The Mathematics of Surfaces IX

Springer-Verlag London Ltd.

Roberto Cipolla and Ralph Martin (Eds)

The Mathematics of Surfaces IX

Proceedings of the Ninth IMA Conference on the Mathematics of Surfaces

 Springer

Roberto Cipolla, BA(Hons), MSE, MEng, DPhil
Department of Engineering, University of Cambridge, Cambridge CB2 1PZ, UK

Ralph Martin, MA, PhD, FIMA, CMath, MBCS, CEng
Department of Computer Science, Cardiff University, PO Box 916, 5 The Parade, Cardiff CF24 3XF, UK

ISBN 978-1-85233-358-4   ISBN 978-1-4471-0495-7 (eBook)
DOI 10.1007/978-1-4471-0495-7

Typesetting: Camera-ready by editors

12/3830-543210 Printed on acid-free paper SPIN 10776750

Preface

These proceedings collect the papers accepted for presentation at the biennial IMA Conference on the Mathematics of Surfaces, held in the University of Cambridge, 4–7 September 2000. While there are many international conferences in this fruitful borderland of mathematics, computer graphics and engineering, this is the oldest, the most frequent and the only one to concentrate on surfaces.

Contributors to this volume come from twelve different countries in Europe, North America and Asia. Their contributions reflect the wide diversity of present-day applications which include modelling parts of the human body for medical purposes as well as the production of cars, aircraft and engineering components. Some applications involve design or construction of surfaces by interpolating or approximating data given at points or on curves. Others consider the problem of 'reverse engineering'—giving a mathematical description of an already constructed object.

We are particularly grateful to Pamela Bye (at the Institue of Mathematics and its Applications) for help in making arrangements; Stephanie Harding and Karen Barker (at Springer Verlag, London) for publishing this volume and to Kwan-Yee Kenneth Wong (Cambridge) for his heroic help with compiling the proceedings and for dealing with numerous technicalities arising from large and numerous computer files. Following this Preface is a listing of the programme committee who with the help of their colleagues did much work in refereeing the papers for these proceedings. Due to their efforts, many of the papers have been considerably improved. Our thanks go to all of them and particularly to our fellow member of the Organising Committee, Malcolm Sabin.

Cambridge,
June 2000

Roberto Cipolla
Ralph Martin

Programme Committee

Contents

Meshless Parameterization
and B-Spline Surface Approximation

Michael S. Floater

SINTEF, P.O. Box 124, Blindern, 0314 Oslo, Norway

Summary. This paper proposes a method for approximating unorganized points in $\mathbb{R}^3$ with smooth B-spline surfaces. The method involves: meshless parameterization; triangulation; shape-preserving reparameterization; and least squares spline approximation.

1 Introduction

The goal of this paper is to describe a new method for approximating a set of distinct points

$$X = \{x_1, \ldots, x_N\}, \qquad x_i \in \mathbb{R}^3, \tag{1}$$

with a tensor-product B-spline surface

$$s(u,v) = \sum_{i=1}^{m_1} \sum_{j=1}^{m_2} B_i(u) C_j(v) c_{ij}, \qquad c_{ij} \in \mathbb{R}^3. \tag{2}$$

Here $B_1, \ldots, B_{m_1}$ and $C_1, \ldots, C_{m_2}$ are B-splines over non-uniform knot vectors with orders (degrees plus 1) K and L respectively. We assume that the points have been sampled from a (simply connected) patch of the surface of some object in $\mathbb{R}^3$. Many sets of measured data in practice are in the form of single patches, even though their geometry can be quite complex. The surface generation method we propose consists of several sequential steps: meshless parameterization; triangulation; reparameterization; and least squares approximation. The key ingredient is the meshless parameterization of [11] which we use to parameterize the points x_i without the need of a given mesh or topological structure. The overall surface approximation method has performed very well in numerical examples, at least when the underlying surface is not too far from being developable.

Surface generation from 'single patch' data sets can be viewed as a simple form of reverse engineering, which, in its widest sense, is usually understood to be the generation of a full B-rep (boundary representation) surface model from points measured from the whole surface of a physical object in $\mathbb{R}^3$. Since such surfaces are closed and can have arbitrary topology, building a B-rep model requires not only topology construction but also segmentation and the sewing together of surface patches. Thus reverse engineering in general is a non-trivial operation and considerable research effort has been invested in developing automatic and semi-automatic methods; for a survey of techniques up to 1997, see [22].

Single-patch data is simpler than general point data in the sense that it offers the possibility of constructing a *parameterization*

$$U = \{u_1, \ldots, u_N\}, \qquad u_i \in \mathbb{R}^2, \tag{3}$$

a set of points in the plane corresponding to the data points x_i. Any standard scattered method [17] can then be applied to find control points c_{ij} in (2) which in some sense minimize the error vectors

$$s(u_i) - x_i. \tag{4}$$

We will minimize these vectors in a least squares sense, drawing on the work of several authors. A major part of this paper however, concerns the construction of a good parameterization. Certainly a simple parameterization can be generated if the data set X is sampled from a surface patch which can be projected 1-1 onto some *base surface*, a simple parametric surface such as a plane, sphere, or cylinder. One can then simply take the projections of the points x_i onto this base surface as parameter points u_i. However, in the absence of a base surface, the only published parameterization method we know of is that of Hoschek and Dietz [16]. The main idea of their approach is to construct the parameterization and the surface simultaneously, and iteratively. The initial parameterization is formed by projecting the points x_i onto some least squares fitting plane: a mapping which may or may not be 1-1. Provided the iteration converges, the method generally leads to very well parameterized surfaces with small error. This is partly due to finding the parameterization through *parameter correction* [17] which aims to yield error vectors in (4) which are perpendicular to the surface. However, it has been pointed out in [16], [5], and [6] that when the geometry of the data set is complex, the method sometimes fails completely, due to foldover in the parameterization.

The method we propose here provides an alternative approach which generates a parameterization in which the parameter point of each data point is forced to lie in the convex hull of the parameter points of neighbouring data points. Though we are not at this stage able to offer any theoretical justification, all the parameterizations in our (numerous) test examples have been free of the foldover effect exhibited by the method of [17].

The basic structure of the paper is built around the three main steps of the surface approximation algorithm: meshless parameterization and triangulation in Section 2, shape-preserving reparameterization in Section 3, and least squares spline approximation in Section 4. Section 5 contains the results of applying the algorithm to a variety of test examples.

2 Meshless Parameterization and Triangulation

Our first step consists of *meshless parameterization*, recently introduced in
[11]. Meshless parameterization determines a sequence of parameter points
u_i from the points x_i *without* the need for any given topological structure.
This almost immediately gives us a triangulation $\mathcal{T}$ of the points set X: we
simply compute a Delaunay triangulation $\mathcal{S}$ of the (planar) parameter points
u_i and define $\mathcal{T}$ to be the corresponding (surface) triangulation of the x_i.
In other words, we take $\mathcal{T}$ to be the set of triangles $[x_i, x_j, x_k]$ for which
$[u_i, u_j, u_k]$ is a triangle in $\mathcal{S}$.

Notice here that the emphasis is not on choosing the parameter points u_i
to yield an optimal triangulation $\mathcal{T}$ for the x_i in some sense. This is because
later, in Section 3, shape-preserving reparameterization will be used to make
a better parameterization of X by using the topological structure of $\mathcal{T}$. For
example, it is not critical if $\mathcal{T}$ contains unnecessarily long thin triangles in
some regions, as long as the shape of $\mathcal{T}$ roughly mimics the shape of X. On
the other hand the parameter points u_i should at least have the property
that the resulting triangulation $\mathcal{T}$ is free of self-intersections. Currently we
are not able to provide a condition which will ensure this, though we have
never experienced this in any of our test cases.

Triangulation by meshless parameterization is relatively fast to both implement and compute compared with existing methods for triangulating unorganized points sets, though it is only applicable to single-patch data sets.
Other triangulation methods include those of [3] and [18], which are based on
the idea of successively removing tetrahedra from a Delaunay tetrahedriza-
tion of the points, and those of [21] and [1], which are based on properties
of the Voronoi diagram. There are also the implicit methods of [14] and [2],
which only approximate the data points.

How do we determine the set of parameter points U without needing a
mesh? We first divide the set X into two disjoint subsets: X_I, the set of
interior points, and X_B, the set of boundary points. Moreover the boundary
points must be ordered. A method was outlined in [11] for first identifying
boundary points and subsequently ordering them using a univariate analog
of meshless parameterization. Without loss of generality we thus assume that
$X_I = \{x_1 \ldots, x_n\}$ for some n, and $X_B = \{x_{n+1} \ldots, x_N\}$, where the points
$x_{n+1} \ldots, x_N$ are ordered consecutively along the boundary.

The method has two steps. In the first step we map the boundary points
$x_{n+1}, \ldots, x_N$ into the boundary of some convex polygon D in the plane. Thus
we choose the corresponding parameter points $u_{n+1}, \ldots, u_N$ to lie around ∂D
in some anticlockwise order. In our numerical examples we take $u_{n+1}, \ldots, u_N$
to lie either on an ellipse or a rectangle and we place the points around these
boundaries according to chord length.

In the second step, we choose for each interior point $x_i \in X_I$, a *neighbourhood* N_i, a set of indexes of points x_j which are in some sense close by.

We also choose a set of (strictly) positive weights λ_{ij}, for $j \in N_i$, such that

$$\sum_{j \in N_i} \lambda_{ij} = 1.$$

Then, in order to find the n parameter points $u_1, \ldots, u_n \in \mathbb{R}^2$ corresponding to the interior points $x_1, \ldots, x_n \in \mathbb{R}^3$, we solve the linear system of n equations

$$u_i = \sum_{j \in N_i} \lambda_{ij} u_j. \qquad i = 1, \ldots, n. \tag{5}$$

These equations demand that each interior u_i be some convex combination of its neighbours $\{u_j : j \in N_i\}$. Thus u_i will be contained in the convex hull of the neighbours.

It was established in [11] that the linear system (5) is nonsingular under a very mild condition, namely that every interior point x_i is *boundary connected*. By boundary connected we mean that x_i can be joined to the boundary X_B by a path of points in X where each point x_{j+1} in the path is a neighbour of the previous point x_j, in the sense that $x_{j+1} \in N_j$. This condition will be fulfilled if the neighbourhoods N_i are chosen large enough. In fact, in practice one can usually choose the neighbourhoods to be both large enough that every interior point is connected to *every* boundary point, and at the same time small enough that each point neighbourhood $\{x_j : j \in N_i\}$ consists of points which are nearby with respect to the geometry of the underlying surface. Certainly we do not want to pick up points from extraneous branches of the surface. It was shown in [11] that due to the convex combinations and the fact that the domain D is convex, all parameter points will lie inside D.

Several choices of neighbourhood N_i were proposed in [11] but a simple and effective choice is to take the 'd nearest neighbours', in other words, we let N_i be the set of indexes of the d points x_j closest to x_i. Going on several numerical tests, it appears that setting $d = 10$ or $d = 20$ is adequate for all but the most extreme data sets. As regards the choice of weights λ_{ij}, the naive choice of uniform weights $\lambda_{ij} = 1/d_i$, where $d_i = |N_i|$, can lead to the result that two data points x_i and x_j end up being mapped to the same points: $u_i = u_j$, as shown in [11]. Our preferred choice in practice is to use the *reciprocal distance weights*

$$\lambda_{ij} = \frac{1}{\|x_j - x_i\|} \Big/ \sum_{k \in N_i} \frac{1}{\|x_k - x_i\|},$$

which depend on the distances between x_i and its neighbours. These have always resulted in distinct parameter points in all the numerical examples we have run. The theoretical best choice of weights still requires further research. Since the matrix A arising from the linear system (5) is sparse and in general non-symmetric, we have used the biconjugate gradient method to solve (5).

3   Shape-preserving Reparameterization

The goal of the meshless parameterization used in the previous section was to generate a (possibly rough) triangulation $\mathcal{T}$ of the data set X so that we have some topological structure. With the triangulation $\mathcal{T}$ in place, we can reparameterize the interior points x_i, but this time use the neighbourhood information given by $\mathcal{T}$. Methods for doing this have been proposed in [7], [8], [12], [20] and we will apply the *shape-preserving parameterization* of [8]. This parameterization has the advantage that it has linear precision and so the triangles in the mapped triangulation tend to mimic the shape of the triangles in $\mathcal{T}$: hence the name. This parameterization also tends to lead to surface approximations $s(u, v)$ in (2) with smooth isocurves. We note that it was shown in [9] that the shape-preserving parameterization of [8] is very similar, both empirically and theoretically, to the harmonic map of [7]. Indeed the harmonic map also has linear precision [9]. However, unlike the shape-preserving parameterization of [8], the harmonic map can fold over, due to negative weights; an example is given in [9].

   To help in understanding the shape-preserving parameterization for triangulations in [8], let us first consider the analogous *chord length* parameterization for polygonal curves. If the vertices of the curve are $x_1, \ldots, x_N \in \mathbb{R}^3$, then the sequence of real values $t_1, t_2, \ldots, t_N$ is called a *chord length* parameterization if

$$(t_{i+1} - t_i) = \rho \| x_{i+1} - x_i \|,$$

for some constant $\rho > 0$, where $\| \cdot \|$ is the Euclidean norm. It was observed in [8] that the sequence $t_1, t_2, \ldots, t_N$ is a chord length parameterization if and only if

$$t_i = \frac{\| x_{i+1} - x_i \| t_{i-1} + \| x_i - x_{i-1} \| t_{i+1}}{\| x_{i+1} - x_i \| + \| x_i - x_{i-1} \|}.$$

This implies that t_i is a convex combination of its two neighbouring parameter values t_{i-1} and t_{i+1}, and the weighting depends on the lengths of the chords, in such a way that if the points x_i lie on a straight line (though not necessarily uniformly), then the whole sequence $t_1, \ldots, t_N$ will be some affine transformation, from $\mathbb{R}^3$ to $\mathbb{R}^1$, of the sequence $x_1, \ldots, x_N$. In other words, chord length parameterization has linear precision and it is this property that is carried over to the shape-preserving parameterization for triangulations in [8].

   In the shape-preserving parameterization for triangulations, we solve the system (5) where we take the neighbourhoods N_i to be the neighbourhoods of the triangulation $\mathcal{T}$. In addition we choose positive weights λ_{ij} which give linear precision.

   For the sake of completeness, we will outline how the weights are determined. For each i, we refer to the set of triangles incident on x_i as the *cell* of x_i. The vertices in the cell are x_i and its neighbours $\{ x_j : j \in N_i \}$. The shape-preserving weights λ_{ij} depend only on x_i and its neighbours and are constructed in two steps.

The first step is to 'flatten out' the cell into the plane, yielding *local* (temporary) parameter points u_i and $\{u_j : j \in N_i\}$ in $\mathbb{R}^2$; see Figure 1. We use an approximation of the geodesic polar map, adapted to triangulations, which, it seems, was first proposed in [23], in the context of computer graphics. We let u_i be arbitrary and choose the neighbours u_j such that for each $j \in N_i$,

$$\|u_j - u_i\| = \|x_j - x_i\|$$

and for each triangle $[x_i, x_j, x_k]$ in the cell of x_i,

$$\text{ang}(u_k, u_i, u_j) = \rho\, \text{ang}(x_k, x_i, x_j),$$

where ρ is a constant. The scaling factor ρ is needed to ensure that the interior angles in the mapped cell sum to 2π.

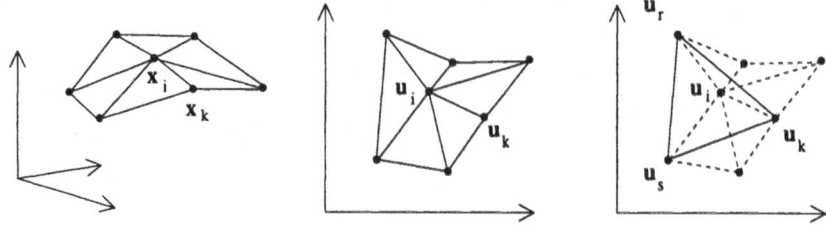

Fig. 1. Calculating the shape-preserving weights

The second step is to express u_i as a convex combination of the neighbouring mapped points $\{u_j : j \in N_i\}$, in order to obtain linear precision. This is always possible since the mapped cell is star-shaped with u_i in its kernel. For each $k \in N_i$, we locate an edge $[u_r, u_s]$, $r, s \in N_i$, in the mapped cell for which

$$u_i \in [u_k, u_r, u_s],$$

and with τ_k^k, τ_r^k, τ_s^k the barycentric coordinates of u_i in this latter triangle, we have

$$u_i = \tau_k^k u_k + \tau_r^k u_r + \tau_s^k u_s.$$

Letting $\tau_j^k = 0$ for all $j \in N_i$, $j \neq k, r, s$ we then have

$$u_i = \sum_{j \in N_i} \tau_j^k u_j.$$

Finally, we take the shape-preserving weights to be averages of the local weights τ_j^k over all $k \in N_i$,

$$\lambda_{ij} = \frac{1}{d} \sum_{k \in N_i} \tau_j^k,$$

and we have

$$u_i = \sum_{j \in N_i} \lambda_{ij} u_j, \quad \text{and} \quad \sum_{j \in N_i} \lambda_{ij} = 1,$$

and $\lambda_{ij} > 0$ for all $j \in N_i$.

4 Least Squares Approximation

Finally, we propose a least squares method for approximating the parameterized data by a tensor-product spline surface (2). Our discussion builds on the theoretical studies of [13] and [19] and follows closely the work of [16], [5], [6], [12], [10]. Least squares approximation of scattered data by tensor-product splines with a smoothing term seems to yield very well behaved, smooth surfaces even when the data set is very large and is certainly an attractive alternative to other scattered data methods such as piecewise polynomials over triangulations or radial basis functions.

Our goal is to determine the control points c_{ij} in (2) to minimize the sum of squared errors

$$\sum_{k=1}^{N} \| s(u_k) - x_k \|^2. \tag{6}$$

We assume here that we have already chosen the degrees K and L of the B-splines in (2). We must also choose the two knot vectors and we ensure that the rectangular domain $[a_1, b_1] \times [a_2, b_2]$ of s contains all the parameter points u_i. If the parameter domain D of the parameter points u_i is chosen to be a rectangle then the knot vectors are chosen so that

$$D = [a_1, b_1] \times [a_2, b_2],$$

and the boundary of the resulting surface approximation s will closely follow the boundary points X_B. If on the other hand, the domain D has a different shape, for example a circle or ellipse, we choose the knot vectors so that

$$D \subset [a_1, b_1] \times [a_2, b_2].$$

The resulting surface approximation s will then need to be trimmed so that the final surface approximation will be a trimmed B-spline surface. One should also choose non-uniform knot vectors which reflect the distributions of the points u_k along each of their two coordinate axes.

It remains to find the control points c_{ij}. One could try to minimize the function (6) directly but since the u_i are unstructured, the minimum will rarely be unique and the usual procedure is to add a smoothing term. There are several smoothing terms one can add such as ones defined in terms of curvature, torsion, etc. Several approaches, linear and nonlinear, are mentioned in [5] and [12]. We have found in practice that provided the shape-preserving parameterization is used to compute the u_i, good surface approximations result from simply adding the thin-plate spline energy term. Specifically we propose minimizing the function

$$F = \sum_{k=1}^{N} ||s(u_k) - x_k||^2 + \lambda J, \qquad (7)$$

where J is the thin-plate spline integral

$$J = \int_{a_1}^{b_1} \int_{a_2}^{b_2} ||s_{uu}||^2 + 2||s_{uv}||^2 + ||s_{vv}||^2 \, dv \, du, \qquad (8)$$

and $\lambda > 0$ is a constant.

We will discuss some computational details of how one minimizes F and the first thing to notice is that the minimization 'decouples' in the sense that since

$$F = \sum_{\alpha=1}^{3} \left(\sum_{k=1}^{N} (s^\alpha(u_k) - x_k^\alpha)^2 + \lambda J^\alpha \right),$$

it is minimized by the independent minimization of its three component functions. Thus for the sake of simplicity, we next replace the points x_i by real values x_i and the control points c_{ij} by real coefficients c_{ij}. The vector valued spline surface s is now replaced by the real valued spline function s.

Our simplified goal is to determine a coefficient vector

$$c = (c_{1,1}, \ldots, c_{m_1,1}, c_{1,2}, \ldots, c_{m_1,m_2})^T$$

of length $m = m_1 m_2$, which minimizes the function

$$F(c) = \sum_{k=1}^{N} (s(u_k) - x_k)^2 + \lambda J(c), \qquad (9)$$

where J is the thin plate spline integral

$$J(c) = \int_{a_1}^{b_1} \int_{a_2}^{b_2} s_{uu}^2 + 2s_{uv}^2 + s_{vv}^2 \, dv \, du.$$

This integral can be re-expressed as

$$\sum_{i=1}^{m_1} \sum_{j=1}^{m_2} \sum_{r=1}^{m_1} \sum_{s=1}^{m_2} E_{ijrs} c_{ij} c_{rs},$$

where

$$E_{ijrs} = A_{ijrs} + 2B_{ijrs} + C_{ijrs}$$

and

$$A_{ijrs} = \int_{a_1}^{b_1} B_i''(u) B_r''(u) du \int_{a_2}^{b_2} C_j(v) C_s(v) dv,$$

$$B_{ijrs} = \int_{a_1}^{b_1} B_i'(u)B_r'(u)du \int_{a_2}^{b_2} C_j'(v)C_s'(v)dv,$$

$$C_{ijrs} = \int_{a_1}^{b_1} B_i(u)B_r(u)du \int_{a_2}^{b_2} C_j''(v)C_s''(v)dv.$$

A minimum of $F(c)$ must occur at a point c where all partial derivatives are zero, a *critical point*. The equations $\partial F/\partial c_i = 0$ are called the *normal equations* of the least squares problem. It will help to write

$$J(c) = c^T E c,$$

where E is the $m \times m$ matrix whose elements are

$$E_{(j-1)m_1+i,\,(s-1)m_1+r} = E_{ijrs},$$

for $i, r = 1, \ldots, m_1$ and $j, s = 1, \ldots, m_2$. By differentiating $F(c)$ in (9) explicitly and rearranging the subsequent expression, the normal equations can thus be rewritten as the single matrix equation

$$(B^T B + \lambda E)c = B^T x, \tag{10}$$

where $x = (x_1, \ldots, x_N)^T$ and B is the $N \times m$ matrix

$$B = \begin{pmatrix} B_1(u_1)C_1(v_1) & B_2(u_1)C_1(v_1) & \cdots & B_{m_1}(u_1)C_{m_2}(v_1) \\ \vdots & \vdots & & \vdots \\ B_1(u_N)C_1(v_N) & B_2(u_N)C_1(v_N) & \cdots & B_{m_1}(u_N)C_{m_2}(v_N) \end{pmatrix},$$

where $u_k = (u_k, v_k)$. Then the solution to minimizing (9) is the solution c to (10). As observed in [13], the $m \times m$ matrix

$$G = B^T B,$$

whose elements are

$$G_{(j-1)m_1+i,\,(s-1)m_1+r} = \sum_{k=1}^{N} B_i(u_k)C_j(v_k)B_r(u_k)C_s(v_k) \tag{11}$$

for $i, r = 1, \ldots, m_1$ and $j, s = 1, \ldots, m_2$, is symmetric and positive semidefinite. So is E and thus the matrix sum

$$A = G + \lambda E \tag{12}$$

is also symmetric and positive semidefinite. We will derive a sufficient condition for when A is *strictly* positive definite and thereby nonsingular. The matrix A is strictly positive definite if the only solution to $c^T A c = 0$ is $c = 0$. First observe that

$$c^T E c = J(c) = 0$$

implies that s must be a linear polynomial $a + bu + cv$. Second, observe that

$$c^T G c = c^T B^T B c = ||Bc||^2 = 0$$

implies that $s(\boldsymbol{u}_k) = 0$ for all $k = 1, \ldots, N$. Thus we have that $c^T A c = 0$ implies that s is a linear polynomial which is zero at every parameter point $\boldsymbol{u}_k$. Clearly then, if there are at least three points $\boldsymbol{u}_k$ which do not lie on a straight line, s would have to be zero and therefore all the coefficients c_{ij} would also have to be zero. Since, due to our parameterization method, the parameter points $\boldsymbol{u}_k$ can never all be collinear, we deduce that A is indeed nonsingular and the minimizer c of (10) is unique.

In order to compute the elements of the matrix A in (12), we have used the Cox de Boor algorithm for the evaluation of B-splines and their derivatives, and the algorithm in [4] for the exact evaluation of inner products of B-splines. When building A, it is very important not to compute each element of G independently, for this can be terribly costly when N is large. The simple trick is to instead process each parameter point $\boldsymbol{u}_k$ in turn and compute all the tensor-product B-splines whose supports contain it, applying the Cox de Boor algorithm just once. For each new pair (u_k, v_k), all non-zero products

$$B_i(u_k) C_j(v_k) B_r(u_k) C_s(v_k)$$

are added to the current value of $G_{(j-1)m_1+i,\,(s-1)m_1+r}$.

The matrices G and E and A are clearly sparse. If the B-splines B_i and C_j have orders K and L respectively, then

$$A_{(j-1)m_1+i,\,(s-1)m_1+r} = 0$$

if either $|i - r| \geq K$ or $|j - s| \geq L$. The non-zero elements of A are shown in Figure 2 for the case when $m_1 = m_2 = 10$ and the spline $s(u, v)$ is bicubic ($K = L = 4$). As illustrated by the figure, the matrix A has $m_2 \times m_2$ blocks, each of which has size $m_1 \times m_1$, and is banded with bandwidth $2K - 1$. If A is viewed as an $m_2 \times m_2$ block matrix then it has bandwidth $2L - 1$. The real bandwidth of A however is $(2L - 1)m_1$.

We have chosen to use the conjugate gradient method to solve each of the two components of (10), which capitalizes on the sparse form of A. The number of non-zero elements of A in any row is at most $(2K - 1)(2L - 1)$ and is independent of m_1 and m_2. For example, in Figure 2, the maximum number of non-zeros per row is 49.

The final point about the least squares is the choice of the smoothing parameter λ in (7). This can be used to vary the emphasis of the approximation between error minimization and smoothing. However, it is useful in practice to a have a good default value. In the absence of any better estimation, we suggest the simple and pragmatic choice of

$$\lambda = ||G|| / ||E||,$$

Fig. 2. Matrix structure when $K = L = 4$, $m_1 = 10$, and $m_2 = 8$

for some matrix norm $||.||$, such as the l_2 norm $||M|| = \sqrt{(\sum_{ij} m_{ij}^2)}$. The effect of this choice of λ is roughly speaking to ensure that the two contributions G and λE to the matrix A in (12) have equal weight. This choice has performed very well in our test examples.

After the surface s has been computed, the error of the approximation can be computed numerically at each point and if the error is unacceptable with respect to some error measure, the knot vectors of the surface can be refined and a new surface computed, keeping the parameterization fixed. Eventually the error will be within a chosen tolerance.

5   Numerical Examples

The first numerical example illustrates all the steps of the whole algorithm from unorganized points to B-spline surface. Figure 3a shows 192 unorganized points for which a boundary is self-evident. Figure 3b shows its meshless parameterization where the points are mapped into the rectangular domain $D = [0,2] \times [0,1]$. The shape of the domain roughly mimics the shape of the point set X. Figure 3c shows the Delaunay triangulation of the parameter points in Figure 3b and Figure 3d shows the corresponding triangulation $\mathcal{T}$ of the original points. Since the points in Figure 3b provide only a rough parameterization, lacking for example linear precision, we then make a shape-preserving reparameterization of the points x_i, which uses the neighbourhood information of the triangulation in Figure 3d. The new parameterization, with the corresponding triangulation, is shown in Figure 3e and is used to compute the least squares surface approximation of Figure 3f. The spline surface is bicubic ($K = L = 4$) and has dimensions $m_1 = m_2 = 10$.

For the sake of comparison, we also show in Figure 3g the Delaunay retriangulation of the final parameter points of Figure 3e and one sees how the corresponding retriangulation $\mathcal{T}'$ of X in Figure 3h has better proportioned triangles than the earlier one, $\mathcal{T}$, in Figure 3d. This would be useful if one simply wanted to triangulate the original data set.

The second example illustrates how well the method performs equally well on data for which there is no obvious base surface to project onto (in which case other methods could be applied). Following the shape of the boundary, the 1800 points of the 'S-shape in Figure 4a were also mapped into a rectangle and following the same steps as for the previous example, the triangulation $\mathcal{T}'$ of the points is shown in Figure 4b (corresponding to Figure 3h). The spline approximation is shown in Figure 4c, together with the corners of the polynomial patches. The surface is again bicubic.

Figure 5a shows a real data set of 24712 points from a sofa. Rather than try to mimic the complex boundary shape with a rectangular parameter domain, we chose here to map the boundary into the unit circle. After making a shape-preserving reparameterization, the resulting triangulation $\mathcal{T}'$ is shown in Figure 5b. Figure 5c shows a bicubic spline surface approximation of dimensions $m_1 = m_2 = 100$ which captures the crease in the back of the sofa. This surface would need to be trimmed by the circle in its parameter domain for it to hug the original data set properly. Note here that in a real reverse engineering application one would probably want to segment the data set according to the 'folds' in the sofa.

The last example illustrates the limitation of our surface approximation technique when the data set is far from being developable. In contrast to the S-shape of Figure 2a, the Spock data set of Figure 6a requires considerable deformation to map into the plane. After mapping it into the unit square and reparameterizing, the distance between the closest pair of parameter points was found to be just 0.0004. This poses no obvious problem for the corresponding triangulation $\mathcal{T}'$ of the original points in Figure 6b, but the 100×100 bicubic least squares approximation in Figure 6c exhibits undesirable oscillations, for example, around the mouth. Here the two knot vectors were constructed so that there are many densely placed knot lines, in both directions, in the middle of the domain, and fewer elsewhere. This was necessary in order to have sufficient degrees of freedom to reproduce the detail at the top of the head and the face. On the other hand these dense knot lines create the unwanted oscillations lower down the head. Despite this, the smooth rows of patch corners in Figure 6d clearly indicate that the parameterization is good. As pointed out in [15], the inevitably high deformation when parameterizing a data set such as the head of Spock suggests that approximating with some kind of multiresolution surface, as in [12], would give better results.

Fig. 3a. Point set.   Fig. 3b. Meshless parameterization.

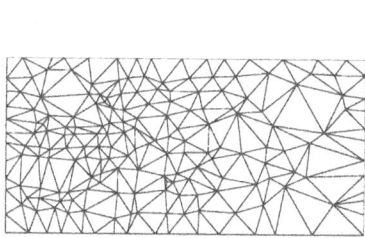

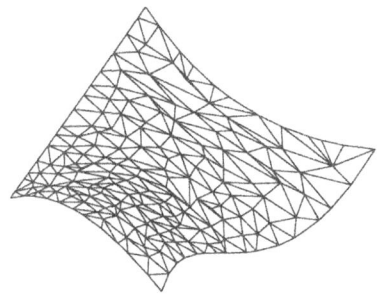

Fig. 3c. Delaunay triangulation.   Fig. 3d. Surface triangulation.

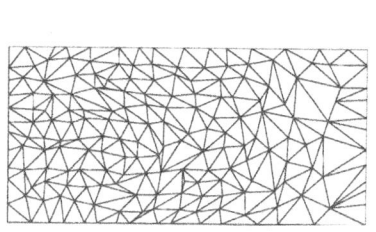

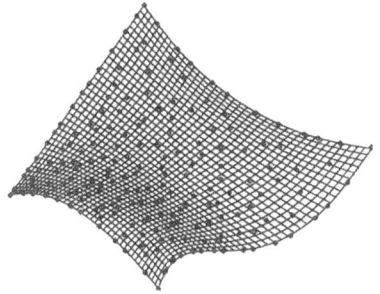

Fig. 3e. Shape-preserving parameterization.   Fig. 3f. Spline surface.

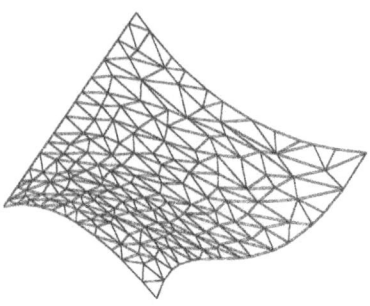

Fig. 3g. Delaunay retriangulation.   Fig. 3h. Surface retriangulation.

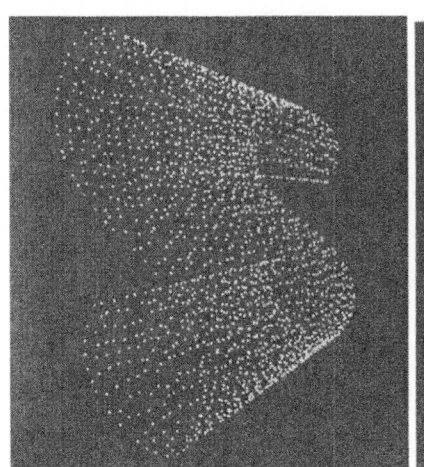

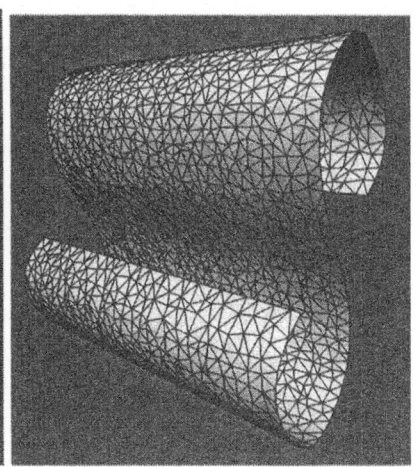

Fig. 4a. Point set.   Fig. 4b. Triangulation.

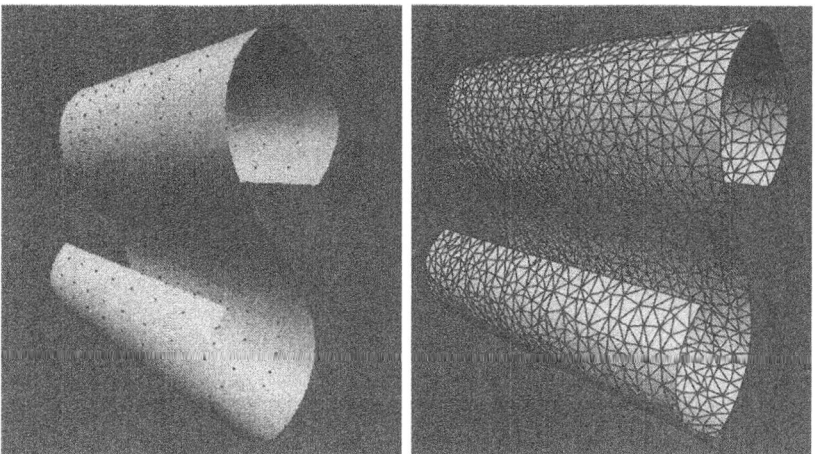

Fig. 4c. Spline surface with patch corners.

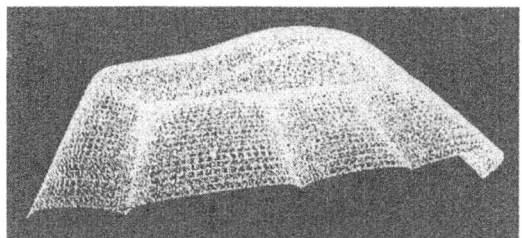

Fig. 5a. Point set.

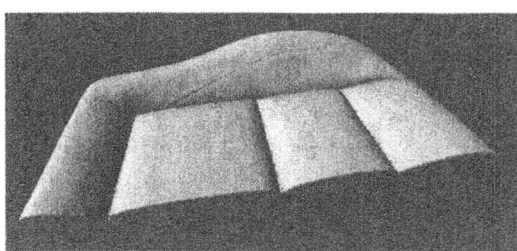

Fig. 5b. Triangulation.

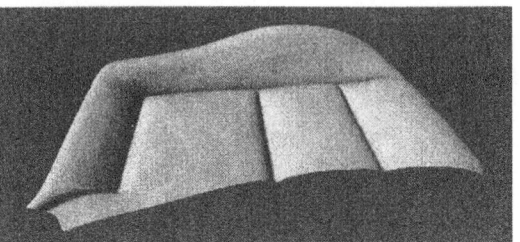

Fig. 5c. Spline surface.

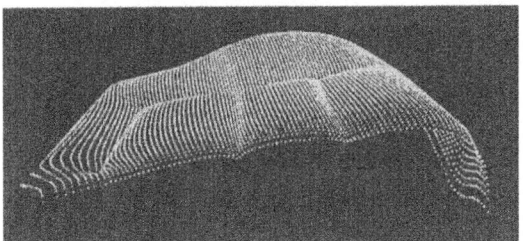

Fig. 5d. Patch corners.

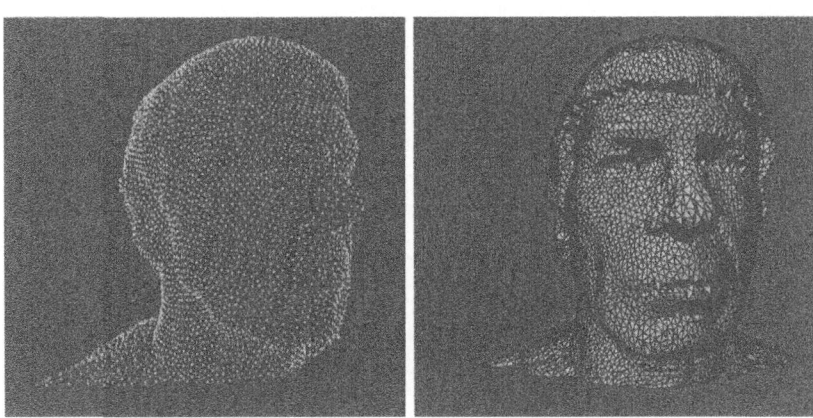

Fig. 6a. Point set.   Fig. 6b. Triangulation.

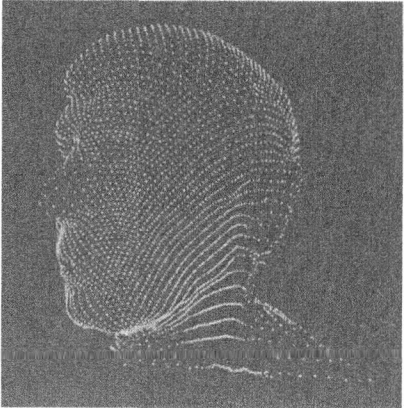

Fig. 6c. Spline surface.   Fig. 6d. Patch corners.

Acknowledgement. I wish to thank Martin Reimers for help with some of the data sets, Ewald Quak for helpful discussions, and Steve Hull of Metrocad GmbH for use of the sofa data set.

References

1. N. Amenta, M. Bern, and M. Kamvysselis, A new Voronoi-based surface reconstruction algorithm, Computer Graphics, Proceedings ACM Siggraph 98 (1998), 415–421.
2. C. L. Bajaj, F. Bernardini, and G. Xu, Automatic reconstruction of surfaces and scalar fields from 3d scans, Computer Graphics Proceedings, SIGGRAPH '95, Annual Conference Series (1995), 109–118.
3. J.-D. Boissonnat, Geometric structures for three-dimensional shape representation, ACM Transactions on Graphics **3(4)** (1984), 266–286.
4. C. de Boor, T. Lyche, and L. Schumaker, On calculating with B-splines, II. Integration, in Proc. Oberwolfach, 1975.
5. U. Dietz, B-spline approximation with energy constraints, in *Advanced Course on Fairshape*, J. Hoschek and P. Kaklis (eds.), Teubner, Stuttgart, (1996), 229–240.
6. U. Dietz, Fair surface reconstruction from point clouds, in *Mathematical Methods for Curves and Surfaces II*, M. Dæhlen, T. Lyche & L. L. Schumaker (eds.), Vanderbilt University Press, Nashville, (1998), 79–86.
7. M. Eck, T. DeRose, T. Duchamp, H. Hoppe, M. Lounsbery, and W. Stuetzle, Multiresolution analysis of arbitrary meshes, SIGGRAPH Comp. Graph. (SIGGRAPH '95 Proceedings) **26**(2) (1995), 173–182.
8. M. S. Floater, Parametrization and smooth approximation of surface triangulations, Comp. Aided Geom. Design **14** (1997), 231–250.
9. M. S. Floater, Parametric tilings and scattered data approximation, International Journal of Shape Modeling **4** (1998), 165–182.

10. M. S. Floater, How to Approximate Scattered Data by Least Squares, SINTEF Report No. STF42 A98013, Oslo, (1998).

11. M. S. Floater and M. Reimers, Meshless parameterization and surface reconstruction, preprint.

12. G. Greiner and K. Hormann, Interpolating and approximating scattered 3D data with hierarchical tensor product B-splines, in *Surface Fitting and Multiresolution Methods*, A. Le Méhauté, C. Rabut, and L. L. Schumaker (eds.), Vanderbilt University Press, Nashville 1997, 163–172.

13. M. von Golitschek and L. L. Schumaker, Data fitting by penalized least squares, in *Algorithms for Approximation II*, J. C. Mason and M. G. Cox (eds.), Chapman & Hall, 1990, 210-227.

14. H. Hoppe, T. DeRose, T. DuChamp, J. McDonald, W. Stuetzle, Surface reconstruction from unorganized points, Computer Graphics, Vol. 26, No. 2 (1992), 71–78.

15. K. Hormann and G. Greiner, MIPS: An efficient global parametrization method, preprint.

16. J. Hoschek and U. Dietz, Smooth B-spline surface approximation to scattered data, in *Reverse Engineering*, J. Hoschek and W. dankwort (eds.), B.G. Teubner, 1996, 143–152.

17. J. Hoschek and D. Lasser, *Fundamentals of Computer Aided Geometric Design*, AKPeters, Wellesley, 1994.

18. F. Isselhard, G. Brunnett, and T. Schreiber, Polyhedral reconstruction of 3d objects by tetrahedral removal, Technical Report No. 288/97, Fachbereich Informatik, University of Kaiserslautern, Germany, 1997.

19. C. L. Lawson and R. J. Hanson, *Solving Least Squares Problems*, SIAM, Philadelphia, 1995.

20. B. Lévy and J. L. Mallet, Non-distorted texture mapping for sheared triangulated meshes, Proceedings of SIGGRAPH 98 (1998), 343–352.

21. T. Schreiber and G. Brunnett, Approximating 3d objects from measured points, in *Proceedings of 30th ISATA*, Florence, Italy, 1997.

22. T. Varady, R. R. Martin, and J. Cox, Reverse engineering of geometric models — an introduction, CAD **29** (1997), 255–268.

23. W. Welch and A. Witkin, Free-form shape design using triangulated surfaces, Computer Graphics, SIGGRAPH 1994, 247–256.

Computation of Local Differential Parameters on Irregular Meshes

Péter Csákány and Andrew M. Wallace

Heriot-Watt University, Scotland EH14 4AS, UK

Summary. The aim of our work was to carry out the calculation of local differential properties on an irregular 3D mesh. We review the existing computational methods and give a simplified version which suits better the practical requirements. Our approach is applicable not only to range images but to 3D objects as well, applying a direct and simple method to calculate Gaussian and mean curvature on ain irregular 3D mesh. This is an important step towards recognition based only on object shape as it enables the application of subsequent higher level processing to data that has been converted to a mesh representation.

1 Introduction

Local differential properties are employed in many areas of computer vision such as segmentation, object recognition and modelling. Most methods of computing these properties assume a regular parameterisation of the 3D space, which is appropriate for many 3D sensors. However, with the emerging number and variety of the different vision systems and applications this limitation is more and more restrictive. The mesh data structure is already very successful in the field of computer graphics. Its flexibility to define an approximation of the surface of objects also makes it important for several areas of computer vision. Unfortunately a 3D mesh does not support regular parameterisation of the 3D space. This paper shows that although there are a number of existing approaches to carry out local differential computations on mesh structures they are closely related and usually computationally expensive. We also give a simplified but still effective solution to the problem.

2 Mean and Gaussian Curvature

Local differential properties like the mean and Gaussian curvature or the principal curvatures serve as a basis for some important vision algorithms, as described memorably by Besl and Jain [1,2]. They applied the sign of the mean and Gaussian curvatures for segmentation, Koenderink and van Doorn [3] showed the importance of the principal curvatures for shape description, and Liang and Todhunter [7] applied differential properties to object modelling. These fundamental results led to many other applications; however, it is essential to show that the basic differential geometric measures can be calculated not only in the case of the regular sampling of the 3D space but also in the case of irregular sampling. In this section we show different methods of this computation.

2.1 Computation in the regular case

There are many definitions of the mean and the Gaussian curvatures depending on which of their properties are considered. In the regular case, the most practical definition of the surface curvatures is based on the partial derivatives of the 3-D surface. If the parameterisation of the graph surface takes the form: $x(u,v) = [u, v, f(u,v)]^T$, then the Gaussian and mean curvatures are:

$$K = \frac{f_{uu}f_{vv} - f_{uv}^2}{(1 + f_u^2 + f_v^2)^2}, \tag{1a}$$

$$2H = \frac{f_{uu} + f_{vv} + f_{uu}f_v^2 + f_{vv}f_u^2 - 2f_u f_v f_{uv}}{(1 + f_u^2 + f_v^2)^{\frac{3}{2}}} \tag{1b}$$

where f_u, f_v refer to the first derivatives in u, v, etc. The calculation of the partial derivatives is straightforward if the graph surface $(x(u,v) = [u, v, f(u,v)]^T)$ is considered as discrete sampled data of a continuous surface. The approach of Besl and Jain finds the best fitting continuous surface, computes the derivatives analytically and evaluates them at the corresponding discrete points. They even show a windowing operator for efficient evaluation. The issue that we address is the problem of the calculation of the first and second partial derivatives if the parameterisation is not that given above.

2.2 Computation in the irregular case

If one can calculate the first and second partial derivatives of a surface at a given point independent of the irregularity of the parameterisation, the mean and Gaussian curvatures can be calculated. Thus, the key to the problem is the derivation of discrete, irregular functions. In 3D the first partial derivatives of a surface at a point specify the normal vector at that point. Therefore, we address firstly the computation of the normal vector.

Surface fitting. One approach to generate the normals on a surface, at points irregularly placed in 3D space, is to fit a surface to a subset of those points and its neighbours that is given by an analytic form. The differentiation in this case can be done analytically. This is the method that Besl and Jain used to generate the derivatives in the regular case, and can be reasonably extended to the irregular case.

The first difficulty with fitting is the selection of points. In the regular case the subset of points is selected from a window with a given size. In the irregular case the definition of such a neighbourhood is more complicated. In both cases the specification of the size of the given neighbourhood is also a problem. At this point we should note that those topological properties are used later for segmentation that are to be calculated after the fitting. This leads to the classic chicken and egg problem, i.e. the surface cannot be fitted as long as the data is not segmented but the data cannot be segmented as long as the surface is not fitted. To make the procedure effective, it is necessary to pre-process depth data by robust fitting of outliers, anisotropic smoothing, before, iteratively re-fitting and re-segmenting the 3D data [11].

Computation without derivatives. There is a way to compute curvatures without direct computation of the derivatives. Besl and Jain [1] briefly discussed the problem of computing Gaussian curvature without partial derivatives on a triangulated surface, stating that they had obtained "amazingly accurate estimation of the Gaussian curvature of a sphere using five points on the surface of a hemisphere". This method is based on the computation of the angle deficit of the N different triangular facets around a node (Fig. 1).

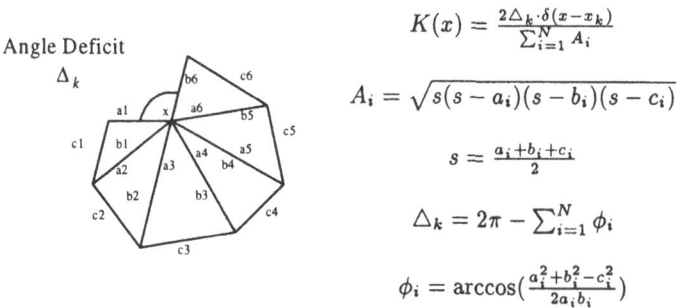

$$K(x) = \frac{2\triangle_k \cdot \delta(x - x_k)}{\sum_{i=1}^{N} A_i}$$

$$A_i = \sqrt{s(s - a_i)(s - b_i)(s - c_i)}$$

$$s = \frac{a_i + b_i + c_i}{2}$$

$$\triangle_k = 2\pi - \sum_{i=1}^{N} \phi_i$$

$$\phi_i = \arccos\left(\frac{a_i^2 + b_i^2 - c_i^2}{2a_i b_i}\right)$$

Fig. 1. Computation of the Gaussian curvature on a triangulated surface

A similar method to compute the mean curvature was presented subsequently [5,6]. This was based on a slightly different definition of the mean curvature. Let $\kappa(\theta, p)$ define the curvature of the surface curve C_θ, which is the intersection of S surface and P_θ, at p; where p is a point of S, and P_θ denotes the orthogonal plane to the tangent plane at p rotated by θ angle around the surface normal at p (Fig. 2).

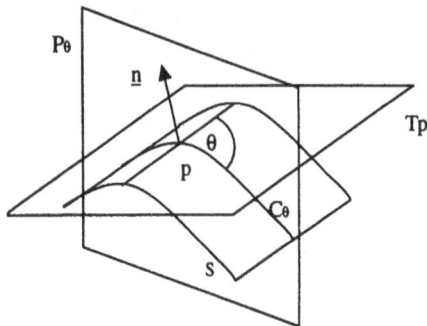

Fig. 2. Mean curvature computation on continuous surface

The mean curvature H(p) of S at p is the average of the curvature $\kappa(\theta, p)$ for $\theta \in [0, 2\pi]$. Thus,

$$H(p) = \frac{1}{2\pi} \int_0^{2\pi} \kappa(\theta, p) d\theta, \tag{2a}$$

$$H(p) = \frac{1}{2}(\kappa(\theta, p) + \kappa(\theta + \pi/2, p)) \tag{2b}$$

which implies that the mean curvature can be calculated at any point p on a surface S by choosing two mutually orthogonal directions at point p. This result can be applied to polyhedral objects. Let q be a point on an edge e between two orthogonal planes. Let the two mutually orthogonal directions be the direction along the edge and the one perpendicular to it. The curvature along the edge is zero, thus the mean curvature at q is equal to the half of the angle between the planes on the two sides of the edge. To calculate the mean curvature at a node of the tessellated surface, we have to compute the integral mean curvature of an area U similar to the one used during the Gaussian curvature calculation. The mean curvature is the integral mean curvature divided by the surface area.

$$H(p) = \frac{\sum\limits_{q \in e} H(q) \cdot Length(e \cap U)}{Size(U)}. \tag{3}$$

This can be reformulated according to the notations of the Gaussian curvature computation above (Fig. 3).

Computation using auto-correlation. A computationally more expensive approach was given by Liang and Todhunter [7] then extended by Berkmann and Caelli [8]. They first calculated the local surface normal on a Monge patch using local covariance

$$C_I = \frac{1}{n} \sum_{i=1}^{n} (\boldsymbol{x}_i - \boldsymbol{x}_m) \cdot (\boldsymbol{x}_i - \boldsymbol{x}_m)^T \tag{4}$$

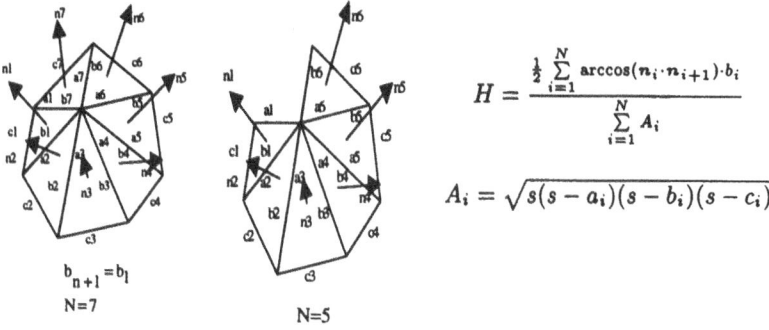

$$H = \frac{\frac{1}{2} \sum\limits_{i=1}^{N} \arccos(n_i \cdot n_{i+1}) \cdot b_i}{\sum\limits_{i=1}^{N} A_i}$$

$$A_i = \sqrt{s(s - a_i)(s - b_i)(s - c_i)}$$

Fig. 3. Mean curvature computation on a triangulated mesh in a closed and in an open neighbourhood

where $x_i = (x_i, y_i, z_i)$ corresponds to the projection plane (x,y) and depth (z) values at position i; $x_m = \frac{1}{n} \sum_{i=1}^{n} x_i$ is the mean position vector, and n is the total number of pixels in the local neighbourhood (Fig. 2.2a). The two eigenvectors of this matrix with the largest eigenvalues correspond to two vectors on the tangent plane. The cross product of the two eigenvectors yields the normal at the middle of the correlation window. The estimation of the tangent plane in this way is optimal in the least squares sense. The estimation of the principal direction is similar to the tangent plane estimation. The second fundamental form can be expressed as:

$$II_C = v^T \cdot C_{II} \cdot v \tag{5}$$

where

$$C_{II} = \frac{1}{n} \sum_{i=1}^{n} (y_i - y_m) \cdot (y_i - y_m)^T \tag{6}$$

$$y_i = [(x_i - x_0)^T \cdot n_i] \cdot \begin{pmatrix} (x_i - x_0)^T \cdot t_{i1} \\ (x_i - x_0)^T \cdot t_{i2} \end{pmatrix} \tag{7}$$

The first part of y_i measures the orthogonal distance from the tangent plane to point x_i. The second part of the expression measures the projection of the difference vector from x_0, the central vertex, to x_i, one of its neighbours, onto the tangent plane. t_{i1}, t_{i2} and n_i are the tangent and normal vectors at point x_0, respectively; v is a chosen unit vector on the tangent plane. The principal directions are directions given by the eigenvectors of this covariance matrix.

Another way of computing the principal directions is to calculate the covariance matrix from the projections of the normal vectors, within a neighbourhood of a point, onto the tangent plane and define the principal directions as the eigenvectors of this matrix.

$$C_P = \frac{1}{n} \sum_{i=1}^{n} (v_i - v_m) \cdot (v_i - v_m)^T \tag{8}$$

$$v_i = \begin{pmatrix} n_i^T \cdot t_{i1} \\ n_i^T \cdot t_{i1} \end{pmatrix} \tag{9}$$

The advantage of this method is that it enables one to differentiate between planar, parabolic and curved regions directly without further computation based on the characteristic clustering of the projected normals onto the tangent plane. The two eigenvalues of this covariance matrix indicate the local surface type as they give a measure of the curvature in the principal directions. In the case of a locally planar surface, both eigenvalues become zero. Locally parabolic surfaces result in one eigenvector being very small, while locally curved surfaces manifest themselves as two non-zero eigenvectors.

Caelli et al [9] extended this method to make it applicable to tessellated surfaces. In their approach, the distance between two surface points was defined as the length of the minimal path on the mesh from one node to the other (Fig. 2.2b). They also applied a weighting factor, thus the correlation matrices C_I, C_P are given as follows:

$$C_I = \frac{1}{n} \sum_{i=1}^{n} w_i^2 (x_i - x_m) \cdot (x_i - x_m)^T \tag{10}$$

$$C_P = \frac{1}{n} \sum_{i=1}^{n} w_i^2 (v_i - v_m) \cdot (v_i - v_m)^T \tag{11}$$

$$x = \frac{1}{n} \sum_{i=1}^{n} x_i w_i + x(1 - w_i) \tag{12}$$

The weight of each point is given as the Gaussian function of x, the centre point, and x_i :

$$w_i = \exp \frac{d(x, x_i)}{2\sigma^2} \tag{13}$$

where $d(x, x_i)$ is the minimum spanning distance between the two points measured along the edges of the triangulated surface.

The difficulty in defining the size of the neighbourhood is similar to the problems of the surface fitting. Because the data is not segmented, in the surrounding of discontinuities the estimated surface normals are distorted and this distortion spreads further in the second step, the computation of principal directions.

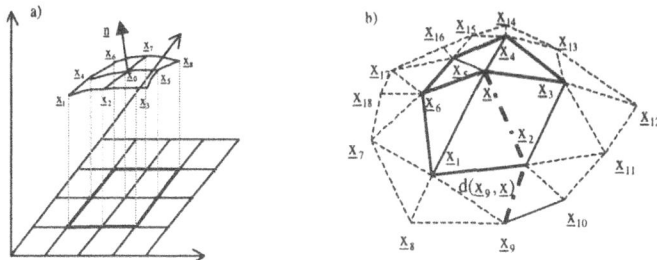

Fig. 4. The distribution of surface points on (**a**) regular Monge patch surface (**b**) tessellated surface of two-connected neighbourhood for correlation based methods of curvature computation. The direct neighbours of the central point are connected with a continuous line while the second connected neighbours are connected with a broken line. The dashed line depicts the distance measured between two nodes of the mesh

Estimation applying a Hessian matrix. A further approach to estimate the Gaussian and mean curvature on a mesh is given in [10]. If the surface on which the curvature computation is required is twice differentiable and the surface normals are close enough to each other, then the normal vector distribution around a point, given in the local frame, can be estimated by its first-order Taylor expression as shown in Fig. 5.

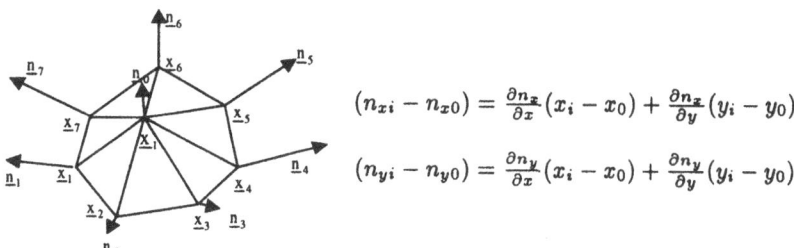

$$(n_{xi} - n_{x0}) = \frac{\partial n_x}{\partial x}(x_i - x_0) + \frac{\partial n_x}{\partial y}(y_i - y_0)$$

$$(n_{yi} - n_{y0}) = \frac{\partial n_y}{\partial x}(x_i - x_0) + \frac{\partial n_y}{\partial y}(y_i - y_0)$$

Fig. 5. First order Taylor approximation of the node normal at the centre of a tessellated patch

where $x_i = (x_i, y_i, z_i)$, $n_i = (n_{xi}, n_{yi}, n_{zi})$ and $n_i - n_0 = (n_{xi} - n_{x0}, n_{yi} - n_{y0}, 0)^T$ measures the change of the surface normal. The displacement between two surface points in the x and y directions is given by $\Delta x_i = x_i - x_0$ and $\Delta y_i = y_i - y_0$, respectively. Putting the partial derivatives in one matrix we get the Hessian matrix.

$$\aleph = \begin{pmatrix} \frac{\partial^2 f}{\partial x^2} & \frac{\partial^2 f}{\partial x \partial y} \\ \frac{\partial^2 f}{\partial x \partial y} & \frac{\partial^2 f}{\partial y^2} \end{pmatrix} = \begin{pmatrix} \frac{\partial n_x}{\partial x} & \frac{\partial n_x}{\partial y} \\ \frac{\partial n_y}{\partial x} & \frac{\partial n_y}{\partial y} \end{pmatrix} = \begin{pmatrix} \alpha & \beta \\ \beta & \gamma \end{pmatrix} \tag{14}$$

The target parameters, the Gaussian and mean curvature, can be expressed with the elements of the Hessian matrix as they are the product and the mean of the eigenvalues of this matrix.

$$H = \frac{1}{2}(\alpha + \beta) \tag{15}$$

$$K = \alpha\gamma - \beta^2 \tag{16}$$

Substituting the elements of the Hessian into the Taylor expansion we get two linear equations in the elements of the Hessian matrix. Applying the expression on N neighbouring points yields 2N linear equations; therefore using the method of least squares the 3 independent elements of the Hessian matrix can be estimated. More explicitly:

$$\Delta n_{xi} = \delta x_i \cdot \alpha + \Delta y_i \cdot \beta + 0 \cdot \gamma \tag{17}$$

$$\Delta n_{yi} = 0 \cdot \alpha + \delta x_i \cdot \beta + \Delta y_i \cdot \gamma \tag{18}$$

As discussed in section 2.2, this approach requires prior estimation of the normal vectors which is not simple in an irregular mesh structure. The importance of this method is that no direct computation of the second derivatives is required. They are yielded as the solutions of a linear equation system.

Discussion of correlation and Hessian estimators The correlation based method, as a first step, fits a plane to the raw points of the surface. The normal vector of this plane serves as the normal vector of the surface at the centre of the correlation window. Two orthogonal, tangent vectors of the plane are also estimated. This must be done for all points on the surface. The second step of the method is to project all normal vectors in the correlation window to the tangent vectors of the centre point. These projections are the second derivatives of the original surface in the local frame of the centre point. The z-axis is parallel to the local normal vector while x and y are parallel to the tangent vectors. The last step is to estimate the second derivatives at the centre point by applying a least-squares based correlation method on them. Substituting the second derivatives into the expression of K and H they can be computed as the first derivatives are zero in the local frame.

The extension of this approach modifies the approximation's linearity by applying higher weights to points closer to the central one, therefore it yields better results, but does not change the basis of the method.

In the Hessian based approach the surface normals were assumed to be available in the form of $n = (n_x, n_y, 1)$ in a local frame. This vector is not of unit length; it can be generated from the unit normal vector $(\tilde{n})$ by dividing all its components by the z one.

$$f = (x, y) \tag{19a}$$

$$\tilde{n} = (\tilde{n}_x, \tilde{n}_y, \tilde{n}_z) \tag{19b}$$

$$n = (n_x, n_y, 1) \tag{19c}$$

$$n_x = \frac{\tilde{n}_x}{\tilde{n}_z} = \frac{\partial f}{\partial x} \tag{19d}$$

$$n_y = \frac{\tilde{n}_y}{\tilde{n}_z} = \frac{\partial f}{\partial y} \tag{19e}$$

As shown, the availability of the normal vectors of this form is dependent on the availability of the first derivatives, the algorithm consists of the estimation of the second derivatives only. The original function describing the object surface is turned into its first partial derivatives in x by changing a point $p = (x, y, z)$ on the original surface to $p' = (x, y, n_x)$. The first order Taylor approximation of the first derivatives forms the base expression of this method.

$$dn_x = \frac{\partial n_x}{\partial x} dx + \frac{\partial n_y}{\partial y} dy \tag{20}$$

Applying the above given linear approximation to the points around a central point in a mesh, the unknown partial derivatives can be estimated. If there are more than two neighbours of the centre point the derivatives can be approximated by the method of least-squares.

As explained, the Hessian based method is identical to the second half of the basic correlation method, but it is applied to a non-regular parameterisation. However, the correlation method can also be applied to a mesh surface without the weighting factor. In that case the approximation is less accurate because it does not take into account that the original surface should diverge further from the local approximating plane at points further from the centre point. Both methods apply the same local co-ordinate frame for the calculation, therefore the first derivatives equal zero, and the results are the determinant and the trace of the Hessian matrix.

In the following sections we give our approximating solution for the calculation of both the Gaussian and mean curvatures directly on an irregular 3D triangular mesh. Like the methods discussed above, we will apply linear approximations to compute derivatives. However, the approximations are simplified by linear averaging of mesh normals.

3   Simplified approximation

The method presented in this section emphasises fast and efficient implementation. We assume the object surface is approximated by an irregular 3D mesh. The mesh nodes are given in an arbitrary 3D co-ordinate system as (x_i, y_i, z_i) triples, where i means the i-th node of the mesh. Mesh triangles are ordered lists of nodes. The ordering of the nodes contains the information about the facing (i.e. inward, outward normal) of the mesh triangle. The first step of the curvature computation is the well-known estimation of the node normals of a mesh. This is applied in computer graphics for shading mesh objects. The surface normal at a node point is generated as the sum of the surrounding triangular facets' normals. The 3D co-ordinates of the nodes of the triangles define a plane. The normal of this plane serves as the normal of the triangular facet.

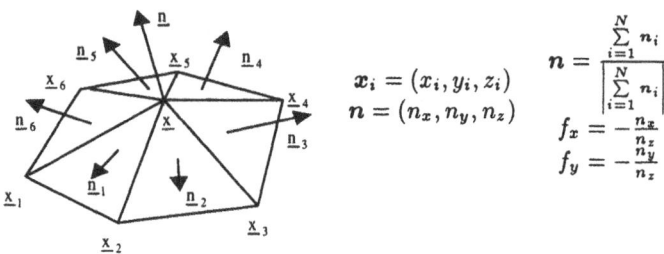

$$x_i = (x_i, y_i, z_i)$$
$$n = (n_x, n_y, n_z)$$

$$n = \frac{\sum\limits_{i=1}^{N} n_i}{\left|\sum\limits_{i=1}^{N} n_i\right|}$$
$$f_x = -\frac{n_x}{n_z}$$
$$f_y = -\frac{n_y}{n_z}$$

Fig. 6. Estimation of surface normals and the first derivatives

The surface normal at a point of a surface uniquely defines the first partial derivatives in any given reference frame in any variable as $f_x(x, y) = -\frac{n_x}{n_z}$ and $f_y(x, y) = -\frac{n_y}{n_z}$. At this point, an arbitrary (x, y) plane can be chosen. The normal can be calculated under all circumstances except for the unique case when the sum of the vectors is zero. This can only happen in the case of very special surfaces with unusually high curvature. If the resultant node normal is parallel to the parameter plane, the partial derivatives become infinite. However, this is unlikely in real faceted objects. It can appear in artificial scenes, though it is still not likely. In our experiments it happens very rarely and in all cases only on individual points and not on connected regions or curves. The easiest way to solve the problem is simply to ignore these points. If it is a non-acceptable loss, then the application of a temporary local frame, in which the normal is no longer parallel to the parameter plane must be considered.

The computation of the second derivatives follows a similar path. One of the co-ordinates of the original node points is replaced by one of the partial derivatives. The re-application of the above-defined process on the modified mesh yields two of the second partial derivatives (f_{uu}, f_{vv}). Replacing the node co-ordinates by the other first derivative, the other two partial derivatives can be generated. The question, which co-ordinate of the three should be changed, should still be answered. Before the first step, that is the generation of the node normals, a parameter plane was assumed. At that stage, the choice of plane is arbitrary. In later computations the replaced co-ordinate must be the one orthogonal to the parameter field.

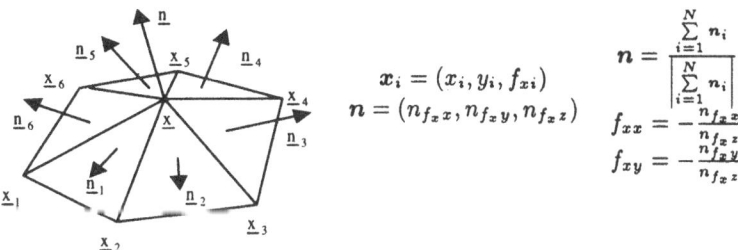

$$x_i = (x_i, y_i, f_{xi})$$
$$n = (n_{f_x x}, n_{f_x y}, n_{f_x z})$$

$$n = \frac{\sum_{i=1}^{N} n_i}{\left| \sum_{i=1}^{N} n_i \right|}$$

$$f_{xx} = -\frac{n_{f_x x}}{n_{f_x z}}$$
$$f_{xy} = -\frac{n_{f_x y}}{n_{f_x z}}$$

Fig. 7. Estimation of surface normals of the first derivative surface and the second derivatives

Although, the surface curvatures are invariant to transformations of the parameter field, they are not entirely invariant to choosing another non-parallel parameter field. For example, the ridge and valley surfaces differ only in their relation to the parameter field. The mean curvature depends on how the object is embedded in the 3D space. To answer this question - as well as to solve the problem of normal vectors parallel to the parameter field - the principal curvatures can always be computed with a parameter field perpendicular to the local surface normal. The original mesh contains the information, which side of the surface is outside and which is inside. This way, the locally concave and convex points can be differentiated depending on their Gaussian curvature.

3.1   Examples

To show the behaviour of the labelling method in a controlled environment under different circumstances, we made experiments on a sphere with unit radius. In the first experiment the sphere consisted of 80 triangles and in the second it contained 1280. The difference in the results shows that in the case of higher tessellation the approximation is more accurate. This is an interesting unpredicted result, since the mesh nodes are in exact position and the tessellation is uniform in each case, therefore the surface normals should be exact. In fact the error arose in the calculation of the second derivatives. As the surface normals are accurate, the first derivative surface contains no more errors than the applied number representation of the machine. According to our algorithm, the tessellation of the second derivative surface is the same as the original one on the (x, y) plane but different on the z axis. This yields a non-uniform tessellation of the second derivative surface, which leads to slight inaccuracies in the second derivatives (Fig. 8).

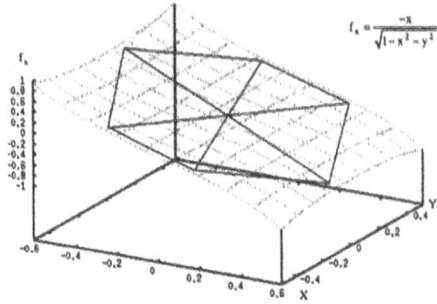

Fig. 8. The tessellation of the first derivative of a unique sphere in a local frame of a node

The discrete lines in Fig. 9 are the results of the rotation of the hexagon around its centre. The (x, y) plane of the co-ordinate frame in Fig. 8 is the perpendicular plane to the surface normal at the point in which the curvatures are to be computed. The orientation of this plane around the z-axis is arbitrary. We aligned the z-axis to the normal vector by rotating the original frame around the x and the y axis. As the second derivative surface is not rotationally symmetric the rotation of the hexagon projected onto the surface yields different approximations of the surface normals, leading to different second derivatives, at the centre point. The orientation and the number of triangles of the polygon applied in the calculation depend on the original tessellation of the surface in question.

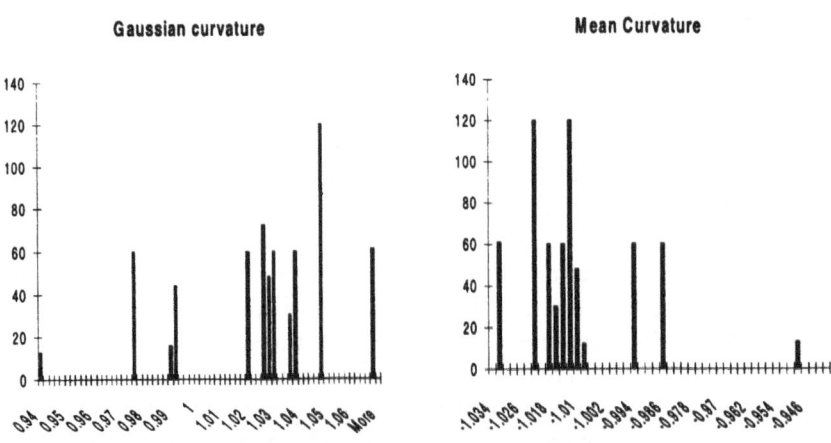

Fig. 9. Histograms of the mean and the Gaussian curvature distributions of points of a unit sphere. The computation was made by the simplified method. The discrete lines of the histogram are the result of the asymmetric second derivative surface

To demonstrate the power of the simplified method we also applied it to an object created by registering synthetically generated range images. Although there is no noise present in the range images, the imperfect registration of the images creates damage to the surface. The image in Fig. 10 is the input image of the segmentation, Fig. 11 shows some of the regions of the object with identical topological labels. The topolocical labels are planar, cylindrical valley, cylindrical ridge, saddle pit, saddle ridge, spherical pit and non-valid. The obviously mistaken non-valid and pit points occur at the edges of the mesh. At these points the mesh does not continue, the object has no back and bottom part, therefore the approximated surface normals are less reliable. There are rough sections along the registered boundaries of the different views, causing points labelled pit and saddle around these areas. The surface deviations affected by the range image registration process are clearly visible on the rendered image.

Fig. 10. Rendered original object

We created a segmentation algorithm for objects given by 3D irregular meshes. This algorithm is based on H, K labelling followed by region growing. We also successfully applied this segmentation method to objects bounded by meshes of closed as well as open surfaces as shown in Fig. 12.

4   Conclusion

In this paper we have addressed the problem of computing differential geometric measures on irregular 3D mesh surfaces. We have shown that this problem is equivalent to the problem of computing first and second derivatives on an irregularly sampled surface. We reviewed the different approaches to solve the problem and introduced our own solution. Our method is a simplified approach to estimate the derivatives, applying fewer and simpler operations. We have shown a segmentation algorithm based on the curvature computation of our simplified method, which shows that our method produces satisfying results to serve as a basis of higher level vision algorithms.

Fig. 11. The first 5 images show some of the identically labeled topologicial regions of the test objects. The bottom right image shows the disturbed area between regions

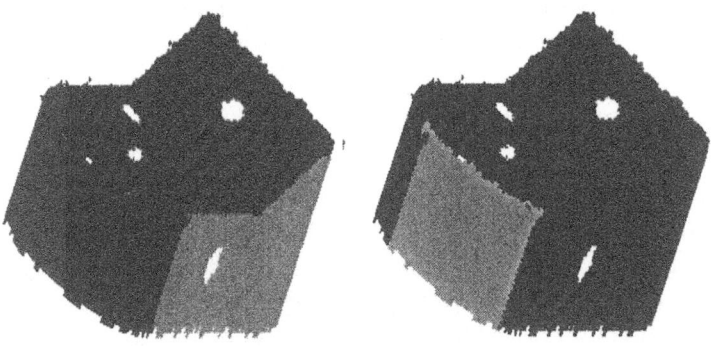

Fig. 12. Segmentation results. The segmentation algorithm is based on the simplified computation of mean and Gaussian curvature

References

1. P.J. Besl and R.C. Jain (1986) Invariant Surface Characteristics for 3D Object Recognition in Range Images. Journal of Computer Vision, Graphics, and Image Processing **33**, 33-80
2. P.J. Besl and R.C. Jain (1988) Segmentation Through Variable-Order Surface Fitting. IEEE Trans. PAMI **10**, No. 2, 167-192
3. J.J Koenderink, A.J van Doorn (1992) Surface shape and curvature scales. Image and Vision Computing **10**, No. 8, 557-565
4. P. Saint-Marc, J.S. Chen and G. Medioni (1991) Adaptive Smoothing: A General Tool for Early Vision. IEEE Trans. PAMI **13**, No. 4, 514-529
5. L. Alboul, R. van Damme (1006) Polyhedral Metrics in Surface Reconstruction. Mathematics of Surfaces VI, Oxford University Press, 171-199
6. P. Krsek, G. Lukács and R.R. Martin (1998) Algorithms for Computing Curvatures from Range Data. The Mathematics of Surfaces VIII, 1-16
7. P. Liang and J. S, Todhunter (1990) Representation and Recognition of Surface Shapes in Range Images: A Differential Geometric Approach. Computer Vision Graphics, and Image Processing **52**, 78-109
8. J. Berkmann and T. Caelli (1994) Computation of Surface Geometry and Segmentation Using Covariance Techniques. IEEE Trans. PAMI **16** No. 11, 1114-1116
9. T. Caelli, G. West, M. Robey and E. Osman (1999) A relational learning method for pattern and object recognition. Image and Vision Computing **17**, 391-401
10. R.C. Wilson, E.R. Hancock (1999) Consistent topographic surface labelling. Pattern Recognition **32**, 1211-1223
11. M.Umasuthan, A.M. Wallace (1996) Outlier removal and discontinuity preserving smoothing of range data. IEE Proc. Vision, Image & Signal Processing **143**, No. 3, 191-200

Remarks on Meshless Local Construction of Surfaces

Robert Schaback

Universität Göttingen
Institut für Numerische und Angewandte Mathematik
Lotzestraße 16–18
D–37083 Göttingen
Germany

Summary. This contribution deals with techniques for the construction of surfaces from N given data at irregularly distributed locations. Such methods should ideally have the properties

- computational efficiency,
- smoothness of the resulting surface, if required, and
- quality of reproduction,

but these goals turn out to be hard to meet by a single algorithm. Methods are split into a single construction or precalculation part and subsequent pointwise evaluations. Both parts are analyzed with respect to their complexity. It turns out that one has to expect the main workload on the side of geometric subproblems rather than within numerical techniques. Furthermore, if exact reconstruction at the data locations is required, and if the user wants to avoid solving non–local linear systems, there is no way around localized Langrange–type interpolation formulae. Thus two instances of such techniques are studied in some detail:

- interpolation by weighted local Lagrangians based on radial basis functions and
- moving least squares.

While the former is much more simple than the latter, it still has some deficiencies in theory and practice. Moving least squares, if equipped with certain additional features, turn out to be widely satisfactory, even in difficult cases.

1 Introduction

Ignoring more general and more subtle definitions, we consider surfaces here as sets Y of points $y \in I\!\!R^3$ that are either

- *implicitly* represented via an equation $g(y) = 1$ for a scalar function g on $I\!\!R^3$ or
- *explicitly* represented as images $y = F(x)$ of a function F defined on a subset Ω of $I\!\!R^2$ called the *parameter domain*.

Implicit representations have the advantage that one can often define a body with surface Y as the set of points y with $g(y) \leq 1$, while all points with $g(y) > 1$ are "outside". This feature is very convenient for ray tracing algorithms, because one has a quick test for points y on the ray for being inside or outside the body. The transition between implicit and explicit representations of the same surface is a difficult problem that we ignore here. An explicit representation is called *nonparametric*, if the defining map F has the simple form $F(x) = (x, f(x))$ with a scalar function f on Ω.

We focus on the construction of surfaces from given data. These can come in different forms. The most standard case is *quantitative point data* as

- a set $\{y_1, \ldots, y_N\} \subset I\!R^3$ of points on (or near) the surface, or
- a set $X = \{x_1, \ldots, x_N\} \subset \Omega \subseteq I\!R^2$ together with a set $\{y_1, \ldots, y_N\} \subset I\!R^3$ such that $y_j = F(x_j)$ for all j, $1 \leq j \leq N$, either exactly or approximately.

Again, the nonparametric setting specializes to the case $y_j = (x_j, f(x_j))$, $1 \leq j \leq N$. In general, derivative values can be specified, but we skip over such extensions here. More serious are *qualitative* data like "smoothness", "good shape" or whatever the user may prescribe. Here, we ignore everything except smoothness, and we shall restrict the latter to the classical mathematical definition.

The construction of explicit representations of surfaces from data of the form $y_j = F(x_j)$ can clearly be done by any multivariate vector–valued scattered data interpolation or approximation technique. This will be the main topic of this paper. But before that, and for completeness, we want to point at the specific problems coming up in the case of unstructured data $\{y_1, \ldots, y_N\} \subset I\!R^3$. Imagine 8 data points to be given, forming the vertices of the unit cube in $I\!R^3$. Whatever method is used to find a surface containing these points, there is an intrinsic ambiguity, described by the following possible solutions:

1. A closed, bounded and connected surface, e.g. a sphere,
2. three solutions consisting of two connected components, each picking up the four points of opposite facets, e.g. two parallel planes,
3. twelve different U–shaped solutions formed by picking up the vertices of a chain of three adjacent facets.

This ambiguity even concerns the global topological structure of the solution, and the arising problem is much more serious than finding a numerical method that actually constructs *some* surface containing the data points. For instance, given an arbitrary multivariate scalar–valued interpolation technique for scattered data, one can easily construct a function g on $I\!R^3$ such that $g(y_j) = 1$ for all j. Now, except for certain degenerations, the set of points y with $g(y) = 1$ will define a surface that picks up the given data, but it is not clear how the method behaves in situations like the one above. In principle, it also does not help to do a local triangulation first, because the same problem arises with the triangulation. We leave this interesting area to future research.

2   Construction, Evaluation and Complexity

For the rest of this paper, we focus on constructing explicit representations of surfaces from given data in the form $y_j = F(x_j)$, $1 \leq j \leq N$ for some unknown function F. In many cases, the actual and final evaluation of surface points will not be based on the given data, but rather on some intermediate data needed for the representation. For instance, many CAD packages evaluate surfaces from Bernstein–Bézier control nets, and then these nets form the intermediate data for the representation. We thus split the process in

$$\text{Input data} \xrightarrow{Construction} \text{Representation data} \xrightarrow{Evaluation} \text{Surface points.}$$

The construction step can contain some data reduction. Typical cases are provided by the lower levels of hierarchical or multilevel schemes for representing surfaces (see [36] for example), or by greedy methods like [33]. We do not consider such methods here. There are also cases where the intermediate data are much larger than the original data. We mentioned an example at the beginning of this section.

The construction step will often be much more complex than the evaluation step, but the evaluation usually has to be performed many times. This is why it is prohibitive to have an $\mathcal{O}(N)$ complexity of evaluation. But if evaluation at a point x is to be done a $\mathcal{O}(1)$ cost and reasonable quality, one needs at least some geometric information about data near to x. This geometric part of the reconstruction process turns out to be much more important than expected, and we conjecture the following:

> In the development of *efficient* techniques for reconstruction of multivariate functions (or surfaces, in particular), the major computational complexity lies within the *geometric* algorithms, not the numerical techniques.

This fact should have been widely recognized in the past, but the scientific focus still is very much on the side of Numerical Analysis than on Computational Geometry. For instance, in any case of univariate spline interpolation, we need for each evaluation point x the smallest knot interval $[x_j, x_{j+1}]$ containing x. This is what we called a "geometric" information. A naive way getting this information is to use a sorting algorithm at cost $\mathcal{O}(N \log N)$ within the construction step, followed by an $\mathcal{O}(\log N)$ search within each evaluation. The actual numerical construction step via solving a banded system will take $\mathcal{O}(N)$ operations, while the numerical evaluation is of $\mathcal{O}(1)$

complexity for a fixed degree. Our statement is valid already in this simple example, but things will naturally be worse in the multivariate case. This is why we deal with geometric issues in the next section.

To fix our efficiency goals somewhat more precisely, let us look at the relative computational complexity of construction and evaluation of surfaces, provided that N data are given.

- We consider a *construction* technique to be efficient, if it produces $\mathcal{O}(N)$ intermediate data at a computational cost of $\mathcal{O}(N)$ operations for a fixed accuracy requirement. This will rule out precalculations involving triangulations, sorting methods, or full–size linear systems, and it will normally require some additional assumptions on the geometry of the data.
- We consider an *evaluation* technique to be efficient, if it takes $\mathcal{O}(1)$ operations to evaluate the surface at a single point. This rules out all nonlocal methods, methods based on the evaluation of sums with more than $\mathcal{O}(1)$ terms, or methods that require nontrivial search techniques for each evaluation.

The rest of the paper is concentrating on techniques that at least promise to meet these goals, together with the ability to yield surfaces of any prescribed smoothness. The reader will wonder how and why we drop the additional $\log N$ complexity factor that already arises in univariate spline algorithms. But we shall show below that this is justified for "reasonable" data geometries, and in the univariate case it turns out that this is possible whenever there is an upper bound ρ on the mesh ratio

$$\frac{\max\limits_{1 \le j < N} |x_{j+1} - x_j|}{\min\limits_{1 \le j < N} |x_{j+1} - x_j|}.$$

3   Efficient Geometric Algorithms

If there are no additional assumptions on the data locations, any geometric algorithm with a complexity of $\mathcal{O}(N \log N)$ within the construction step and $\mathcal{O}(\log N)$ for each evaluation must be considered to be efficient, as we are taught by univariate spline theory. But if the set $X = \{x_1, \ldots, x_N\} \subset \Omega \subseteq I\!\!R^2$ of data locations is not too badly distributed, we hope to get away with $\mathcal{O}(N)$ and $\mathcal{O}(1)$, respectively. The first basic idea is to assume *quasi-uniformity* of the data locations x_j, $1 \le j \le N$ on a bounded domain Ω which at least contains the convex hull of the data. This property requires that the quotient of the *fill distance*

$$h := h(X, \Omega) := \max_{y \in \Omega} \min_{x_j \in X} \|y - x_j\|_2 \tag{1}$$

and the *separation distance*

$$q := q(X) := \frac{1}{2} \min_{x_j \neq x_k \in X} \|x_j - x_k\|_2 \leq h(X, \Omega)$$

is bounded above by a constant $\rho > 1$.

The second basic idea is to ignore sorting and triangulations in favour of the *k nearest neighbor problem*. The goal is to do some geometric preprocessing at $\mathcal{O}(N)$ cost such that for every given evaluation point x it takes only $\mathcal{O}(k)$ operations to get the k nearest neigbours from the data set X.

The standard folklore recipe, described in d dimensions here, implements a space decomposition technique like those used in Computer Graphics. By a first $\mathcal{O}(N)$ scan over the given N data locations, a bounding box for the whole data set is constructed, defined by maximal and minimal coordinates. Then there are several possible strategies for splitting the global box into $\mathcal{O}(N)$ smaller boxes, hopefully containing only $\mathcal{O}(1)$ data points each.

A standard grid–type decomposition of the global bounding box does the job for quasi–uniform data sets. To see this, let us first prove that h^{-d}, q^{-d}, and N have the same asymptotics for $N \to \infty$. In fact, since each data point has a ball of radius q around it such that the ball does not contain any other data point, these balls are disjoint and the sum of their volumes must be bounded above by a constant. Thus $N = \mathcal{O}(q^{-d})$. On the other hand, the union of the balls of radius h around the data points must cover the domain Ω, and thus the sum of their values is bounded below by a constant, proving $h^{-d} = \mathcal{O}(N)$.

Now let n_B be the maximum number of data points in each box. The balls of radius q around these points will be disjoint and contained in the box of volume $\mathcal{O}(1/N) = \mathcal{O}(q^d)$ plus a surrounding volume that can be bounded by $\mathcal{O}(q^d)$, too. Therefore n_B is bounded above by a constant.

If the data distribution is not quasi–uniform, a decomposition via median splits into a binary tree of boxes will work at the price of $\mathcal{O}(N \log N)$ operations. We prefer the former case and suggest to drop excess points of clusters, keeping the number of points in each grid box at $\mathcal{O}(1)$ by brute force. The treatment of details of surfaces related to data clusters can always be postponed to a second problem, working locally at a finer scale, and having the residuals of the first step as input data. As a byproduct, the above strategy provides a simple "thinning" algorithm along the lines of papers by Floater and Iske [14–16].

Anyway, it takes $\mathcal{O}(N)$ or $\mathcal{O}(N \log N)$ operations to distribute the given N data into $\mathcal{O}(N)$ boxes with $\mathcal{O}(1)$ points in each box. For any given point x, it takes $\mathcal{O}(1)$ operations to find the box containing x, if the data are quasi-uniform. In the general case, however, one has to go down the binary tree at $\mathcal{O}(\log N)$ cost.

The basic data structure will consist of a list of point indices for each box. The implementation of such a structure can use standard techniques from sparse matrices. To cope with allocation problems, we prefer to use a second scan over all data points that just counts the number of points in each box. Then allocation can be done once and precisely, and the actual placement of points into the correct boxes is done by a third scan over all data points. The overall storage requirement is $\mathcal{O}(N)$

Since the number of points in each box is $\mathcal{O}(1)$, one can then easily use the data structure to solve nearest neighbour problems for any point x with a complexity of $\mathcal{O}(k)$ (in the quasi-uniform case, or $\mathcal{O}(k \log N)$ in general) for a fixed number k of required neighbours of a point x. The idea is to go into the box of x first and then into all neighbouring boxes with increasing distance, picking up all the data points in those boxes. Whenever one has finished the boxes covering a full ball of radius r around x, one can be sure that at least all neighbours of x at distance at most r are found. The process is stopped if one has found at least k such points, and these are then sorted with respect to their distance to x. No more that $\mathcal{O}(k^d) = \mathcal{O}(1)$ boxes need to be checked in the quasi-uniform case, because the k nearest neighbours cannot be further away from x than $h + 2(k-1)q$.

Note that a univariate simplification of this algorithm allows to sort N real numbers in $\mathcal{O}(N)$ operations, provided that they are quasi-uniformly distributed in a bounded interval. Such algorithms are called "sorting by distribution", and their prototype is the well-known *radix sort*. Furthermore, a subsequent search algorithm can then be implemented at $\mathcal{O}(1)$ cost.

Methods for constructing good triangulations will cost at least $\mathcal{O}(N \log N)$ operations in the two-dimensional case, but they work for general point distributions. Whether they can be reduced to $\mathcal{O}(N)$ computational complexity for quasi-uniform data, is beyond the knowledge of the author.

4   Localization and Oversampling

We now go back to numerical techniques and introduce some more notation. If both the calculation and the evaluation step are linear, one can write

$$b_k = \sum_{\ell=1}^{N} \beta_{k\ell} y_\ell, \ 1 \leq k \leq M$$
$$F(x) = \sum_{k=1}^{M} u_k(x) b_k \tag{2}$$

with certain evaluation functions u_k and intermediate data b_k, $1 \leq k \leq M$, starting from input data y_ℓ at x_ℓ for $1 \leq \ell \leq N$. The number M can be much larger that N, and the index k of intermediate data b_k and basis functions u_k need not have any relation to the index ℓ of the data. For example, if we define the u_k as basis functions of some finite element space or as Bernstein–Bézier or NURBS basis functions with respect to some representation of the surface by many standard patches, we require the intermediate data to be nodal data for finite elements or to be control points with respect to the various standard surface patches. In such a case, the value of M is much larger than N, and it may be not at all obvious that the constructed surface has sufficient smoothness, unless certain linear equations for the control points b_k are satisfied. We want to ignore the "patching problem" in this contribution, but we shall see later how it arises unexpectedly.

The resulting surface mapping is

$$
\begin{aligned}
F(x) &= \sum_{k=1}^{M} u_k(x) \sum_{\ell=1}^{N} \beta_{k\ell} y_\ell \\
&= \sum_{\ell=1}^{N} y_\ell \sum_{k=1}^{M} u_k(x) \beta_{k\ell} \\
&= \sum_{\ell=1}^{N} y_\ell L_\ell(x),
\end{aligned}
\tag{3}
$$

and the final form uses Lagrange–type functions

$$
L_\ell(x) := \sum_{k=1}^{M} u_k(x) \beta_{k\ell}, \ 1 \leq \ell \leq N
$$

that have to satisfy $L_\ell(x_j) = \delta_{j\ell}$ if exact reproduction of the data is required.

In principle, one could fix the evaluation functions u_k beforehand, depending on the final application, and maybe even in a very convenient local form, via finite elements, bicubic spline patches or NURBS. The matrix B with entries $\beta_{k\ell}$ should then be a one–sided inverse to the matrix U with entries $u_k(x_j)$. For $M \geq N$, and if U has full rank N, the determination of such an inverse is possible in theory, but we already mentioned the additional conditions on the intermediate data that will be required to guarantee smoothness, if the u_k are not automatically smooth enough.

At this point, the reader should have understood why we do not want to get into serious trouble with smoothness conditions defined indirectly via additional conditions on the intermediate data. We restrict ourselves to cases where the functions u_k or L_ℓ have the required smoothness, and then we are free to find convenient linear mappings to generate the intermediate data we need.

But there is an important point to be noted at this stage. If we want to make both steps local and carry them out as they are in (2), i.e. without reformulation of the first equation as a linear system for the b_k, the matrices U and B should be sparse, and at the same time be one–sided inverses of each other. This is a very serious obstacle. In general, matrices with a fixed, but irreducible sparsity structure can have full inverses, if perturbations of the matrix entries are allowed ([9], p. 271). This rules out the case $M = N$ except for the standard situation $U = I$ that we analyze below. For $M \gg N$ the chances are better, but there are just a few results on such "oversampling" techniques. Roughly speaking, the above argument amounts to the following:

> If locality or sparsity of both the construction and the evaluation process for exact reconstruction of surfaces from parametric data is required, and if both steps are carried out by linear formulae without solving linear systems, one has to do oversampling or to stick to a variation of Lagrange interpolation.

We refrain from casting this guideline into the shape of a theorem, but we shall follow it throughout the paper.

For $M = N$, most of the well–known U matrices (e.g. from splines or radial basis functions) are non–sparse. Note that they are the inverses of the (possibly sparse) matrices of the linear systems for calculating the intermediate data. The pitfall of the above principle is avoided by not using the inverses as linear mappings. Instead, one solves the (possibly sparse) linear systems.

Let us describe a univariate example. Imagine a standard scalar univariate interpolation problem with data $y_j = f(x_j)$, $1 \le j \le N$ for nodes $x_1 < x_2 < \ldots < x_N$. We already used this example for pointing at the bulk of work induced by generating the necessary geometric information. We now focus on numerical techniques for construction of intermediate data and evaluation. Linear splines have a Lagrange formulation without any construction step. This follows the above principle via local Lagrange interpolation. Splines of higher degree are usually treated via a nonlocal construction step involving a sparse system with a non–sparse inverse. This follows the principle by resorting to solving a system. There is no local formula that allows circumvention of solving a system in case of $M = N$ and higher–degree splines. This is what the above principle enforces, if neither Lagrange interpolation, nor solving a system, nor oversampling is done. But oversampling can possibly avoid both the nonlocal evaluation and the linear system. In fact, if sufficiently many derivatives at the knots are approximated using the point data by any of the standard techniques, and if piecewise odd–degree Hermite interpolation is done on the oversampled data, we get away without any system, using local construction and evaluation.

The general trick is to use oversampling in such a way that sufficiently many local intermediate data are constructed, such that a subsequent local construction step finds all the data it needs. Finding good multivariate oversampling strategies is a major open problem. But note that the example also shows that we are back to a situation that we did not want to pursue here: the introduction of the "patching problem" through the back door via oversampling. This is easy in the univariate case, but serious in multivariate settings. We close this section with the remark that there may be sparse *approximate* inverses. Examples are in [35]. Transition from interpolation to approximation will thus be another feasible workaround, but note that in this context approximation coincides with quasi–interpolation.

5 Local Lagrange Interpolation

Let us go back to interpolation and consider the simple case $U = I$ implied by Lagrange interpolation on the original data, and look at localized techniques. In such a situation there are no intermediate data, and there is no preprocessing required and no system to be solved. On the downside, we now need Lagrange–type evaluation functions which have a prescribed global smoothness and a cheap $\mathcal{O}(1)$ local evaluation. Such functions do exist, but the race for practically good functions is open. The early Shepard–type techniques were nonlocal, and their localized extensions were nonsmooth. On the other hand, any sufficiently smooth and sufficiently localized peak function u_k which is one at x_k will do the job, but at the price of a useless resulting surface, looking like a bed of nails. The approximation quality comes in as a third criterion, besides smoothness and computational complexity.

But there are simple and cheap methods that do better than localized peaks. A good class of methods with limited smoothness is provided by *natural neighbour coordinates*. Originally due to Sibson [37,38] as a method yielding a continuous surface, there was an extension by Farin [12] to a continuously differentiable interpolant. If implemented naively, natural neighbour coordinates require a preprocessed Dirichlet tesselation at a cost of at least $\mathcal{O}(N \log N)$, which violates our efficiency goals. If a preprocessing at $\mathcal{O}(N)$ is done for solving the k nearest neighbour problem as described in section 3, one can possibly calculate the natural neighbour coordinates locally within each evaluation step, getting a $\mathcal{O}(1)$ cost per evaluation. But since smoothness is limited, we do not pursue natural neighbour techniques in this paper. We also skip over extensions of Shepard–type techniques and refer the reader to [1] and subsequent papers.

Let us describe a rather general recipe for calculating smooth local interpolants. Around any of the data points $x_j \in \Omega$ we consider a ball $B_r(x_j)$ of some fixed radius $r > 0$. Then we take all points x_k in this ball and construct a local Lagrange function L_j^{loc} with respect to these points by an arbitrary method for local scattered data interpolation, provided that the solution has the required smoothness. Thus we have

$$L_j^{loc}(x_k) = \delta_{jk} \text{ for all } x_k \text{ with } \|x_j - x_k\| < r, \ 1 \leq j, k \leq N,$$

but we cannot use these functions globally, because they fail to work on far-away points. But there is an easy remedy. Take any nonnegative scalar function w on $\mathbb{R}$ with $w(0) = 1$ and support $[-r, r]$ such that $w(\|x - x_j\|_2)$ has the required smoothness. Then

$$L_j(x) := L_j^{loc}(x)w(\|x - x_j\|_2) \tag{4}$$

will be a global Lagrange function, and the surface construction can proceed via (3). Note that this recipe allows for a wide range of possible cases, and the contest for good examples is open. We shall show some cases after we have described how to solve the local scattered data interpolation problems.

If the data distribution is quasi–uniform, the calculation of a local Lagrangian function at x_j will require only $\mathcal{O}(1)$ operations. Precalculation of all Lagrangians can be done at $\mathcal{O}(N)$ complexity, and local evaluation at a single point x will only require $\mathcal{O}(1)$ Lagrangians. Thus we have an efficient method in the sense of section 2, independent of the type of local interpolation used. The race for cases with good reproduction qualities is open.

6   Radial Basis Functions

Now it is time to explicitly describe the tools we want to use for local interpolation to scattered data. The presentation can be brief, because there are many survey articles on the subject (in chronological order: [18,17,10,24,11,20,26,8,31,35]). By a fundamental observation of Mairhuber [21], nontrivial spaces for multivariate scattered data interpolation must necessarily depend on the data locations. To make this dependence as simple as possible, one uses functions of the form

$$s(x) = \sum_{j=1}^{N} \alpha_j \phi(\|x - x_j\|_2) + \sum_{i=1}^{Q} \beta_i p_i(x)$$
$$0 = \sum_{j=1}^{N} \alpha_j p_i(x_j), \ 1 \leq i \leq Q \tag{5}$$

with a *radial basis function* ϕ on $[0, \infty)$ and a basis $p_1, \ldots, p_Q$ of the space $\mathbb{P}_m$ of bivariate polynomials of degree up to $m - 1$, where $Q = m(m + 1)/2$. The function ϕ and the number m are related by the requirement that ϕ

must be (strictly) conditionally positive definite of some order $m_0 \leq m$, and this property makes sure that the systems

$$s(x_k) = \sum_{j=1}^{N} \alpha_j \phi(\|x_k - x_j\|_2) + \sum_{i=1}^{Q} \beta_i p_i(x_k) = y_k, \ 1 \leq k \leq N$$

$$0 = \sum_{j=1}^{N} \alpha_j p_i(x_j), \ 1 \leq i \leq Q \tag{6}$$

arising for arbitrary scattered data problems are uniquely solvable, if there is no nontrivial polynomial in $I\!P_m, m \geq m_0$, that vanishes at all data locations $x_1, \ldots, x_N$. The coefficients α_j and β_i are scalars in the case of nonparametric data, and vectors in the general case. We shall ignore this in the sequel, restricting ourselves to the scalar case without loss of generality.

The most prominent examples of radial basis functions are

$$\begin{array}{lll}
\phi(r) = r^\beta, & \beta > 0, \beta \notin 2I\!N_0, & m_0 = \lceil \beta/2 \rceil \\
\phi(r) = r^{2k} \log(r), & k \in I\!N \quad \text{(thin-plate splines)} & m_0 = k+1 \\
\phi(r) = (c^2 + r^2)^\beta, & \beta < 0, \quad \text{(inverse multiquadrics)} & m_0 = 0 \\
\phi(r) = (c^2 + r^2)^\beta, & \beta > 0, \beta \notin I\!N_0 \quad \text{(multiquadrics)} & m_0 = \lceil \beta \rceil \\
\phi(r) = e^{-\alpha r^2}, & \alpha > 0 \quad \text{(Gaussians)} & m_0 = 0 \\
\phi(r) = (1-r)_+^4 (1+4r) & & m_0 = 0
\end{array}$$

together with their orders m_0 of conditional positive definiteness. A comprehensive presentation of these functions together with full proofs of their fundamental properties is in [34].

Note that in the context of section 2 we have representation data consisting of two vectors $\alpha \in I\!R^N$, $\beta \in I\!R^Q$, which already is a weak form of oversampling in case of $Q > 0$. But, except for trivial choices of scaling, the system (6) has no sparse inverse, even if a compactly supported function like $(1-r)_+^4(1+4r)$ is used. The latter function is C^2 on R^2 when written as a radial function of two variables, and it can act as a weight function in (4). Other reasonable weight functions are

$$w_{\delta,k}(r) = \begin{cases} 1 & r \leq 1-\delta \\ \delta^{-2k}(1-r)^k(r - (1-2\delta))^k & 1-\delta < r \leq 1 \\ 0 & r > 1 \end{cases} \tag{7}$$

for $\delta \in (0,1)$ and $k > 0$, yielding prescribed degrees of smoothness.

7 Global Interpolation by Radial Basis Functions

We do not consider global solutions of large–scale scattered data interpolation problems in detail here. For completeness, we only point out the two current lines of research and mark their fundamental differences. The starting point is the behaviour of radial basis function interpolants with respect to scaling. It is a standard technique, arising already in finite elements and being the background of the convergence theory initiated by Strang and Fix [39], to scale the interpolants in a way that is proportional to the data density, using "narrow" basis functions for dense data and "wide" basis functions for coarse data. For historical reasons this is called a *stationary* setting, while the *nonstationary* case uses the same radial basis function for all possible interpolation problems, irrespective of the data density.

Let us first look at computational issues. In the stationary setting, the arising matrices will have a condition that is basically independent of the data density. For compactly supported basis functions, the sparsity structure is fixed and the evaluation of approximants will be cheap due to localization. In the nonstationary setting the condition will dramatically increase when the data get dense, because rows and columns of the system matrix tend to be more and more similar. Furthermore, sparse matrices arising from compactly supported radial basis functions get filled up, and the complexity of evaluation increases.

But the situation is different, if we look at approximation properties. In the nonstationary setting, all radial basis functions have good approximation properties which are closely related to the numerical condition: the better the approximation properties, the worse the condition [30]. On the other hand, in the stationary case there is no convergence for interpolation problems based on integrable radial basis functions [7], while thin–plate splines and multiquadrics show good approximation properties. But the latter do not share the advantage of the stationary setting with respect to the matrix structure: the systems will always be non–sparse.

Thus there is no fully satisfying way out, if users look at problems on varying scales. Using the stationary setting with global radial basis functions like thin–plate splines, powers or multiquadrics will cause no convergence problems, but the user is forced to add strategies for dealing with large full matrices and a costly evaluation process. The groups around M.J.D. Powell [27,29] and R. Beatson [6,4,5,3] have made great progress in this direction. A second approach uses compactly supported radial basis functions and exploits sparsity as much as possible. If fill–in is to be limited, one is bound to a stationary setting, but then there are problems getting good approximation quality, because there is no convergence in theory. As long as the data are not too dense, the stationary technique improves with data density, but there is a small final error level that cannot be improved by adding more data. This phenomenon was called *approximate approximation* my Mazỳa and Schmidt [22], and it deserves further study. The approximation quality of the final level is mainly determined by the admitted amount of fill–in [32], but the natural way out of this is to go over to multiscale techniques [13,14,23] applying the steps of a stationary setting recursively to residuals. This is quite successful, but still needs theoretical work. First steps are in [19].

The method of section 5, using local weighted Lagrangians, avoids solving large systems and guarantees locality without using compactly supported basis functions. Its properties will be discussed in the next section.

8 Local Weighted Interpolation by Radial Basis Functions

Let us now look at some specific cases, implementing the cut–off Lagrangian technique of section 5 via local interpolants based on radial basis functions from section 6. For ease of publication in printed form, we confine ourselves here to simple 2D graphics and present much more sophisticated 3D images at the conference. It is a rather convenient rule–of–thumb to use about 50 local data around each evaluation point, and thus we start in Figure 1 with presenting a one–dimensional cross–section of the Lagrangian calculated via thin–plate splines and linear polynomials for 49 local neigbours on a two–dimensional grid. These 49 neighbors are within a circle of radius 0.5 on a grid with spacing $1/8$, and thus the cross section of the Lagrange function along an axis has 8 symmetric zeros in $[-0.5, 0.5]$, being regularly distributed at distance $1/8$, if zero is added. To see the behaviour outside $[-0.5, 0.5]$, we replaced the values inside by zero to get the second plot in Figure 1. The outside peaks have a maximum height of 0.000717, and this is a coarse upper bound of the relative deviation between the global and local Lagrangian. This unexpected behaviour of thin–plate spline Lagrangians was first observed by Powell [25]. The decay for arguments tending to infinity is exponential, and thus a weighted cutoff does no serious harm. In our figures, we have not yet multiplied the local Lagrangian with a weight function, but we prefer to use weights that are equal to 1 for most of $[0, 1]$, because otherwise the peak of the Lagrangian gets too sharp. A good strategy for the Lagrangian based on 49 points was (7) for $\delta = 0.1$.

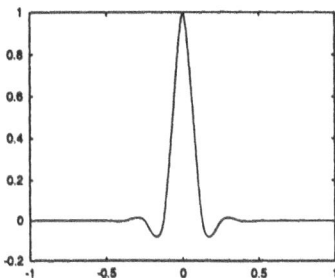 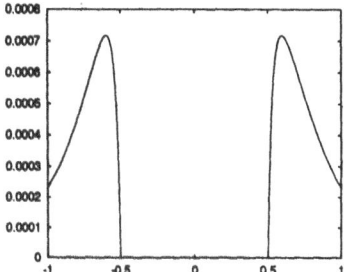

Fig. 1. Local Lagrange Function for Thin–plate Splines

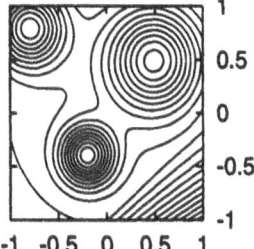 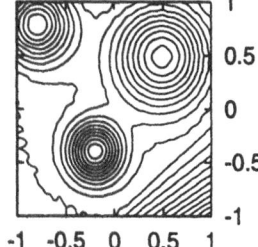

Fig. 2. Contour Plots

More sophisticated examples reveal that at high graphical resolutions the smooth cut–off induced by the weight function shows up considerably, though it is quantitatively of a small order of magnitude. Thus there is quite some work to be done on methods of this kind. An example is provided by Figure 2, where contours of the reproduced Franke–type surface, plotted at high resolution on the right–hand side, get rough in comparison to the original function on the left.

We now want to focus on the reproduction quality and start with the remark that the classical error bounds for radial basis function interpolation in the nonstationary setting are local. This is not directly stated in the literature, but can be read between the lines of the various proof techniques, e.g. [41,28]. In principle, if the fill distance $h := h(X, \Omega)$ of (1) is small enough, and if local reconstruction is to be done at some point $x \in \Omega$, one can confine the local interpolant to data at points x_j with $\|x - x_j\|_2 \leq ch$ with a suitable constant $c > 1$. Thus the number of locally required data points can be bounded independent of h for reasonably distributed data sets, but due to the nonstationary setting the condition will not be bounded above. However, the numerically feasible range is much larger than in global problems. In cases that are scale invariant (powers and thin–plate splines), there is no difference between the stationary and nonstationary settings, and then the local systems have no serious stability problems.

But, unfortunately, there is a subtle difference to the techniques of section 5. In local radial basis function interpolation as described by the localized standard convergence analysis, the selection of data points depends on a selection of nearest neighbours of the evaluation point x, while in section 5 we used precomputed selections based on neighbours of each x_k. The local Lagrangians of the two cases will not be comparable, and the proof of local convergence orders does not cover the situation of section 5. Furthermore, the local interpolant in case of radial basis functions is a true linear combination of the $\phi(\|x - x_j\|_2)$ with $\|x - x_j\|_2 \leq ch$, while in case of section 5 such functions are multiplied with the weight function. Thus, unfortunately, there is no easy way to carry standard results on radial basis functions over to this situation.

9   Fully Local Methods with Polynomial Reproduction

We now look generally at localized techniques in the sense of the previous section. We depart from radial basis functions for a while and describe a folklore argument proving m–th order of convergence for stable local methods with local reproduction of polynomials of order up to m. At a point $x \in \Omega$ we want to take only a subset $X(x) := \{x_j \in X \ : \ \|x - x_j\|_2 \leq ch\} \subseteq X$ of the data set x with fill distance h as in (1), where $c > 1$ is a constant. We simply assume that we have a linear local process at x that is based on data $X(x)$ and that locally reproduces polynomials of order at most m. In particular, we keep x fixed and write

$$R_f(x) := \sum_{x_j \in X(x)} f(x_j) u_j(x) \tag{8}$$

with certain real numbers $u_j(x)$ such that

$$R_p(x) = p(x)$$

holds for all polynomials p up to order m. Note that the reproduction of polynomials is confined to the single point x. Now we assume that f has continuous derivatives up to order m around x, and thus the Taylor expansion $T_{x,f,m}$ of f at x of order at most m satisfies

$$|f(x_j) - T_{x,f,m}(x_j)| \leq Ch^m$$

for all $x_j \in X(x)$, where C depends on c and the derivatives of f near x. Now we can bound the local error via

$$
\begin{aligned}
|f(x) - R_f(x)| &= |T_{x,f,m}(x) - R_f(x)| \\
&= |R_{T_{x,f,m}}(x) - R_f(x)| \\
&= \left| \sum_{x_j \in X(x)} (T_{x,f,m}(x_j) - f(x_j)) u_j(x) \right| \\
&\leq C h^m \sum_{x_j \in X(x)} |u_j(x)|,
\end{aligned}
$$

and we see that the "Lebesgue constants"

$$
L(x) := \sum_{x_j \in X(x)} |u_j(x)|
$$

should be bounded independent of h, which is the stability condition we mentioned at the outset.

Let us look at simple examples first. For $m = 1$, we can get local reproduction of constants by always picking the function value at the nearest neighbor. The Lebesgue constant is 1. If f is continuously differentiable on Ω, and because any $x \in \Omega$ has a nearest neighbour from X at distance at most h, we get a method of order 1. The reconstruction is piecewise constant on the Dirichlet tesselation induced by X, though the tesselation is never actually calculated. For $m = 2$ and if Ω is the convex hull of X, we can use barycentric coordinates with respect to triangles containing x, taking the data at the vertices. Any triangulation of Ω via X will then lead to a piecewise linear and continous reconstruction by linear finite elements. The Lebesgue constant is 1 again.

Natural neighbour interpolation is another case fitting into this framework, The original version by Sibson [37,38] is continuous and reproduces linear polynomials with Lebesgue constant 1, while the C^1 extension by Farin [12] even reproduces quadratic polynomials.

Of course, the general approach above can be combined with radial basis function techniques and a sufficiently large order m of polynomial reproduction. By an argument in [30], the quantity

$$
\sum_{x_j \in X(x)} u_j^2(x)
$$

can be bounded above in all relevant cases, even in the nonstationary situation. This is not precisely what we require for the above line of argumentation, but if the local data sets $X(x)$ consist of $\mathcal{O}(1)$ points, which is what we can assume for quasi–uniform data distributions, the Lebesgue constants are uniformly bounded. However, it always has to assume that the local data do not allow a vanishing nontrivial polynomial of order m, and under this assumption one can go back to m-th order polynomials right away. This is why we do not pursue this setting any further.

Here is a little digression. One is tempted to consider the optimization problem

$$\sum_{j=1}^{N} |u_j(x)| = \text{Minimum}$$
$$\sum_{j=1}^{N} u_j(x)x_j^{\alpha} = x^{\alpha}, \ 0 \le |\alpha| < m$$

to hope for a reasonable method with *automatic* localization near x. The standard split of the variables $u_j(x) = u_j^+(x) - u_j^-(x)$ into nonnegative parts leads to a linear programming problem of simple form. But due to reproduction of constants via

$$1 = \sum_{j=1}^{N} (u_j^+(x) - u_j^-(x)),$$

we have

$$\sum_{j=1}^{N} u_j^+(x) \ge 1$$

and the objective function always satisfies

$$\text{Minimum} = \sum_{j=1}^{N} (u_j^+(x) + u_j^-(x)) \ge 1.$$

Thus for $m = 2$ all cases with interpolation via local barycentric coordinates in a triangle containing x will be optimal, irrespective of the size or position of the triangle. There is no automatic selection of local neighbours via this optimization problem.

Things are even worse when the point x is outside the convex hull of the data. If the problem is solvable at all, linear programming tells us that there always is an optimal solution based on three points for $m = 2$, and the solution must be determined by barycentric coordinates again, at least one of which must now be negative. The optimum is attained for choices of triangles where the sum of negative barycentric coordinates is minimal in absolute value. Closer inspection reveals that those optimal triangles are geometrically awful, because negative barycentric coordinates of a point x outside a triangle are small in absolute value, if the vertices "antipodal" to x are far away from x.

Similarly bad results are obtained if we replace the L_1 objective function by L_2 or L_∞, and we conclude that optimal stability does not imply locality, finishing our digression.

For upsampling of gridded data, there are simple and useful folklore formulae obtainable via the arguments of this section. For linear precision, upsampling at the midpoint of edges or at the center of a square should use the arithmetic mean of the data values. Again, we have Lebesgue constants bounded by 1, and the process will be of second order in terms of the meshwidth. Of course, such a process yields the bilinear local interpolant when started on four values at the vertices of a square and repeated indefinitely. Note that though the order is 2 for data from C^2 functions or surfaces, the resulting function or surface will not be C^2. Schemes with quadratic precision in two variables should use 6 points in general. A simple recipe can be obtained from looking at quadratic polynomials in Dernstein Bóoior representation, but the result will not yield a smooth surface.

10   Moving Least Squares

The examples above had the disadvantage that they generate surfaces with little smoothness, because the local schemes depend on the evaluation point x and the point selection $X(x)$ in a nontrivial and possibly noncontinuous way. We now look at a general recipe that overcomes this drawback and allows arbitrary smoothness and approximation order, at least in theory.

For a fixed evaluation point $x \in \Omega$ we consider the weighted least–squares problem

$$\text{Minimize} \sum_{j=1}^{N} (f(x_j) - p(x_j))^2 \, \phi(\|x - x_j\|_2)$$

over all polynomials $p \in \mathbb{P}_m$. Here, the weight function is a smooth nonnegative radial basis function ϕ with compact support, and this is how the above problem turns out to be localized. The resulting process, if well–defined, will reproduce polynomials up to order m, but we still have to write it in the form (8) and show that the functions u_j come out to be smooth.

Since the resulting linear system has a right–hand side that is a linear function of the data $f(x_j)$, we get (8) without further arguments, but we have to find a representation of the $u_j(x)$. To this end, we introduce self–explanatory matrix notation to write the objective function as $\|D_x f - D_x A a\|_2^2$ with a diagonal $N \times N$ matrix D_x having entries $\sqrt{\phi(\|x - x_j\|_2)}$ and an $N \times Q$ matrix A with entries $p_k(x_j)$ for a basis $p_1, \ldots, p_Q$ of $\mathbb{P}_m$. The solution vector $a_x \in \mathbb{R}^Q$ with respect to the data vector $f = (f(x_1), \ldots, f(x_N))^T$ is uniquely determined by the system

$$A^T D_x D_x A a_x = A^T D_x D_x f,$$

provided that the coefficient matrix $A^T D_x D_x A$ has full rank $Q \leq N$. We assume this for a moment, and we proceed to construct a vector $u(x) \in \mathbb{R}^N$

such that for $p(x) := (p_1(x), \ldots, p_Q(x))^T$ we can write $R(x) := a_x^T p(x) = u(x)^T f$. This is easy, if we look at

$$A^T D_x D_x A v(x) = p(x)$$
$$u(x) = D_x D_x A v(x) \tag{9}$$

and solve the first system for $v(x)$, putting the solution into the second equation. Thus we get $A^T u(x) = p(x)$ for free, which is the polynomial reproduction property at x. The entries of $A^T D_x D_x A$ are

$$\sum_{i=1}^N \phi(\|x - x_i\|_2) p_j(x_i) p_k(x_i),$$

and the matrix has full rank, if we define

$$X(x) := \{x_j \in X : \phi(\|x - x_i\|_2) > 0\}$$

and assume that there is no nontrivial polynomial in $I\!P_m$ that vanishes on $X(x)$. One can see clearly how the weight function localizes the least–squares problem if it is of compact support, but the support must be large enough to host at least a set of points near x that are in general position with respect to $I\!P_m$.

Since we can write the reconstruction in the form $R(x) = u^T(x)f$ without taking care of the localization explicitly, we see from the system (9) that the smoothness of the overall approximation is completely determined by the smoothness of the weight function. Thus we are left with the highly nontrivial problem of bounding the Lebesgue constants. A thorough treatment of this, giving all constants in explicit form, is due to Wendland [40]. Thus moving least squares are a technique that satisfies all requirements: it is effective in the sense of section 2, and it can produce surfaces with any prescribed smoothness. However, in its standard form it is an approximation rather than an interpolation.

One of the main computational problems of moving least squares is the proper determination of the local point selection $X(x)$. In particular, there may be great variations in the data density, and these variations should be flexibly incorporated into the algorithm. We propose to use all data points in a ball with varying radius around the evaluation point x, i.e.

$$X(x) := X \cap B_{\delta(x)}(x) := \{x_j \in X : \|x - x_j\|_2 \le \delta(x)\}$$

where δ is a smooth function that is calculated beforehand, preferably by another moving least squares apppproximation. For instance, one can generate $\mathcal{O}(N)$ regularly distributed points $y_1, \ldots, y_N$ in the domain Ω and find a "good calculation radius" δ_j for $X(y_j) := B_{\delta_j}(x)$ for each of these points.

Then $\delta(x)$ is constructed via an intermediate moving least squares algorithm, and the result is inserted into the actual surface construction technique.

We finish the paper with examples provided by R. Baule [2], illustrating the use of a varying calculation radius. We pick the glacier data ($N = 8345$) from R. Franke's website http://www.math.nps.navy.mil/~rfranke/, because it has a very inhomogeneous data distribution (see Figure 3). The main problem of any reconstruction method is to produce good results where the data are scarce, while keeping a good overall reproduction quality of the data. Naive and direct application of moving least squares can either result in a staircase or an overdose of smoothing (see Figures 4,5 and the examples from [40]). If the calculation radius varies as in Figure 6, one gets the much more realistic results of Figures 7 and 8. In fact, the L_∞ error on the data goes down from 81 to 21 when variable radii are used. A further variation, not described here in detail, includes interpolation via infinite weights, and then we get the same visual appearance as in Figure 8, but with zero error on the data.

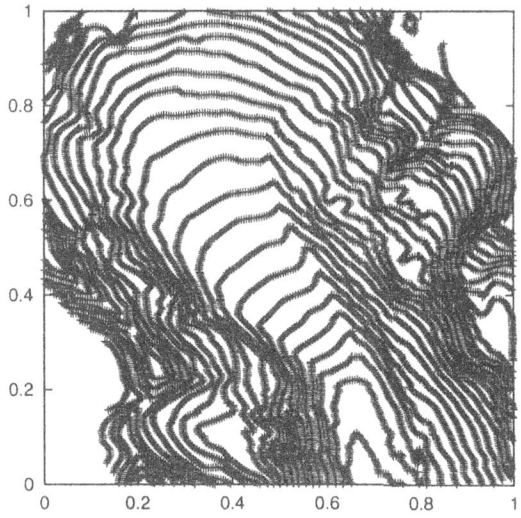

Fig. 3. Glacier data

Acknowledgement

Special thanks go to Rainer Baule and Holger Wendland for proofreading.

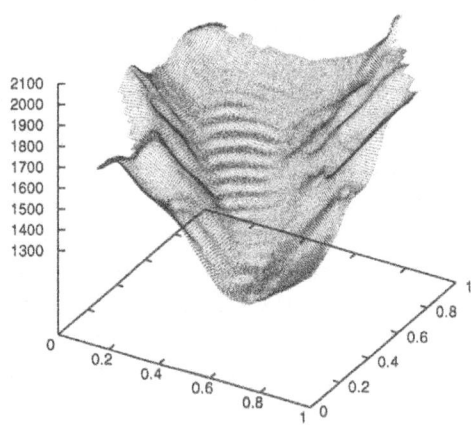

Fig. 4. Moving least squares with fixed radius 0.06

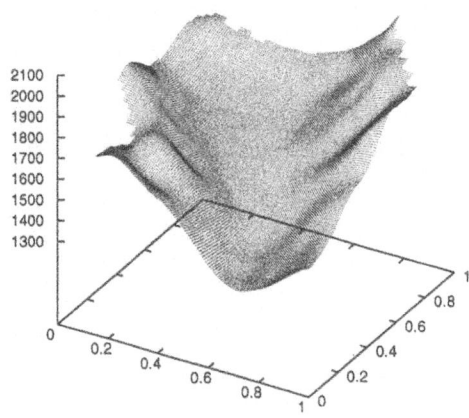

Fig. 5. Moving least squares with fixed radius 0.12

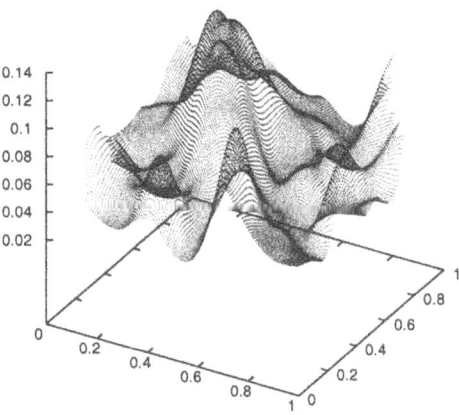

Fig. 6. Variable radius used in Figure 7

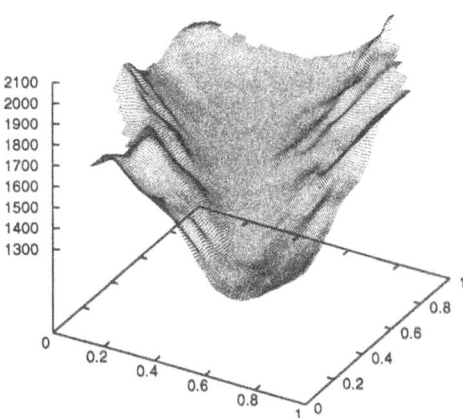

Fig. 7. Moving least squares with variable radius, degree zero

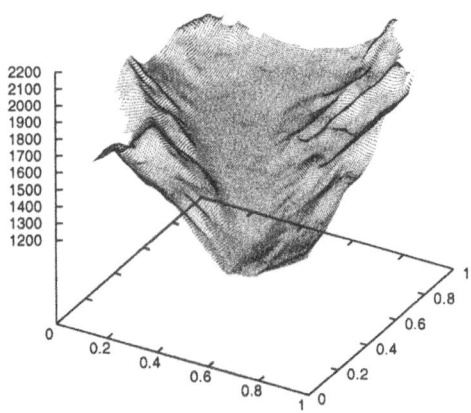

Fig. 8. Moving least squares with variable radius, degree 2

References

1. G. Allasia. A class of interpolating positive linear operators: theoretical and computational aspects. In S.P. Singh, editor, *Approximation Theory, Wavelets, and Applications*, pages 1–36. Kluwer, 1995.
2. R. Baule. Moving Least Squares Approximation mit parameterabhängigen Gewichtsfunktionen. Diplomarbeit, Universität Göttingen, 2000.
3. R.K. Beatson, J.B. Cherrie, and C.T. Mouat. Fast fitting of radial basis functions: Methods based on preconditioned GMRES iteration. *Advances in Computational Mathematics*, 11:253–270, 1999.
4. R.K. Beatson and L. Greengard. A short course on fast multipole methods. In M. Ainsworth, J. Levesley, W. Light, and M. Marletta, editors, *Wavelets, Multilevel Methods and Elliptic PDEs*, pages 1–37. Oxford University Press, 1997.
5. R.K. Beatson and W.A. Light. Fast evaluation of radial basis functions: Methods for two–dimensional polyharmonic splines. *IMA Journal of Numerical Analysis*, 17:343–372, 1997.
6. R.K. Beatson and G.N. Newsam. Fast evaluation of radial basis functions: I. Advances in the theory and applications of radial basis functions. *Comput. Math. Appl.*, 24(12):7–19, 1992.
7. M.D. Buhmann. *Multivariable interpolation using radial basis functions*. PhD thesis, University of Cambridge, 1989.
8. M.D. Buhmann. New developments in the theory of radial basis function interpolation. In K. Jetter and F.I. Utreras, editors, *Multivariate Approximations: From CAGD to Wavelets*, pages 35–75. World Scientific, Singapore, 1993.
9. Iain S. Duff, Albert M. Erisman, and John K. Reid. *Direct Methods for Sparse Matrices*. Monographs on Numerical Analysis. Clarendon Press, Oxford, 1986.

10. N. Dyn. Interpolation of scattered data by radial functions. In C.K. Chui, L.L. Schumaker, and F.I. Utreras, editors, *Topics in Multivariate Approximation*, pages 47–61. Academic Press, Boston, 1987.

11. N. Dyn. Interpolation and approximation by radial and related functions. In C.K. Chui, L.L. Schumaker, and J.D. Ward, editors, *Approximation Theory VI*, volume 1, pages 211–234. Academic Press, New York, 1989.

12. G. Farin. Surfaces over Dirichlet tessellations. *Comput. Aided Geom. Design*, 7:281–292, 1990.

13. M.S. Floater and A. Iske. Multistep scattered data interpolation using compactly supported radial basis functions. *J. Comp. Appl. Math.*, 73:65–78, 1996.

14. M.S. Floater and A. Iske. Thinning and approximation of large sets of scattered data. In F. Fontanella, K. Jetter, and P.-J. Laurent, editors, *Advanced Topics in Multivariate Approximation*, pages 87–96. World Scientific Publishing, Singapore, 1996.

15. M.S. Floater and A. Iske. Thinning, inserting, and swapping scattered data. In A. Le Mehaute, C. Rabut, and L. L. Schumaker, editors, *Surface Fitting and Multiresolution Methods*, pages 139–144. Vanderbilt University Press, Nashville, 1996.

16. M.S. Floater and A. Iske. Thinning algorithms for scattered data interpolation. *BIT*, 38:4:705–720, 1998.

17. R. Franke. Scattered data interpolation. Test of some methods. *Mathematics of Computation*, pages 181–199, 1982.

18. R.L. Hardy. Multiquadric equations of topography and other irregular surfaces. *J. Geophys. Res.*, 76:1905–1915, 1971.

19. S. Hartmann. *Multilevel-Fehlerabschätzung bei der Interpolation mit radialen Basisfunktionen*. Dissertation, Universität Göttingen, 1998.

20. W. Light. Some aspects of radial basis function approximation. In S.P. Singh, editor, *Approximation Theory, Spline Functions and Applications*, volume 356, pages 163–190. Kluwer, 1992.

21. J. C. Mairhuber. On Haar's theorem concerning Chebychev approximation problems having unique solutions. *Proc. Amer. Math. Soc.*, 7:609–615, 1956.

22. V. Maz'ya and G. Schmidt. On approximate approximations using Gaussian kernels. *IMA Journal of Numerical Analysis*, 16:13–29, 1996.

23. F. J. Narcowich, R. Schaback, and J. D. Ward. Multilevel interpolation and approximation. *Appl. Comput. Harmonic Anal.*, 7:243–261, 1999.

24. M.J.D. Powell. Radial basis functions for multivariable interpolation: a review. In D.F. Griffiths and G.A. Watson, editors, *Numer. Algorithms*, pages 223–241. Longman Scientific & Technical (Harlow), 1987.

25. M.J.D. Powell. Tabulation of thin plate splines on a very fine two–dimensional grid. In *Numerical Methods of Approximation Theory*, pages 221–244. Birkhäuser, Basel, 1992.

26. M.J.D. Powell. The theory of radial basis function approximation in 1990. In W.A. Light, editor, *Advances in Numerical Analysis II: Wavelets, Subdivision Algorithms, and Radial Basis Functions*, pages 105–210. Oxford Univ. Press, Oxford, 1992.

27. M.J.D. Powell. Truncated Laurent expansions for the fast evaluation of thin plate splines. *Numer. Algorithms*, 5(1–4):99–120, 1993.

28. M.J.D. Powell. The uniform convergence of thin–plate spline interpolation in two dimensions. *Numer. Math.*, 67:107–128, 1994.

29. M.J.D. Powell. A review of algorithms for thin plate spline interpolation in two dimensions. In F. Fontanella, K. Jetter, and P.-J. Laurent, editors, *Advanced Topics in Multivariate Approximation*, pages 303–322. World Scientific Publishing, 1996.

30. R. Schaback. Error estimates for approximations from control nets. *Comput. Aided Geom. Design*, 10:57–66, 1993.

31. R. Schaback. Multivariate interpolation and approximation by translates of a basis function. In Charles K. Chui and L.L. Schumaker, editors, *Approximation Theory VIII*, volume 1: Approximation and interpolation, pages 491–514. World Scientific Publishing, 1995.

32. R. Schaback. On the efficiency of interpolation by radial basis functions. In A. LeMéhauté, C. Rabut, and L.L. Schumaker, editors, *Surface Fitting and Multiresolution Methods*, pages 309–318. Vanderbilt University Press, Nashville, TN, 1997.

33. R. Schaback and H. Wendland. Adaptive greedy techniques for approximate solution of large RBF systems. Preprint, 2000.

34. R. Schaback and H. Wendland. Characterization and construction of radial basis functions. In N. Dyn, D. Leviatan, and D Levin, editors, *Eilat proceedings*. Cambridge University Press, 2000.

35. R. Schaback and H. Wendland. Numerical techniques based on radial basis functions. In Albert Cohen, Christophe Rabut, and Larry Schumaker, editors, *Curve and Surface Fitting*. Vanderbilt University Press, Nashville, TN, 2000.

36. H.-P. Seidel. Hierarchical methods in computer graphics. In László Szirmay Kalos, editor, *14th Spring Conference on Computer Graphics*, pages 1–16. Comenius University, Bratislava, Slovakia, 1998.

37. R. Sibson. A vector identity for the Dirichlet tesselation. *Mathematical Proceedings of the Cambridge Philosophical Society*, 87:151–155, 1980.

38. R. Sibson. A brief description of natural neighbor interpolation. In V. Barnett, editor, *In Interpreting Multivariate Data*, pages 21–36. John Wiley, Chichester, 1981.

39. G. Strang and G. Fix. A Fourier analysis of the finite element variational method. In G. Geymonat, editor, *Constructive Aspects of Functional Analysis*, C.I.M.E. II Ciclo 1971, pages 793–840. ???, 1973.

40. H. Wendland. Local polynomial reproduction and moving least squares approximation. to appear in IMA Journal of Numerical Analysis, 2000.

41. Z. Wu and R. Schaback. Local error estimates for radial basis function interpolation of scattered data. *IMA Journal of Numerical Analysis*, 13:13–27, 1993.

Gaussian and Mean Curvature of Subdivision Surfaces

Jörg Peters and Georg Umlauf

CISE, University of Florida, P.O. Box 116120, Gainesville, FL 32611-6120, USA,
e-mail: [jorg|umlauf]@cise.ufl.edu

Summary. By explicitly deriving the curvature of subdivision surfaces in the extraordinary points, we give an alternative, more direct account of the criteria necessary and sufficient for achieving curvature continuity than earlier approaches that locally parametrize the surface by eigenfunctions.

The approach allows us to rederive and thus survey the important lower bound results on piecewise polynomial subdivision surfaces by Prautzsch, Reif, Sabin and Zorin, as well as explain the beauty of curvature continuous constructions like Prautzsch's. The parametrization neutral perspective gives also additional insights into the inherent constraints and stiffness of subdivision surfaces.

1 Introduction

Almost all subdivision algorithms in the current literature achieve tangent continuity but not curvature continuity. We give a simple characterization of the causes underlying this phenomenon by explicitly expressing Gaussian and mean curvature in the minimally smooth extraordinary points. This allows us to rederive and thereby survey the important lower bound results of [18,14,19,8] and constructions for curvature continuous piecewise polynomial subdivision algorithms by [6,10,16]. Beyond this we get additional insights into the inherent constraints and stiffness of such subdivision algorithms. Since a subdivision surface consists of an infinite collection of polynomial pieces around every extraordinary point one might expect such surfaces to be more flexible than spline surfaces. However, we will see that the infinite application of the *same* subdivision rule enforces strict rules on the piecewise polynomial rings converging towards extraordinary points. For example, the Jacobian of the subdominant eigenfunctions of a curvature continuous subdivision algorithm must have lower degree than the Jacobian of the subdivision surface itself.

The paper is organized as follows. With the notation of Section 2, we express in each subdivision step m the curvatures of the innermost spline ring around a given extraordinary point as $K_m = (\mu/\lambda^2)^{2m} f_K^m(u, v)$, respectively, $H_m = (\mu/\lambda^2)^m f_H^m(u, v)$ for scalar constants $\mu < \lambda$ and rational functions f. The factor μ/λ^2 immediately implies necessary constraints on curvature continuous subdivision surfaces and the weakest form of curvature smoothness: the principal curvatures of piecewise polynomial C^1 subdivision algorithms are square integrable. Section 4 derives and reviews necessary constraints on subdivision surfaces to be curvature continuous by observing that f should be constant in the limit and equating the degree of numerator and denominator of f at the extraordinary point. Section 5 reviews Prautzsch's sufficient condition and his unique construction of a curvature continuous, linear, stationary subdivision algorithm by projection.

2   Notation and basic facts

In this section we define just the basic notation and facts needed for our analysis; for a formal, more general and abstract setting, the reader is referred to [15] and [19].

While our analysis applies to a larger class of subdivision algorithms, we focus in the following on *generalized box-spline subdivision algorithms*, that is on affine invariant, symmetric, linear, local, stationary algorithms that generalize box-spline subdivision and generate (regular) C^1 surfaces. In particular, the limit surface has a piecewise polynomial parametrization and the parametric smoothness between the pieces is C^2 except at a finite number of extraordinary points. An extraordinary point is the limit point of a minimal subnet of the initial control net under repeated application of the subdivision algorithm. Such a subnet consists of an n-valent vertex (for a primal subdivision algorithm) or an n-sided facet (for a dual subdivision algorithm) and just those neighboring control points that determine a surface ring x_0 around an n-sided hole by the regular box-spline subdivision rules (Figure 1 *left*).

If we arrange the points of the subnet into the column vector C_0 then each subdivision step transforms the subnet by applying the same square, stochastic subdivision matrix A. After m applications the result is the subnet

$$C_m = A^m C_0.$$

The mth subdivision step adds the surface ring $x_m(u, v)$ inside the hole left by the $(m - 1)$st subdivision step (Figure 1 *right*). The surface rings are box-splines and can therefore be represented as

$$x_m : \{0, \ldots, n - 1\} \times \Omega \to \mathbb{R}^3, \qquad x_m(u, v) = B(u, v) C_m,$$

where $B(u, v)$ is the row vector of the box-spline basis functions and the domain Ω is either $2\square \setminus \square$ or $2\triangle \setminus \triangle$, with $\square$ the unit square and $\triangle$ the unit triangle.

For simplicity we assume that A is diagonalizable with eigenvalues

$$1 = \lambda_0 > \underbrace{\lambda_1 = \lambda_2}_{=:\lambda} > \underbrace{\lambda_3 = \lambda_4 = \lambda_5}_{=:\mu} > \cdots \geq 0,$$

such that $\lambda_1 = \lambda_2$ correspond to the 1st and $(n - 1)$st block, $\lambda_3 = \lambda_4$ (for $n > 3$) to the 2nd and $(n - 2)$nd block and λ_5 to the 0th block of the Fourier

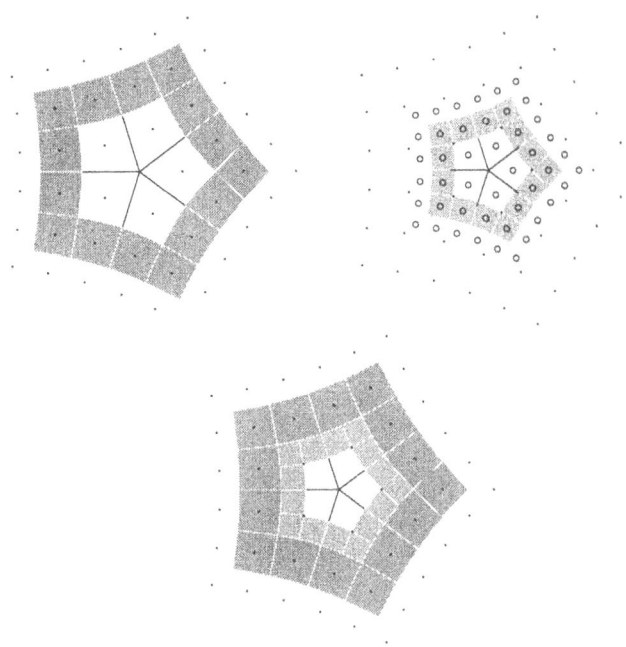

Fig. 1. Inserting a surface ring x_m (light grey) into the hole left by x_{m-1} (dark grey). x_m consists of 5 images of $2\square \setminus \square$. The vector of control points C_m of x_m (circles) is obtained by one subdivision step applied to C_{m-1} (dots)

decomposition of A (c.f. [5]). The corresponding eigenvectors are denoted by v_i, i.e. $Av_i = \lambda_i v_i$ for all i. For subdivision algorithms with more general subdivision matrices A the reader is referred to [15,19].

In terms of multiples p_i of the eigenvectors the subnet can be expressed as

$$C_m = \sum_i \lambda_i^m v_i p_i, \qquad p_i \in \mathbb{R}^3.$$

Expanded in the *eigenfunction* $e^i : \{0, \dots, n-1\} \times \Omega \to \mathbb{R}, (u, v) \mapsto B(u, v) v_i$ associated with v_i the surface ring x_m is of the form

$$x_m(u, v) = \sum_i \lambda_i^m B(u, v) v_i p_i = \sum_i \lambda_i^m e^i(u, v) p_i.$$

A well-known fact of differential geometry (see e.g. [2]) is that for *any* regular surface parametrization x the Gauss curvature K and the mean curvature H are

$$K(u,v) = \frac{e(u,v)g(u,v) - f(u,v)^2}{E(u,v)G(u,v) - F(u,v)^2},$$

$$H(u,v) = \frac{e(u,v)G(u,v) - 2f(u,v)F(u,v) + g(u,v)E(u,v)}{2(E(u,v)G(u,v) - F(u,v)^2)}, \tag{1}$$

where x_u is the partial derivative of $x(u,v)$ with respect to u

$$E = x_u x_u^t, \quad F = x_u x_v^t, \quad G = x_v x_v^t,$$
$$e = n x_{uu}^t, \quad f = n x_{uv}^t, \quad g = n x_{vv}^t,$$

and $n = (x_u \times x_v)/\|x_u \times x_v\|$ is the normal. Since x is assumed to be regular, the denominators of (1), $EG - F^2 = \|x_u \times x_v\|^2$, are nonzero and we have

$$K = \frac{1}{\|x_u \times x_v\|^4}(\det(x_u, x_v, x_{uu})\det(x_u, x_v, x_{vv})$$
$$- \det(x_u, x_v, x_{uv})^2),$$

$$H = \frac{1}{2\|x_u \times x_v\|^3}(\det(x_u, x_v, x_{uu})(x_v x_v^t)$$
$$- 2\det(x_u, x_v, x_{uv})(x_u x_v^t) + \det(x_u, x_v, x_{vv})(x_u x_u^t)).$$

3   Gauss curvature and mean curvature

In this section, we derive the Gauss curvature and the mean curvature of the limit surfaces of generalized box-spline subdivision algorithms at extraordinary points. We expand each of the surface rings x_m in terms of the eigenfunctions e^i. This approach goes back at least to [12]. However, in contrast to [12] we do not analyze the curvature by parametrizing the limit surface locally as a function over the subdominant eigenfunctions e^1 and e^2 but rather compute the curvature expansion explicitly. That is, we determine the curvatures $K_m(u,v)$ and $H_m(u,v)$ of $x_m(u,v)$ and then take the limit as $m \to \infty$.

For simplicity we write x instead of x_m in the following. Since the basis functions form a partition of unity, $e^0 \equiv 1$ and

$$x_u = \lambda^m\left(e_u^1 p_1 + e_u^2 p_2\right) + \mu^m\left(e_u^3 p_3 + e_u^4 p_4 + e_u^5 p_5\right) + o(\mu^m),$$
$$x_{uv} = \lambda^m\left(e_{uv}^1 p_1 + e_{uv}^2 p_2\right) + \mu^m\left(e_{uv}^3 p_3 + e_{uv}^4 p_4 + e_{uv}^5 p_5\right) + o(\mu^m).$$

Symmetry in u and v yields the analogous terms for x_v, x_{uu} and x_{vv}. With the abbreviations

$$\Delta_{ij} := e_u^i e_v^j - e_u^j e_v^i,$$
$$D_{st}^i := \Delta_{12} e_{st}^i - \Delta_{1i} e_{st}^2 + \Delta_{2i} e_{st}^1, \quad s,t \in \{u,v\},$$
$$P_{ij} := \det(p_1, p_2, p_i)\det(p_1, p_2, p_j),$$

it is now easy to see that

$$x_u \times x_v = \lambda^{2m} \Delta_{12}(p_1 \times p_2) + o(\lambda^{2m}),$$

$$\det(x_u, x_v, x_{uu}) = \lambda^{2m} \mu^m \sum_{i=3,4,5} \det(p_1, p_2, p_i) D^i_{uu} + o(\lambda^{2m} \mu^m).$$

Symmetry yields the analogous terms for $\det(x_u, x_v, x_{uv})$ and $\det(x_u, x_v, x_{vv})$ and

$$K_m = \left(\frac{\mu}{\lambda^2}\right)^{2m} \frac{\sum_{i,j=3,4,5} P_{ij}\left(D^i_{uu}D^j_{vv} - D^i_{uv}D^j_{uv}\right) + o(1)}{\Delta^4_{12}\|p_1 \times p_2\|^4 + o(1)}. \tag{2}$$

All dependencies on m in the equality are either explicit or hidden in the $o(1)$ terms. We note that Δ_{12} is the Jacobi determinant of the characteristic map [12,13], which is non-zero if the characteristic map is regular, and that $\|p_1 \times p_2\|$ is positive for almost all initial control nets C_0. The leading factor of the expression (2) for the Gauss curvature readily yields the following basic characterization of the curvature at extraordinary points (c.f. [12, page 25])

Observation 1 *Let A be the subdivision matrix of a regular C^1 generalized box-spline subdivision algorithm as defined in Section 2 with eigenvalues $1 > \lambda > \mu > \ldots \geq 0$.*

(a) *If $\mu > \lambda^2$ then the Gauss curvature at the extraordinary point is infinite.*
(b) *If $\mu < \lambda^2$ then the Gauss curvature at the extraordinary point is zero.*
(c) *If $\mu = \lambda^2$ then the Gauss curvature at the extraordinary point is bounded by the second factor of (2), but is possibly non-unique.*

Examples for (a) are [1,4,11], for (b) are [Prautzsch & Umlauf '98a, Prautzsch & Umlauf '98b] and examples for (c) are [18,3]. Note the curious combination of tangent continuity and infinite curvature for the standard algorithms in (a). In case (c), the limit for $m \to \infty$ yields at the extraordinary point

$$K = \sum_{i,j=3,4,5} \frac{P_{ij}}{\|p_1 \times p_2\|^4} \frac{D^i_{uu}D^j_{vv} - D^i_{uv}D^j_{uv}}{\Delta^4_{12}}. \tag{3}$$

Recall that the factor $(D^i_{uu}D^j_{vv} - D^i_{uv}D^j_{uv})/\Delta^4_{12}$ is a rational function in u and v. In order for the Gauss curvature to be well-defined at the extraordinary point rather than multi-valued or divergent, K must be constant. Since the P_{ij} depend on the initial net, they can be arbitrary except for $P_{ij} = P_{ji}$ and each of the six resulting summand has to be constant. We conclude that the eigenfunctions $e^1, \ldots, e^5$ must satisfy the six partial differential equations:

$$D^i_{uu}D^j_{vv} - 2D^i_{uv}D^j_{uv} + D^i_{vv}D^j_{uu} = \Delta^4_{12} \cdot \mathrm{const}_{ij}, \text{ for } i,j \in \{3,4,5\}, j > i,$$
$$D^i_{uu}D^i_{vv} - (D^i_{uv})^2 = \Delta^4_{12} \cdot \mathrm{const}_{ii}, \text{ for } i = 3,4,5.$$

$$\tag{4}$$

Lemma 2 *The limit surface of a generalized box-spline subdivision algorithm with $\Delta_{12} \neq 0$ has continuous, for almost all initial nets non-zero, Gauss curvature at the extraordinary point if and only if $\mu = \lambda^2$ and the differential equations (4) hold.*

Similarly, with $\tilde{P}_{ikl} := \det(\boldsymbol{p}_1, \boldsymbol{p}_2, \boldsymbol{p}_i)(\boldsymbol{p}_k \boldsymbol{p}_l^t)$ and $\varepsilon_{kl} = 1/2$ for $k = l$ and 1 otherwise, the mean curvature is

$$H_m = \left(\frac{\mu}{\lambda^2}\right)^m \frac{1}{\Delta_{12}^3 \|\boldsymbol{p}_1 \times \boldsymbol{p}_2\|^3 + o(1)} \Big(\sum_{\substack{i=3,4,5 \\ k,l=1,2,\, k \geq l}} \tilde{P}_{ikl} \varepsilon_{kl} (D_{uu}^i(e_v^k e_v^l)$$

$$- D_{uv}^i(e_u^k e_v^l + e_v^k e_u^l) + D_{vv}^i(e_u^k e_u^l)) + o(1)\Big). \tag{5}$$

The expression for H_m yields Observation 1, with Gauss curvature replaced by mean curvature. For $\mu = \lambda^2$ we get bounded, not forcibly zero, but possibly non-unique mean curvature

$$H = \sum_{\substack{i=3,4,5 \\ k,l=1,2,\, k \geq l}} \frac{\varepsilon_{kl} \tilde{P}_{ikl}}{\Delta_{12}^3 \|\boldsymbol{p}_1 \times \boldsymbol{p}_2\|^3} (D_{uu}^i(e_v^k e_v^l)$$

$$- D_{uv}^i(e_u^k e_v^l + e_v^k e_u^l) + D_{vv}^i(e_u^k e_u^l)). \tag{6}$$

In analogy to Lemma 2 the sufficient conditions for a mean-curvature continuous limit surface require that nine partial differential equations hold:

$$D_{uu}^i(e_v^k e_v^l) - D_{uv}^i(e_u^k e_v^l + e_v^k e_u^l) + D_{vv}^i(e_u^k e_u^l) = \Delta_{12}^3 \cdot \text{const}_{ikl}, \tag{7}$$

for $i = 3, 4, 5, k, l = 1, 2, k \geq l$.

Since the *principal curvatures* at a point on the mth surface ring are

$$\kappa_{1,2}^m = H_m \pm \sqrt{H_m^2 - K_m},$$

(2) and (5) imply that κ_1^m and κ_2^m converge like $O(\mu^m/\lambda^{2m})$ for $m \to \infty$.

Observation 3 *The limit surface of a generalized box-spline subdivision algorithm with $\Delta_{12} \neq 0$ is curvature continuous at the extraordinary point if $\mu = \lambda^2$ and the differential equations (4) and (7) hold.*

Since $\int d\boldsymbol{x}_m = O(\lambda^{2m})$ and $\mu < \lambda$

$$\sum_m \int_{\boldsymbol{x}_m} |\kappa_{1,2}^m|^2 d\boldsymbol{x}_m = \sum_m O(\mu^{2m}/\lambda^{2m}) < \infty.$$

This immediately implies an interesting fact derived for general L^p spaces in [Reif & Schröder '00].

Observation 4 *The principal curvatures of the limit surface of a generalized box-spline subdivision algorithm are square integrable.*

4   Lower bounds on the degree

We now take a look at the important lower bound results of [Sabin '91, Reif '96, Zorin '98, Prautzsch & Reif '99]. For this it is crucial to distinguish between the apparent or *formal degree* of box-spline eigenfunctions, possibly the result of degree-raising, and the *true degree* denoted by "deg". The true degree is defined to be the minimal number of non-vanishing derivatives. We focus on the differential equations resulting from the Gaussian curvature – the analysis of the mean curvature yields the same results.

We recall that the left hand side of the differential equations (4) are for $i, j = 3, 4, 5$,

$$G_{ij} := D_{uu}^i D_{vv}^j - 2 D_{uv}^i D_{uv}^j + D_{vv}^i D_{uu}^j, \qquad j > i,$$
$$G_{ii} := D_{uu}^i D_{vv}^i - (D_{uv}^i)^2.$$

Let $d = \deg(x_0)$ denote the total degree, respectively, the bi-degree of a regular box-spline parametrization. A straightforward count yields for all $i, j = 3, 4, 5, j \geq i$, that

- for the total degree $\deg(G_{ij}) \leq 2(2(d-1) + d - 2) = 6d - 8$ and
- for the bi-degree $\deg(G_{ij}) \leq 2(2d - 1 + d - 1) = 6d - 4$,

whereas for the right hand side of (4)

- the formal total degree of Δ_{12}^4 is $4(2d - 2)$ and
- the formal bi-degree of Δ_{12}^4 is $4(2d - 1)$.

This degree mismatch implies the following observation.

Observation 5 *The limit surface of a generalized box-spline subdivision algorithm with total degree*

$$\deg(x_0) = d \quad and \quad \deg(\Delta_{12}) = 2(d - 1)$$

is curvature continuous at an extraordinary point P if and only if P is a flat point, i.e. $\mu < \lambda^2$. The limit surface of a generalized tensor-product subdivison algorithm with bi-degree

$$\deg(x_0) = d \quad and \quad \deg(\Delta_{12}) = 2d - 1$$

is curvature continuous at an extraordinary point P if and only if P is a flat point.

In other words, a generalized box-spline subdivision algorithm can only have a curvature continuous limit surface for $\mu = \lambda^2$, if the true degree of the Jacobian Δ_{12} is less than its formal degree. This is the case, if either one or both of the following conditions hold:

(i) The true degree of e^1 or e^2 is less than d.

(ii) The leading terms in the Jacobian Δ_{12} cancel.

Since we assume symmetric masks,

$$\deg(e^1) = \deg(e^2) =: d'.$$

If the subdivision surface is curvature continuous and not flat in the extraordinary point and if condition (ii) does not apply then $d' < d$ must hold by condition (i). In fact, we compute

- for the total degree $\deg(G_{ij}) = 2(2d' + d - 4)$ and $\deg(\Delta_{12}^4) = 4(2d' - 2)$ and
- for the bi-degree $\deg(G_{ij}) = 2(2d' + d - 2)$ and $\deg(\Delta_{12}^4) = 4(2d' - 1)$.

Comparing degrees we find in either case that $2d' = d$ and arrive at the following observation:

Observation 6 *If the leading terms in the Jacobian Δ_{12} do not cancel then the limit surface of a generalized box-spline subdivision algorithm is curvature continuous and not flat in an extraordinary point only if the true (bi-)degree of the surface is at least twice the true (bi-)degree of the subdominant eigenfunctions e^1 and e^2.*

This is consistent with the degree estimate of [14,19]. The central idea of these proofs appears already as a parting sentence in [18]: the surface is viewed as a function over the tangent plane parametrized by e^1 and e^2. To have non-zero curvature, it is necessary that the non-tangential component of the surface is at least quadratic in e^1 and e^2, i.e. $d \geq 2d'$. More generally, if the non-tangential component of the surface is at least of degree r in e^1 and e^2 then the surface representation has to be at least of degree rd'. Since e^1 and e^2 have to have a minimal degree to form C^k rings, e.g. $d' \geq k+1$ in the tensor-product case, a lower bound, say $r(k+1)$, is deduced [8].

If, on the other hand, the subdivision surface is curvature continuous and not flat in the extraordinary point and if condition (i) does not apply then the leading terms of Δ_{12} must cancel by condition (ii). Now we compute

- for the total degree $\deg(G_{ij}) = 2\max\{\deg(\Delta_{12})+d-2, 2(d-1)+d-2\} = 6d - 8$ and
- for the bi-degree $\deg(G_{ij}) = 2\max\{\deg(\Delta_{12}) + d - 1, 2d-1+d-1\} = 6d - 4$.

Comparing against $\deg(\Delta_{12}^4) = 4\deg(\Delta_{12})$ we obtain a counterpart to Observation 6.

Observation 7 *If the true degree of e^1 and e^2 is not less than d then the limit surface of a generalized box-spline subdivision algorithm is curvature continuous and not flat in an extraordinary point only if the total degree $\deg(\Delta_{12}) \leq 3d/2 - 2$, respectively, the bi-degree $\deg(\Delta_{12}) \leq 3d/2 - 1$.*

From these observations, it is evident that the key to curvature continuous subdivision surfaces is the answer to the following question.

Central Question *For what choices of eigenfunctions e^1 and e^2 is $\deg(\Delta_{12})$ less than $2\deg(x_0) - 2$ for total degree, respectively, $2\deg(x_0) - 1$ for bi-degree generalized box-spline subdivision algorithms?*

5   Curvature continuous subdivision constructions

If we interpret curvature smoothness in the weak sense of L^2 integrability, then Observation 4 guarantees that almost all C^1 subdivision algorithms qualify as curvature smooth. If we allow flat spots, then [Prautzsch & Umlauf '98a, Prautzsch & Umlauf '98b] yield low degree, small mask, curvature continuous subdivision algorithms. If we want non-zero bounded curvature, we can adapt the leading eigenvalues as in [18,3]. However, if we want curvature continuity without flat spots the stringent constraints of Lemma 2 apply and a degree-reduced Jacobian in the sense of Observations 6 or 7 is necessary. A trivial example that satisfies these constraints is the regular case of any C^2 box-spline: here e^1 and e^2 are linear. The only non-trivial constructions reported to date subdivide a polynomial filling of an n-sided hole obtained by projection [6,16].

To see in our newly acquired framework why these constructions yield curvature continuous subdivision surfaces without forced flat points we re-state the sufficient conditions derived by Prautzsch [7]. (Parametrizing the limit surface over the characteristic map, [15] concludes also the necessity of these conditions.)

The *sufficient conditions* of [7] state, that in order to be able to solve the differential equations (4) it suffices that the eigenfunctions e^3, e^4 and e^5 be quadratic polynomials in e^1 and e^2:

$$e^i = a_i(e^1)^2 + b_i e^1 e^2 + c_i(e^2)^2, \quad a_i, b_i, c_i \in \mathbb{R}, \quad \text{for } i = 3, 4, 5. \tag{8}$$

Indeed the derivatives of $e^i, i = 3, 4, 5$, are then of the form

$$
\begin{aligned}
e^i_u &= 2a^i e^1 e^1_u + b^i(e^1_u e^2 + e^1 e^2_u) + 2c^i e^2 e^2_u, \\
e^i_{uu} &= 2a^i\left((e^1_u)^2 + e^1 e^1_{uu}\right) + b^i\left(e^1_{uu} e^2 + 2e^1_u e^2_u + e^1 e^2_{uu}\right) \\
&\quad + 2c^i\left((e^2_u)^2 + e^2 e^2_{uu}\right)
\end{aligned}
$$

and analogously for e^i_v, e^i_{vv} and e^i_{uv}. This yields

$$
\begin{aligned}
\Delta_{1i} &= \Delta_{12}(b^i e^1 + 2c^i e^2), \\
\Delta_{2i} &= -\Delta_{12}(2a^i e^1 + b^i e^2) \qquad \text{and} \\
D^i_{uu} &= 2\Delta_{12}(a_i(e^1_u)^2 + b_i e^1_u e^2_u + c_i(e^2_u)^2).
\end{aligned}
$$

Substituting D^i_{uu}, D^i_{vv} and D^i_{uv} in (3) and (6), the limit terms for the Gauss and the mean curvatures simplify to

$$K = \sum_{i,j=3,4,5} \frac{P_{ij}}{\|p_1 \times p_2\|^4} \cdot f_{ij} \quad \text{and} \quad H = \sum_{\substack{i=3,4,5 \\ k,l=1,2,\ k \geq l}} \frac{\bar{P}_{ikl}}{\|p_1 \times p_2\|^3} \cdot \tilde{f}_{ikl}$$

where

$$f_{ij} = \begin{cases} 4(a_ic_j + a_jc_i) - 2b_ib_j & \text{for } i \neq j \\ 4a_ic_i - (b_i)^2 & \text{for } i = j \end{cases}, \quad \tilde{f}_{ikl} = \begin{cases} c_i & \text{for } k = l = 1 \\ a_i & \text{for } k = l = 2 \\ -b_i/2 & \text{for } k \neq l \end{cases}.$$

Since these last expressions for the Gauss and mean curvature are constant, the limit surface is curvature continuous at the extraordinary point.

Subdivision algorithms that satisfy the sufficient conditions (8) were derived by [Prautzsch '97, Reif '98b]. We analyze Prautzsch's approach in detail in terms of our necessary and sufficient conditions of Lemma 2. Here v_1 and v_2 are the eigenvectors to the subdominant eigenvalue λ of the Catmull-Clark algorithm. Then e^1 and e^2 have true bi-degree 3. Now set $e^3 = (e^1)^2, e^4 = e^1e^2$ and $e^5 = (e^2)^2$ with control nets $w_i, i = 3, 4, 5$. Furthermore, w_1 and w_2 are the control nets of e^1 and e^2, respectively, in a degree-doubled representation. Then the subdivision matrix of Prautzsch's construction is given by

$$A = MDM^+$$

where

$$M := [1, w_1, w_2, w_3, w_4, w_5], \qquad D := \text{diag}(1, \lambda, \lambda, \lambda^2, \lambda^2, \lambda^2),$$
$$M^+ := (M^tM)^{-1}M^t.$$

In this construction the only non-zero eigenvalues of A are $1, \lambda(2\text{-fold}), \lambda^2(3\text{-fold})$ corresponding to the eigenvectors $1, w_1, \ldots, w_5$.

6 Conclusion

We surveyed and restated a number of important recent results concerning the curvature continuity of the limit surfaces of generalized box-spline subdivision algorithms. The direct computation of K and H simplifies the matter and yields new insights into why the limit surfaces of most subdivision algorithms are not curvature continuous and what criteria need to be enforced by new constructions.

References

1. E. Catmull and J. Clark. Recursive generated B-spline surfaces on arbitrary topological meshes. *Computer-Aided Design*, 10(6):350–355, 1978.

2. M.P. do Carmo. *Differential geometry of curves and surfaces.* Prentice-Hall Inc., Englewood Cliffs, N.J., 1976.
3. F. Holt. Towards a curvature-continuous stationary subdivision algorithm. *Z. Angew. Math. Mech.*, 76(Suppl. 1):423–424, 1996.
4. C.T. Loop. Smooth Subdivision Surfaces Based on Triangles. Master's thesis, Department of Mathematics, University of Utah, August 1987.
5. J. Peters and U. Reif. Analysis of algorithms generalizing B-spline subdivision. *SIAM J. Numer. Anal.*, 35(2):728–748, 1998.
6. H. Prautzsch. Freeform splines. *Comput. Aided Geom. Design*, 14(3):201–206, 1997.
7. H. Prautzsch. Smoothness of subdivision surfces at extraordinary points. *Advances in Comp. Math.*, 9:377–389, 1998.
8. H. Prautzsch and U. Reif. Degree estimates for C^k-piecewise polynomial subdivision surfaces. *Advances in Comp. Math.*, 10(2):209–217, 1999.
9. H. Prautzsch and G. Umlauf. A G^2-subdivision algorithm. In G. Farin, H. Bieri, G. Brunnett, and T. DeRose, editors, *Geometric Modelling*, pages 217–224. Dagstuhl 1996, Computing Supplement 13, Springer-Verlag, 1998.
10. H. Prautzsch and G. Umlauf. Improved triangular subdivision schemes. In F.-E. Wolter and N.M. Patrikalakis, editors, *Computer Graphics International 1998*, pages 626–632, Hannover, 22.-26. June 1998. IEEE Computer Society.
11. R. Qu. *Recursive subdivision algorithms for curve and surface design.* PhD thesis, Department of Mathematics and Statistics, Burnel University, Uxbridge, Middlesex, England, August 1990.
12. U. Reif. *Neue Aspekte in der Theorie der Freiformflächen beliebiger Topologie.* PhD thesis, Mathematisches Institut A, Universität Stuttgart, Stuttgart, August 1993.
13. U. Reif. A unified approach to subdivision algorithms near extraordinary vertices. *Comput. Aided Geom. Design*, 12:153–174, 1995.
14. U. Reif. A degree estimate for subdivision surfaces of higher regularity. *Proc. Amer. Math. Soc.*, 124(7):2167–2174, 1996.
15. U. Reif. Analyse und Konstruktion von Subdivisionsalgorithmen für Freiformflächen beliebiger Topologie. Habilitationsschrift, Mathematisches Institut A, Universität Stuttgart, Stuttgart, December 1998. Verlag Shaker, Aachen.
16. U. Reif. TURBS – topologically unrestricted rational B-splines. *Constr. Approx.*, 14(1):57–78, 1998.
17. U. Reif and P. Schröder. Curvature smoothness of subdivision surfaces. Technical report TR-00-03, Caltech, Pasadena, 2000.
18. M.A. Sabin. Cubic recursive division with bounded curvature. In P.J. Laurent, A. Le Méhauté, and L.L. Schumaker, editors, *Curves and Surfaces*, pages 411–414. Academic Press, Boston, 1991.
19. D. Zorin. *Stationary Subdivision and Multiresolution Surface Representations.* PhD thesis, California Institut of Technology, Pasadena, 1998.

Best Fit Translational and Rotational Surfaces for Reverse Engineering Shapes

Pál Benkő and Tamás Várady

Computer and Automation Research Institute, Hungarian Academy of Sciences

Summary. An algorithm is presented to approximate a given point set by either a translational or a rotational surface. The algorithm has two main parts: in the first part the translational direction or the rotational axis; in the second part the sweeping profile curve are approximated. The latter requires fitting planar curves to 'thick' point sets: either free form curves or constrained straight-circular contours are generated using a so called guiding polygon. In the last section the possible application of this technique is discussed in a region growing context.

1 Introduction

This paper focuses on extracting exact translational and rotational parts of complex shapes given by a dense set of measured data points. There are several applications where roughly *approximating* models are acceptable or preferable: for example, triangular meshes or models with small gaps between adjacent faces should not cause any problem for visualization or rough collision checking. The majority of mechanical engineering applications, however, requires CAD models with *exact* geometry, due to traditions, the standards of data exchange and last, but not least, due to the existing NC-machining technology. Recognising translational and rotational symmetries generally improves surface quality and the efficiency of NC processing.

The most critical phases of reverse engineering are segmentation and surface fitting [9]. We want to partition the whole point cloud into geometrically homogeneous parts, for which a good approximation by a single surface exists. The difficulty of *segmenting point clouds* is the following. If we knew all the points of a region, it would be relatively easy to find a good surface description by least squares fitting. On the other hand, if we knew the description, we could collect all the corresponding points. Unfortunately none of these holds: this is why either there are strong assumptions on the model what we want to reconstruct, or no assumptions, but then iterative methods need to be applied. A good example for the first is *direct segmentation* [8], where based on the 'regularity' of the object, it is possible to subdivide the point cloud into smaller regions, and fit simple or constrained, multiple surfaces over the individual regions. Concerning iterative solutions, a widely applied, general segmentation technique is *region growing*, where the connected point set and the corresponding surface description are built up simultaneously, step by step [4,7].

Traditional mechanical engineering parts are mainly composed of regular and semi-regular surfaces, i.e. planes, natural quadrics and tori; or surfaces created by translating or rotating a planar profile curve. There is copious literature on detecting simple algebraic surfaces; but only a few on the other. In this paper we address two related problems. Problem one: how to fit the best possible translational or rotational surface to a given, previously seg-mented, smooth point cloud. Problem two: how to detect the largest possible (translational or rotational) point region starting from a given, small initial point set called *seed*, and how to determine the best corresponding profile curve, which is also unknown. Region growing has been applied using various surface types, for example Sapldis & Desl [7] used B-splines, Leonardis et al [4] used superquadrics. In our context, we grow regions based on their special shape property.

The *smooth, sweeping profile curve* can be either a free-form B-spline or for the majority of conventional mechanical engineering parts it is composed of a sequence of straight segments and circular arcs. In the latter case, the translational surface is built of planes and cylinders, or the rotational surface of cones and tori, sharing a common translational direction or rotational axis, respectively, and we enforce additional constraints for smooth connection. Without these steps it would be quite hard to precisely locate the smooth boundaries and reconstruct these surface elements.

We would like to emphasize that the following algorithms are based on *triangulated* point clouds. The triangulation of the original point set is often decimated for computational efficiency, when only a sparse set of the data points are used for the triangles, and the points left out are assigned to the closest decimated triangle. Another important feature of our algorithm is that for each point a *normal vector* estimation is needed. Our experience shows that the best normal vector estimates are based on fitting a second order algebraic surface to surrounding points in the neighbourhood (the standard technique of fitting the best least squares plane proved to be inaccurate). This means that first we translate all the points in the neighbourhood so that the given point $\mathbf{p}$ goes to the origin (the image point set is denoted by P'), then we look for a symmetric matrix A and unit vector $\mathbf{d}$ which minimise

$$\sum_{\mathbf{x} \in P'} \left(\mathbf{x}^T A \mathbf{x} + \langle \mathbf{d}, \mathbf{x} \rangle \right)^2.$$

This leads to a three dimensional eigenvalue problem and the normal we look for is just $\mathbf{d}$.

The outline of the paper is the following. After giving the formulae to compute the best direction or axis we deal with generating optimal profile curves using the concept of guiding polygons. In the second half of the pa-per we investigate geometric region growing. Several examples illustrate the difficulties and the results of the proposed algorithms.

2 Accurate Reconstruction of Translational and Rotational Surfaces

To approximate a given point set by a translational or rotational surface, we perform a two step fit. First we determine the translational direction, then project all points to its ortho-complement. In case of rotational surfaces: first we determine the rotational axis, then rotate all points to a fixed half-plane containing the axis. In the second step a planar curve is fitted for the profile using a so-called *guiding polygon*, which is the basis of creating both free-form and straight-circular profiles.

2.1 Determining the Translational Direction

The normal vectors of a translational surface must be perpendicular to a common translational direction. We may use the deviation of these angles from $\pi/2$ as the error measurement, but the simplest is to minimise the cosine of the angles, which is just the scalar product of the estimated normal vectors $\{n_i\}$ and the unknown direction d. The translational characteristic of a shape can be visualised with the help of the Gaussian sphere: the normal vectors of a translational surface lie on a great circle, that is, in a plane which goes through the origin and is perpendicular to the translational direction.

Formally: we look for the unit vector d which minimises

$$\sum \langle n_i, d \rangle^2.$$

This is a well-known three dimensional eigenvalue problem.

2.2 Determining the Rotational Axis

The normal lines of a rotational surface must intersect a common axis (in projective sense, i.e. the lines may also be parallel to the axis). The first problem is how to measure the error of a noisy normal line. Using the distance of the two lines has two problems: first, a normal line parallel to the axis does not have zero error; second, the noise of the direction of a line near to the axis is less penalised than the same deviation of a line farther from the axis. Another idea is using the angle of the normal line and the plane spanned by the axis and the data point corresponding to the normal line. Now we do not face problem one as above, but have the opposite of problem two: the noise of a data point near to the axis is more penalised than the same deviation of a farther point. Using the product of the distance and the sine of the angle, however, eliminates the above problems (and is fairly easy to compute as will be shown below). This measure was suggested by Pottmann and Randrup in [5].

We cannot go into details here, just the basic concept is presented. Lines are represented using the so-called Plücker coordinates. A line through point

p_i in direction d_i is represented by a sextuple $(d_i, d_i \times p_i)$ (the latter term, the *momentum vector* is denoted by $\bar{d}_i$ — note that it is independent of the choice of p_i). The axis la is also given in the same form as $(da, \bar{da})$, where $\bar{da} = da \times pa$. da denotes the direction of the axis, pa is an arbitrary point on it.

Using this formalism, our error function is linear in the coordinates of the unknown axis: we minimise

$$\sum [\langle \bar{d}_i, da \rangle + \langle d_i, \bar{da} \rangle]^2.$$

We solve the above least squares expression for $(da, \bar{da})$ having the constraints $\|da\| = 1$, $\langle da, \bar{da} \rangle = 0$. First the second constraint is omitted, and the remaining system (which is a generalised eigenvalue problem) is solved. Then the full system is solved iteratively, the result of the first solution is used as initial value. Generally just a few (three or four) iterations are needed. For a geometric interpretation of the initial value see [5].

2.3 Guiding Polygon

The *guiding polygon* is the key element in the forthcoming profile fitting procedures. Approaches, which are based on point data only, may find it difficult to approximate the point set after projection. Due to the inaccuracies of the translational direction or the rotational axis we may obtain a point cloud, which is 'thick' (see Figures 8 and 9), thus it is hard to order the points, which is generally needed for the different approximating methods. An approach to thin point clouds was reported recently in [3]. Our advantage of working over a triangulation is that not only the point information, but their connectivity can be transformed into the plane of the profile.

We create the guiding polygon by intersecting the triangulation with planes, which are either orthogonal to the direction of translation or contain the axis. The problem is that these often do not contain the entire profile — even all the planes can have this property — see for example Figure 1. This is why we apply a gradual approach. Start at an arbitrary triangle — possibly in the middle of the point set. Intersect it with a plane through its barycentre and extend the polygon taking the adjacent triangles in that plane. Of course, this may be just a part of the guiding polygon and generally it will be extended later. After building the initial polygon we check the remaining triangles. Each triangle is projected (or rotated) into the profile plane and its image is projected onto the polygon. If this image is not fully covered by the current polygon, the polygon is extended, as shown in Figure 1. In order to build up the polygon correctly, we do not traverse the triangles in an arbitrary order: always a triangle adjacent to a previously processed one is taken.

If the distance between the two endpoints of the polygon gets too small (say, smaller than the average distance of adjacent vertices), we close the

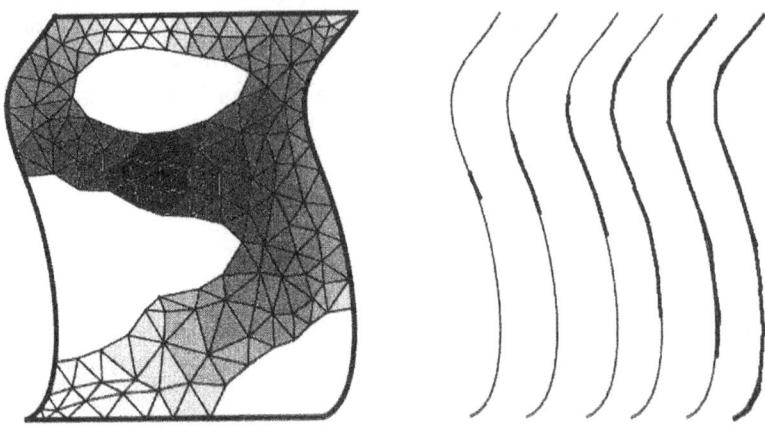

Fig. 1. Step by step build up of the guiding polygon

polygon and stop the building process. (A closed profile can be seen in Figure 15.) Note that by this procedure we always get a manifold profile. In fact, we cannot detect if the profile is non-manifold, just notice that the error of the curve fit is too large, see later fast and safe fitting in Section 3.

2.4   Free-form Profiles

Free-form profiles — see for example Figures 8, 9 and 15 — are usually represented by parametric curves. In order to approximate a point cloud by a parametric curve a parametrisation of the points is needed, which is particularly difficult for thick point data. Standard curve fitting methods combine least squares terms of distances and smoothness terms. For example maybe the simplest is:

$$\sum (\mathbf{p}_i - \mathbf{r}(t_i))^2 + \lambda \int \ddot{\mathbf{r}}^2(t)dt,$$

where t_i denotes an assigned parameter value to the 2D point $\mathbf{p}_i$. We do not go into details of curve approximation here; see related literature in [2], but emphasize that projecting the points onto the above guiding polygon provides a 'natural' parametrisation, since the guiding polygon is a good approximation of the final profile curve.

2.5   Straight-circular Profiles

Of course, every profile can be approximated by a free-form curve, but many engineering applications require an explicit line/circle profile representation.

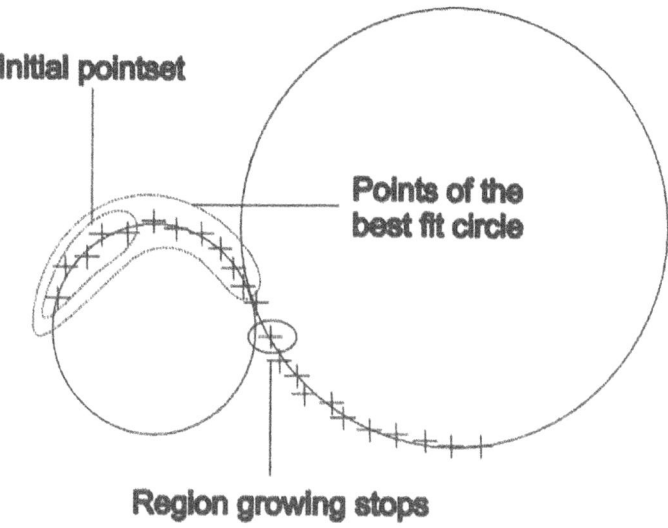

Fig. 2. Segmenting the guiding polygon

While in the previous case the guiding polygon provided the parametrisation of the projected points, here it helps the segmentation of the profile into circular arcs and straight segments. One may apply a 'recover & select' [4] procedure, but we preferred a simple sequential heuristic method. As it will be shown, after segmentation a *constrained least squares fitting* will be performed for the whole point set, which guarantees the smooth connection of the consecutive elements.

The basic concept of the algorithm is illustrated in Figure 2. Given the points of the planar polygon, what we want to segment. First assume that the polygon is open. We search for the largest segment which locally can be fitted by a circle within a given tolerance. We start from a small initial point set, the seed region and fit a circle. Then check whether the next point of the guiding polygon is still within tolerance, if yes we refit. Otherwise, we terminate the current segment, and step back for a reduced point set, where we obtained the best average least squares error. Then we create a new seed region and look for a new segment. Note the seed region cannot be too short, because then the initial fit may fail being far from the 'ideal' circle. For the same reasons, if at the end of the process fewer points remain than a seed we do not fit a circle, but a straight line, and either throw these points out or force them to belong to the previous segment. Segmenting a closed polygon is similar, the only difference is that at the end we need to check if it is the same circle as the initial one or not.

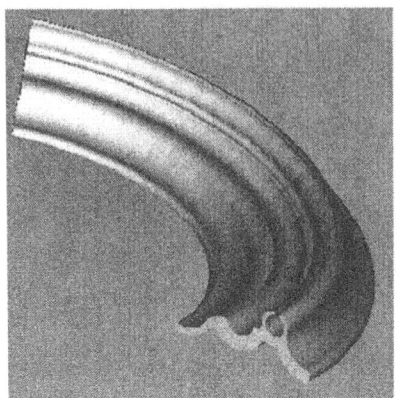 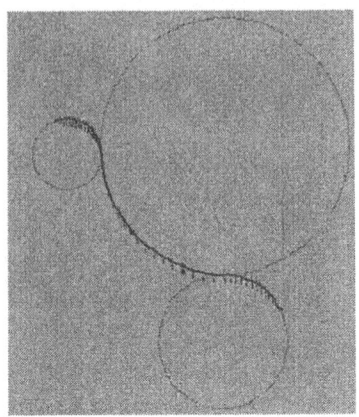

Fig. 3. a rotational surface   **Fig. 4.** profile — top right

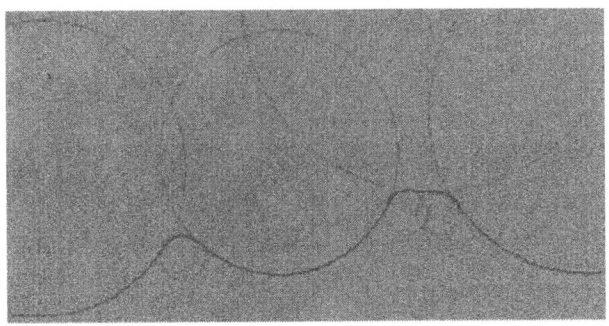

Fig. 5. profile composed of tangentially joining circles — bottom

As an example consider the three circles of the profile in Figure 4. We use Pratt's formula for the circle [6],

$$A(x^2 + y^2) + Bx + Cy + D = 0, \quad where \quad B^2 + C^2 - 4AD = 1.$$

which has the advantage that a straight line is a special (not singular) case of a circle. After fitting, if the radius of the circle is greater than a given threshold, the fitted object is interpreted as a straight line.

Denote the parameters of the i-th circle by A_i, B_i, C_i and D_i respectively. Assume that the point set has been segmented, and P_i is the set of points, which corresponds to the i-th circle: $(x_{ij}, y_{ij}) \in P_i, j = 1, \ldots, n_i$. Now we want to perform a least squares minimisation for

$$\sum_i \sum_j \left(A_i(x_{ij}^2 + y_{ij}^2) + B_i x_{ij} + C_i y_{ij} + D_i \right)^2.$$

In our example $i = 3$ and we have five constraints:

$$B_1^2 + C_1^2 - 4A_1D_1 = 1$$
$$B_2^2 + C_2^2 - 4A_2D_2 = 1$$
$$B_3^2 + C_3^2 - 4A_3D_3 = 1$$
$$(A_1B_2 - A_2B_1)^2 + (A_1C_2 - A_2C_1)^2 - (A_1 \pm A_2)^2 = 0$$
$$(A_3B_2 - A_2B_3)^2 + (A_3C_2 - A_2C_3)^2 - (A_3 \pm A_2)^2 = 0$$

The first three equations are Pratt's normalisation conditions, the last two express the tangency.

In the above case we may say that each circle is represented proportionally to the number of corresponding points. Alternatively, we may use uniform weighting, and minimise

$$\sum_i \frac{1}{n_i} \sum_j \left(A_i(x_{ij}^2 + y_{ij}^2) + B_i x_{ij} + C_i y_{ij} + D_i \right)^2$$

The most preferred solution we have found, is to weight the individual terms by their arc lengths l_i, which can be well approximated by means of the guiding polygon. Thus we minimise

$$\sum_i \frac{l_i}{n_i} \sum_j \left(A_i(x_{ij}^2 + y_{ij}^2) + B_i x_{ij} + C_i y_{ij} + D_i \right)^2$$

Based on the above equation a system is set up to be minimised at the zeros of another system of equations. One may apply the general method of Lagrangian multipliers, or its special variant for constrained fitting, as described in details in a companion paper [1]. This solution is of course an iterative process starting from the initial values, which were computed during segmentation of the guiding polygon by independent unconstrained fitting. Our basic assumption is that the initial unconstrained fit provides reasonably good initial values, not far from the final constrained solution.

A simple rotational object can be seen in Figure 3. Our approach guarantees that the profiles are smooth and are represented by consecutive arcs. Three arcs define the top right profile in Figure 4; while the bottom profile consists of seven such arcs (Figure 5).

3   Region Growing

In this section, we deal with surface reconstruction, where unlike in the previous situation, the extent of the point region, which needs to be represented is not known. We assume that there is an initial *seed region* selected by a user

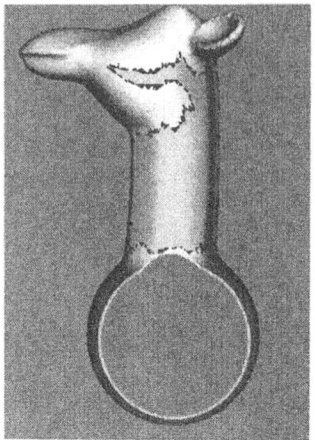

Fig. 6. fast fit went wrong

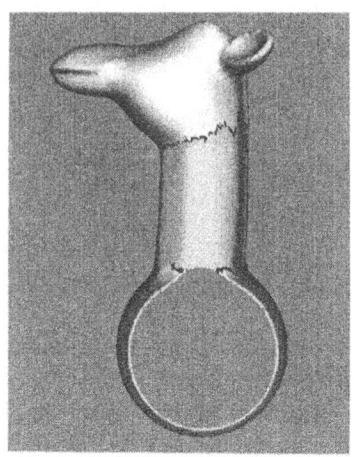

Fig. 7. safe fit stops in time

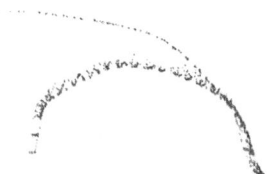

Fig. 8. profile not manifold

Fig. 9. manifold profile

or some high level module, which contains a relatively small set of triangles based on which our initial estimates can be computed. Our task is to gradually enlarge the region until all points, which belong to the given translational (or rotational) surface, are found. The main difficulty of the whole process is that the sweeping profile curve is unknown, nevertheless, the points being included need to be consistent with this unknown curve.

Region growing is composed of two alternating phases of iteration: in phase one we fit a surface to a given point set, in phase two we extend this point set by collecting those points from the neighbourhood, which are *close*. This closeness is not necessarily taken in strict geometrical sense, but a partial characterisation of the surface, which is needed for the second phase, is often sufficient. Neighbourhood information is based on the decimated triangulation; a triangle is considered to be a candidate for inclusion, if all its points are 'close' (or alternatively, if the average error of these points is small). For smooth point regions, a useful parameter to exclude undesirable triangles is the *search angle*, i.e. if there is an abrupt change of the surface normal between two adjacent triangles, the latter one will not be processed.

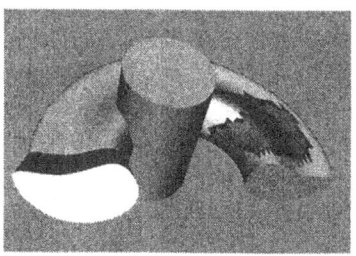

Fig. 10. Region growing: phase 1

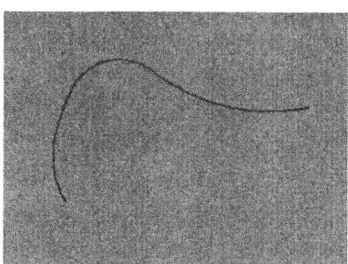

Fig. 11. profile 1

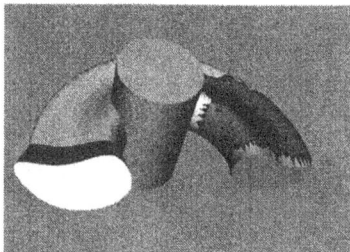

Fig. 12. Region growing: phase 2

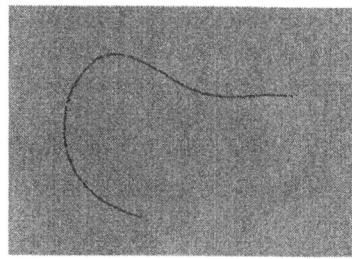

Fig. 13. profile 2

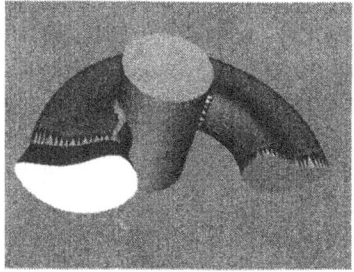

Fig. 14. Region growing: the result

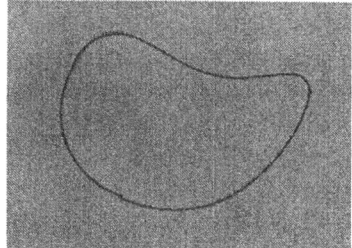

Fig. 15. final profile

3.1   Fast Versus Safe Region Growing

There are two alternatives of our region growing called *fast* and *safe* method, respectively. Since we have to apply the above fitting many times, we often want to apply the fastest possible method. As a minimal, sufficient surface description we can use just the translational direction or the rotational axis. For closeness measure, we can use the magnitude what we minimise: in the translational case the scalar product of the estimated normal and the translational direction, in the rotational case either the angle γ between the normal lines and the axis plane, its sine, or Pottmann's measure, as it was explained in Section 2.2.

This approach is computationally very efficient, however, it does not take into account possible branching of the profile. In Figures 6 and 8 an example is shown with a non-manifold profile set using measured data. As a result the current region embeds two subregions, which should be represented separately. If we had an *a priori* assumption, that there are no branching profiles, the fast method is sufficient and this should be applied. If this assumption does not hold, we have to apply the *safe* approach, as follows.

At each step of safe region growing the complete surface description is computed, which makes it possible to use geometric distances from the *current profile* as error measure. For including new points outside the current profile, either the profile needs to be extended and a distance check with a loose tolerance need to be performed, or we have to apply the error measure of the fast method. Safe region growing was applied to the object previously shown; as a result only the region with the first branch of the profile was found, which goes through the seed region, compare Figures 8 and 9.

Finally, we present a sequence of region growing for a free-form rotational surface part (see Figures 10 – 15). Three stages of the growing process are shown, as it can be seen the profile becomes closed in the third picture. It is represented in B-spline form.

4   Conclusion

Problems and algorithms for detecting and reconstructing translational and rotational surfaces have been presented. This is particularly important for reverse engineering CAD models for mechanical engineering applications, where fair and accurate surface representations are required with a complete and consistent boundary representation. The best translational direction or rotational axis was computed by least-squares methods. Different approaches were used for free-form or straight-circular profile curves. In the latter case constrained fitting assured the smoothness of the profile. In both cases a guiding polygon was used, which was extracted from the underlying, decimated triangulation of the point cloud. Algorithm variants, prioritising computational efficiency or safety, were also discussed.

5   Acknowledgments

This research was partly supported by MetroCad GmbH, Saarbrücken, special thanks are due for useful discussions and interesting test objects. The project was also supported by the National Science Foundation of the Hungarian Academy of Sciences (OTKA, No. 26203).

References

1. Benkő P., Andor L., Lukács G., Kós G., Várady T. (1999) Constrained Fitting in Reverse Engineering, Geometric Modelling Studies. GML 1999/4, Computer and Automation Research Institute, Hungarian Academy of Sciences
2. Hoschek J., Lasser D. Fundamentals of Computer Aided Geometric Design, A K Peters, 1993
3. Lee I. K. (2000) Curve reconstruction from unorganised points. Computer Aided Geometric Design **17**, 161–177
4. Leonardis A., Gupta A., Bajcsy R. (1995) Segmentation of Range Images as the Search for Geometric Parametric Models. International Journal of Computer Vision **14** 253–277
5. Pottmann H., Randrup T. (1998) Rotational and Helical Surface Approximation for Reverse Engineering. Computing **60**, 307–322
6. Pratt V. (1987) Direct least squares fitting of algebraic surfaces. Computer Graphics (Proceedings of SIGGRAPH87), **21** 145–152
7. Sapidis N. S., Besl P. J. (1995) Direct Construction of Polynomial Surfaces from Dense Range Images Through Region Growing. ACM TGO **14** 171–200
8. Várady T., Benkő P., Kós G. (1998) Reverse Engineering Regular Objects: Simple Segmentation and Surface Fitting Procedures. Int. Journal of Shape Modeling **4**, 127–141
9. Várady T., Martin R. R., Cox J. (1997) Reverse Engineering of Geometric Models — an Introduction. Computer Aided Design **29**, 255–268

Problem Reduction to Parameter Space

Myung-Soo Kim[1] and Gershon Elber[2]

[1] Seoul National University, Seoul 151-742, South Korea
[2] Technion (Israel Institute of Technology), Haifa , Israel

Summary. This paper presents a problem reduction scheme that converts geometric constraints in work space to a system of equations in parameter space. We demonstrate that this scheme can solve many interesting geometric problems that are usually considered quite difficult to deal with using conventional techniques. An important advantage of our approach is that equations represented in the parameter space have degrees significantly lower than those required for a solution in the original work space.

1 Introduction

In computer-aided geometric design, many algorithms have been developed for the design of freeform curves and surfaces [7,16,24]. Given a set of control points and other input parameters, curves and surfaces can be generated explicitly by the appropriate construction algorithms. Mathematically speaking, there are many well-defined unary and binary geometric operations that can be applied to these curves or surfaces. For example, the convex hull, offset, sweep, and bisector are geometric operations that have a clear mathematical meaning. Nevertheless, in practice their implementation is non-trivial.

This paper proposes a problem reduction scheme that converts a variety of geometric operations on freeform curves and surfaces to problems of finding zero-sets in the parameter space of the original curves and surfaces.

The zero-set of an m-variate function $F(u_1, \ldots, u_m) = 0$ is essentially the same as the intersection of the function graph surface $u_{m+1} = F(u_1, \ldots, u_m)$ and the hyperplane $u_{m+1} = 0$. Thus it is a special case of the hypersurface intersection problem. Some operations can be converted to the solution of a set of simultaneous equations in m variables: $F_i(u_1, \ldots, u_m) = 0$, for $i = 1, \ldots, k$. Implementing an operation of this sort becomes a question of intersecting k hypersurfaces in m-dimensional space. Though still quite challenging, the surface intersection problem is now relatively well understood, with a considerable amount of research reported in the literature [13,16].

To illustrate the effectiveness of our approach, we apply this problem-reduction scheme to some important geometric operations: Minkowski sums in three dimensions, general sweeps of freeform surfaces, bisectors of two freeform surfaces, intersections of two ruled surfaces, and intersections of two ringed surfaces. Some of these results are reported in recent work by the authors and their colleagues. In this paper, we give only a brief summary of these results; more details can be found in References [5,6,11,12].

An approach based on parameter space has an important advantage: representing constraints in parameter space requires equations of significantly lower degree than the same constraints represented in the initial work space. To illustrate the basic idea, we consider a simple two-dimensional sweep problem in the plane. When a curve $C_1(u_1) = (x_1(u_1), y_1(u_1))$ moves (with a fixed orientation) along another curve $C_2(u_2) = (x_2(u_2), y_2(u_2))$ in xy-plane, the sweep of C_1 generates a planar region. This region is bounded by *envelope curve*, which is defined by

$$x = x_1(u_1) + x_2(u_2), \tag{1}$$
$$y = y_1(u_1) + y_2(u_2), \tag{2}$$
$$0 = x_1'(u_1)y_2'(u_2) - y_1'(u_1)x_2'(u_2). \tag{3}$$

The third equation is the parallel constraint $C_1'(u_1) \parallel C_2'(u_2)$. Eliminating u_1 and u_2 from the three equations above produces an algebraic equation $e(x, y) = 0$ of very high degree [2,10]. In contrast, the parallel constraint of the third equation is of considerably lower degree.

Let $F(u_1, u_2) = x_1'(u_1)y_2'(u_2) - y_1'(u_1)x_2'(u_2)$; then $F(u_1, u_2) = 0$ is a necessary condition for a point $(x(u_1, v_1), y(u_1, v_1)) = C_1(u_1) + C_2(u_2)$ to be located on the envelope curve. After computing solutions of $F(u_1, u_2) = 0$ in the u_1u_2-plane, the corresponding points on the envelope curve $e(x, y) = 0$ are constructed by simple addition: $(x, y) = C_1(u_1) + C_2(u_2)$.

In computing an envelope surface for two freeform surfaces $S_1(u_1, v_1)$ and $S_2(u_2, v_2)$, we need to consider five equations in seven variables [3]. After eliminating four variables, we will end up with an algebraic surface $e(x, y, z) = 0$ of extremely high degree [3]. Therefore, instead of eliminating u_1, u_2, v_1, v_2, our approach considers two simultaneous equations in the four parameters: $F_i(u_1, u_2, v_1, v_2) = 0$, for $i = 1, 2$. After computing solutions of these two equations, the points on the envelope surface $e(x, y, z) = 0$ are constructed by addition: $(x, y, z) = S_1(u_1, v_1) + S_2(u_2, v_2)$.

It is interesting to compare our approach with the *dimensionality paradigm* of Hoffmann [14], who considers the solution-set in $xyzu_1u_2v_1v_2$-space. In the above example of an envelope surface, the solution-set is contained in two cylindrical hypersurfaces: $F_i(u_1, u_2, v_1, v_2) = 0$, for $i = 1, 2$. Representing only two out of a total of five hypersurfaces under consideration, these equations have degrees significantly lower than the solution-set in $xyzu_1u_2v_1v_2$-space, where the solution-set is itself the result of intersecting all five hypersurfaces. In contrast, the projection of the solution-set to an xyz-subspace has a considerably higher degree. Thus the selection of an appropriate subspace on which to project is very important. In this paper, we consider a few examples, where the x, y, z terms or other parameters are linear and thus can be eliminated very easily.

As a simple two-dimensional example, where the x and y terms are linear, we consider the bisector curve of two rational curves $C_1(u_1)$ and $C_2(u_2)$ in the plane. Each point (x, y) on the bisector curve is at equal distance from $C_1(u_1)$ and $C_2(u_2)$; thus each point satisfies the following constraint equations:

$$\langle (x, y) - C_1(u_1), C_1'(u_1) \rangle = 0, \tag{4}$$

$$\langle (x, y) - C_2(u_2), C_2'(u_2) \rangle = 0, \tag{5}$$

$$\left\langle (x, y) - \frac{C_1(u_1) + C_2(u_2)}{2}, C_1(u_1) - C_2(u_2) \right\rangle = 0. \tag{6}$$

There are four variables in these three equations. By eliminating two variables, we obtain one equation in two variables, which represents an implicit curve in the plane of the two remaining parameters.

Conventional approaches eliminate the parameters u_1 and u_2 to generate an implicit equation $b(x, y) = 0$ for the bisector curve in xy-plane [15], or trace the curve $b(x, y) = 0$ numerically [8,9]. Unfortunately, the algebraic degree of $b(x, y) = 0$ is very high. (Using Mathematica [26], the authors observe [5] that the degree of $b(x, y) = 0$ becomes $7d_1 d_2 - 3(d_1 + d_2) + 1$, for $1 \leq d_1 \leq 3$ and $1 \leq d_2 \leq 6$, when each $C_i(u_i)$ is a polynomial curve of degree d_i.) Note that equations (4)–(6) are linear in x and y. Thus it is much easier to eliminate the variables x and y than the other curve parameters, u_1 and u_2. Moreover, the degree of the resulting implicit equation $F(u_1, u_2) = 0$ is significantly lower than that of $b(x, y) = 0$; in this case the degree of $F(u_1, u_2)$ is $2(d_1 + d_2) - 2$. When both $C_i(u_i)$ are cubic polynomial curves, $F(u_1, u_2) = 0$ has degree 10, whereas $b(x, y) = 0$ has degree 46. When we consider the bisector of two rational surfaces in three dimensions, the difference will be even greater. The authors therefore reformulate [6] the bisector surface problem as a common zero-set finding problem: $F_i(u_1, u_2, v_1, v_2) = 0$, for $i = 1, 2$. A brief summary of this result is given in Section 4.

A ruled surface has a linear parameter along each ruling direction. Thus, in computing the intersection of two ruled surfaces, formulated as three equations in four variables, two linear variables can be eliminated very easily. Based on this observation, Heo et al. [12] reformulate the intersection of two ruled surfaces as a zero-set finding problem: $F(u_1, u_2) = 0$.

Johnstone [19] defines a *ringed surface* as the sweep of a moving circle under translation, rotation, and scaling. The intersection of two ringed surfaces can be decomposed into the intersection of two moving circles. Heo [11] reformulates the intersection of two circles $C_1^{u_1}$ and $C_2^{u_2}$ as the intersection of three planes and a sphere. Three planes in general position have a unique intersection point $(x(u_1, u_2), y(u_1, u_2), z(u_1, u_2))$, where the x, y, z coordinates are given as rational functions of u_1 and u_2. Substituting these rational representations into the sphere equation, Heo [11] derives an implicit equation, $F(u_1, u_2) = 0$, which is a necessary condition for two circles to intersect.

The rest of this paper is organized as follows. In Sections 2–4, we consider the Minkowski sum of two freeform shapes, the general sweep of a freeform shape, and the bisector of two freeform surfaces. The intersection of two ruled surfaces and that of two ringed surfaces are briefly reviewed in Sections 5 and 6 respectively. Section 7 concludes this paper.

2 Minkowski Sum Computation

Given two objects O_1 and O_2, their Minkowski sum $O_1 \oplus O_2$ is defined as the set of all vector sums generated by all pairs of points in O_1 and O_2 respectively:

$$O_1 \oplus O_2 = \{a + b \mid a \in O_1,\ b \in O_2\}. \tag{7}$$

The Minkowski sum of two objects considers all points in the interiors as well as on the boundaries of the two objects.

The Minkowski sum has been used as an important tool for computing collision-free paths in robot motion planning [2,20]. Figure 1(a) shows the Minkowski sum $O_1 \oplus O_2$ of two planar curved objects O_1 and O_2. In Figure 1(b), we compute the Minkowski sum $O_1 \oplus (-O_2)$, where $-O_2$ is the symmetric object of O_2 with respect to the local reference point (which is located at the origin in this case). There is no collision between O_1 and O_2 as long as the reference point of O_2 does not penetrate $\partial(O_1 \oplus (-O_2))$ (see Figure 1(c)). The object $O_1 \oplus (-O_2)$ is called the *Configuration-space (C-space) obstacle* of O_1 with respect to the moving object O_2.

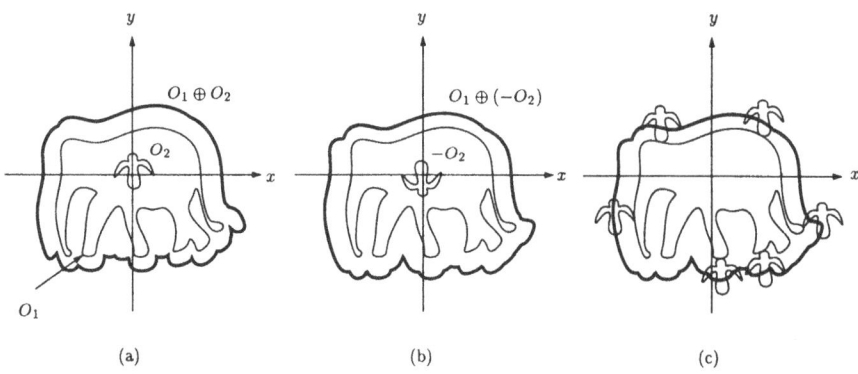

Fig. 1. A Minkowski sum and a *C-space* obstacle.

Assume that O_1 and O_2 are bounded by the surfaces $S_1(u_1, v_1)$ and $S_2(u_2, v_2)$, respectively. The problem of computing the Minkowski sum boundary, denoted as $\partial(O_1 \oplus O_2)$, can then be transformed into the problem of computing the convolution of S_1 and S_2, denoted as $S_1 * S_2$ [3]. In the convolution operation, the vector sums are applied only to the pairs of surface points that have the same normal direction.

Definition 1. Let $S_1(u_1, v_1)$ and $S_2(u_2, v_2)$ be two regular parametric surfaces. Moreover, let $N_i(u_i, v_i) = \frac{\partial S_i}{\partial u_i}(u_i, v_i) \times \frac{\partial S_i}{\partial v_i}(u_i, v_i)$ denote the normal vector of $S_i(u_i, v_i)$, for $i = 1, 2$. The convolution surface $S_1 * S_2$ is defined by

$$(S_1 * S_2)(u_1, v_1) = S_1(u_1, v_1) + S_2(u_2(u_1, v_1), v_2(u_1, v_1)), \tag{8}$$

where

$$N_1(u_1, v_1) \parallel N_2(u_2, v_2) \quad \text{and} \quad \langle N_1(u_1, v_1), N_2(u_2, v_2) \rangle > 0, \tag{9}$$

for a reparametrization $(u_2, v_2) = (u_2(u_1, v_1), v_2(u_1, v_1))$.

When both O_1 and O_2 are convex objects, the convolution surface $S_1 * S_2$ is exactly the same as the Minkowski sum boundary $\partial(O_1 \oplus O_2)$. In general, $\partial(O_1 \oplus O_2)$ is only a subset of $S_1 * S_2$ [2]. Thus the convolution surface $S_1 * S_2$ may have some redundant patches which do not contribute to the Minkowski sum boundary. To construct $\partial(O_1 \oplus O_2)$, we need to follow two steps: (i) compute the convolution surface $S_1 * S_2$, and (ii) eliminate the redundant patches of $S_1 * S_2$ which do not contribute to $\partial(O_1 \oplus O_2)$. The second step, called trimming, essentially reduces to a surface intersection problem. A robust implementation of trimming is a non-trivial task even for the simple case of curve convolution [22]. This paper deals only with the generation of the convolution surface.

The convolution surface $S_1 * S_2$ is an envelope which is obtained by sweeping one surface S_1 (with a fixed orientation) along the other surface S_2 [2]. Each point on the convolution surface $S_1 * S_2$ satisfies

$$\left\langle N_1(u_1, v_1), \frac{\partial S_2}{\partial u_2}(u_2, v_2) \right\rangle = 0, \tag{10}$$

$$\left\langle N_1(u_1, v_1), \frac{\partial S_2}{\partial v_2}(u_2, v_2) \right\rangle = 0, \tag{11}$$

which is also equivalent to the following two equations:

$$F_1(u_1, u_2, v_1, v_2) = \begin{vmatrix} \frac{\partial S_1}{\partial u_1}(u_1, v_1) & \frac{\partial S_1}{\partial v_1}(u_1, v_1) & \frac{\partial S_2}{\partial u_2}(u_2, v_2) \end{vmatrix} = 0, \tag{12}$$

$$F_2(u_1, u_2, v_1, v_2) = \begin{vmatrix} \frac{\partial S_1}{\partial u_1}(u_1, v_1) & \frac{\partial S_1}{\partial v_1}(u_1, v_1) & \frac{\partial S_2}{\partial v_2}(u_2, v_2) \end{vmatrix} = 0. \tag{13}$$

Now the problem has essentially been reduced to that of computing the intersection of two implicit hypersurfaces in $u_1 u_2 v_1 v_2$-space, the parameter space of the two original surfaces. For each solution (u_1, u_2, v_1, v_2) of two equations $F_i(u_1, u_2, v_1, v_2) = 0$, for $i = 1, 2$, the corresponding point (x, y, z) in the work space is given by the following well-defined mapping:

$$(x(u_1, u_2, v_1, v_2), y(u_1, u_2, v_1, v_2), z(u_1, u_2, v_1, v_2))$$
$$= S_1(u_1, v_1) + S_2(u_2, v_2). \tag{14}$$

Figure 2 shows an example of a surface-surface convolution. Dots on the convolution surface are solutions of Equations (12) and (13). The convolution surface (shown in the leftmost) is approximated by interpolating these discrete sample points with a spline surface.

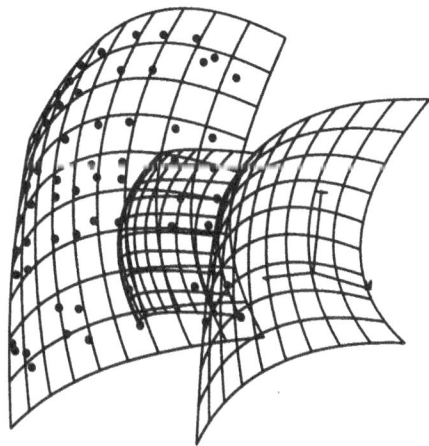

Fig. 2. Convolution surface of two freeform surfaces.

3   General Sweep Computation

Let O be a three-dimensional object bounded by a freeform surface $S(u, v)$, and let $A(t)$ denote an affine transformation represented by a 4×4 matrix,

$$\begin{bmatrix} a_{11}(t) & a_{12}(t) & a_{13}(t) & t_x(t) \\ a_{21}(t) & a_{22}(t) & a_{23}(t) & t_y(t) \\ a_{31}(t) & a_{32}(t) & a_{33}(t) & t_z(t) \\ 0 & 0 & 0 & 1 \end{bmatrix},$$

where $(a_{ij}(t))_{3 \times 3}$ represents a linear transformation (e.g. rotation or shearing) and $(t_x(t), t_y(t), t_z(t))$ denotes a translation of the coordinate system.

The swept volume of the object O under the affine transformation $A(t)$ is given by $\cup_t A(t)[O]$. Assuming $a \leq t \leq b$, the boundary surface of the swept volume consists of some patches of $A(a)[S(u, v)]$, $A(b)[S(u, v)]$, and the envelope surface, which is the set of points $A(t)[S(u, v)]$ that satisfies the following equation [1,18,23]:

$$F(u, v, t) = \left| A'(t)[S(u, v)] \quad A(t) \left[\frac{\partial S}{\partial u}(u, v) \right] \quad A(t) \left[\frac{\partial S}{\partial v}(u, v) \right] \right| = 0.$$

Figures 3 and 4 show two examples of a general sweep. An ellipsoid is moving and also changing its shape, but its orientation is fixed. These examples show the potential of our approach in effectively generating complex three-dimensional shapes based on a few simple motion and shape parameters.

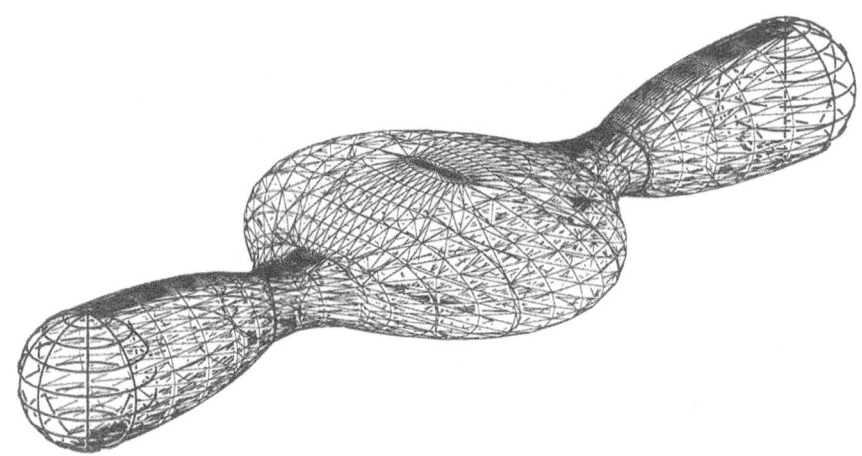

Fig. 3. General sweep of an ellipsoid along an open trajectory.

4   Bisector Surface Computation

Let $S_1(u_1, v_1)$ and $S_2(u_2, v_2)$ be two rational surfaces. We consider the bisector surface of $S_1(u_1, v_1)$ and $S_2(u_2, v_2)$, which consists of points (x, y, z) satisfying the following five constraint equations:

$$\left\langle (x, y, z) - S_1(u_1, v_1), \frac{\partial S_1(u_1, v_1)}{\partial u_1} \right\rangle = 0, \tag{15}$$

$$\left\langle (x, y, z) - S_1(u_1, v_1), \frac{\partial S_1(u_1, v_1)}{\partial v_1} \right\rangle = 0, \tag{16}$$

$$\left\langle (x, y, z) - S_2(u_2, v_2), \frac{\partial S_2(u_2, v_2)}{\partial u_2} \right\rangle = 0, \tag{17}$$

$$\left\langle (x, y, z) - S_2(u_2, v_2), \frac{\partial S_2(u_2, v_2)}{\partial v_2} \right\rangle = 0, \tag{18}$$

$$\langle (x, y, z), 2(S_2(u_2, v_2) - S_1(u_1, v_2)) \rangle$$
$$+ \|S_1(u_1, v_1)\|^2 - \|S_2(u_2, v_2)\|^2 = 0. \tag{19}$$

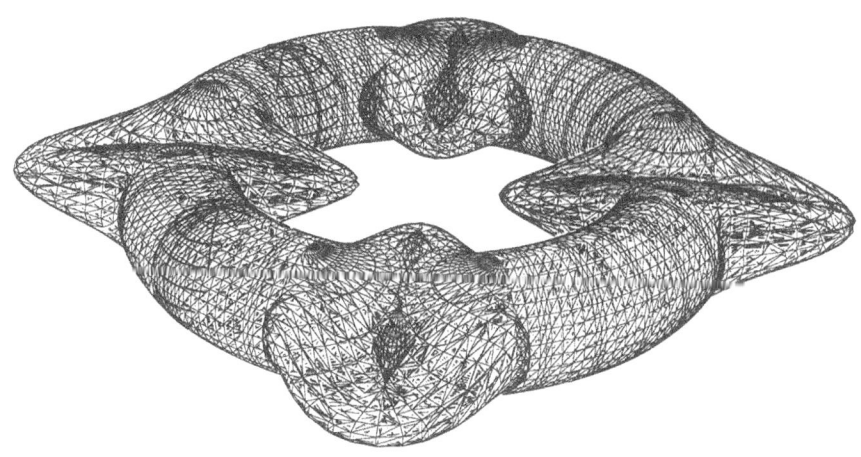

Fig. 4. General sweep of an ellipsoid along a closed trajectory.

Equations (15) and (16) mean that the bisector point (x, y, z) is located on the normal line of $S_1(u_1, v_1)$, and Equations (17) and (18) imply that the point (x, y, z) is on the normal line of $S_2(u_2, v_2)$. Additionally, Equation (19) constrains the point (x, y, z) to the symmetry plane of $S_1(u_1, v_1)$ and $S_2(u_2, v_2)$. Equations (15)–(19) are all linear in (x, y, z).

Let $(x, y, z) = S_1(u_1, v_1) + \alpha n_1(u_1, v_1)$, $\alpha \in \mathbf{R}$, be a bisector point located on the normal line of $S_1(u_1, v_1)$, where $n_1(u_1, v_1) = \frac{\partial S_1(u_1, v_1)}{\partial u_1} \times \frac{\partial S_1(u_1, v_1)}{\partial v_1}$ is an unnormalized normal vector of $S_1(u_1, v_1)$. Equations (15) and (16) are automatically satisfied. By substituting (x, y, z) into Equation (19), we get the equation

$$\langle S_1(u_1, v_1) + \alpha n_1(u_1, v_1), 2(S_2(u_2, v_2) - S_1(u_1, v_2))\rangle$$
$$+ \|S_1(u_1, v_1)\|^2 - \|S_2(u_2, v_2)\|^2 = 0$$

or, equivalently,

$$\alpha(u_1, u_2, v_1, v_2) = \frac{\|S_1(u_1, v_1) - S_2(u_2, v_2)\|^2}{2\langle n_1(u_1, v_1), S_2(u_2, v_2) - S_1(u_1, v_1)\rangle}. \tag{20}$$

Finally, we substitute this representation of $\alpha(u_1, u_2, v_1, v_2)$ into $(x, y, z) = S_1(u_1, v_1) + \alpha(u_1, u_2, v_1, v_2)n_1(u_1, v_1)$, and update Equations (17) and (18) with the resulting representation of (x, y, z).

Let $\Delta(u_1, u_2, v_1, v_2)$ be defined as follows:

$$\Delta(u_1, u_2, v_1, v_2)$$

$$= 2(S_2(u_2, v_2) - S_1(u_1, v_1)) \langle n_1(u_1, v_1), S_1(u_1, v_1) - S_2(u_2, v_2) \rangle$$
$$+ n_1(u_1, v_1) \| S_1(u_1, v_1) - S_2(u_2, v_2) \|^2.$$

Then, Equations (17) and (18) reduce to

$$F_1(u_1, u_2, v_1, v_2) = \left\langle \Delta(u_1, u_2, v_1, v_2), \frac{\partial S_2(u_2, v_2)}{\partial u_2} \right\rangle = 0, \tag{21}$$

$$F_2(u_1, u_2, v_1, v_2) = \left\langle \Delta(u_1, u_2, v_1, v_2), \frac{\partial S_2(u_2, v_2)}{\partial v_2} \right\rangle = 0. \tag{22}$$

The common zero-set of Equations (21) and (22) satisfies all five constraints of Equations (15)–(19). Hence, we have reduced the surface-surface bisector problem of R^3 into the problem of finding a common zero-set for the two four-variate functions of Equations (21) and (22). When the zero-set has been found, the bisector point corresponding to each point (u_1, u_2, v_1, v_2) in the zero-set can be computed by the rational map

$$(x(u_1, u_2, v_1, v_2), y(u_1, u_2, v_1, v_2), z(u_1, u_2, v_1, v_2))$$
$$= S_1(u_1, v_1) - \frac{\| S_1(u_1, v_1) - S_2(u_2, v_2) \|^2}{2 \langle n_1(u_1, v_1), S_1(u_1, v_1) - S_2(u_2, v_2) \rangle} n_1(u_1, v_1). \tag{23}$$

Figure 5 shows two examples of a surface-surface bisector. Dots on the bisector surfaces are solutions of Equations (21) and (22). Each bisector surface is approximated by interpolating these discrete sample points with a spline surface.

5   Intersecting Two Ruled Surfaces

Let $S_1(u_1, t_1)$ and $S_2(u_2, t_2)$ be two ruled surfaces defined by

$$S_1(u_1, t_1) = p_1(u_1) + t_1 d_1(u_1), \tag{24}$$
$$S_2(u_2, t_2) = p_2(u_2) + t_2 d_2(u_2), \tag{25}$$

where $p_1(u_1)$ and $p_2(u_2)$ are position curves, and $d_1(u_1)$ and $d_2(u_2)$ are direction vector curves. When the two surfaces $S_1(u_1, t_1)$ and $S_2(u_2, t_2)$ intersect, we have

$$S_1(u_1, t_1) = S_2(u_2, t_2), \tag{26}$$

and equivalently,

$$p_1(u_1) - p_2(u_2) = -t_1 d_1(u_1) + t_2 d_2(u_2). \tag{27}$$

That is, the difference vector $p_1(u_1) - p_2(u_2)$ is given as a linear combination of $d_1(u_1)$ and $d_2(u_2)$. Consequently, three vectors $p_1(u_1) - p_2(u_2)$, $d_1(u_1)$, and $d_2(u_2)$ are linearly dependent and the following determinant vanishes:

$$F(u_1, u_2) = \left| p_1(u_1) - p_2(u_2) \quad d_1(u_1) \quad d_2(u_2) \right| = 0. \tag{28}$$

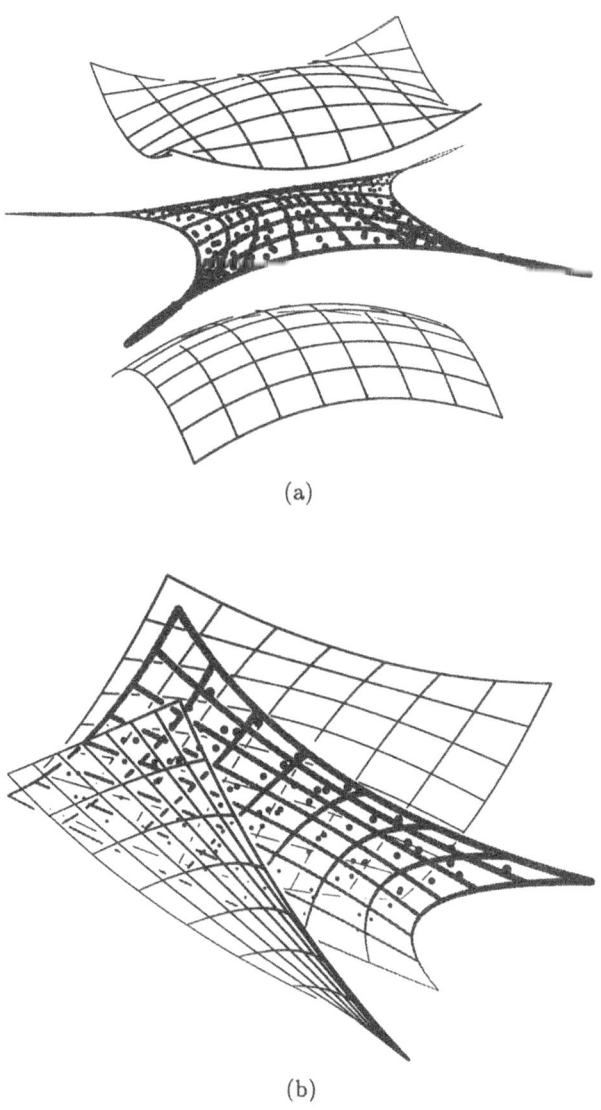

(a)

(b)

Fig. 5. Bisector surfaces (in gray) between two freeform surfaces.

For each point (u_1, u_2) on the zero-set of $F(u_1, u_2) = 0$, the corresponding intersection point (x, y, z) can be computed by the equation (see Equation (24))

$$(x, y, z) = \boldsymbol{p}_1(u_1) + t_1(u_1, u_2)\boldsymbol{d}_1(u_1), \tag{29}$$

where $t_1(u_1, u_2)$ is the solution of Equation (27). We consider how to represent $t_1(u_1, u_2)$ as a rational bivariate function of u_1 and u_2. By taking inner products of Equation (27) with the vectors $\boldsymbol{d}_1(u_1)$ and $-\boldsymbol{d}_2(u_2)$, we obtain the following linear system of equations for t_1 and t_2:

$$\begin{bmatrix} -\|\boldsymbol{d}_1(u_1)\|^2 & \langle \boldsymbol{d}_1(u_1), \boldsymbol{d}_2(u_2)\rangle \\ \langle \boldsymbol{d}_1(u_1), \boldsymbol{d}_2(u_2)\rangle & -\|\boldsymbol{d}_2(u_2)\|^2 \end{bmatrix} \begin{bmatrix} t_1 \\ t_2 \end{bmatrix} = \begin{bmatrix} \langle \boldsymbol{d}_1(u_1), \boldsymbol{p}_1(u_1) - \boldsymbol{p}_2(u_2)\rangle \\ \langle \boldsymbol{d}_2(u_2), \boldsymbol{p}_2(u_2) - \boldsymbol{p}_1(u_1)\rangle \end{bmatrix}$$

Let $\Delta_1(u_1, u_2)$ be defined as follows:

$$\begin{aligned} \Delta_1(u_1, u_2) = &\|\boldsymbol{d}_2(u_2)\|^2 \langle \boldsymbol{d}_1(u_1), \boldsymbol{p}_2(u_2) - \boldsymbol{p}_1(u_1)\rangle \\ &+ \langle \boldsymbol{d}_1(u_1), \boldsymbol{d}_2(u_2)\rangle \langle \boldsymbol{d}_2(u_2), \boldsymbol{p}_1(u_1) - \boldsymbol{p}_2(u_2)\rangle . \end{aligned}$$

Then we have

$$t_1(u_1, u_2) = \frac{\Delta_1(u_1, u_2)}{\|\boldsymbol{d}_1(u_1)\|^2 \|\boldsymbol{d}_2(u_2)\|^2 - \langle \boldsymbol{d}_1(u_1), \boldsymbol{d}_2(u_2)\rangle^2}. \tag{30}$$

Consequently, we have a well-defined rational mapping from the zero-set of $F(u_1, u_2) = 0$ to the intersection curve

$$(x(u_1, u_2), y(u_1, u_2), z(u_1, u_2)) = \boldsymbol{p}_1(u_1) + t_1(u_1, u_2)\boldsymbol{d}_1(u_1). \tag{31}$$

Figure 6(a) shows a simple example of the intersection of two ruled surfaces that meet transversally. The corresponding bivariate function $F(u_1, u_2)$ is shown in Figure 6(b). Figures 7(a)–7(d) show a sequence of examples that intersect two almost coaxial cylinders with angles of $10°$, $1°$, $0.1°$, $0.01°$ between the two cylinders, in that order. Figures 7(e) and 7(f) are the bivariate functions $F(u_1, u_2)$ that correspond to the examples shown in Figures 7(a) and 7(d) respectively. It is very difficult to distinguish the two intersecting cylinders as they appear almost to overlap in Figures 7(c)–7(d). Moreover, the function as shown in Figure 7(f) is almost flat. Nevertheless, the computational results are numerically stable and they produce reasonable solutions, which demonstrates the robustness of our intersection algorithm for two ruled surfaces.

6   Intersecting Two Ringed Surfaces

Let $S_1(u_1, t_1)$ and $S_2(u_2, t_2)$ be two ringed surfaces, each of which is defined as the union of a one-parameter family of circles,

$$S_1(u_1, t_1) = \cup_{u_1} C_1^{u_1}(t_1) \quad \text{and} \quad S_2(u_2, t_2) = \cup_{u_2} C_2^{u_2}(t_2), \tag{32}$$

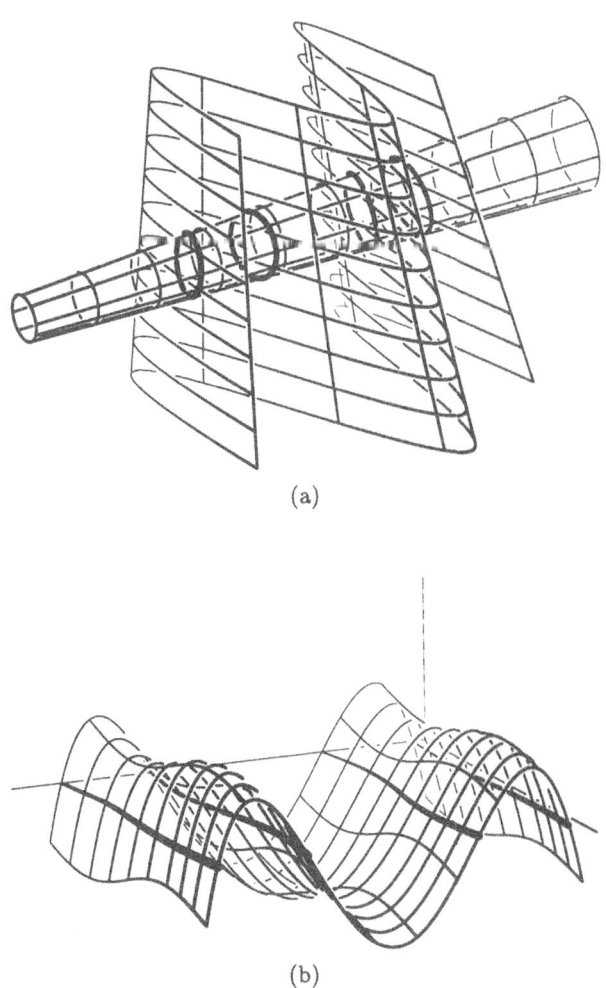

(a)

(b)

Fig. 6. Intersection of two ruled surfaces. (a) Two ruled surfaces. (b) Function graph $(u_1, u_2, F(u_1, u_2))$ and the zero-set $F(u_1, u_2) = 0$.

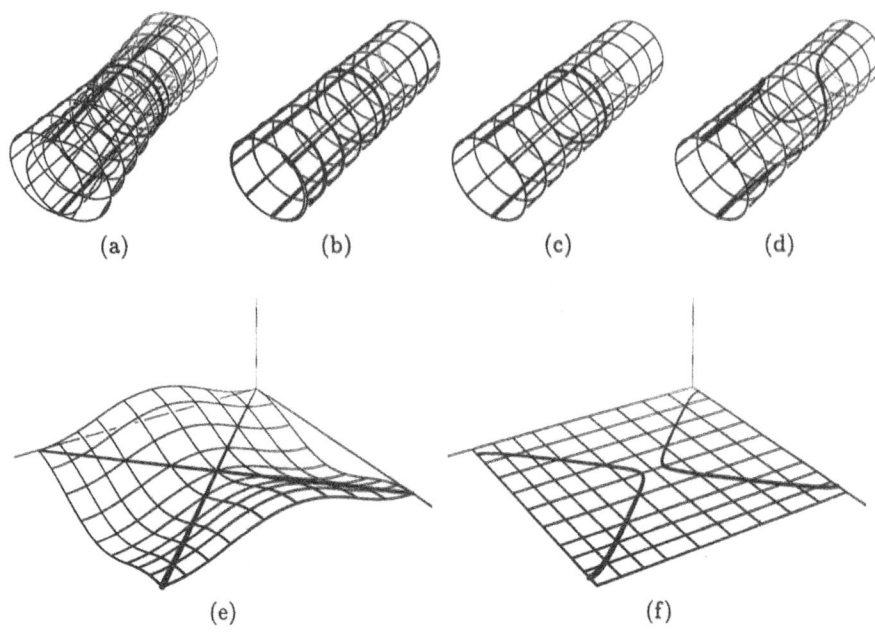

Fig. 7. Intersection of two cylinders. (a)–(d) Two almost coaxial cylinders, meeting with angles of 10°, 1°, 0.1°, 0.01°, respectively. (e)–(f) Function graphs $(u_1, u_2, F(u_1, u_2))$ and the zero-sets $F(u_1, u_2) = 0$ for cases (a) and (d), respectively.

where $C_1^{u_i}(t_i)$ is a circle parameterized by t_i, for $i = 1, 2$. Assume that the circle $C_i^{u_i}$ has center $p_i(u_i)$ and radius $r_i(u_i)$ and, moreover, that it is contained in the plane $P_i^{u_i}$ with normal $n_i(u_i)$. Let $O_i^{u_i}$ denote the sphere with center $p_i(u_i)$ and radius $r_i(u_i)$. Then we have

$$C_1^{u_1} = P_1^{u_1} \cap O_1^{u_1} \quad \text{and} \quad C_2^{u_2} = P_2^{u_2} \cap O_2^{u_2}. \tag{33}$$

The intersection $S_1 \cap S_2$ is the union of a two-parameter family of circle-circle intersections

$$\begin{aligned}
S_1 \cap S_2 &= \cup_{u_1} \cup_{u_2} C_1^{u_1} \cap C_2^{u_2} \\
&= \cup_{u_1} \cup_{u_2} (P_1^{u_1} \cap O_1^{u_1}) \cap (P_2^{u_2} \cap O_2^{u_2}) \\
&= \cup_{u_1} \cup_{u_2} (P_1^{u_1} \cap P_2^{u_2}) \cap (O_1^{u_1} \cap O_2^{u_2}) \\
&= \cup_{u_1} \cup_{u_2} (P_1^{u_1} \cap P_2^{u_2}) \cap C_{12}^{u_1 u_2} \\
&= \cup_{u_1} \cup_{u_2} (P_1^{u_1} \cap P_2^{u_2}) \cap (P_{12}^{u_1 u_2} \cap O_1^{u_1}) \\
&= \cup_{u_1} \cup_{u_2} (P_1^{u_1} \cap P_2^{u_2} \cap P_{12}^{u_1 u_2}) \cap O_1^{u_1} \tag{34}
\end{aligned}$$

$$= U_{u_1} U_{u_2} (P_1^{u_1} \cap P_2^{u_2}) \cap (P_{12}^{u_1 u_2} \cap O_2^{u_2})$$
$$= U_{u_1} U_{u_2} (P_1^{u_1} \cap P_2^{u_2} \cap P_{12}^{u_1 u_2}) \cap O_2^{u_2}, \tag{35}$$

where $C_{12}^{u_1 u_2}$ is the intersection circle of two spheres $O_1^{u_1}$ and $O_2^{u_2}$, and $P_{12}^{u_1 u_2}$ is the plane containing the circle $C_{12}^{u_1 u_2}$.

A simple calculation shows that the plane $P_{12}^{u_1 u_2}$ contains the following point:

$$\frac{p_1(u_1) + p_2(u_2)}{2} + \frac{r_1^2(u_1) - r_2^2(u_2)}{2\|p_1(u_1) - p_2(u_2)\|^2} (p_2(u_2) - p_1(u_1)). \tag{36}$$

Moreover, the difference vector $p_1(u_1) - p_2(u_2)$ is a normal vector of the plane $P_{12}^{u_1 u_2}$. Note that the inner product of this normal vector and the position vector of Equation (36) produces

$$\frac{\|p_1(u_1)\|^2 - \|p_2(u_2)\|^2 + r_2^2(u_2) - r_1^2(u_1)}{2}. \tag{37}$$

Consequently, the intersection point (x, y, z) of three planes $P_1^{u_1}$, $P_2^{u_2}$, and $P_{12}^{u_1 u_2}$ satisfies the following matrix equation:

$$\begin{bmatrix} n_1(u_1) \\ n_2(u_2) \\ p_1(u_1) - p_2(u_2) \end{bmatrix} \begin{bmatrix} x \\ y \\ z \end{bmatrix} = \begin{bmatrix} \langle n_1(u_1), p_1(u_1) \rangle \\ \langle n_2(u_2), p_2(u_2) \rangle \\ \frac{\|p_1(u_1)\|^2 - \|p_2(u_2)\|^2 + r_2^2(u_2) - r_1^2(u_1)}{2} \end{bmatrix}. \tag{38}$$

When the three vectors $n_1(u_1)$, $n_2(u_2)$, and $p_1(u_1) - p_2(u_2)$ are linearly independent, the above matrix equation has a unique solution $(x(u_1, u_2), y(u_1, u_2), z(u_1, u_2))$. The coordinate functions $x(u_1, u_2)$, $y(u_1, u_2)$, $z(u_1, u_2)$ are rational if the position curve $p_i(u_i)$ and the radius function $r_i(u_i)$ are all rational, for $i = 1, 2$. The condition for this intersection point to be located on the sphere $O_i^{u_i}$ can be formulated as follows:

$$F_i(u_1, u_2) = \|(x(u_1, u_2), y(u_1, u_2), z(u_1, u_2)) - p_i(u_i)\|^2 - r_i^2(u_i) = 0.$$

When the vectors $n_1(u_1)$, $n_2(u_2)$, and $p_1(u_1) - p_2(u_2)$ are linearly independent, the equation $F_1(u_1, u_2) = 0$ (or $F_2(u_1, u_2) = 0$) is a necessary condition for two circles $C_1^{u_1}$ and $C_2^{u_2}$ to have a non-empty intersection point. For each point (u_1, u_2) on the zero-set of $F_1(u_1, u_2) = 0$ (or $F_2(u_1, u_2) = 0$), the corresponding intersection point can be computed by the rational map $(x(u_1, u_2), y(u_1, u_2), z(u_1, u_2))$, which is the solution of Equation (38).

Figure 8 shows two examples of intersecting ringed surfaces. As we can see from these examples, ringed surfaces can represent tube-like surfaces with ease; thus an efficient algorithm for intersecting two arbitrary ringed surfaces has much potential.

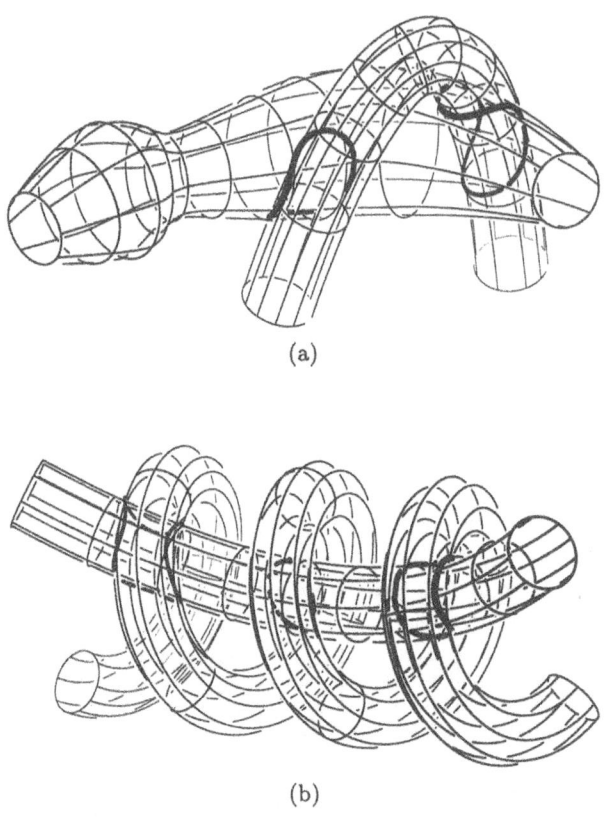

(a)

(b)

Fig. 8. Intersection of two ringed surfaces

7   Conclusions

In this paper we present a problem reduction scheme that converts geometric constraints in work space to a system of equations in parameter space. The effectiveness of this approach has been demonstrated through some important geometric operations such as computing Minkowski sums, general sweeps, bisector surfaces, and intersecting ruled surfaces and ringed surfaces. These examples are taken from the result of our own recent work, and also from joint work with colleagues.

Our problem reduction scheme is not limited to the operations discussed in this paper. Future work will introduce more problems that can be solved using the scheme presented here. New techniques also need to be developed for reducing geometric constraints to those formulated with lower degrees and fewer variables.

An important question that remains is how to compute the zero-set of the reduced equations in the parameter space. A subdivision-based approach, using Bernstein-Bézier basis functions, has been used in implementing the examples shown in this paper [5,6,12]. The basic approach is similar to conventional techniques for computing the solutions of multivariate equations [17,21,25].

Zhang and Martin [27] apply affine arithmetic to algebraic curve drawing. Their test results show plane algebraic curves drawn with higher quality, while using less subdivisions, than those generated by interval arithmetic. However, each subdivision takes more time than that of interval arithmetic. Thus speeding up the subdivision step of affine arithmetic is a possible way of improving the overall performance of our approach based on the zero-set finding.

8   Acknowledgements

The authors thank In-Kwon Lee and Hee-Seok Heo for collaboration on the work reported in this paper. This research was supported in part by the Korean Ministry of Information and Communication.

References

1. Abdel-Malek, K., Yeh, H.-J. (1997) Geometric Representation of the Swept Volume Using Jacobian Rank-Deficiency Conditions. *Computer-Aided Design* **29**, 457–468
2. Bajaj, C., Kim, M.-S. (1989) Generation of Configuration Space Obstacles: The Case of Moving Algebraic Curves. *Algorithmica* 4:2, 157–172
3. Bajaj, C., Kim, M.-S. (1990) Generation of Configuration Space Obstacles: The Case of Moving Algebraic Surfaces. *The Int'l J. of Robotics Research* **9**, 92–112
4. Elber, G. (1996) *IRIT 7.0 User's Manual*, Technion, http://www.cs.technion. ac.il/~irit.
5. Elber, G., Kim, M.-S. (1998) Bisector Curves of Planar Rational Curves. *Computer-Aided Design* **30**, 1089–1096
6. Elber, G., Kim, M.-S. (2000) A Computational Model for Nonrational Bisector Surfaces: Curve-Surface and Surface-Surface Bisectors. To appear in *Proc. of Geometric Modeling and Processing 2000*, Hong Kong, April 10-12, 2000
7. Farin, G. (1997) *Curves and Surfaces for Computer Aided Geometric Design: A Practical Guide*, Fourth Ed., Academic Press, San Diego
8. Farouki, R., Johnstone, J., (1994) Computing Point/Curve and Curve/Curve Bisectors. *Design and Application of Curves and Surfaces: Mathematics of Surfaces V*, R.B. Fisher (Ed.), Oxford Univ. Press, New York, 327–354
9. Farouki, R., Ramamurthy, R. (1998) Specified-precision Computation of Curve/Curve Bisectors. *Int'l J. of Computational Geometry & Applications* **8**, 599–617
10. Kaul, A., Farouki, R. (1995) Computing Minkowski Sums of Plane Curves. *Int'l J. of Comp. Geom. and Appl.* **5**, 413–432

11. Heo, H.-S. (2000) *The Intersection of Ruled and Ringed Surfaces*, Ph.D. Thesis, Dept. of Computer Science, POSTECH, February, 2000
12. Heo, H.-S., Kim, M.-S., Elber, G. (1999) The Intersection of Two Ruled Surfaces. *Computer-Aided Design* **31**, 33–50
13. Hoffmann, C. (1989) *Geometric and Solid Modeling: An Introduction*, Morgan Kaufmann, San Mateo, CA
14. Hoffmann, C. (1990) A Dimensionality Paradigm for Surface Interrogations. *Computer Aided Geometric Design* **7**, 517–532
15. Hoffmann, C., Vermeer, P. (1991) Eliminating Extraneous Solutions in Curve and Surface Operation. *Int'l J. of Comp. Geom. and Appl.* **1**, 47–66
16. Hoschek, J., Lasser, D. (1993) *Fundamentals of Computer Aided Geometric Design*, A.K. Peters, Wellesley, MA
17. Hu, C.-Y., Maekawa, T., Sherbrooke, E., Patrikalakis, N. (1996) Robust Interval Algorithm for Curve Intersections. *Computer-Aided Design* **28**, 495–506
18. Joy, K., Duchaineau, M. (1999) Boundary Determination for Trivariate Solid. *Proc. of Pacific Graphics '99*, Seoul, Korea, October 5–7, 1999, 82–91
19. Johnstone, J. (1993) A New Intersection Algorithm for Cyclides and Swept surfaces using Circle Decomposition. *Computer Aided Geometric Design* **10**, 1–24
20. Kohler, M., Spreng, M. (1995) Fast Computation of the C-space of Convex 2D Algebraic Objects. *The Int' J. of Robotics Research* **14**, 590–608
21. Lane, J., Riesenfeld, R. (1981) Bounds on a Polynomial. *BIT* **21**, 112–117
22. Lee, I.-K., Kim, M.-S., Elber, G. (1998) Polynomial/Rational Approximation of Minkowski Sum Boundary Curves. *Graphical Models and Image Processing* **60**, 136–165
23. Martin, R., Stephenson, P. (1990) Sweeping of Three-Dimensional Objects. *Computer-Aided Design* **22**, 223–234
24. Piegl, L., Tiller, W. (1995) *The NURBS Book*, Springer, Berlin Heidelberg
25. Sherbrooke, E., Patrikalakis, N. (1993) Computation of the Solutions of Nonlinear Polynomial Systems. *Computer Aided Geometric Design* **10**, 379–405
26. Wolfram, S. (1991) *Mathematica*, 2nd Ed., Addison-Wesley
27. Zhang, Q., Martin, R. (2000) Polynomial Evaluation using Affine Arithmetic for Curve Drawing. *Eurographics UK 2000 Conference Proceedings*, Abingdon, UK, 49–56

Higher Order Singularities
in Piecewise Linear Vector Fields

Xavier Tricoche, Gerik Scheuermann, Hans Hagen

Computer Science Department, University of Kaiserslautern, Germany

Summary. Piecewise linear interpolation of 2D scattered vector data is a classical, simple and fast scheme to process the discrete information provided by experiments or numerical simulations. Nevertheless, its major drawback is its low order that prevents satisfying approximation of non linear behaviors. For topology-based methods in particular, commonly applied in vector field visualization, it often restricts the structural features found to very few possible configurations which may be insufficient for interpretation. In this paper, on the contrary, we consider piecewise linear vector fields from the modeling viewpoint, showing that they can exhibit arbitrary complex topological features.

1 Introduction

Piecewise linear vector fields are used in many applications. Their simple mathematical description ensures a very good insight into the topological structure. For scientific visualization in particular, it is the most suited way to handle 2D scattered vector data for the depiction of the corresponding flow: the given positions are first associated with a triangulation and a linear interpolation is then processed in each triangle resulting in a vector field continuous over the domain. The computation of streamlines (or integral curves) can then be done very accurately in the phase plane by using analytic formulas. Unfortunately, this simplicity usually restricts the encountered features to very few possible behaviors (locally linear) which deprives the extracted structure of most topological configurations existing in analytic vector fields. The qualitative analysis of vector fields on the plane has been a subject of major interest in pure and applied mathematics in this century. In [4], Poincaré laid the foundations of this field. A major contribution was then the work of Andronov ([1]). Topology concepts were next introduced in scientific visualization in [2]: It was shown that focusing on the singularities of a vector field leads to a synthetic depiction of the corresponding flow. The deficiencies of the piecewise linear interpolant for that purpose is a problem that has been considered in Scientific Visualization in the last years. In [6], an approximation method was used to enable the appearance of higher order singular points in piecewise linear vector fields by transforming the field in domains made up of neighboring triangles containing several critical points. The mathematics involved referred to Clifford algebra. An higher order method was also introduced in [5] to automatically detect flow features called vortex core lines.

Furthermore, the lack of smoothness of piecewise linear interpolation motivated the application of Nielson's C^1-interpolant for vector field topology depiction (see [7]).

In this paper, we propose an overview of 2D vector field topology taken from the qualitative theory of second order dynamic systems, then we show that complicated singularities may also occur in piecewise linear vector fields and propose a method to model and properly analyze them. The paper is thus structured as follows. First, we introduce the basic notions required for the qualitative analysis of vector fields. Fundamental theorems are given that enable the definition of possible sector types for a singular point in the general case. Second, we focus on linear vector fields, exposing the few singularity configurations that may appear in this special case. Third, we consider piecewise linear interpolation: we show how to model any type of topological sector and conversely how to depict the structure of complicated topological features.

2   Topology of 2D Vector Fields

2.1   Definitions and Fundamental Theorems

In the following, we consider a steady vector field defined on the plane.

Definition 1. A *steady planar vector field* is a map

$$v : I\!R^2 \longrightarrow T I\!R^2 \simeq I\!R^2$$
$$X \longmapsto v(X)$$

that is, a map that associates a 2D-vector value with each point on the plane.

Practically, the vector field will be analyzed through its integral curves (also called *orbits* or *paths*).

Definition 2. An *integral curve through a point* $X_0 \in I\!R^2$ of a vector field $v : I\!R^2 \longrightarrow I\!R^2$ is a map

$$\alpha_{X_0} : I\!R \supset I \longrightarrow I\!R^2$$

where

$$\begin{cases} \dot{\alpha}_{X_0}(t) = v(\alpha_{X_0}(t)), \ \forall t \in I \\ \alpha_{X_0}(0) = X_0 \end{cases}$$

One supposes that the considered vector field is continuous and satisfies the Lipschitz condition around each point of the plane.

Definition 3. The continuous vector field $v : I\!R^2 \longrightarrow I\!R^2$ is said to satisfy the *Lipschitz condition* around each point of $I\!R^2$ if their exists $K > 0$ such that

$$\forall (X, Y) \in I\!R^2 \times I\!R^2, \ \|v(X) - v(Y)\| < K\|X - Y\|$$

If this condition is true, then K is called the *Lipschitz constant* of v.

Remark 1. In the case of a piecewise linear vector field defined on a bounded triangulation, the Lipschitz condition is satisfied. As a matter of fact, the field is C^∞ (being affin linear) over each triangle and thus satisfies the Lipschitz condition over each triangle. Let K_i be a Lipschitz constant corresponding to the ith triangle and consider the following configuration: One has:

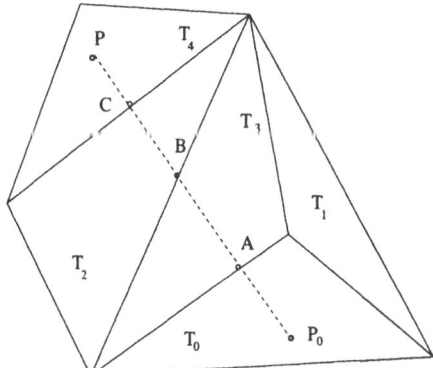

Fig. 1. Lipschitz constant in the piecewise linear case

$$\|v(P) - v(P_0)\| \leq \|v(P) - v(C)\| + \|v(C) - v(B)\| + \|v(B) - v(A)\|$$
$$+\|v(A) - v(P_0)\|$$
$$\leq K_4 \times \|P - C\| + K_2 \times \|C - B\| + K_3 \times \|B - A\|$$
$$+K_0 \times \|A - P_0\|$$
$$\leq (K_0 + K_2 + K_3 + K_4)\|P - P_0\|$$
$$\leq (\sum_{i=0}^{N} K_i)\|P - P_0\|$$

where N is the total number of triangles. So if one sets $K = \sum_{i=0}^{N} K_i$, K is a Lipschitz constant of the vector field for the whole triangulated domain and one verifies the hypotheses in the following theorems.

Theorem 1. *Let $v : \mathbb{R}^2 \longrightarrow \mathbb{R}^2$ be a continuous vector field satisfying the Lipschitz condition around each point of $\mathbb{R}^2$. Then there exists a unique integral curve through any $X_0 \in \mathbb{R}^2$. Furthermore, every integral curve is defined over $\mathbb{R}$.*

Actually, in the theorem above, two different curves through a position $X_0 \in \mathbb{R}^2$ may only differ by the choice of their parameterization. One can then define the *phase portrait* of the vector field as the family of all its paths over the plane. A fundamental property for the study of the structure of a vector field is given in the following theorem.

Theorem 2. *Let v be as before and let α and β be two integral curves of v on the closed interval $[t_0, t_1]$. Then, for all $t \in [t_0, t_1]$ one has*

$$\|\alpha(t) - \beta(t)\| \leq \|\alpha(t_0) - \beta(t_0)\| \exp(K(t - t_0))$$

That is, one has continuity of integral curves with respect to initial conditions. In the domain of study, one distinguishes two types of point:

Definition 4. A *singular point* (also called *equilibrium state*) of a vector field v is a point at which the field is zero.

It results from the uniqueness of the integral curve of v through a point that singular points are the only locations on the plane where two different integral curves can meet. A point which is not singular is said to be *regular*.

Now, one focuses on the topology of the vector field, that is, geometrically, the structure of its integral curves. As shown in [1], the knowledge of the singular points of the field provides a very good characterization of the phase portrait topology.

2.2   Singular Points Analysis

All the results in this section are taken from [1]. One should consult this reference for a more rigorous understanding of these topics.

Center Type First consider the case of singular points that are approached by no integral curve. Such singularities are said to be of *center type*. In this case, on can find a neighborhood of the singular point where all paths are closed, inside one another, and contain the singular point in their interior.

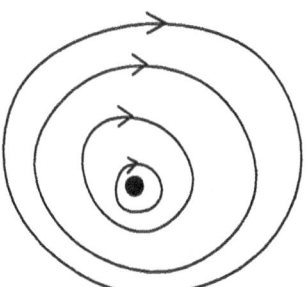

Fig. 2. center

Non Center Type In this case, one has not only a single path converging to the singular point but actually at least two. To analyze the local structure of such a point, consider the following configuration: where L_M^* and $L_{M'}^*$ are

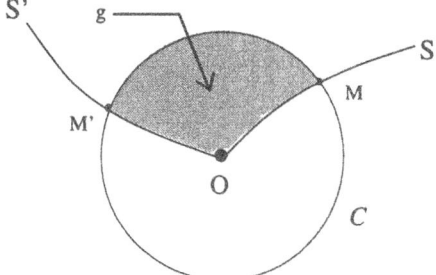

Fig. 3. curvilinear sector

(positive or negative) semi-paths tending to O. C is a circle with arbitrary small fixed radius and g is the (open) region bounded by the integral curves OM and OM' and the arc of MM' on C, called *curvilinear sector*.

The structure of a singular point of non center type is characterized by the behavior of the integral curves passing through points of one of its sectors. In fact, there exist three different types of curvilinear sectors.

- Case 1. If L_M^* tends to 0 for $t \longrightarrow \infty$ and if $L_{M'}^*$ tends to 0 for $t \longrightarrow -\infty$ and if every integral curve passing through the open g leaves g for both $t \longrightarrow \infty$ and $t \longrightarrow -\infty$, the sector will be called a *hyperbolic* or *saddle sector*. In this case, both L_M^* and $L_{M'}^*$ are called *separatrices of*

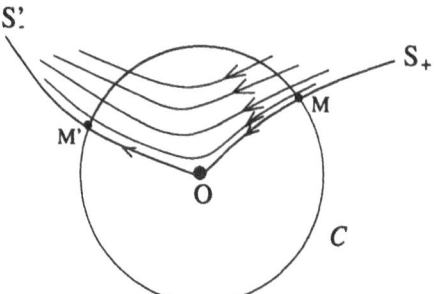

Fig. 4. hyperbolic sector

the singular point O (more precisely, L_M^* is an *ω-separatrix* and $L_{M'}^*$ is an *α-separatrix*).

- Case 2. If L_M^* and $L_{M'}^*$ both tend to O for $t \longrightarrow \infty$ (resp. $t \longrightarrow -\infty$) and if every integral curve through the open g tends to 0 for $t \longrightarrow \infty$ (resp. $t \longrightarrow -\infty$) without leaving g and leaves g for $t \longrightarrow -\infty$ (resp. $t \longrightarrow \infty$), the sector is known as an *ω-* (resp. *α-*) *parabolic sector*.

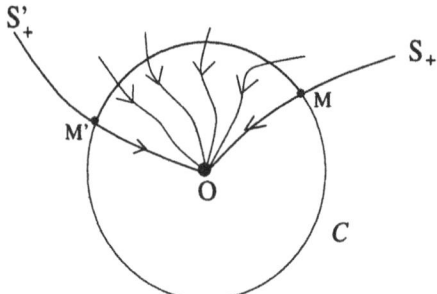

Fig. 5. parabolic sector

- Case 3. If L_M^* and $L_{M'}^*$ are two semi-paths on the same path and if all the paths through a point inside this loop form nested loops tending to O for both $t \longrightarrow \infty$ and $t \longrightarrow -\infty$, the sector is called an *elliptic sector*.

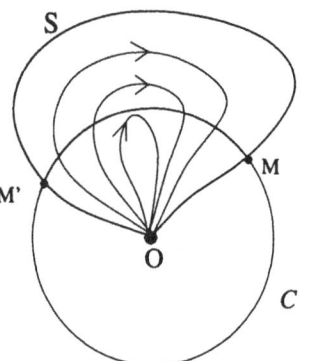

Fig. 6. elliptic sector

Consequently, any singular point may be characterized by the type, angular location, and number of its curvilinear sectors. The precise meaning of this characterization is described in the following.

Theorem 3. *If the structures of two singular points are related through a one-to-one correspondence between their respective ω-separatrices, α-separatrices and elliptic regions then there exists a path-preserving topological mapping of a neighborhood of the first onto a neighborhood of the second preserving orientation and direction of t.*

3   Piecewise Linear Interpolation

After having considered singular points from a general viewpoint, we now focus on their possible structure in the special case of a piecewise linear vector field. We start with a review of the few basic topological features of a linear vector field before presenting a method that makes use of the "flexibility" introduced by the piecewise linearity to model any topological structure for a singular point.

3.1   Linear Interpolation

An affin linear vector field v is described in the following way

$$v(X) = AX + b$$

where

$$X = \begin{pmatrix} x \\ y \end{pmatrix}$$

$$A = \begin{pmatrix} \alpha_X & \beta_X \\ \alpha_Y & \beta_Y \end{pmatrix} \quad \text{and} \quad b = \begin{pmatrix} \gamma_X \\ \gamma_Y \end{pmatrix}$$

(If v has a zero, then one takes its location as new coordinates origin and thus considers the linear field $v'(X) = AX$).

An affin linear vector field is uniquely determined by its Jacobian (or gradient matrix) at the location of its possible zero. That is, depending on the eigenvalues of the matrix A, integral curves of v may have different aspects over the plane. The following classification is taken from [3].

- Case 1. A has real eigenvalues of opposite signs. The zero is called a *saddle point*.

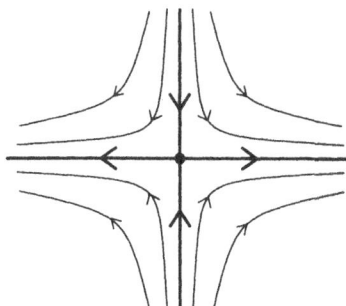

Fig. 7. saddle point

- *Case 2.* All eigenvalues have negative real parts. The zero is called a *sink*, because any integral curve tends to O for $t \longrightarrow \infty$.
  - *Case 2a.* A is diagonalizable and its eigenvalues are different. The zero is called a *node sink*. The special case where the eigenvalues are equal is called a *focus sink*.

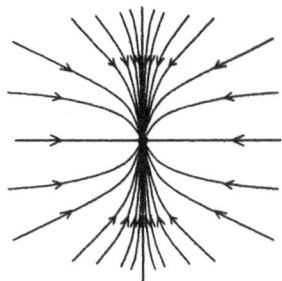

Fig. 8. node sink

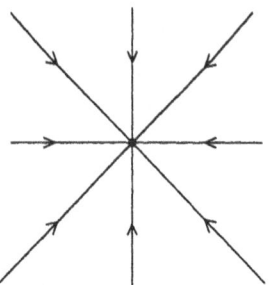

Fig. 9. focus sink

- Case 2b. *A* is not diagonalizable but has one real negative eigenvalue. The zero is called an *improper node sink*.
- Case 2c. *A* has two complex conjugate eigenvalues with negative real parts. The zero is called a *spiral sink*.

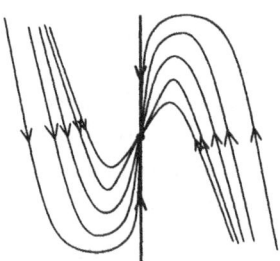

Fig. 10. improper node sink

Fig. 11. spiral sink

- Case 3. All eigenvalues have positive real parts. The zero is called a *source*, because any integral curve tends to it for $t \longrightarrow -\infty$.
  - Case 3a. *A* is diagonalizable and its eigenvalues are different. The zero is a *node source*. If both eigenvalues are equal, the zero is a *focus source*.

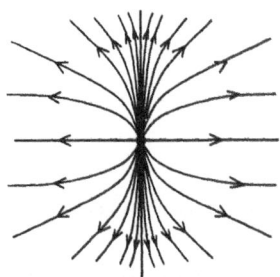

Fig. 12. node source

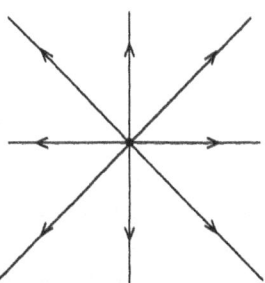

Fig. 13. focus source

- Case 3b. A is not diagonalizable but has one real positive eigenvalue. The zero is called an *improper node source*.
- Case 3c. A has two complex conjugate eigenvalues with positive real parts. The zero is called a *spiral source*.

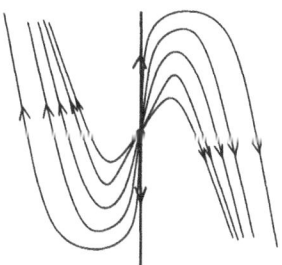

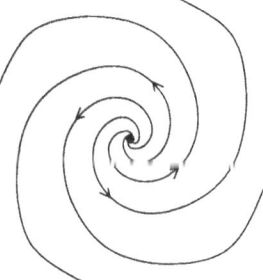

Fig. 14. improper node source   **Fig. 15.** spiral source

- *Case 4.* A has pure imaginary eigenvalues. The zero is then called a *center* (see Fig. 2).

Thus, in a linear vector field, the only singularities that can be found are (possibly after a coordinate change) of the one of the eight configurations above. Such singular points are said to be *simple* or of *first order*.

3.2 Higher Order Singular Points

So what is new, from the topological viewpoint, with piecewise linear interpolation?
One is given a triangulation of planar points associated with vector values and inside each triangle cell, one builds a linear combination of the three vectors at the vertices. So inside each triangle, the vector field is linear and if a singularity exists in its interior, it has the topological structure of one of the cases above. But the situation becomes fundamentally different when the singularity is to be found on a triangle vertex. Indeed, in this case, there is no neighborhood of the singular point lying in the definition domain of a linear field (i.e. one single triangle). Consequently, the analysis of the Jacobian matrix is of no help in finding out the topological structure of such a singular point. Fundamentally, one has to face an arbitrary singular point (i.e. a singular point in the neighborhood of which a linear approximation may no longer be possible). In analytic vector fields, such singular points correspond to points on the plane at which the Jacobian matrix has not full rank. Simple examples are given by the so-called *monkey-saddle*$(v(x,y) = (x^2 - y^2, -2xy))$ and *dipole*$(v(x,y) = (x^2 - y^2, 2xy))$. In the following, we first show how to model any type of singularity (that is the number, positions and natures of

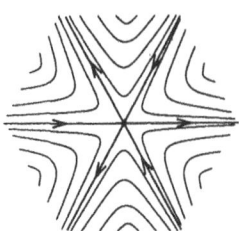

Fig. 16. monkey saddle

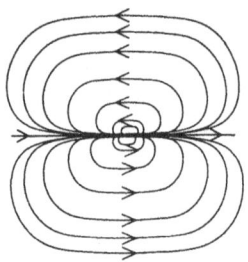

Fig. 17. dipole

the curvilinear sectors are arbitrary defined) and then we focus on the converse problem, detecting the local topological structure of a given singularity lying at a triangulation vertex.

Modeling of Singular Points In this section, one is dealing with the following problem (see theorem 3): Given a list of angles $(\omega_i)_{i=1,..,n1}$ of ω-separatrices, a list of angles $(\alpha_i)_{i=1,..,n2}$ of α-separatrices, and a list $(g_i)_{i=1,..,n3}$ of elliptic sectors, build a piecewise linear vector field that presents an equivalent singular point. Here, the problem should be subdivided into modeling a single curvilinear sector starting and stopping at prespecified angular positions, given by the sorted values of the angles introduced above.

Consequently, one considers the three possible sector type cases.

Remark: In the following, the neighborhood of the singular point is actually the set of triangles that are incident to the considered "singular" vertex.

Hyperbolic Sector A hyperbolic sector is bounded by two separatrices of opposite kind (an ω- and an α-separatrix). Consider an ω-separatrix located at $\theta = \omega$ and an α-separatrix located at $\theta = \alpha$. Building a triangle as in Fig. 18, and setting $v_1 = -u_\omega$ and $v_2 = +u_\alpha$, one gets the expected hyperbolic behavior for all integral curves starting inside the triangle ABC.

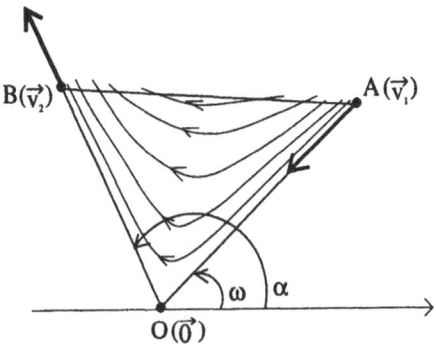

Fig. 18. piecewise linear hyperbolic sector

Parabolic Sector According to the definition above, a parabolic sector is bounded by two separatrices of the same kind (both ω- or both α-separatrices). Consider two ω-separatrices located at $\theta = \omega_1$ and $\theta = \omega_2$ respectively. Building a triangle as in Fig. 19, and setting $v_1 = -u_{\omega_1}$ and $v_2 = -u_{\omega_2}$, one gets the expected parabolic behavior for all integral curves starting inside the triangle ABC.

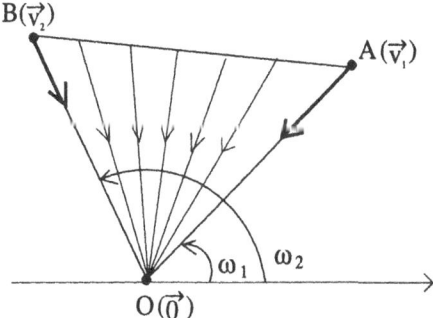

Fig. 19. piecewise linear parabolic sector

Elliptic Sector In this case, one is not given two angle coordinates of two bounding separatrices but a curvilinear sector bounded by a loop integral curve that tends to O for both $t \longrightarrow \infty$ and $t \longrightarrow -\infty$. The modeling of such a sector by a piecewise linear vector field requires the curve itself to be described in terms of its tangential directions for $t \longrightarrow \infty$ and $t \longrightarrow -\infty$. Consider Fig. 20.

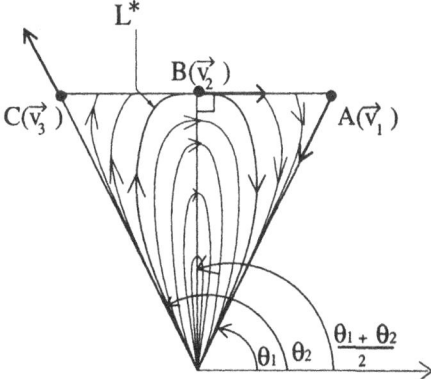

Fig. 20. piecewise linear elliptic sector

The angle coordinates θ_1 and θ_2 are the tangential direction of the loop curve L^* that bounds the modeled elliptic sector. The vector values at A and C are set as in the hyperbolic case. From the former case, one knows that the linear sector defined by A, B and O is hyperbolic. This means that an elliptic sector cannot be met in a linear field. To build such a sector, one has thus to split the triangle into two sub-triangles. By setting the vector value at the corresponding additional point B as shown above, one gets the expected elliptic behavior for all integral curves through points located inside the loop L^*.

Fig. 21 illustrates the possible cases.

Fig. 21. modeled singular point

Local Topology Detection Conversely, suppose that a "piecewise linear" singular point is given, that is a zero vector located at a vertex of a piecewise linear interpolated triangulation. One wants to locate and analyze the different topological sectors of such a singular point. According to what precedes, it consists in seeking the boundary curves of hyperbolic sectors (called separatrices) and the different sets of nested loop curves tending to the singular point for both increasing and decreasing time.

The following lemma will be very useful for the processing.

Lemma 1. *In the neighborhood of a singular point lying on a vertex of a piecewise linear interpolated triangulation, the angle coordinate of the vector field does not depend on the distance to the singular point.*

Proof: Consider the situation in Fig. 22. In the triangle ABC, the vector value at P is linear interpolated between O and Q. One thus gets

$$v(P) = x\, v(Q) + (1 - x)\, v(O)$$
$$= x * Q$$

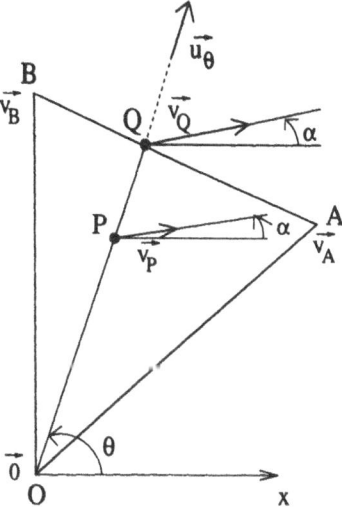

Fig. 22. Angular coordinate

That is, $v(P)$ is collinear to $v(Q)$ and thus, taking O as coordinate origin, both vectors have the same angle coordinate. Q.E.D.
In the remaining, one calls their common angular coordinate $v(\theta)$.

Using this property, one can locate the angular positions of all separatrices by looking for angles where the vector field is collinear to the coordinate vector and checking the type of the sectors on both sides to find out if one has actually found a separatrix. Seeking the angular coordinates where the vector field is orthogonal to the coordinate vector enables next the distinction between a hyperbolic and an elliptic sector (see section 20). To simplify the results, one adopts the following notations: at the angular coordinates where the vector field is orthogonal to the coordinate vector, one distinguishes angles where the cross-product $u_\theta \wedge v(\theta)$ is positive (called *orthogonal+*) from those where it is negative (called *orthogonal-*). At the angular coordinates where the vector field is parallel to the coordinate vector, one distinguishes angles where the scalar product $u_\theta . v(\theta)$ is positive (called *parallel+*) from those where it is negative (called *parallel-*). (see Fig. 23).
One gets then the graph proposed in Fig. 24 for the determination of a sector type.

4   Conclusion

A piecewise linear interpolation of 2D vector data is a simple and fast way to reconstruct the information provided by simulations or experiments into a continuous vector field. Nevertheless, this scheme is often regarded as insufficient for the detection of complex topological features because of its low

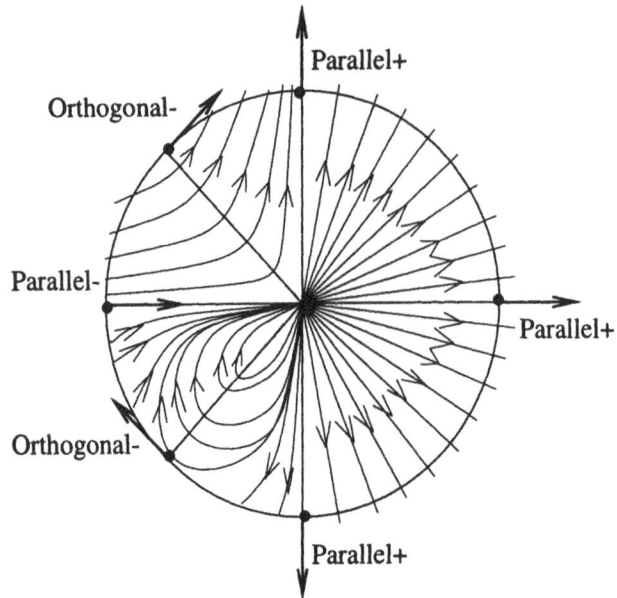

Fig. 23. remarkable positions

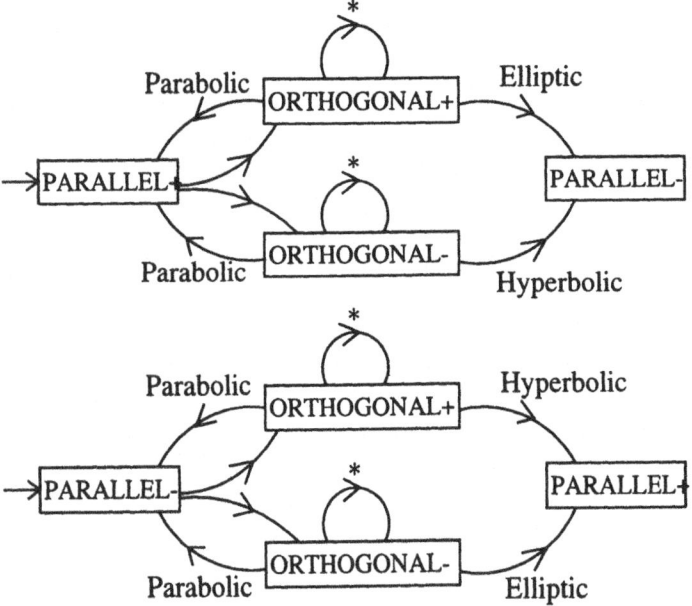

Fig. 24. sector type determination graph

degree. Yet, we have shown in this paper that a zero vector value located at a vertex position results in a complex topological structure in its neighborhood. Actually, one has shown that any possible topological behavior of a vector field may be thus encountered and not only the few basic features existing in linear fields. An interesting application of this result has been considered which is the modeling of singularities of any type and complexity in piecewise linear vector fields.

References

[1]   Andronov A.A., Leontovich E.A., Gordon I.I., Maier A.G., *Qualitative Theory of Second-Order Dynamic Systems* Israel Program For Scientific Translation, Halsted Press, 1973.

[2]   Helman J.L., Hesselink L., *Representation and Display of Vector Field Topology in Fluid Flow Data Sets.* IEEE Computer, 1989.

[3]   Hirsch M.W., Smale S., *Differential Equations, Dynamical Systems and Linear Algebra.* Academic Press, 1974.

[4]   Poincaré H., *Sur les courbes définies par une équation différentielle* Oeuvres, Vol. I., Paris, Gauthier-Villars, 1928.

[5]   Roth M., Peikert R., *A Higher-Order Method For Finding Vortex Core Lines.* IEEE Visualization, 1998, Research Triangle Park, NC.

[6]   Scheuermann G., Krüger H., Menzel M., Rockwood A.P., *Visualizing Non-Linear Vector Field Topology.* IEEE Transactions On Visualization & Computer Graphics, Vol. 4, No. 3, July 1998.

[7]   Scheuermann G., Tricoche X., Hagen H., *C1-Interpolation for Vector Field Topology Visualization.* IEEE Visualization, 1999, San Francisco, CA.

Landmarks of a Surface

Ian R. Porteous[1] and Mike J. Puddephat[2]

[1] Department of Mathematical Sciences, University of Liverpool, Liverpool, L69 7ZL (porteous@liv.ac.uk)
[2] SimCorp Swallow Business Systems Ltd, Cambridge House, 100 Cambridge Grove, Hammersmith, London, W6 0LE (mike@easymeasure.co.uk)

Summary. The second-order robust features of a deforming surface in Euclidean space such as its umbilics and parabolic line, are well-known. This paper is concerned with establishing properties of third-order robust features of a deforming surface, such as its *ridges* and *flexcords* or *sub-parabolic lines*, and the behaviour of those features under perestroikas, such as the birth of umbilics.

1   Introduction

Landmarks of a smooth surface in **R** are features of the surface that preserve some sort of individual identity if the surface is subject to smooth deformation. They are *robust* features of the surface. Familiar examples are the parabolic line of a surface separating the regions of positive Gaussian curvature and negative Gaussian curvature, and the umbilics of a surface, where the principal curvatures of the surface coincide. Generically the parabolic line is a smooth non-singular curve on the surface, while the umbilics are isolated points. However under deformation perestroikas of these may occur. For example new components of the parabolic line may be born or old ones die, while also new umbilics may be born or old ones die. Less familiar are the *ridges* and what we prefer to call the *flexcords* of surfaces, and associated features. By contrast the lines of curvature and geodesics of a surface are not robust. Under deformation of the surface they do not *deform* – they *reform*.

In defining ridges and flexcords we find it convenient to think of the two families of lines of curvature on a surface, at least locally, as being of different colours, namely blue and red, the corresponding leaves of the focal surface likewise being coloured blue and red. Then it is the case that if one travels along a blue line of curvature one crosses a *blue ridge* where the blue curvature of the surface is critical, while one crosses a *blue flexcord* where the red curvature of the surface is critical. Likewise one defines *red ridges* and *red flexcords*. Ridges are extensively studied in Porteous (1994). They lie under the *ribs* of the surface – the cuspidal edges of the focal surface – just as critical points of curvature of plane curves lie under the cusps of the focal curve or evolute of the given curve. Flexcords also make their appearance there, but under the name of *subparabolic lines*, as they were originally introduced by Wilkinson (1991) as the curves on the surface lying under parabolic lines on the focal surface. They also are the loci of geodesic inflections of lines of curvature of the surface, though it should be noted that a *blue* flexcord is the locus of the geodesic inflections of the *red* lines of curvature, and vice versa. The name subparabolic line is not entirely satisfactory for the following reason. Consider an ellipsoid with unequal semi-axes. Then the three principal sections are all lines of curvature that are also geodesics of the ellipsoid, and therefore are *a fortiori* loci of geodesic inflections of lines of curvature. Yet the lifts to the relevant sheets of the focal surface are not parabolic lines in the regular sense of the term, but rather are cuspidal edges of the focal surface. It is for this reason that we prefer the term *flexcord*.

The characterisation of flexcords in terms of the criticality of the principal curvature of the opposite colour seems first to have been observed by Jean-Philippe Thirion of INRIA at Sophia Antipolis (near Nice) (1994, personal communication). As a proof is not to be found in Porteous (1994) we provide one here. The notational conventions are those of Porteous (1994). In particular differentiation is indicated by subscripts and not by the letter d or by primes.

Let us suppose that the surface is given in intrinsic form as the zeros of a smooth map $F : \mathbf{R}^3 \to \mathbf{R}$. Suppose that s is a non-singular point of such a surface, let α and β be mutually orthogonal unit principal vectors at s, the corresponding principal curvatures being $\kappa = \rho^{-1}$ and $\lambda = \sigma^{-1}$, and let $\mathbf{n}$ be unit normal vector at s. Then, in particular,

$$F_1(s)\beta = 0$$
$$\text{and} \quad F_2(s)\beta^2 + \sigma F_1(s)\mathbf{n} = 0.$$

Differentiating these in the direction α we have first of all that

$$0 = F_2(s)\beta\alpha + F_1(s)\beta_1\alpha = F_1(s)\beta_1\alpha.$$

So $\beta_1\alpha$, being necessarily orthogonal to β, is a multiple of α.
Secondly,

$$F_3(s)\alpha\beta^2 + 2F_2(s)\beta(\beta_1\alpha) + \sigma_1\alpha F_1(s)\mathbf{n} + \sigma F_2(s)\mathbf{n}\alpha + \sigma F_1(s)\mathbf{n}_1\alpha = 0.$$

The relevant curvature condtion is that $\sigma_1\alpha = 0$, when the third term on the left hand side is zero. Moreover, since $\beta_1\alpha$ is a multiple of α, $F_2(s)\beta(\beta_1\alpha) = 0$, and since $\mathbf{n} \cdot \mathbf{n}_1 = 0$, $F_1(s)\mathbf{n}_1\alpha = 0$. So

$$F_3(s)\alpha\beta^2 + \sigma F_2(s)\mathbf{n}\alpha = 0.$$

Eliminating σ we then have that the condition is satisfied if and only if

$$(F_3(s)\alpha\beta^2)(F_1(s)\mathbf{n}) = (F_2(s)\beta^2)(F_2(s)\mathbf{n}\alpha),$$

which is a condition for the point s to lie on a flexcord. (Cf. Porteous (1994), Exercise 13.11.)

For another proof of this theorem see Bruce and Tari (1996). For a detailed treatment of the generic perestroikas of ridges and subparabolic lines from the point of view of singularity theory see Bruce, Giblin and Tari (1996) and (1999).

2   Ridges and flexcords at umbilics

It has been known for over 100 years (Darboux (1894)) that there are through a generic umbilic either three directions or one for lines of curvature, though there are not two but rather three typical configurations, known today as the *lemon*, *monstar* and *star* configurations. Cf. Figures 12.1–12.3 of Porteous (1994). It is only much more recently that it has been realised that through any generic umbilic of a surface there pass both ridges and flexcords. For flexcords there is the remarkable theorem of Wilkinson (1991) or Bruce and Wilkinson (1991) that these lines generally pass through an umbilic in the same directions as the lines of curvature. The reader is referred to Porteous (1994) for details.

Wilkinson's theorem however fails at birth points of umbilics. For ridges there is no problem. A generic birth-point is always located on a single ridge. However the situation for flexcords is quite complicated. As we shall see there is always a flexcord tangential to the ridge, but that is not the whole story.

It will greatly simplify the exposition if we choose our origin in $\mathbf{R}^3$ to be some specific point of the surface, and to choose as x, y-plane the tangent plane to the surface there. Then the surface locally takes the *Monge form*

$$z = \frac{1}{2}(rx^2 + 2sxy + ty^2) + \frac{1}{6}(Ax^3 + 3Bx^2y + 3Cxy^2 + Dy^3) + \text{h.o.t.}.$$

The first fundamental form of the surface at the origin is just $x^2 + y^2$, while the second fundamental form is $rx^2 + 2sxy + ty^2$. The eigenlines of the second form are just the root lines of the *Jacobian quadratic form* of these two forms, and if these lines are taken to be the x and y axes then the Monge form takes the form

$$z = \frac{1}{2}(\kappa x^2 + \lambda y^2) + \frac{1}{6}(Ax^3 + 3Bx^2y + 3Cxy^2 + Dy^3) + \text{h.o.t.},$$

the eigenvalues of the form, κ and λ, being the principal curvatures of the surface there. There is a ridge passing through the origin if either A or D is equal to zero and a flexcord passing through the origin if either B or C is equal to zero. At an umbilic the eigenvalues coincide, and the second form is just κ times the first. In that case the cubic form $Ax^3 + 3Bx^2y + 3Cxy^2 + Dy^3$ is the *classifying cubic* of the umbilic.

To determine whether the umbilic is generic or not we require the first and second fundamental forms in the neighbourhood of the origin also. To

keep things simple we consider first the special case where the umbilic is flat
– that is the sectional curvature κ is equal to zero, and the Monge form is

$$z = \frac{1}{6}(Ax^3 + 3Bx^2y + 3Cxy^2 + Dy^3) + \frac{1}{4!}(ax^4 + 4bx^3y + 6cx^2y^2 + 4dxy^3 + ey^4)$$

$$+ \frac{1}{5!}(\alpha x^5 + 5\beta x^4y + 10\gamma x^3y^2 + 10\delta x^2y^3 + 5\epsilon xy^4 + \zeta y^5) + \text{h.o.t.}.$$

It is then the case that near the origin the first fundamental form is to
order three just $x^2 + y^2$, with matrix the identity matrix, while the second
fundamental form has matrix

$$\begin{pmatrix} Ax + By + \text{h.o.t} & Bx + Cy + \text{h.o.t} \\ Bx + Cy + \text{h.o.t} & Cx + Dy + \text{h.o.t} \end{pmatrix}.$$

For the latter matrix to be a multiple of the identity matrix one must have
$Ax + By + \text{h.o.t.} = Cx + Dy + \text{h.o.t}$ and $Bx + Cy + \text{h.o.t.} = 0$. The origin is
a solution of multiplicity 1 unless $(A - C)C = (B - D)B$, that is unless the
root lines of the Hessian determinant

$$\begin{vmatrix} Ax + By & Bx + Cy \\ Bx + Cy & Cx + Dy \end{vmatrix},$$

the *Hessian lines* of the cubic, are mutually orthogonal, in which case by
taking these now as our x and y axes the cubic terms reduce to $Ax^3 + Dy^3$,
with $B = C = 0$. The multiplicity of the umbilic is then greater than 1.

This now being the case, the matrix of the second fundamental form
becomes

$$\begin{pmatrix} Ax + \text{h.o.t} & \frac{1}{2}(bx^2 + 2cxy + dy^2) + \text{h.o.t.} \\ \frac{1}{2}(bx^2 + 2cxy + dy^2) + \text{h.o.t} & Dy + \text{h.o.t.} \end{pmatrix},$$

and the multiplicity is clearly 2 unless $Ax - Dy$ is a factor of the quadratic
form $bx^2 + 2cxy + dy^2$, that is unless $bD^2 + 2cDA + dA^2 = 0$, in which case
the multiplicity is at least equal to 3. In fact it is well known that the line
with equation $Ax = Dy$ is the tangent line to the unique ridge that passes
through the umbilic, and which after the birth of twin umbilics links these
two umbilics, the one a monstar of index $\frac{1}{2}$ and the other a star of index $-\frac{1}{2}$.

In the case that $Ax - Dy$ is a factor of $bx^2 + 2cxy + dy^2$ let the quotient be
$lx + my$. Then the multiplicity of the umbilic at the origin is the multiplicity
of the intersection there of the curves

$$Ax - Dy + \frac{1}{2}((a - c)x^2 + 2(b - d)xy + (c - e)y^2) + \text{h.o.t.} = 0$$

and

$$\frac{1}{2}(bx^2 + 2cxy + dy^2) + \frac{1}{6}(\beta x^3 + 3\gamma x^2y + 3\delta xy^2 + \epsilon y^3) + \text{h.o.t.} = 0.$$

The second equation may be replaced by what we obtain by subtracting from it the first equation multiplied by $lx + my$, namely

$$\frac{1}{6}(\beta x^3 + 3\gamma x^2 y + 3\delta x y^2 + \epsilon y^3) - \frac{1}{4}(lx + my)((a-c)x^2 + 2(b-d)xy + (c-e)y^2)$$

$$+\text{h.o.t.} = 0,$$

starting on the left-hand side with a homogeneous cubic form in x and y. From this it follows that the required multiplicity is 3 unless $Ax = Dy$ is a factor of that cubic form, when it must be at least 4. We conjecture that in the case that the multiplicity is 3 the overall index of the umbilic is either $\frac{1}{2}$ or $-\frac{1}{2}$, according to the sign one obtains on substituting D for x and A for y in this cubic form.

There is a simple formula for determining the tangent line at the umbilic of a regular curve passing through a generic umbilic in terms of the limiting positions of the mutually orthogonal principal tangent lines there. Let C_3 denote the *symmetric thrice linear form* that is the fully polarised form of the classifying cubic at the umbilic, and let $\mathbf{t}$, $\mathbf{u}$ and $\mathbf{v}$ be, respectively, a tangent vector to the curve and limiting principal tangent vectors to the surface. Then $C_3\mathbf{tuv} = 0$. This clearly fails to determine $\mathbf{t}$ in the special case that $C_3\mathbf{uv} = 0$, which is exactly the case where $\mathbf{u}$ and $\mathbf{v}$ are Hessian vectors of the cubic form.

For a ridge passing through an umbilic one of the limiting principal vectors, say $\mathbf{u}$, has to be a root vector of the classifying cubic, that is $C_3\mathbf{u}^3 = 0$, but then $\mathbf{u}$ cannot be either of the Hessian vectors in the case that these are distinct. In that case, taking the cubic to be $Ax^3 + Dy^3$, this has the single real linear factor $A^{\frac{1}{3}}x + D^{\frac{1}{3}}y$, from which it follows that the limiting value along the ridge of the principal vector $\mathbf{u}$ is, up to a non-zero multiple, $(D^{\frac{1}{3}}, -A^{\frac{1}{3}})$, with $\mathbf{v} = (A^{\frac{1}{3}}, D^{\frac{1}{3}})$, so that the equation for $\mathbf{t} = (x, y)$ becomes

$$AD^{\frac{1}{3}}A^{\frac{1}{3}}x - DA^{\frac{1}{3}}D^{\frac{1}{3}}y = 0,$$

that is $Ax - Dy = 0$, as previously asserted.

For a flexcord passing through the umbilic one of the limiting principal vectors, say $\mathbf{u}$, has to satisfy the equation $C_3\mathbf{u}^2\mathbf{v} = 0$, where $\mathbf{v}$ is orthogonal to $\mathbf{u}$, and in the case that the Hessian roots are orthogonal two of the roots of this equation are these Hessian roots, the third being tangent to the ridge. In fact the possible limiting directions of the principal vector $\mathbf{u}$ are given by setting to zero the Jacobian of the forms $Ax^3 + By^3$ and $x^2 + y^2$, namely

$$\begin{vmatrix} Ax^2 & Dy^2 \\ x & y \end{vmatrix} = xy(Ax - Dy) = 0.$$

It follows that $Ax - Dy = 0$ is tangent to a flexcord, as well as to the unique ridge present at the moment of birth. The remaining pattern has been resolved by Richard Morris, who has proved that, if one approaches the birth-point along a flexcord with limiting principal vector **u**, one of the Hessian vectors there, then the tangent vector $\mathbf{t} = (x, y)$ to the flexcord at the birth-point satisfies a *quadratic equation*, namely

$$A(bx^2 + 2cxy + dy^2) = 2(Ax - Dy)(bx + cy),$$

that is

$$-Abx^2 + 2Dbxy + (Ad - 2Dc)y^2 = 0.$$

This has discriminant $b(A^2d + 2ADc + D^2b)$, zero if either $b = 0$ or $A^2d + 2ADc + D^2b = 0$. When $b = 0$, but the bracket is non-zero, the equation for **t** reduces to $y^2 = 0$, implying that the flexcord is cuspidal with limiting tangent the x-axis, while if $b \neq 0$, but the bracket is zero, then it reduces to $(Ax - Dy)^2 = 0$, implying that the flexcord is cuspidal with limiting tangent the line with equation $Ax - Dy = 0$, the tangent line at the birth-point to the ridge through the birth-point.

For flexcords of the opposite colour the discriminant is $d(A^2d + 2ADc + D^2b)$, zero if either $d = 0$ or $A^2d + 2ADc + D^2b = 0$. When $d = 0$, but the bracket is non-zero, the equation for **t** reduces to $x^2 = 0$, implying that the flexcord is cuspidal with limiting tangent the y-axis, while if $d \neq 0$, but the bracket is zero, then it reduces to $(Ax - Dy)^2 = 0$, implying, as before, that the flexcord is cuspidal with limiting tangent the line with equation $Ax - Dy = 0$.

That in either case the factor $A^2d + 2ADc + D^2b = 0$ turns up is no surprise, as this corresponds to the case that the multiplicity of the umbilic is at least 3, as we have already remarked.

The case that the multiplicity is 2 is the generic case of the birth or death of a pair of umbilics on a pre-existing ridge, one a star and the other a monstar, these being joined not only by the ridge but also by a flexcord having in the limit the same tangent direction at the double umbilic as the ridge. The overall index in this case is 0.

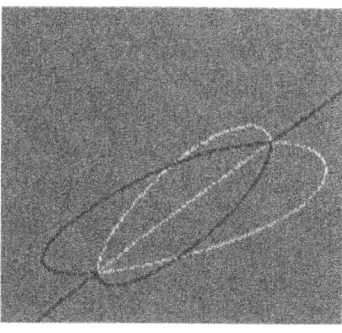

Fig. 1. Birth of a pair of umbilics — the elliptic/elliptic case.

This case is illustrated first by Fig. 1, where $\kappa = -0.04$, $\lambda = 0.04$, $A = 2$, $B = C = 0$, $D = 3$, $a = 0$, $b = 1.5$, $c = -4$, $d = 5$ and $e = 0$, the higher order terms all being zero. Then $b > 0$, while $A^2d + 2ADc + D^2b = -14.5 < 0$, showing that we have in this case close to a flat umbilic of multiplicity 2 two umbilics lying on a flexcord close to the line $2x = 3y$, the lower one to the left being a monstar and the upper one to the right being a star, the other two of the flexcord directions there being close to being horizontal and vertical.

Secondly, consider Fig. 2, where in the first row at the moment of birth $\kappa = 0$, $\lambda = 0$, $A = 2$, $B = C = 0$, $D = 3$, $a = 0$, $b = 0$, $c = 2$, $d = -3$ and $e = 0$, the higher order terms all being zero, except for $\beta = 2$, the *horizontal perestroika* being *parabolic* since $b = 0$, and the *vertical* one *elliptic*, while in the second row at the moment of birth the only change is that $d = 3$, the horizontal perestroika still being parabolic, but the vertical one *hyperbolic*.

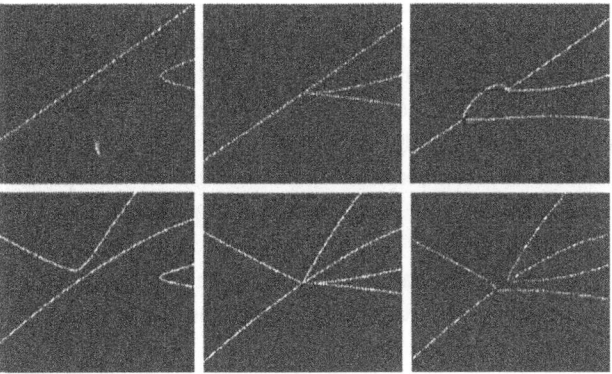

Fig. 2. Birth of a pair of umbilics – in the top row the parabolic/elliptic case and in the bottom row the parabolic/hyperbolic case.

In each case a star and a monstar are born on the flexcord approximating the line $2x = 3y$, with the horizontal new flexcord exhibiting a cusp and undergoing a parabolic perestroiks, while the other exhibits an elliptic perestroika in the first case and a hyperbolic perestroika in the second.

In the case that the multiplicity of the singular umbilic is 3 we have a perestroika from a single umbilic to three, strung out both along a ridge and along a flexcord. The overall index may be either $-\frac{1}{2}$, there being in one direction a single star on the ridge and flexcord and in the other direction a sequence along the ridge and along a flexcord of star, monstar and star, or it may be $\frac{1}{2}$, there being in one direction a single monstar on the ridge and the flexcord and in the other direction a sequence along the ridge and along the flexcord of monstar, star and monstar.

Some possibilities are illustrated in Fig. 3 and Fig. 4.

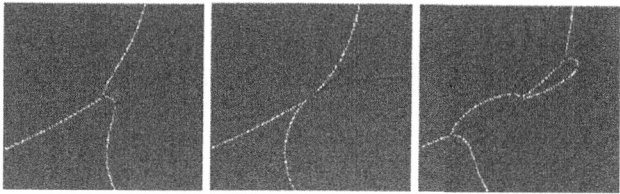

Fig. 3. One umbilic becomes three.

In Fig. 3, in the central position, $A = 2$, $B = C = 0$, $D = 3$, $a = 0$, $b = 4$, $c = 0$, $d = -9$, $e = 0$, with $A^2 d + 2ADc + D^2 b = 0$. Both the horizontal and vertical perestroikas feature a cubic curve having a single component before the birth, acquiring a cusp, with limiting tangent the line $Ax = Dy$ at birth and changing colour there, and acquiring a loop after birth.

In this example the cusps point in opposite directions, but it is possible for them both to point in the same direction.

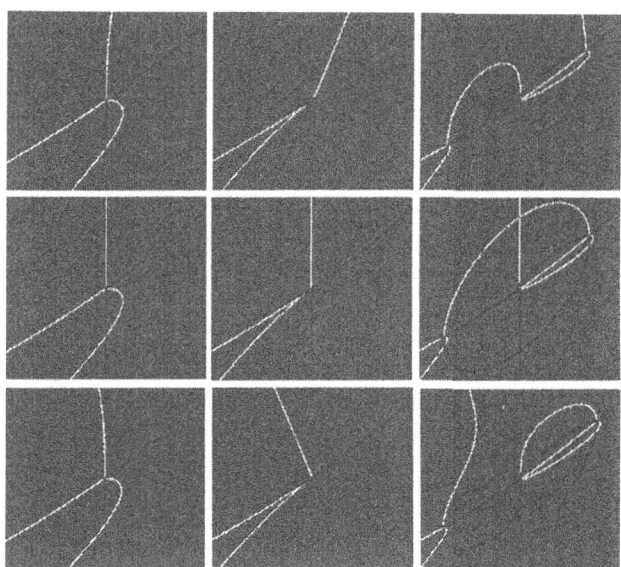

Fig. 4. One umbilic becomes three, giving birth to a snake.

In Fig. 4, in the central picture, $A = 2$, $B = C = 0$, $D = 3$, $a = 0$, $b = 3$, $c = -2.25$, $d = 0$, and $e = 0$, with $A^2 d + 2ADc + D^2 b = 0$, all the higher order terms being zero. In the top row $\epsilon = -4$ and in the bottom row $\epsilon = 4$, while in the left-hand column $B = -0.1$ and the right-hand column $B = 0.01$.

The horizontal perestroika features a cubic curve, acquiring an ordinary cusp and then a loop, as in Fig. 3, while the vertical perestroika features a cubic curve with a *triple* point (two of the tangents there in this case being complex conjugate), which in the middle row becomes an almost vertical line and an ellipse, but in the other two rows becomes a 'snake', passing through each of the three umbilics after birth vertically.

The triple point arises because in this case, since $d = 0$ as well as $A^2d + 2ADc + D^2b = 0$, all the coefficients of the quadratic equation determining the directions of the 'vertical' flexcords at the transition vanish, to be replaced by a cubic equation for the transitional directions, which we have not attempted to determine.

In both these examples a star develops into a star followed by a monstar followed by a star, but it is also possible to have a monstar developing into a monstar, followed by a star, followed by a monstar.

Until now we have assumed that both A and D are non-zero – in all our examples we have chosen $A = 2$ and $D = 3$. The remaining case, where the cubic terms of the Monge from are not all zero is when $A = 0$ and $D \neq 0$ or when $D = 0$ and $A \neq 0$. Suppose that $A = 0$ but that $D \neq 0$. Then, provided that b, the coefficient of x^2y, is non-zero, the umbilic at the origin is of multiplicity 2. T. R. Barrick has shown that in that case there are two ridges passing through the umbilic, the one tangential to the line $y = 0$ and the other to the line $ax + by = 0$, each changing colour there, the latter not being orthogonal to the former, since $b \neq 0$. If $b = 0$, with $a \neq 0$, the multiplicity of the umbilic at the origin is 3, the ridges passing through the umbilic being tangential to the axes. The vertical one changes colour, but the horizontal one does not. A particular case is where the surface has reflectional symmetry about the vertical direction, in which case only even powers of x occur in the Monge form. The further details of the ridge and flexcord patterns that can arise in this case are still under investigation.

In all the above we have assumed that the umbilic of central interest was flat, with all the sectional curvatures equal to zero. It is not difficult to carry through the analysis where $\kappa \neq 0$. For example, in that case the condition for the umbilic to be of multiplicity at least 2 reduces to $A^2d + 2ADc' + D^2b = 0$, where $c' = c - \kappa^3$. Otherwise everything is just as before.

3   Bumpy Spheres

In his thesis Stelios Markatis (1980) studied the ridges of bumpy spheres, surfaces of the form

$$x^2 + y^2 + z^2 + \tfrac{1}{3}\epsilon C_3(x,y,z)^3 = 1,$$

where C_3 is a thrice linear form on $\mathbf{R}^3$, ϵ is small, and faraway non-compact components are disregarded. His pictures are reproduced as Figures 16.11–16.20 of Porteous (1994).

The final exercise in that book is

16.10   Determine and plot the flexcords of the bumpy spheres.
The answer is somewhat surprising!

On any bumpy sphere of the form

$$x^2 + y^2 + z^2 + \tfrac{1}{3}\epsilon C_3(x,y,z)^3 = 1,$$

where C_3 is a thrice linear form on $\mathbf{R}^3$, ϵ is small, and faraway non-compact components are disregarded, the umbilics occur in mutually antipodal pairs, and the flexcords are in the limit the great circles on S^2 cut out by the planes that bisect the diameters of the sphere with end points a pair of umbilics.

To prove this, let $\mathbf{r}$ be an umbilic of the bumpy sphere, $\mathbf{s}$ some point of the sphere lying on the plane bisecting orthogonally the diameter whose end-points are the umbilic $\mathbf{r}$ and its antipodal companion, and let $\mathbf{t}$ be a unit vector that completes a triple of orthogonal vectors. To a first approximation all are unit vectors, with $\mathbf{s}$ and $\mathbf{t}$ tangent to the surface at the umbilic and $\mathbf{r}$ and $\mathbf{t}$ tangent to the surface at $\mathbf{s}$.

On differentiating three times the defining equation of the bumpy sphere we have the following holding at any point $\mathbf{r}$:

$$F(\mathbf{r}) = \mathbf{r} \cdot \mathbf{r} + \epsilon C_3 \mathbf{r}^3 - 1 = 0,$$

$$F_1(\mathbf{r}) = 2\mathbf{r} \cdot + 3\epsilon C_3 \mathbf{r}^2,$$

$$F_2(\mathbf{r}) = 2 \cdot + 6\epsilon C_3 \mathbf{r},$$

$$F_3(\mathbf{r}) = 6\epsilon C_3.$$

For $\mathbf{r}$ an umbilic, the second fundamental form, $\mathrm{II}_2(\mathbf{r})$, is a multiple of the first fundamental form, $\mathrm{I}_2(\mathbf{r}) = \cdot$. Since the restriction of $F_2(\mathbf{r})$ to the tangent plane to the surface at $\mathbf{r}$ is a multiple of $\mathrm{II}_2(\mathbf{r})$ it then follows, in particular, that $C_3\mathbf{r}\mathbf{s}^2 = C_3\mathbf{r}\mathbf{t}^2$ and $C_3\mathbf{r}\mathbf{s}\mathbf{t} = 0$. Indeed these equations hold if and only if $C_3\mathbf{r}$ is a multiple of $\cdot$, that is, if and only if $\mathbf{r}$ is an umbilic of the surface.

By the symmetry of C_3 in all three slots, $C_3\mathbf{s}\mathbf{t}\mathbf{r} = 0$. Since also $\mathbf{t} \cdot \mathbf{r} = 0$ it follows that the vectors $\mathbf{t}$ and $\mathbf{r}$ are principal tangent vectors to the surface at the point $\mathbf{s}$.

Finally again from Exercise 13.11 of Porteous (1994), $\mathbf{s}$ is on a flexcord of the surface if and only if, for the principal tangent vectors $\mathbf{u}$ and $\mathbf{v}$ at $\mathbf{s}$ in some order or other,

$$(F_1(\mathbf{s})\mathbf{n})(F_3(\mathbf{s})\mathbf{u}^2\mathbf{v}) = (F_2(\mathbf{s})\mathbf{n}\mathbf{v})(F_2(\mathbf{s})\mathbf{u}^2),$$

where $\mathbf{n}$ is normal to the surface. So

$$(2\mathbf{s} \cdot \mathbf{n} + 3\epsilon C_3\mathbf{s}^2\mathbf{n})(6\epsilon C_3\mathbf{u}^2\mathbf{v}) = (2\mathbf{n} \cdot \mathbf{v} + 6\epsilon C_3\mathbf{s}\mathbf{n}\mathbf{v})(2\mathbf{u} \cdot \mathbf{u} + 6\epsilon C_3\mathbf{s}\mathbf{u}^2).$$

Now, as $\epsilon \to 0$, $\mathbf{n} \to \mathbf{s}$, and $\mathbf{u}$ and $\mathbf{v}$ tend to $\mathbf{r}$ and $\mathbf{t}$ in some order or other. Suppose that $\mathbf{u} \to \mathbf{t}$ and $\mathbf{v} \to \mathbf{r}$. Then, ignoring terms in ϵ^2 the flexcord condition reduces to

$$(2\mathbf{s} \cdot \mathbf{s})(6\epsilon C_3\mathbf{t}^2\mathbf{r}) = (6\epsilon C_3\mathbf{s}^2\mathbf{r})(2\mathbf{t} \cdot \mathbf{t}),$$

that is, cancelling 12ϵ,

$$C_3 \mathbf{t}^2 \mathbf{r} = C_3 \mathbf{s}^2 \mathbf{r}.$$

But this is indeed the case, since $\mathbf{r}$ is an umbilic of the surface, as we observed earlier. So $\mathbf{s}$ lies on a flexcord of the surface.

Conversely, if $\mathbf{s}$ lies on a flexcord of the surface with $\mathbf{t}$ and $\mathbf{r}$ principal unit tangent vectors at $\mathbf{s}$, so that $C_3 \mathbf{srr} = 0$ and $C_3 \mathbf{t}^2 \mathbf{r} = C_3 \mathbf{s}^2 \mathbf{r}$, then $\mathbf{r}$ is an umbilic of the surface.

This completes the proof.

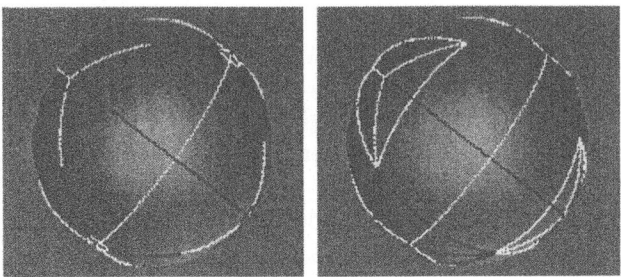

Fig. 5. Tne bumpy cube.

As an example, Fig. 5 shows the ridges and flexcords of the 'bumpy cube', where $C_3(x, y, z) = xyz$, ¿From the flexcord picture it is clear that the umbilics near the mid-points of the twelve sides of the cube are all monstars, each of index $\frac{1}{2}$. The eight at the vertices of the cube are of course all elliptic stars, each of index $-\frac{1}{2}$, the total index of the 20 umbilics being 2. Further examples are to be found in Puddephat (1994).

With suitable software, for example the Liverpool Surface Modeling Package of Richard Morris (1997), one may illustrate many more general perestroikas of flexcords by taking ϵ no longer small.

References

1. Bruce, J. W., Giblin, P. J. and Tari, F. (1996) Ridges, crests and sub-parabolic lines of evolving surfaces, *Int J. Computer Vision* **18**, 195-210.
2. Bruce, J. W., Giblin, P. J. and Tari, F. (1999) Families of surfaces: focal sets, ridges and umbilics, *Math. Proc. Camb. Phil. Soc.* **125**, 243-268.
3. Bruce, J. W. and Tari, F. (1996) Extrema of principal curvatures and symmetry, *Proc. Edinburgh Math. Soc.* **39**, 397-402.
4. Bruce, J. W. and Wilkinson, T. C. (1991) Folding maps and focal sets, *Proc. of Warwick Symposium on Singularities*, Lecture Notes in Math. 1462, Springer, Berlin, pp 63-72.
5. Darboux G. (1894) *Leçons sur la théorie générale des surfaces*, 4me partie, Gauthiers-Villars et Fils, Paris.

6. Markatis, S. (1980) *Some generic phenomena in families of surfaces in* $\mathbf{R}^3$, Thesis, University of Liverpool.

7. Morris R. J. (1997) *The use of computer graphics for solving problems in singularity theory*, in Visualisation and Mathematics, ed. H. -C. Hege and K. Polthier, Heidelberg: Springer-Verlag, 53–66. For details of the Liverpool Surface Modeling Package, for Silicon Graphics and X Windows, contact Richard Morris (rjm@amsta.leeds.ac.uk).

8. Porteous, I. R. (1994) *Geometric Differentiation*, C.U.P.

9. Puddephat, M. J. (1994) Robust features of bumpy spheres, MSc Thesis, University of Liverpool.

10. Wilkinson, T. C. (1991) The geometry of folding maps, Thesis, University of Newcastle.

Time-Optimal Paths Covering a Surface

Taejung Kim and Sanjay E. Sarma

Massachusetts Institute of Technology, Cambridge MA 02139, USA.
MIT ROOM 35-014B, TEL:(617)253-1925; FAX:(617)253-7549;
taejung@mit.edu, sesarma@mit.edu

Summary. A new point of view is presented for tool path generation for free-form surfaces. The task of tool path generation is treated as a time-optimization problem. We seek the path that covers an entire surface as quickly as possible while respecting both geometric tolerance requirements and the kinematic limits of the machine tool. This is an important consideration in high speed machining of dies and molds. We first frame the optimization problem mathematically. The problem turns out to be difficult to solve, and a closed form solution does not exist. We instead describe approaches towards for finding numerical solutions. We present preliminary results of our algorithms. The resulting tool paths tend to take 10-30% less time.

1 Introduction

NC machining is a fundamental but time-consuming process in modern manufacturing. In the automotive industry, for example, dies often require hundreds of hours to machine. Reducing machining time is of considerable economic importance. The insatiable need for performance of machine tools has spawned the area of high speed machining in recent years. However, there is a limit to how fast a machine can move. After a point, it becomes necessary to optimize the tool path for maximum performance. In this paper, we seek the machining path that covers an entire surface as quickly as possible. We describe how the geometry of tool paths (as opposed to just the feed rates) can be optimized for a specific machine tool just as computer programs are compiled for specific computers. To answer this global question, we will deal with **families** of paths rather than individual paths. Specifically, our final set of machining paths is a finite selection from infinitely-many streamlines of a **vector field** derived from the kinematics of the particular machine tool under consideration.

Kinematic limits and high-speed machining: Every motor has a torque-speed curve that determines the maximum toque as a function of velocity. The motors drive the cutting tool through a kinematic mechanism. This has two implications. First, as one would expect, there are limits on the feed rate and acceleration achievable in any configuration. Second, the kinematic transformations skew the velocity and acceleration envelope of a machine tool and make them extremely direction dependent. This means that motion in some directions is considerably faster than motion in other directions. Furthermore, improperly selected tool paths will also have significant overlaps— unnecessarily machining certain areas repeatedly. The preferable tool paths must, therefore, not only avoid unnecessary overlaps between sweeping strips, but also avoid the directions in which, for example, the acceleration and velocity requirements are high. Generating optimal tool paths for general freeform surfaces is a difficult problem, beyond human intuition. The current work addresses this problem in the context of high-speed machining.

2   Background

There is a massive body of previous work in the area of surface machining. Much previous work has dealt with basic issues of geometric feasibility such as: techniques for path generation including iso-parametric, iso-planar, iso-cusp-height, iso-phote and iso-engagement schemes [2, 3, 4, 5, 6, 7 and 8]; gouge-free tool paths and gouge correction [9 and 10]; issues of accessibility and global interference [11 and 12]; tool shape, tool orientation and surface finish prediction [13]. Once feasibility is ensured, it is possible to improve the efficiency of machining, and there has been work in this area as well. Elber and Cohen, for example, try to reduce unnecessary tool path overlaps following iso-parametric curves [14]. In iso-cusp-height scheme, one of the families of the curves, instead of iso-parametric curves, is selected not to generate the overlaps [4, 5 and 6]. Some researchers have also considered more global issues. Elber divides a surface into convex, concave and parabolic regions and suggests the use of a flat-end cutter for convex regions in order to sweep much larger area during the same time [15]. Choi et al. and Mullins et al. show how to reduce cusp height for a given tool path to reduce grinding or manual polishing time [16 and 17]. Comprehensive lists of references are available in [2, 3, 18 and 19].

The idea of adjusting the feed rate for considerations related to cutting physics has been studied by some researchers [20, 21 and 22]. A typical application is the reduction of the feed rate at points on a given path where the cut is too aggressive. The idea of choosing preferable directions is discussed by Lim et. al. [23]. The authors suggest following the directions that cause the minimum cutting force and tool deflection. Their work is very pertinent in high-force situations, where machine deflection is an important issue. We, however, seek tool paths for surface finishing (in the context of high-speed machining), where speed is usually the bottleneck.

Time-optimal control (for robots) is a well-established theory [24, 25, 26, 27, 28 and 29]. However, there is one fundamental difference between path planning for machining and for robots: while robotic paths seek at most to follow a given path, or go from point to point, machining paths almost always seek to fill or cover an area.

The strategy of the tool path of today is decided purely from a geometric viewpoint and it is up to the operator to recognize and harness any advantages from a particular strategy. It is here that we aim to contribute with our research: we seek to generate, automatically, (near)-optimal tool paths for a particular surface for a given machine tool.

3   A Formulation of the Problem

In this section we formulate the optimal tool path problem as a mathematical problem.

For simplicity, we consider only the case of **ball end-mills** and assume that **the tool orientation is pre-defined** at every point on the surface. We can extend this formulation to other mill geometries and also ask the more general question—what is the most preferable set of orientations in 5-axis machining?

3.1   Description of Tool Paths: Kinematics on a Surface

1. **Designed Surface:** A designed surface S is assumed to be given as a *regular* parametric form, $r(u,v)$. In other words, $r : (\mathcal{P} \subset \mathcal{R}^2) \to (\mathcal{W} \subset \mathcal{R}^3)$, $S = r(\mathcal{P}) \subset \mathcal{W}$ and $r_u \times r_v \neq 0$ for a compact set $\mathcal{P}$ and an open set $\mathcal{W}$. We call the domain $\mathcal{P}$ the *parameter space* and the co-domain $\mathcal{W}$ the *workpiece space*. The independent variables, u and v, are called the parameters of the surface. For compactness, the parameters are grouped into a column vector $\boldsymbol{u}$, i.e., $\boldsymbol{u} = [u\ v]^T \in \mathcal{P}$. The unit outside normal vector of the surface at (u,v) is denoted by $n(u,v)$ or $n(\boldsymbol{u})$. Without loss of generality, we assume that the surface is parameterized such that $n(\boldsymbol{u}) = (+)r_u \times r_v / \|r_u \times r_v\|$. Thus, $det[r_u r_v n] > 0$. We denote partial derivatives by subscripts, *e.g.*, $\eta_u = \partial \eta / \partial u$.

A *change of frame* between $\mathcal{W}$ and an *inertial reference frame* should be established for mechanical analysis. While the designed surface remains stationary in $\mathcal{W}$ during the entire machining process, the workpiece itself can be in movement with respect to the inertial reference frame. We choose an orthonormal basis $\{i\ j\ k\}$ which is stationary in $\mathcal{W}$ as shown in Fig. 1. This is a basis moving with the surface. When we represent a vector as a matrix, the vector will be treated as a column vector and it will be projected into the basis $\{i\ j\ k\}$, *e.g.* $r = [r \bullet i\ r \bullet j\ r \bullet k]^T$.

2. **Typical NC Code:** In 5-axis machining, we need 3 translational and 2

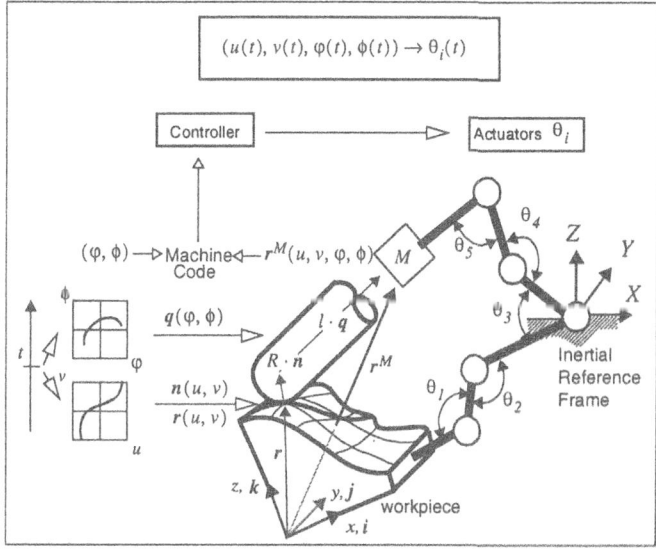

Fig. 1. Surface inverse kinematics

rotational coordinates to describe the cutter configuration in the workpiece space $\mathcal{W}$. A classic way to describe a rigid body motion is to apply a translation and a rotation about a specified point on the body, which we call the *reference point* of the motion [*c.f.* 30]. It is convenient to choose a point M on the centerline shown in Fig. 1 as the reference point of the tool motion, whose position vector is denoted by $r^M(= r_x^M i + r_y^M j + r_z^M k)$. The orientation in 5-axis machining is represented by a *unit* vector q along the centerline as shown in Fig. 1.[1] We parameterize the orientation q with 2 parameters (ψ, ϕ). For example, we can utilize the spherical coordinate system.

The 5-tuple $(r_x^M, r_y^M, r_z^M, \psi, \phi)$ determines the configuration of a machine tool as well as the joint angles θ_i $(i = 1...N)$, where N is the number of actuators. Typically, CNC G codes for a 5-axis machine tool consist of a series of 5-tuple location descriptors and feed rates.

3. Motion restricted by the designed surface: To prevent the tool from gouging into the surface, we require the tool to touch the surface tangentially at every instant. The trajectory $r(u(t), v(t))$ of such a point of contact will be called the *(surface) tool path*, which is a surface curve, where t is time. (Of course, we can use any other parameters.) For example, in the case of a ball-end cutter, the *requirement of tangency* results in

---

[1] The tuple (r^M, q) is called cutter location data as described in [16].

$r^M = r(u,v) + R \cdot n(u,v) + l \cdot q(\psi,\phi)$, where R is the radius of the tool and l is the length to the reference point as shown in Fig. 1. Then, the configuration of a machine tool $(r_x^M, r_y^M, r_z^M, \psi, \phi)$, restricted by this condition, is described by 4 parameters (u, v, ψ, ϕ).

The 4-tuple,(u, v, ψ, ϕ), is transformed into machine code by a planning system using the requirement of tangency. The machine code is further transformed into the joint angles θ_i by the controller based on the kinematics of the mechanical system. In other words, we can establish a mapping f between the parameters and the joint angles, $i.e.$

$$\theta_i = f_i(u, v, \psi, \phi) \tag{1}$$

which we call *surface inverse kinematics*. It is possible to develop the *equations of motion* of the machine tool with joint angles being taken as the *generalized coordinates*. [2] We would have to assume that the cutting forces could be derived in an analytical form. By plugging the surface inverse kinematics into the equations of motion, it is possible to derive joint torques in the following form: $\tau_i = g_i(u, v, \psi, \phi, \dot{u}, ..., \ddot{u}, ...)$.

4. Field description: Trajectories of the form $r(u(t), v(t)) = r(t)$ on a surface are the candidate tool paths of a cutting tool to machine the surface. We postulate a vector field $v(r)$ on the surface S which corresponds to the velocity of such candidate trajectories: $v(r(t)) = \dot{r}(t)$. By this, we also postulated a corresponding vector field $\dot{u}(u)$ ($= [\dot{u}(u,v) \quad \dot{v}(u,v)]^T$) on the parameter space $\mathcal{P}$ s.t. $v(r(u)) = \dot{u} \cdot r_u + \dot{v} \cdot r_v$ ($= r_* \dot{u}$) . This field, along with "start points" from which tool paths begin machining the surface, contains all the information required to construct the tool paths $r(t)$ or $u(t)$. This is because the differential equation $\dot{r} = v(r)$ (or $\dot{u} = \dot{u}(u)$) can be integrated from the start points or initial conditions.

The field v or $\dot{u}$ is conveniently specified by its magnitude and direction on the surface as well as by its Cartesian components. The magnitude is the cutting speed ϑ and the direction is the angle η with one of the iso-parametric lines as shown in Fig. 2-(a): $v = \vartheta \cdot (\cos\eta \cdot r_u/\|r_u\| + \sin\eta \cdot n \times r_u/\|r_u\|)$. The field η is called a *direction angle*. The speed $\vartheta(u)$ and the direction angle $\eta(u)$ are related to the vector field $\dot{u}(u)$ on $\mathcal{P}$ by the following relation:

$$\vartheta = \sqrt{v^T v} = \sqrt{\dot{u}^2 \cdot r_u^T r_u + 2\dot{u}\dot{v} \cdot r_u^T r_v + \dot{v}^2 \cdot r_v^T r_v} = \sqrt{\dot{u}^T G \dot{u}}$$

$$\eta = \text{atan}_2((\dot{v} \cdot r_v^T(n \times r_u)), (\dot{u} \cdot r_u^T r_u + \dot{v} \cdot r_v^T r_u)) \tag{2}$$

where G is the *1st fundamental matrix* of the surface in differential geometry:

---

[2] For establishing equations of motion, we refer to standard texts on multi-body dynamics, *e.g.* [31].

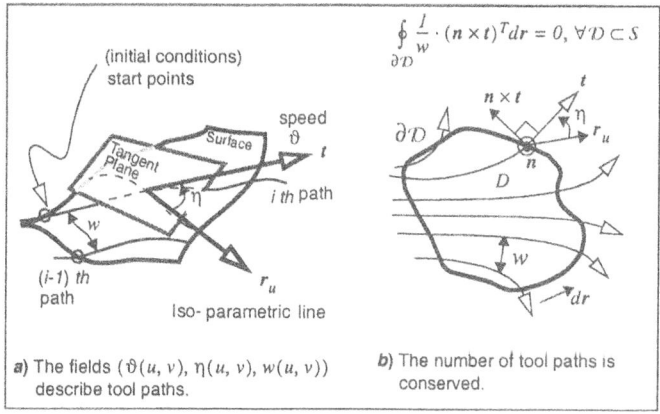

a) The fields $(\vartheta(u, v), \eta(u, v), w(u, v))$ describe tool paths.

b) The number of tool paths is conserved.

Fig. 2. Field description and an equality constraint of tool paths

$$G \equiv \begin{bmatrix} r_u^T r_u & r_u^T r_v \\ r_u^T r_v & r_v^T r_v \end{bmatrix} = [g_{ij}]. \tag{3}$$

And its inverse is

$$\dot{u} = \vartheta \cdot h(\eta, u, v) \tag{4}$$

where $h(\eta, u, v) \equiv P(u, v) \cdot a(\eta)$, $a(\eta) \equiv [\ \cos\eta\ \ \sin\eta\ ]^T$ and P is defined in terms of the *1st fundamental form coefficients*, g_{ij}:

$$P \equiv \frac{1}{\sqrt{g_{11}}} \left[\left\{ \begin{matrix} 1 \\ 0 \end{matrix} \right\} \middle| \frac{1}{\sqrt{det(G)}} \left\{ \begin{matrix} -g_{12} \\ g_{11} \end{matrix} \right\} \right]. \tag{5}$$

5. Streamlines: The vector field specified by $(\vartheta(u, v), \eta(u, v))$ is integrated to derive streamlines (integral curves) on the surface. In the parameter space $\mathcal{P}$, Equation (4) forms a system of ODEs for a known $(\vartheta(u, v), \eta(u, v))$. It is integrated to produce streamlines. We refer to these as *parameter streamlines*. These parameter streamlines are mapped to surface curves $r(u(t), v(t))$, which we call *surface streamlines*. The unit tangent vector of a surface streamline is denoted by $t = \dot{r}/\|\dot{r}\|$.

6. Sidestep field: The streamlines by themselves do not completely define tool paths. After all, tool paths are separate swathes, not a continuum. The missing element is how to choose the tool paths among the streamlines, which is effectively described by the start points or initial conditions of these paths in the field. We now take a step further so as to eliminate even the need for specifying the initial conditions. We define the *side step* w which represents the distance between two neighboring tool paths on the surface as shown in Fig. 2-(a). Instead of the initial condition for each tool path, we specify the side step as a function of (u, v), *i.e.* $w = w(u, v)$. By treating the side-step as a field, we make the path-planning problem amenable to useful mathematical tools such as calculus and differential geometry. Postulating the side-step as a field is analogous to *continuum hypothesis* in continuum mechanics. And, in the outset, we assumed that the orientation is also fixed as a known field, *i.e.* $\psi = \psi(u, v)$ and $\phi = \phi(u, v)$.

7. Summary: Simply put, tool paths are described by the unknown functions: $(\vartheta(u, v),\ \eta(u, v),\ w(u, v))$. It is general enough to require such participating functions to be **piece-wise smooth** and **piece-wise continuous**.

3.2   An Equality Constraint: Compatibility

1. The side step $w(u, v)$ and the direction angle $\eta(u, v)$ are not independent: As shown in Fig. 2-(b), the net number of tool paths across any closed region on the surface must vanish. We express this approximately in the following form of equation:

$$\oint_{\partial \mathcal{D}} \tfrac{1}{w} \cdot (\boldsymbol{n} \times \boldsymbol{t})^T d\boldsymbol{r} = 0, \quad \forall \mathcal{D} \subset \mathcal{S}.$$

We invoke *generalized Stokes' theorem*, and say that $\int_{\mathcal{D}} d[(\boldsymbol{n} \times \boldsymbol{t})^T d\boldsymbol{r}/w] = 0$, where by the symbol d we denote the *exterior derivative of a differential form*.[3] By the arbitrariness of the domain $\mathcal{D}$, the differential 2-form, $d[(\boldsymbol{n} \times \boldsymbol{t})^T d\boldsymbol{r}/w]$, itself should vanish. By simply applying the rules for the exterior derivative and wedge product on a manifold, the following can be derived:

$$h_1(\eta, u, v) \cdot \frac{\partial w}{\partial u} + h_2(\eta, u, v) \cdot \frac{\partial w}{\partial v} = h_3(\eta, \eta_u, \eta_v, u, v) \cdot w \quad (a.e.), \quad (6)$$

where

$$\boldsymbol{h} = [\ h_1\ \ h_2\ ]^T = \boldsymbol{P}\boldsymbol{a}(\eta) \text{ (as defined earlier,) } \boldsymbol{R} \equiv \begin{bmatrix} 0 & -1 \\ 1 & 0 \end{bmatrix}, \text{ and}$$

$$h_3 \equiv \left[[\ \eta_u\ \ \eta_v\ ]\boldsymbol{P} + ([\ 0\ 1\ ]\tfrac{\partial}{\partial u}(\boldsymbol{GP}) - [\ 1\ 0\ ]\tfrac{\partial}{\partial v}(\boldsymbol{GP}))/\sqrt{\det(\boldsymbol{G})} \right]\boldsymbol{R}\boldsymbol{a}(\eta).$$

This compatibility only applies in regions where there are **no intermittent cuts**. It is an analog of the principle of conservation of mass assuming no sources or sinks in fluid mechanic terms.

2. Compatibility as a 1st order linear partial differential equation: The above equation can be thought of as a 1st order (linear) partial differential equation (PDE) about w when the direction angle $\eta(u, v)$ is known. It is

---

[3] Differential forms and generalized Stokes' theorem are described in several books including [32,33]. We used exterior calculus here. This approach is more natural for a curved space than vector and tensor analysis [34].

a classic result that a 1st order PDE can be integrated along what are called (*base*) *characteristic curves*[4] which are, in this case, expressed as:

$$\frac{du}{h_1} = \frac{dv}{h_2} = \frac{dw}{h_3 \cdot w} = ds. \tag{7}$$

In fact, it can be shown that the (*base*) *characteristics* are the (parameter) streamlines themselves and the parameter s in the above equation is the *arc length* parameter of a surface streamline.

3. Integration of the side step following streamlines: Let us integrate Equation (7) along a streamline, one of the characteristic curves. Of course, this is only possible after the direction angle $\eta(u, v)$ is given. By integrating the 1st two equations ($d\boldsymbol{u}/ds = \boldsymbol{h}$) of Equation (7), we get the streamlines $(u(s), v(s))$ and then by the last equation we have

$$w(s) \equiv w(u(s), v(s)) = w(0) \cdot \exp\left(\int_0^s h_3 ds\right), \tag{8}$$

where $w(0)$ is the initial side step for a streamline which is not yet decided. Therefore, if we are given the direction angle $\eta(u, v)$, we can extract all the information on the side step $w(u, v)$ except for the initial value $w(0)$ of each start point.

3.3   Cost Functional: Cutting Time

Cutting time T_c can be divided into two parts, *effective cutting time* T_e and *non-effective cutting time* T_{ne}, i.e. $T_c = T_e + T_{ne}$. Effective cutting time is the time during which the cutting tool stays in contact with the surface of the part and removes material. During this period, the machine executes NC commands such as G01, G02 and G03. The rest of the time is referred to as non-effective cutting time. Examples of non-effective cutting include G00 rapid motions between contiguous tool paths.

Effective cutting time is modeled as follows:

$$T_e = \sum_i \int_{C_i} \frac{ds}{\vartheta} = \sum_i \int_{C_i} \frac{w \cdot ds}{w \cdot \vartheta} \approx \int_S \frac{dA}{\vartheta \cdot w} = \iint_{\mathcal{P}} \frac{\|\boldsymbol{r}_u \times \boldsymbol{r}_v\| \, dudv}{\vartheta(u, v) \cdot w(u, v)}, \tag{9}$$

where C_i is the *i*th tool path, s is its arc length parameter as mentioned earlier, ϑ is the cutting speed and w is the side step. Of course, the integration domain must include enough tool paths to make the approximation be valid. As in continuum mechanics, the domains we look at must be of much larger scale than the features.

---

[4] A background on partial differential equations is available in [35,36].

Non-effective cutting time can be assumed to be proportional mainly to the **number of tool paths** in the domain. This is because at the end of each tool path, the tool must be lifted, moved and lowered again for the new path. If there are *no tool path loops* and *no intermittent cuts* in the domain, one of the simplest model of the non-effective cutting time is

$$T_{ne} = \frac{\tau_o}{2} \cdot \oint_{\partial S} \left| \frac{1}{w} \cdot (n \times t)^T dr \right|, \tag{10}$$

where τ_o is a constant of the proportionality which characterizes the time consumed by each non-effective movement and the integration is done along the boundary curve, ∂S. If tool paths have any loops (vortices) or intermittent cuts (sources), we need to add a correction term. In addition, the constant τ_o itself is to become a functional as we detail the modeling.

In summary, we minimize the functional $T_c = T_e + T_{ne}$ as above explained by finding the functions $\vartheta(u, v)$, $\eta(u, v)$ and $w(u, v)$ which are subject to the equality constraint, Equation (6), and some inequality constraints which follow.

3.4   Inequality Constraints: Cusp Height Limit, Actuator Limits, Collision Avoidance etc.

We consider the constraints that make this a bounded optimization problem.

1. Cusp Height Limit: Due to the mismatch between the curvatures of the surface and the tool, cusps are unavoidable in general surface machining as shown in Fig. 3-(a). It is a usual practise to *control cusp height h* during machining. In our optimization problem, the height h of cusps is required to be below a certain constant h_o all over a surface.

Fig. 3-(b) shows that there is a maximum allowable tool spacing w_o, for a given limit h_o depending on to which direction η the tool proceeds [4]. We call such allowed spacing *coverage*. In order to ensure that cusps are smaller than the allowable limit, we approximately require that the side step should be less than the coverage:

$$w(u, v) \leq w_o(\eta, u, v). \tag{11}$$

2. Actuator Limits: As discussed in Section 1, motors have torque and speed limits which are coupled. The feasible combinations of torque and speed are in some empirical set: $(\tau_i, \omega_i) \in \mathcal{A}_i \subset \mathcal{R}^2$, as shown in Fig. 4-(a). From the *surface inverse kinematics* $(\theta_i = f_i(u, v, \psi, \phi))$ of a machine tool, which we covered in Equation (1), the joint velocity is derived in the following form of equations:

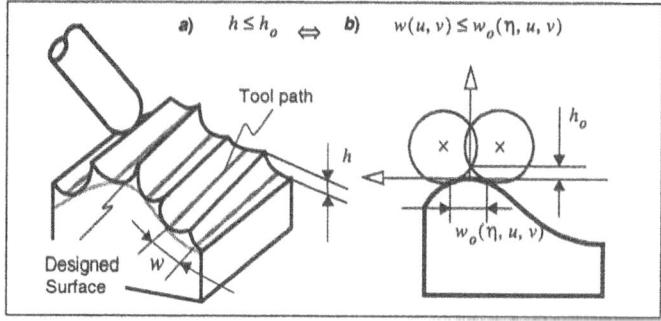

Fig. 3. An inequality constraint: Cusp height or side step is regulated.

$$\omega_i = \dot{\theta}_i = \left(\frac{\partial f_i}{\partial u} + \frac{\partial f_i}{\partial \psi}\frac{\partial \psi}{\partial u} + \frac{\partial f_i}{\partial \phi}\frac{\partial \phi}{\partial u}\right)\dot{u} + \left(\frac{\partial f_i}{\partial v} + \frac{\partial f_i}{\partial \psi}\frac{\partial \psi}{\partial v} + \frac{\partial f_i}{\partial \phi}\frac{\partial \phi}{\partial v}\right)\dot{v}$$
$$= a_i \cdot \dot{u} + b_i \cdot \dot{v}, \qquad (i = 1...N.) \quad (12)$$

In a similar manner, from *the equations of motion* of the mechanical system, the joint toques τ_i can be derived. And recall that $\dot{u} = \vartheta \cdot h(\eta, u, v)$. Therefore, the actuator limits, $(\tau_i, \omega_i) \in \mathcal{A}_i \subset \mathcal{R}^2$, is expressed in terms of ϑ, η and their partial derivatives, in general.

In high speed machining, we assume that speed limits are critical because the objective is to move as fast as possible while taking light cuts. In such situations, the actuator limit reduces to a very simple form. The speed limit in Fig. 4-(a) is shown as the right-extreme of the torque-speed curve. We can say then that

$$|\omega_i| = |a_i \cdot \dot{u} + b_i \cdot \dot{v}| \leq \omega_o^i \qquad (13)$$

where ω_o^i is the maximum allowable speed of joint i, $(i = 1...N.)$ If we now map these actuator velocity constraints back on to the $(\dot{u}, \dot{v})$-space[5] at every point (u, v), we create a symmetric polygon. This is shown in Fig. 4-(b). We call it a *velocity polygon*.[6]

3. Other Constraints: Excessive cutting force can cause the tool to deflect more than accuracy requirements permit. This also form a feasible area in the $(\dot{u}, \dot{v})$-space [23]. In the context of high speed machining, we assume that this is less critical. There are several other constraints that reflect the reality of machining: structural stiffness limits, system band-width, tool wear considerations and so on. We ignore these considerations in our initial analysis.

[5] In fact, $(\dot{u}, \dot{v})$-space is the tangent space $T_r\mathcal{S}$ of the surface in mathematical terms.

[6] A similar concept is found in the context of robot manipulability [38].

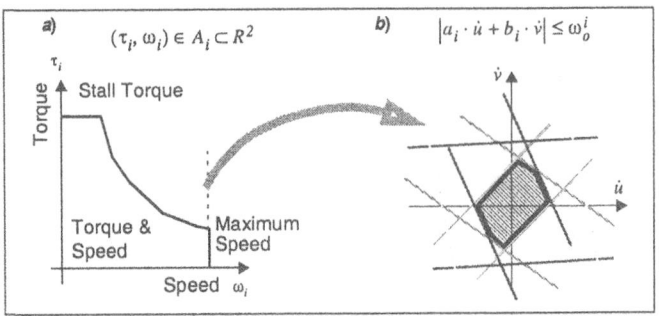

Fig. 4. An inequality constraint: Velocity Polygon

And in the outset, we assumed that the geometric feasibility is satisfied by defining the feasible orientation field.

3.5   The Formulated Problem

We have now derived the final form of the optimization problem. The final formulation is summarized below in its simplest form:

Find the (piece-wise continuous and piece-wise smooth) functions of the parameters, $(u, v) \in \mathcal{P} \subset \mathcal{R}^2$:

> *Unknowns:* $\vartheta(u, v)$, $\eta(u, v)$ and $w(u, v)$,   covered in Section 3.1

which minimize the cutting time (*cost functional*):

$$T_c = \iint_{\mathcal{P}} \frac{\|\mathbf{r}_u \times \mathbf{r}_v\|}{\vartheta \cdot w} \cdot du\,dv + T_{ne}[\vartheta, \eta, w], \qquad \text{covered in Section 3.3}$$

and satisfy the following *constraints:*

Compatibility: $h_1(\eta, u, v) \cdot \frac{\partial w}{\partial u} + h_2(\eta, u, v) \cdot \frac{\partial w}{\partial v} = h_3(\eta, \eta_u, \eta_v, u, v) \cdot w$ (*a.e.*),
covered in Section 3.2
Cusp Height Limit: $w(u, v) \leq w_0(\eta, u, v)$,   covered in Section 3.4.1
Actuator Velocity Limits: $\vartheta \cdot |a_i(u, v) \cdot h_1(\eta, u, v) + b_i(u, v) \cdot h_2(\eta, u, v)| \leq \omega_o^i$,
$(i = 1...N)$ covered in Section 3.4.2

where w_0, h_1, h_2, a_i and b_i are usual functions which are known. ω_o^i are positive constants. $T_{ne}[...]$ is a functional by which we mean that we can decide its value at least procedurally with the functions in the square brackets being specified. We denote partial derivatives by subscripts, e.g., $\eta_u = \partial \eta / \partial u$. N is the number of actuators of a machine tool.

---

4   Solving the Problem: Approaches and Preliminary Results

We now look at some approaches we have considered to solve the problem.

4.1   Minimization of Effective Cutting Time Subject to Local Constraints: Greedy Approach

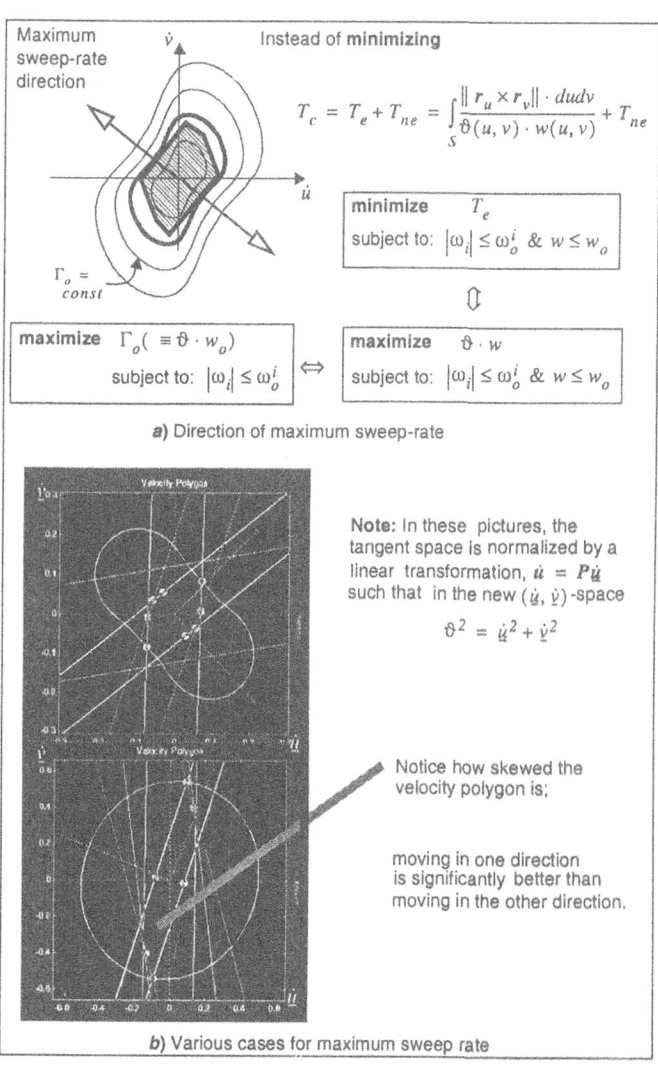

Fig. 5. Direction of Maximum Sweep Rate

Local constraints: The general problem we have formulated is difficult to solve. As an initial analysis, we take a local approach, which we could also refer to as "greedy." The greedy approach is to pick a direction in which the tool path performs best, and to generate streamlines by integrating the direction field. The "local" constraints are (i) the cusp height limit ($w \leq w_o$) and (ii) the actuator speed limits ($|\omega_i| \leq \omega_o^i$.) Consequently, we ignore constraints that affect the system globally such as compatibility and torque limits of actuators.

Direction of Maximum Sweep-rate: Instead of T_c, the effective cutting time T_e is minimized subject to the two local constraints. Consequently, the denominator $\Gamma(\equiv \vartheta \cdot w)$ of its integrand should be maximized because there are no differential inequality constraints. The denominator Γ is called the *effective sweep-rate*. From the cusp height limit $w \leq w_o$, we necessarily find the direction which maximize the *sweep-rate*, $\Gamma_o(\equiv \vartheta \cdot w_o)$ subject to only the velocity limits. The problem can be represented in $(\dot{u}, \dot{v})$-space as shown in Fig. 5-(a): we should find a point on the velocity polygon, where the largest iso-sweep rate contour passes. The iso-sweep rate contour lines can be convex or concave depending on the curvatures of the surface as shown in Fig. 5-(b). If it is convex, only the vertices are candidate points. If not, the point can be on an edge of the polygon as shown in Fig. 5-(a). By the symmetry, there is always a pair of directions of maximum sweep rate as shown in Fig. 5-(a).

Field fitting: We sample some number of points on the surface and evaluate the greedy direction in each point. And then, we fit a vector field to the data with a least-square method. This resolves the difficulty in the ambiguity of the two directions of maximum sweep rate.

Tool path integration: It is possible to perform a Runge-Kutta integration and generate streamlines of the fitted field. Finally, these streamlines can be merged to form a template for tool paths. Using the *compatibility equation*, Equation (8), we also integrate the side step and the information can be used to space the tool paths within the cusp height limit. In this example, rather large cusp height limit, $h_o = 1mm$, was set for visualization. The resulting tool path is shown in Fig. 6-(a). In some regions, there is shown significant overlap between tool paths. In such regions with high overlap, some sections of tool paths are removed using a criterion as shown in Fig. 6-(b). In the same figure, the non-effective movement is shown. Fig. 6-(c) shows a designed surface and the machine tool we took as an example.

This tool path is compared with the two iso-parametric tool paths, as shown in Fig. 6-(d). 36% and 4% of cutting time can be saved. In this example, the non-effective cutting time of the greedy path is higher than the one of the other two paths, which hinders the greedy approach from taking more

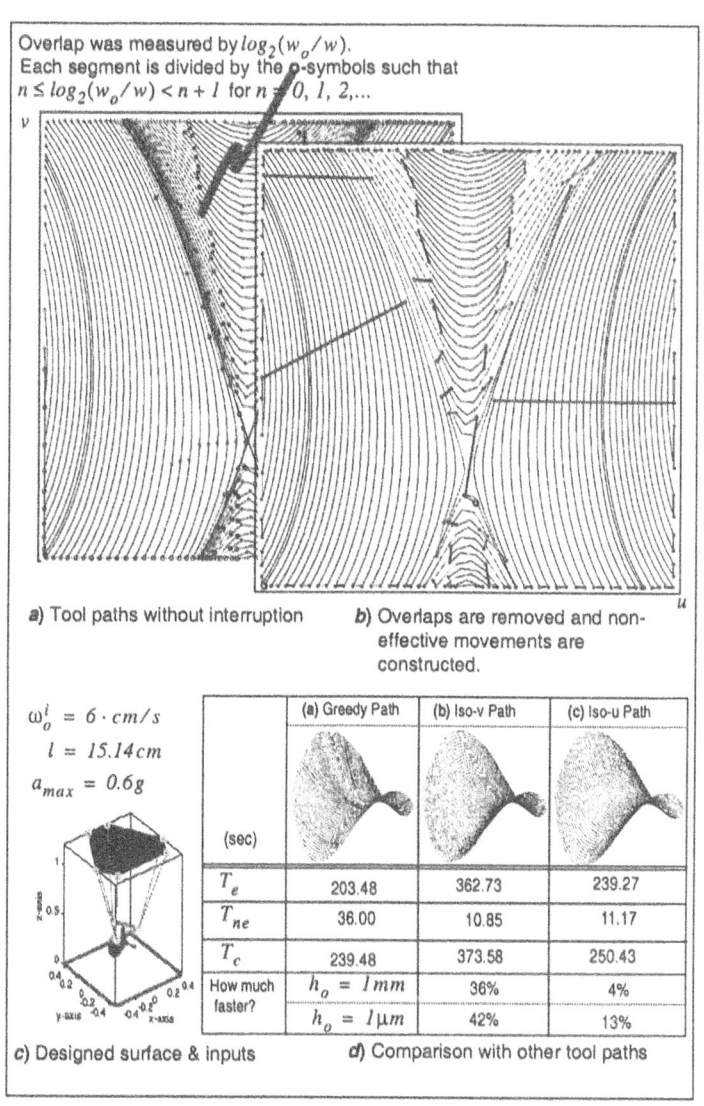

Fig. 6. Greedy Approach with Fitting

advantage. However, if we set the cusp height limit in the realistic range, the ratio is reduced significantly, (roughly, $T_{ne}/T_e \propto h_o^{1/4}$, $T_e \propto h_o^{-1/2}$ and $T_{ne} \propto h_o^{-1/4}$) and the greedy approach becomes more advantageous. For example, if we set the cusp height limit to be $1\mu m$, the advantage is expected to be increased upto 42% and 13%, respectively.

4.2   Minimizing the Entire Integral: General Search

In the greedy approach, global constraints were ignored and non-effective cutting time was optimized. Now we consider solving the general problem. In spite of the complexity, it is possible to construct a straightforward algorithm for minimizing the cutting time for a given direction field $\eta(u, v)$ (with the positions of sinks/sources.) Here, we sketch the algorithm. This is equivalent to actually constructing tool paths: (1) With a given direction field $\eta(u, v)$, a streamline $\{r(u(s), v(s)) : 0 \leq s \leq L\}$ is constructed from a start point by $du/ds = h(\eta, u, v)$. (2) Finding the optimal velocity distribution $\vartheta(s)$ along a given trajectory is a well-studied problem in the field of robotics [28 and 29] and if we consider only velocity limits, the problem becomes even easier. (3) To set the side-step as wide as possible, we choose the side step $w(0)$ at the start point $s.t.$ $w(0)=\inf \{w_o(s)/\exp(\int_o^s h_3 ds) : 0 \leq s \leq L\}$. By Equation (8), now the distribution $w(s)$ of a path is known. (4) We repeat the same procedure for other start points until the surface is properly covered. By collecting the data $\{(\eta(s), \vartheta(s), w(s), u(s), v(s))\}$, we are ready to evaluate our cost functional minimized for a given direction field $\eta(u, v)$.

We therefore need to consider the variations of the direction field $\eta(u, v)$. This input function will be approximated with a finite number of basis functions. We can then use various optimization schemes such as the steepest descent method. In the general sense, this is likely to be a computation-intensive problem f perhaps prohibitively so. However, we are considering several ways to simplify the problem. For example, (1) we can use coarser grids first and then finer grids and (2) we can make some approximate analytic formula for the cutting time. Additionally the local approaches described in preceding sections can also potentially be used to prune the search process.

5   Conclusions

We presented a fairly general formulation of the tool path optimization problem. A tool path was modeled as a streamline of a vector field on a surface. And we have theoretically considered various aspects of machining—especially the kinematics.

In practical terms, we have described approaches that show how it is possible to make progress in solving this rather difficult problem. Greedy approach is invaluable when a machine tool approaches its singularities. Near singularities, behavior of the machine tool becomes extremely anisotropic as shown in Fig. 5, and trajectory planning becomes invaluable.

References

1. Kim, T. and Sarma, S., Towards Machine-optimal NC Tool Path Generation, CAGD, submitted.
2. Dragomatz, D. and Mann, S., A classified bibliography of literature on NC milling path generation, Computer-Aided Design, Vol. 29, No. 3, pp. 239-247, 1997.
3. Marshall, S. and Griffiths, J. G., A survey of cutter construction techniques for milling machines, International Journal of Production Research, Vol. 32, No. 12, pp. 2861-2877, 1994.
4. Suresh, K. and Yang, D. C. H., Constant scallop-height machining of free-form surfaces, ASME Journal of Engineering for Industry, Vol. 116, pp.253-259, May 1994.
5. Lin, R.S. and Koren, Y., Efficient tool-path planning free-form surfaces, ASME Journal of Engineering for Industry, Vol. 118, pp.20-28, Feb 1996.
6. Lo, C-C., Efficient cutter-path planning for five-axis surface machining with a flat-end cutter, Computer-Aided Design, Vol. 31, pp.557-566, 1999.
7. Han, Z. and Yang, D. C. H., Isophote based machining for feature intensive surfaces, Proceedings of the ASME, MED-Vol. 8, pp.483-495, 1998.
8. Stori, J. A. and Wright, P. K., A constant engagement offset for 2-1/2D tool path generation, Proceedings of the ASME, MED-Vol. 8, pp.475-481, 1998.
9. Choi, B. K. and Jun, C. S., Ball-end cutter interference avoidance in NC machining of sculptured surfaces, Computer-Aided Design, Vol. 21, No. 6, pp. 371-378, 1989.
10. Lee, Y. S., Admissible tool orientation control of gouging avoidance for 5-axis complex surface machining, Computer-Aided Design, Vol. 29, No. 7, pp. 507-521, 1997.
11. Lee, Y. S., Chang, T. C., 2-Phase approach to global tool interference avoidance in 5-axis machining, Computer-Aided Design, Vol. 27, No. 10, pp. 715-729, 1995.
12. Elber, G., Accessibility in 5-axis milling environment, Computer-Aided Design, 26(11):796-802, 1994.
13. Vickers, G. W. and Quan, K. W., Ball-Mills Versus End-Mills for Curved Surface Machining, ASME Journal of Engineering for Industry, Vol. 111, pp. 22-26, Feb 1989.
14. Elber, G. and Cohen, G., Toolpath generation for free-form surface models, Computer-Aided Design, Vol. 26, No. 6, pp. 490-496, 1994.
15. Elber, G., Freeform surface region optimization for 3-axis and 5-axis milling, Computer-Aided Design, Vol. 27, No. 6, pp. 465-470, 1995.
16. Choi, B. K., Park, J. W. and Jun, C. S., Cutter-location optimization in 5-axis surface machining, Computer-Aided Design, Vol. 25, No. 6, pp. 377-385, 1993.
17. Mullins, S. H., Jensen, C. G. and Anderson, D. C., Scallop elimination based on precise 5-axis tool placement, orientation, and stepover calculations, Advances in Design Automation ASME 65-2, pp.535-544, 1993.

18. Choi, B. K. and Jerard, B. J., Sculptured Surface Machining: Theory and Application, Kluwer Academic Publishers, 1998.
19. Marciniak, K., Geometric modeling for numerically controlled machining, Oxford University Press, 1991.
20. Yazar, Z., Koch, K. F., Merrick, T. and Altan, T., Feed rate optimization based on the cutting force calculations in three-axis milling of dies and molds with sculptured surfaces, Int. J. Mach. Tools Manufact, 34(3), 365, 1994.
21. Chu, C. N., Kim, S. Y., Lee, J. M. and Kim, B. H., Feed-rate optimization of ball end milling considering local shape features, Annals of CIRP, Vol. 46/1/1997.
22. Park, S., Jun, Y. T., Lee C. W. and Yang, M. Y., Determining the cutting conditions for sculptured surface machining, Int J. Adv Manuf Technol, 8:61-70, 1993.
23. Lim, E. and Menq, C., Integrated planning for precision machining of complex surfaces. Part I: Cutting-path and feedrate optimization, Int. J. Mach. Tools Manufact, 37(1), pp. 61-75, 1996.
24. Yang, H. S. and Slotine, J.-J. E., Fast algorithms for near-minimum-time control of robot manipulators, The International Journal of Robotics Research, Vol. 13, No. 6, pp. 521-532, Dec 1994.
25. Kahn, M. E. and Roth, B., The near-minimum-time control of open-loop articulated kinematic chains, Journal of Dynamic Systems, Measurement, and Control, 93(3):164-172, 1971.
26. Meier, E-B. and Bryson, A. E. Jr., Efficient algorithm for time-optimal control of a two-link manipulator, J. Guidance Control Dynam. 13(5):859-866, 1990.
27. Rajan, V. T., Minimum-time trajectory planning, Proc. of IEEE Conference on Robotics and Automation, pp. 759-764, 1985.
28. Bobrow, J. E., Dubowsky, S. and Gibson, J. S., Time-optimal control of robotic manipulators along specified paths, The International Journal of Robotics Research, Vol. 4, No. 3, Fall 1985.
29. Shin, K. G. and McKay, N. D., Minimum-time control of robotic manipulators with geometric path constraints, IEEE Transactions on Automatic Control, Vol. AC-30, No. 6, June 1985.
30. Synge, J. L., Classical Dynamics, pp.14, Principles of Classical Mechanics and Field Theory, in Flugge, S. (ed.), Handbuch der Physik, Vol. III/1, Springer Verlag, Berlin, 1960.
31. Crandall, Dynamics of mechanical and electromechanical systems, McGraw-Hill, 1968.
32. Weintraub, S. H., Differential forms: a complement of vector calculus, Academic Press, 1997.
33. Buck, R. C., Advanced Calculus, pp. 366-436, 1965.
34. Flanders, H., Differential forms with applications to the physical science, pp. 3-4, Dover Pub., 1989.
35. Hildebrand, F. B., Advanced Calculus for Applications, 2nd ed., p.387, Englewood Cliffs, 1976.
36. Arnold, V. I., Chapter 2, First-Order Partial Differential Equations, Geometrical Methods in the Theory of Ordinary Differential Equations, 2nd ed., p. 59-88, Springer-Verlag, 1988.
37. Marciniak, K., Influence of surface shape on admissible tool positions in 5-axis face milling, Computer-Aided Design, Vol. 19, No. 5, pp. 233-236, June 1987.

38. Huynh, P. and Arai, T., Maximum velocity analysis of Parallel manipulators, Proc. IEEE International Conference on Robotics and Automation, pp.3268-3273, April 1997.

Surface Evolution and Representation using Geometric Algebra

Anthony Lasenby[1] and Joan Lasenby[2]

[1] Astrophysics Group, Cavendish Laboratory, Madingley Road, Cambridge, CB3 OHE, UK
[2] Department of Engineering, University of Cambridge, Trumpington Street, Cambridge CB2 1PZ, UK

Summary. Recent developments in geometric algebra have shown that by moving from a projective to a conformal representation (5d representation of 3d space), one is able to extend the range of geometrical operations that can be carried out in an efficient and elegant way. For example, while in projective space one is able to intersect lines and planes in a simple fashion, in conformal space one is able to intersect and represent spheres, lines, circles and planes. In addition, all the operations of Euclidean geometry (dilations, translations, rotations and inversions) are smoothly integrated with the projective representation.

The paper will use the conformal representation to look at the problems of surface representation and evolution, and of wavefront propagation from such surfaces.

1 Introduction

The mathematical language we will use throughout will be that of geometric algebra (GA). This language is based on the algebras of Clifford and Grassmann and the form we follow here is that developed by David Hestenes [1]. There are now many texts and useful introductions to GA, [2–5] so we do no more here than outline some aspects used in the problems we will discuss.

In a geometric algebra of n-dimensions, we have the standard *inner* product which takes two vectors and produces a scalar, plus an outer or *wedge* product that takes two vectors and produces a new quantity we call a *bivector* or oriented area. Similarly, the outer product between three vectors produces a *trivector* or oriented volume etc. Thus the algebra has basic elements which are oriented geometric objects of different orders. The highest order object in a given space is called the *pseudoscalar* with the unit pseudoscalar denoted by I, e.g. in 3d I is the unit trivector $e_1 \wedge e_2 \wedge e_3$ for basis vectors $\{e_i\}$. Multivectors are quantities which are made up of linear combinations of these different geometric objects. More fundamental than the inner or wedge products is the *geometric product* which can be defined between any multivectors – the geometric product, unlike the inner or outer products, is invertible. For vectors the inner and outer products are the symmetric and antisymmetric parts of the geometric product;

$$ab = a{\cdot}b + a{\wedge}b \qquad (1)$$

In effect the manipulations within geometric algebra are keeping track of the objects of different grades that we are dealing with (much as complex number arithmetic does). For a general multivector M, we will use the notation $\langle M \rangle_r$ to denote the rth grade part of M.

In what follows we shall use the convention that vectors will be represented by non-bold lower case roman letters, while we use non-bold, upper case roman letters for 5d vectors and certain multivectors — exceptions to this are stated in the text. Unless otherwise stated, repeated indices will be summed over.

1.1   Rotations

If, in 3d, we consider a rotation to be made up of two consectutive reflections, one in the plane perpendicular to a unit vector m and the next in the plane perpendicular to a unit vector n, it can easily be shown [4] that we can represent this rotation by a quantity R we call a **rotor** which is given by

$$R = nm$$

Thus a rotor in 3D is made up of a scalar plus a bivector and can be written in one of the following forms

$$R = e^{-B/2} = \exp\left(-I\frac{\theta}{2}n\right) = \cos\frac{\theta}{2} - In\sin\frac{\theta}{2}, \tag{2}$$

which represents a rotation of θ radians about an axis parallel to the unit vector n in a right-handed screw sense. Here the bivector B represents the plane of rotation. Rotors act two-sidedly, ie. if the rotor R takes the vector a to the vector b then

$$b = Ra\tilde{R}$$

where $\tilde{R} = mn$ is the reversion of R (i.e the order of multiplication of vectors in any part of the multivector is reversed). We have that rotors must therefore satisfy the constraint that $R\tilde{R} = 1$. One huge advantage of this formulation is that rotors take the same form, i.e. $R = \pm \exp(B)$ in any dimension (we can define hyperplanes or bivectors in any space) and can rotate any objects, not just vectors; e.g.

$$R(a\wedge b)\tilde{R} = \langle Rab\tilde{R}\rangle_2 = \langle Ra\tilde{R}Rb\tilde{R}\rangle_2$$
$$= Ra\tilde{R}\wedge Rb\tilde{R} \tag{3}$$

gives the formula for rotating a bivector.

2   Conformal Geometry in Geometric Algebra

It has long been known that going to a 4d, projective, description of 3d Euclidean space can have various advantages – particularly when intersections of planes and lines are required. Such projective descriptions are used extensively in computer vision and computer graphics where rotations and translations can be described by a single 4×4 matrix and non-linear projective transformations become linear. Projective geometry fits very nicely into the geometric algebra framework and applications are given in [6,7]. In [1] conformal geometry was briefly discussed and recently, [8,9], the application of these ideas in which a 5d conformal space is used as the representation of 3d Euclidean space has been the subject of renewed interest. In this conformal space we have, as a subset, the projective geometry, but also the ability to extend to circles and spheres. Below we describe the basics of this conformal representation and outline how it can be of use in specific problems, e.g. reflecting a general wavefront from a spherical surface.

We start with the simplest formulation of Hestenes' conformal geometry work in 3d. However, unlike the treatments given in [1,8,9], we will use *reflection* as the key to *inversion* which will enable us to treat circles and spheres very easily and efficiently.

The notation we use will follow the original notation given in [1]. Let x be a vector in a space $\mathcal{A}(p,q)$, where the *signature* (p,q) implies that the space has a basis $\{e_i\}$, $i = 1, \ldots, n = p + q$ where $e_i^2 = +1$ for $i = 1, \ldots, p$ and $e_i^2 = -1$ for $i = p + q + 1, \ldots, n$ – i.e. we take a general mixed signature space. Now extend this to a space $\mathcal{A}(p + 1, q + 1)$ via the inclusion of two additional basis vectors, e and $\bar{e}$, such that

$$e^2 = +1, \quad \bar{e}^2 = -1, \quad e \cdot \bar{e} = 0$$

Note that if $x \in \mathcal{A}(p,q)$, then $e \cdot x = \bar{e} \cdot x = 0$ since $e_i \cdot e = e_i \cdot \bar{e} = 0$ for $i = 1, \ldots, n$. We now introduce the vectors n and $\bar{n}$ where

$$n = e + \bar{e} \qquad \bar{n} = e - \bar{e} \tag{4}$$

$$\implies e = \frac{1}{2}(n + \bar{n}) \qquad \bar{e} = \frac{1}{2}(n - \bar{n}) \tag{5}$$

n and $\bar{n}$ are **null vectors** since

$$n^2 = (e + \bar{e}) \cdot (e + \bar{e}) = e^2 + 2e \cdot \bar{e} + \bar{e}^2 = 1 + 0 - 1 = 0$$
$$\bar{n}^2 = (e - \bar{e}) \cdot (e - \bar{e}) = e^2 - 2e \cdot \bar{e} + \bar{e}^2 = 1 - 0 - 1 = 0$$

Note also that

$$n \cdot \bar{n} = (e + \bar{e}) \cdot (e - \bar{e}) = e^2 - \bar{e}^2 = 2$$
$$x \cdot n = 0 \quad \text{and} \quad x \cdot \bar{n} = 0$$

for $x \in \mathcal{A}(p,q)$. We now map a point x in $\mathcal{A}(p,q)$ to a point $F(x)$ in $\mathcal{A}(p + 1, q + 1)$ via the Hestenes ([1], p.302) representation

$$F(x) = -(x - e)n(x - e) \tag{6}$$

Substituting for $n = e + \bar{e}$ and using the fact that $\bar{e} \cdot x = 0 = \bar{e} \cdot e = n \cdot x$, it is not hard to rewrite this equation in terms of the null vectors as follows:

$$F(x) = x^2 n + 2x - \bar{n} \tag{7}$$

which is precisely the form which is used in the more recent 'horosphere' formulations of the conformal framework [8,9].

Now for any $x_i \in \mathcal{A}(p,q)$ we evaluate $[F(x)]^2$

$$[F(x)]^2 = (x^2 n + 2x - \bar{n})\cdot(x^2 n + 2x - \bar{n})$$
$$= -x^2 n\cdot\bar{n} + 4x^2 = -4x^2 + 4x^2 = 0 \qquad (8)$$

Thus we have mapped vectors in $\mathcal{A}(p,q)$ into **null vectors** in $\mathcal{A}(p+1,q+1)$ – this is precisely the horosphere construction.

More generally, it can be shown that any *null* vector in $\mathcal{A}(p+1,q+1)$ can be written as

$$X = \lambda(x^2 n + 2x - \bar{n}) \qquad (9)$$

with *lambda* a scalar. We can now use this to provide a projective mapping between $\mathcal{A}(p,q)$ and $\mathcal{A}(p+1,q+1)$: the family of null vectors, $\lambda(x^2 n + 2x - \bar{n})$, in $\mathcal{A}(p+1,q+1)$ are taken to correspond to the single point $x \in \mathcal{A}(p,q)$.

At this point it is interesting to see what happens when we take the inner product of any two such null vectors:

$$A\cdot B = \{a^2 n + 2a - \bar{n}\}\{b^2 n + 2b - \bar{n}\} = -2a^2 - 2b^2 - 4a\cdot b = -2(a-b)^2 \qquad (10)$$

We see therefore that taking the inner product of two 5d representative vectors gives a scalar which is proportional to the distance between the 3d vectors. This is where the formulation can be related to the study of *distance geometry* [10].

We will be especially interested in **Conformal Transformations** in $\mathcal{A}(p,q)$, and we shall see later that these are represented by *rotors* and *reflections* in $\mathcal{A}(p+1,q+1)$. We now look at the operations of rotation and reflection more closely.

1. **Rotations**

If $x \mapsto Rx\tilde{R}$ with $x \in \mathcal{A}(p,q)$ and R a rotor in $\mathcal{A}(p,q)$, then what happens when R acts on $F(x)$?

$$RF(x)\tilde{R} = R(x^2 n + 2x - \bar{n})\tilde{R} = x^2 Rn\tilde{R} + 2Rx\tilde{R} - R\bar{n}\tilde{R}$$

Since R is a rotor, it contains only even blades and therefore commutes with n and $\bar{n}$ ($e_i n = -ne_i$, so if we have an even number of *e*s we have commutation), so that $Rn\tilde{R} = R\tilde{R}n = n$ and $R\bar{n}\tilde{R} = \bar{n}$. Thus we have

$$RF(x)\tilde{R} = x'^2 n + 2x' - \bar{n} \qquad (11)$$

where $x' = Rx\tilde{R}$. That is, rotors in $\mathcal{A}(p,q)$ remain rotors in $\mathcal{A}(p+1, q+1)$, i.e.

$$x \mapsto Rx\tilde{R} \qquad \Longrightarrow \qquad F(x) \mapsto F(Rx\tilde{R}) \tag{12}$$

2. Inversions

Here we have $x \mapsto \frac{x}{x^2}$ or equivalently, $x \mapsto x^{-1}$ (since $x^{-1} = \frac{x}{x^2}$). Firstly, we look at the properties of relection in e:

$$-ene = -ee\bar{n} = -\bar{n}$$

using the fact that $ne = \frac{1}{2}(e + \bar{e})e = \frac{1}{2}(e^2 + \bar{e}e) = \frac{1}{2}(e^2 - e\bar{e}) = e\bar{n}$. Similarly, we can show that the following reflection properties hold: $-ene = -\bar{n}$, $-e\bar{n}e = -n$ and $-exe = x$. Now we look at what happens to $F(x)$ under reflection in e

$$-eF(x)e = -e(x^2 n + 2x - \bar{n})e = x^2\bar{n} + 2x + n$$

$$= x^2 \left[\frac{1}{x^2} n + 2\frac{x}{x^2} - \bar{n} \right] = x^2 F(\frac{x}{x^2}) \tag{13}$$

Therefore, we have that inversion in $\mathcal{A}(p,q)$ is brought about by reflection in e in $\mathcal{A}(p+1, q+1)$, i.e.

$$x \mapsto \frac{x}{x^2} \qquad \Longrightarrow \qquad F(x) \mapsto -eF(x)e = x^2 F(\frac{x}{x^2}) \tag{14}$$

Note here that it is irrelevant whether we take $-e(...)e$ or $e(...)e$ as the reflection – henceforth we will use $e(...)e$ for convenience.

3. Translations

Here we wish to achieve a translation $x \mapsto x + a$; we will show that this is performed by a rotor $R = T_a = e^{\frac{na}{2}}$, where $a \in \mathcal{A}(p,q)$;

$$R = T_a = e^{\frac{na}{2}} = 1 + \frac{na}{2} + \frac{1}{2}\left(\frac{na}{2}\right)^2 + \ldots = 1 + \frac{na}{2} \tag{15}$$

since n is null and $an = -na$. Firstly we see how R acts on n, $\bar{n}$ and x.

$$R n \tilde{R} = \left(1 + \frac{na}{2}\right) n \left(1 + \frac{an}{2}\right)$$

$$= n + \frac{1}{2}nan + \frac{1}{2}nan + \frac{1}{4}nanan = n \tag{16}$$

again using $an = -na$ and $n^2 = 0$. Similarly we can show that

$$R\bar{n}\tilde{R} = \bar{n} - 2a - a^2 n \tag{17}$$

$$Rx\tilde{R} = x + n(a{\cdot}x) \tag{18}$$

We can now see how the rotor acts on $F(x)$

$$RF(x)\tilde{R} = \left(1 + \frac{na}{2}\right)(x^2n + 2x - \bar{n})\left(1 + \frac{an}{2}\right)$$
$$= x^2n + 2(x + n(a{\cdot}x)) - (\bar{n} - 2a - a^2n)$$
$$= (x + a)^2 n + 2(x + a) - \bar{n}$$
$$= x'^2 n + 2x' - \bar{n} = F(x + a) \qquad (19)$$

where $x' = x + a$. Thus, translations in $\mathcal{A}(p, q)$ can be performed by the $\mathcal{A}(p + 1, q + 1)$ rotor $R = T_a$, $a \in \mathcal{A}(p, q)$ so that

$$x \mapsto x + a \qquad \Longrightarrow \qquad F(x) \mapsto F(x + a) \qquad (20)$$

4. Dilations

To investigate how dilations are formed we start by considering the rotor $R = D_\alpha = e^{\frac{\alpha}{2}e\bar{e}}$ and the following relations which can easily be verified:

$$e\bar{e}n = -\bar{n} = -ne\bar{e} \quad \text{and} \quad e\bar{e}\bar{n} = \bar{n} = -\bar{n}e\bar{e} \qquad (21)$$

Using these relations it is straightforward to show that $RF(x)\tilde{R}$ gives

$$D_\alpha F(x)\tilde{D}_\alpha = e^{\frac{\alpha}{2}e\bar{e}}\{x^2n + 2x - \bar{n}\}e^{-\frac{\alpha}{2}e\bar{e}}$$
$$= e^\alpha \left\{\hat{x}^2 n + 2x' - \bar{n}\right\} \qquad (22)$$

where $x' = e^{-\alpha}x$. The above can be verified by expanding $e^{-\frac{\alpha}{2}e\bar{e}}$ as $1 - \frac{\alpha}{2}e\bar{e} + \frac{1}{2!}\left(\frac{\alpha}{2}e\bar{e}\right)^2 + \dots$ etc. and using the relations given in equation (21). Thus, dilations in $\mathcal{A}(p, q)$ can be performed by the $\mathcal{A}(p + 1, q + 1)$ rotor $R = D_\alpha$, so that

$$x \mapsto e^{-\alpha}x \qquad \Longrightarrow \qquad F(x) \mapsto e^\alpha F(e^{-\alpha}x) \qquad (23)$$

2.1   Special Conformal Transformations

We have seen above that we are able to express rotations, inversions, translations and dilations in $\mathcal{A}(p, q)$ by rotations and reflections in $\mathcal{A}(p+1, q+1)$. This now leads us to consider *special conformal transformations* of the form

$$x \mapsto x\frac{1}{1 + ax} \qquad (24)$$

which is actually a combination of inversion, translation and inversion again:

$$x \xrightarrow{\text{inversion}} \frac{x}{x^2}$$
$$\xrightarrow{\text{translation}} \frac{x}{x^2} + a$$
$$\xrightarrow{\text{inversion}} \frac{\frac{x}{x^2} + a}{\left(\frac{x}{x^2} + a\right)\left(\frac{x}{x^2} + a\right)}$$
$$= \frac{x + ax^2}{1 + 2a{\cdot}x + a^2x^2} = x\frac{1}{1 + ax} \qquad (25)$$

The final line in the above expression shows us that $x\frac{1}{1+ax}$ is indeed a vector. As we have built up the special conformal transformation via inversions and translations, we know exactly how to construct the $\mathcal{A}(p+1,q+1)$ operator that performs such a transformation – the required rotor is

$$K_a = eT_ae = 1 - \frac{\bar{n}a}{2}, \qquad \text{so that} \qquad x \mapsto K_axK_a \tag{26}$$

and

$$K_ax\tilde{K}_a = e\left\{T_a(exe)\tilde{T}_a\right\}e \tag{27}$$

Now, when we act on $F(x)$ with K_a we can use our previous results to obtain

$$K_aF(x)\tilde{K}_a = (1 + 2a{\cdot}x + a^2x^2)F\left(x\frac{1}{1+ax}\right) \tag{28}$$

which therefore tells us that

$$x \mapsto x\frac{1}{1+ax} \qquad \Longrightarrow \qquad F(x) \mapsto (1 + 2a{\cdot}x + a^2x^2)F\left(x\frac{1}{1+ax}\right) \tag{29}$$

2.2   Projective geometry in the conformal space

In this section we first consider the part $\lambda(2x - \bar{n})$ of the conformal representation of a point x and show it enables us to deal with projective geometry and the incidence of planes, lines etc. Indeed it is very similar to forming $\lambda(x + \gamma_o)$ (x in Euclidean 3-space) – in the conformal representation the signature of the bit we add on is irrelevant for projective geometry. Indeed it is generally better, if dealing only with projective geometry, that we do not have null structures present, which implies adding a 4th basis vector which gives a $\mathcal{A}(4,0)$ space. So, we need to ask ourselves if there is any advantage in going to a $\lambda(x^2n + 2x - \bar{n})$ representation.

The key fact here is that by enlarging the representation and employing the reflection formula to do inversions, we can now study the incidence relations of spheres, circles, lines and planes and not just lines and planes. A second, linked, advantage, is that in the conformal representation we can represent incidence relations via wedge products just as we can in the GA version of projective geometry. For example, in projective geometry, if a line, L, passes through two points a, b, whose (4d) homogeneous representations are A, B, we can write the line as a bivector, $L = A{\wedge}B$. Then, any X lying on the line L will satisfy

$$X{\wedge}L = 0$$

In the conformal representation, rotations, translations, dilations, inversions are all represented by rotors or reflections, which tells us that any incidence relations remain invariant in form under such operations – we can see this explicitly as follows.

Suppose we have the incidence relation

$$X \wedge Y \wedge \ldots \wedge Z = 0$$

where $X, Y, ..., Z \in \mathcal{A}(p+1, q+1)$. Then under reflections in e we have

$$X \wedge Y \wedge \ldots \wedge Z \mapsto (eXe) \wedge (eYe) \wedge \ldots \wedge (eZe) = e(X \wedge Y \wedge \ldots \wedge Z)e \qquad (30)$$

where we have used the fact that $(eXe) \wedge (eYe) = \frac{1}{2}(eXeeYe - eYeeXe) = \frac{1}{2}e(XY - YX)e = e(X \wedge Y)e$, since $e^2 = 1$. Thus, if $X \wedge Y \wedge .... \wedge Z = 0$ then so too does $(eXe) \wedge (eYe) \wedge .... \wedge (eZe)$.

Similarly, under rotations we have

$$X \wedge Y \wedge \ldots \wedge Z \mapsto (RX\tilde{R}) \wedge (RY\tilde{R}) \wedge \ldots \wedge (RZ\tilde{R}) = R(X \wedge Y \wedge \ldots \wedge Z)\tilde{R} \qquad (31)$$

where again we use $(RX\tilde{R}) \wedge (RY\tilde{R}) = \frac{1}{2}(RX\tilde{R}RY\tilde{R} - RY\tilde{R}RX\tilde{R}) = \frac{1}{2}R(XY - YX)\tilde{R} = R(X \wedge Y)\tilde{R}$, since $R\tilde{R} = 1$. Thus, if $X \wedge Y \wedge .... \wedge Z = 0$ then so too does $(RX\tilde{R}) \wedge (RY\tilde{R}) \wedge .... \wedge (RZ\tilde{R})$.

We see therefore that translations, rotations, dilations and inversions can now be brought into the context of projective geometry – a significant increase in the usefulness of the projective representation. It is now possible to build up a set of useful results in this projective conformal representation.

2.3   The equation of a line

Because the incidence relations are invariant under rotations and translations in the $\mathcal{A}(p, q)$ space, wlog we can consider a line in the direction e_1 passing through the origin.

Let three points on this line be x_1, x_2, x_3 with $\mathcal{A}(p+1, q+1)$ representations X_1, X_2, X_3. Now, the $\{X_i\}$ contain only the vectors $n, \bar{n}$ and e_1 (since $x_i = \lambda_i e_1$). Thus, if X is the representation of any other point on the line we have

$$X \wedge X_1 \wedge X_2 \wedge X_3 = 0 \qquad (32)$$

This is because each of the above 5d vectors contains only the three vectors $n, \bar{n}$ and e_1 and therefore the wedge of four of them must be zero since each term will involve a wedge of two identical vectors. By invariance of the incidence relations under rotations and translations, we see that (32) is the equation of a line for *any* three general points X_1, X_2 and X_3 on the line. It is interesting to see how this parallels the projective case and also to note that we appear to need 3 points in this conformal representation to describe a line – we will return to this later.

2.4   The equation of a plane

Exactly the same sort of thing goes through here. By translational and rotational invariance we can, wlog, take the plane as that spanned by e_1 and e_2 and passing through the origin. If x lies in this plane then we can write

$$x = \lambda e_1 + \mu e_2$$

and its conformal representation, X, therefore only contains the vectors $n, \bar{n}, e_1, e_2$, i.e.

$$X = x^2 n + 2(\lambda e_1 + \mu e_2) - \bar{n}$$

Take $\Phi = X_1 \wedge X_2 \wedge X_3 \wedge X_4$, where X_i, $i = 1, ..., 4$ lie in the plane; following the same reasoning as given for the line, we see that for any X on Φ we must have $X \wedge \Phi = 0$, therefore

$$X \wedge X_1 \wedge X_2 \wedge X_3 \wedge X_4 = 0 \tag{33}$$

is the equation of the plane passing through points X_i, $i = 1, ..., 4$. Once again we note here that in the conformal representation we appear to require 4 points rather than 3 to specify the plane.

Extending this to higher dimensions we see that to specify an r-d hyperplane (where a line is $r = 1$, a plane is $r = 2$ etc) the equation is

$$X \wedge X_1 \wedge X_2 \wedge \ldots X_{r+1} \wedge X_{r+2} = 0 \tag{34}$$

where X_i, $i = 1, ..., r + 2$ are conformal representations of the $r + 2$ points x_i lying in the hyperplane.

2.5   The role of inversion

It may be thought strange that we need to specify $r + 2$ points in order to determine an r-d hyperplane. For example, 2 points clearly suffice to determine a line, 3 points for a plane etc. So what is the role of the extra points?

We can best understand this, and the role inversion plays, by considering a simple example. Let the $\mathcal{A}(p, q)$ space be $\mathcal{A}(2, 0)$, i.e. the ordinary Euclidean plane with basis (e_1, e_2), $e_1^2 = 1$, $e_2^2 = 1$.

Let the line L be $x = 1$ i.e. $(1, y) : -\infty \leq y \leq +\infty$ and let $a = (x, y)$. Suppose we want to invert points on this line – we then obtain the set of points

$$a \mapsto \frac{a}{a^2} \quad \Longrightarrow \quad L \mapsto \left(\frac{1}{1 + y^2}, \frac{y}{1 + y^2} \right) \tag{35}$$

Parameterizing the original line as $x = 1, y = t$; $-\infty \leq t \leq +\infty$, the inversion produces $(x', y') = \left(\frac{1}{1+t^2}, \frac{t}{1+t^2} \right)$ – it is then easy to show that

$$(x' - \tfrac{1}{2})^2 + y'^2 = \left(\frac{1}{2} \right)^2$$

Hence the inversion produces a circle, centre $(\frac{1}{2}, 0)$ radius $\frac{1}{2}$.

straight line $\xrightarrow{\text{inversion}}$ circle

Any three points on this line, X_i, $i = 1, 2, 3$ (conformal representation) therefore invert to give three points, X_i', $i = 1, 2, 3$ on this circle. Let the general point on the line be X; we know that

$$X \wedge X_1 \wedge X_2 \wedge X_3 = 0$$

Thus, if X' is a general point on the circle,, we know that

$$X' \wedge X_1' \wedge X_2' \wedge X_3' = 0$$

Recall that we see this by performing an inversion via reflection in e; i.e.

$$e(X \wedge X_1 \wedge X_2 \wedge X_3)e = eXe \wedge eX_1e \wedge eX_2e \wedge eX_3e$$

This gives a very useful form for the equation of a circle. Here, we derived it for a special case but since we know that we can dilate and translate as we wish, it must in fact be true for a completely general circle. Thus if X_i, $i = 1, 2, 3$ are *any* three points, the equation of the circle passing through these points is

$$X \wedge X_1 \wedge X_2 \wedge X_3 = 0 \tag{36}$$

If we now invert this equation via $e(...)e$ we will in general obtain another circle since $X_i' = eX_ie$ will be another three general points in the plane. This only fails if X_1', X_2', X_3' are collinear and this will occur if the original circle passes through the origin (as in the case we started with here). Recall that if $x = 0$, $X \equiv x^2 n + 2x - \bar{n} = -\bar{n}$, so that $\bar{n}$ represents the origin – by inversion, $e\bar{n}e = n$, we see therefore that we can associate n with the point at infinity – the inversion of the origin. What happens for the original circle passing through the origin is that the representation of the origin, any multiple of $\bar{n}$, is transformed by inversion to a multiple of n, the point at infinity. Thus, the equation of a line can always be written as

$$X \wedge n \wedge X_1 \wedge X_2 = 0 \tag{37}$$

where X_1 and X_2 are any two (finite) points on the line – we can see this by choosing the origin as one of the 3 points on the circle before inverting. This therefore explains the extra point we appeared to need in describing a line earlier – what is really going on is that

$$X \wedge X_1 \wedge X_2 \wedge X_3 = 0$$

describes a *circle* and therefore genuinely requires 3 points – while a line is just a special case of a circle which passes through the point at infinity.

2.6 Extension to higher dimensions

All of the previous section transfers immediately to higher dimensions and different signatures – although for indefinite metrics, hyper-hyperboloids have to be considered as well as hyperspheres. Here we just illustrate the extension from $\mathcal{A}(2,0)$ to $\mathcal{A}(3,0)$, i.e. Euclidean 2-space to Euclidean 3-space.

The special case we start from this time – from which everything else can be derived – is the $x = 1$ plane, i.e. the set of points $(1, y, z)$: $-\infty \leq y \leq +\infty$, $-\infty \leq z \leq +\infty$. Inverting this plane we have

$$(1, y, z) \mapsto (x', y', z') = \left(\frac{1}{1 + y^2 + z^2}, \frac{y}{1 + y^2 + z^2}, \frac{z}{1 + y^2 + z^2} \right)$$

It is then not difficult to show that x', y', z' satisfy the following equation

$$\left(x' - \frac{1}{2} \right)^2 + y'^2 + z'gho^2 = \left(\frac{1}{2} \right)^2$$

which is the equation of a sphere, radius $\frac{1}{2}$ and centre $(\frac{1}{2}, 0, 0)$. We already know that the equation of a plane is

$$X \wedge X_1 \wedge X_2 \wedge X_3 \wedge X_4 = 0 \tag{38}$$

where X_i, $i = 1, .., 4$ are any 4 points on the plane. By inversion, translation, dilation and rotation we can now see that the equation of a sphere is given by the same equation

$$X \wedge X_1 \wedge X_2 \wedge X_3 \wedge X_4 = 0$$

for X_i, $i = 1, .., 4$ any 4 points on the sphere. The arguments are precisely as before – since we can always translate and rotate our plane to the plane $x = 1$ and we have shown that under inversion the plane $x = 1$ gives a sphere, then we know that we must have the general equation for a sphere. Thus, as expected, it really does take 4 points to describe a sphere.

We shall now show how a plane is a special case of a sphere. Its inverse is a sphere passing through the origin – again, we see this using the arguments we have used previously. Equation (38) is the equation of a sphere passing through the 4 points X_i, $i = 1, .., 4$. We now invert this to give

$$X' \wedge X_1' \wedge X_2' \wedge X_3' \wedge X_4' = 0$$

where $X_i' = eX_i e$. The X_i' are generally another set of general points, so we get another *sphere* through these new points. However, if the X_i' are coplanar then the above construction will not give a sphere – this occurs if the original sphere passes through the origin. In this case the equation of the original sphere can be written as

$$X \wedge n \wedge X_2 \wedge X_3 \wedge X_4 = 0$$

so that when we transform to give a *plane* we will get (since $ene = \bar{n}$)

$$X \wedge \bar{n} \wedge X_2' \wedge X_3' \wedge X_4' = 0 \tag{39}$$

That is, any plane passing through the points X_2', X_3', X_4' is given by equation (39). So indeed we only need 3 points to describe a plane – we can think of a plane as a sphere passing through the point at infinity.

Finally we consider a plane passing through the origin – we know that we can write this as

$$X \wedge n \wedge \bar{n} \wedge X_1 \wedge X_2 = 0$$

with X_1, X_2 lying on the plane. Now invert this to get

$$X \wedge \bar{n} \wedge n \wedge X_1' \wedge X_2' = 0$$

since $ene = \bar{n}$, $e\bar{n}e = n$. Under inversion we know that $x_1 \mapsto \frac{x_1}{x_1^2}$ and similarly for x_2; we therefore see that the plane is mapped onto itself under this inversion operation. Thus a plane passing through the origin is its own inverse since the null vectors $n, \bar{n}$ are just swapped under inversion. These results are all well known using a conventional approach of course, [11]. The novelty here is to show how easy they are to derive in the conformal approach using the key idea of reflection to perform inversion.

3   Intersections of Surfaces

In problems in computer graphics, robotics and inverse kinematics, large parts of the tasks involve intersecting lines, planes, circles, spheres, and indeed more general surfaces. In this section we will begin to put the formalism described so far to work in particular problems involving such intersections and hope to show that it provides a very elegant framework for carrying out these tasks.

Before looking at particular examples we will look briefly at representations of linear combinations of points. In order to take full advantage of the projective representation we should be able to consider linear combinations of points in the $\mathcal{A}(p,q)$ space. We know that a linear combination of two points a, b of the form

$$\lambda a + \mu b \quad \text{where} \quad \lambda + \mu = 1$$

gives another point on the line joining a and b. Similarly, a linear combination of 3 points, a, b, c of the form

$$\lambda a + \mu b + \nu c \quad \text{where} \quad \lambda + \mu + \nu = 1$$

gives another point on the plane containing a, b and c. The usual projective representation, where we go up just one dimension, has the advantage of still being linear in the representatives of the points e.g. if $x = \lambda a + \mu b + \nu c$ ($\lambda + \mu + \nu = 1$) then its 4d projective representation, X, can also be written in the form

$$X = \lambda A + \mu B + \nu C$$

Here we note that we have insisted that the point X is 'normalised', i.e. that $X = x + e$ rather than some multiple of this. With the conformal representation, working in $\mathcal{A}(p+1, q+1)$, we appear to have lost this advantage of linearity. For example, if A and B are the $\mathcal{A}(p+1, q+1)$ representatives of a and b, then in general

$$\lambda A + (1 - \lambda)B \neq \quad \text{a multiple of} \quad F(\lambda a + (1 - \lambda)b)$$

This is due to the presence of the $x^2 n$ term in the representation, which removes linearity. However, the following is true and is easy to show from the definition of $F(x)$:

$$F(\lambda a + (1 - \lambda)b) = \lambda A + (1 - \lambda)B + \frac{1}{2}\lambda(1 - \lambda)A \cdot Bn \tag{40}$$

We therefore see that the departure from linear behaviour is given by the addition of a multiple of the point at infinity. This is relatively benign behaviour and means that many of the techniques we use in the GA version of projective geometry will still work here. For example, this gives us another way of seeing that the equation for a line passing through points a and b is $X \wedge n \wedge A \wedge B = 0$ – the wedging with n knocks out the non-linear term $\frac{1}{2}\lambda(1 - \lambda)A \cdot Bn$ and we are left with the usual GA projective geometry result.

Precisely the same sort of thing goes through for a plane: let a, b, c define a plane and let

$$x = \alpha a + \beta b + \gamma c \quad \text{where} \quad \alpha + \beta + \gamma = 1$$

be a general point on the plane. Then it is easy to show that the representative of x, $X = F(x)$ satisfies

$$X = \alpha A + \beta B + \gamma C + \delta n \quad \text{where} \quad \delta = \frac{1}{2}(\alpha\beta A \cdot B + \alpha\gamma A \cdot C + \beta\gamma B \cdot C) \tag{41}$$

– again making it clear why the equation of the plane can be written as

$$X \wedge n \wedge A \wedge B \wedge C = 0$$

Note that in this section, and subsequent sections where we use the same multiples (α, β, γ etc) in $\mathcal{A}(p+1, q+1)$ space as in $\mathcal{A}(p, q)$ space, it is important that the representatives are taken as $F(a)$, $F(b)$ etc and not arbitrary multiples of these. What it amounts to is that we have to use **normalised** representatives (as referred to previously) satisfying

$$X \cdot n = -n \cdot \bar{n} = -2 \tag{42}$$

Working with these normalised points will also turn out to be useful shortly when we consider an alternative representation for spheres, circles etc.

3.1   Intersection of a line and a sphere

For reflection of a wavefront from a spherical surface, we need to find the intersection points of a sphere, S, and a line, L. Let the line be specified by $\mathcal{A}(3,0)$ points a and b and the sphere by $\mathcal{A}(3,0)$ points p, q, r and s. A general point, x, on the line is therefore given by

$$x = \lambda a + (1 - \lambda)b$$

where λ is a scalar. Writing as usual the $\mathcal{A}(4,1)$ representations of the points as the corresponding captial letters, e.g. $x \mapsto X$ etc., the 4-vector representing the sphere can be written as

$$\Sigma = P \wedge Q \wedge R \wedge S$$

Thus, taking $x = \lambda a + (1 - \lambda)b$ as a general point on the line, we can use equation (40) to write the intersection of the line and the sphere as

$$[\lambda A + (1 - \lambda)B + \frac{1}{2}\lambda(1 - \lambda)A \cdot Bn] \wedge \Sigma = 0 \tag{43}$$

If we write $\lambda = \frac{1}{2}\mu$ we obtain the following, symmetric form for the intersection equation

$$\left[-\frac{1}{2}\mu^2 A \cdot Bn + \mu(A - B) + \frac{1}{2}(A + B + \frac{1}{4}A \cdot Bn) \right] \wedge \Sigma = 0 \tag{44}$$

Multiplying this by $I = e_1 e_2 e_3 e \bar{e}$, we get the scalar quadratic in μ that we desire – this time with explicit coefficients.

3.2   Alternative representation for spheres and circles

We know that $X \wedge \Sigma = 0$ can be rewritten as

$$X \cdot (I\Sigma) = 0 \quad \implies \quad X \cdot \Sigma^* = 0$$

where $\Sigma^* = \Sigma I^{-1}$ is the **dual** to Σ and is a vector. This therefore suggests a very useful alternative representation for a sphere (or a circle in one dimension down), which we now discuss.

We know that for any two normalised points A and B

$$A \cdot B = -2(a-b)^2 \tag{45}$$

Thus, if X is a point on a sphere and C is its centre we know that we can write

$$X \cdot C = -2(x-c)^2 \equiv -2\rho^2$$

where ρ is the radius of the sphere. For a normalised point X this therefore implies that

$$X \cdot (C - \rho^2 n) = 0$$

since $X \cdot n = -2$. Comparing this with $X \cdot \Sigma^*$ we see that provided we normalise Σ^* after taking the dual, then we will find

$$\Sigma^* = C - \rho^2 n \tag{46}$$

Thus, the vector Σ^* encodes, in a very neat fashion, the centre and radius of the sphere. As an immediate application, writing equation (44) in dual form and multiplying out, gives the following explicit equation for the intersection points of a line through a and b with the sphere centre c and radius ρ:

$$\mu^2 A \cdot B + \mu(A-B) \cdot C + \frac{1}{2}(A+B) \cdot C - \frac{1}{4} A \cdot B + 2\rho^2 = 0 \tag{47}$$

Whether one wishes to use this form or the form given in equation(44) depends on whether it is most useful to specify the sphere by 4 points lying on it or by its centre and radius. Note also that given a Σ^* (via taking the normalised form of the dual of $\Sigma = P \wedge Q \wedge R \wedge S$) we can immediately get the radius from

$$\begin{aligned}(\Sigma^*)^2 &= (C - \rho^2 n)^2 \\ &= -2\rho^2 C \cdot n = 4\rho^2\end{aligned} \tag{48}$$

using the facts that $C^2 = 0$, $n^2 = 0$ and $C \cdot n = -2$. From this it then follows that $C = \Sigma^* + \frac{1}{4}(\Sigma^*)^2 n$. To summarise, from the vector form of the sphere, we can easily obtain the centre and radius as follows

$$(\Sigma^*)^2 = 4\rho^2 \tag{49}$$

$$C = \Sigma^* \left[1 + \frac{1}{4} \Sigma^* n \right] \tag{50}$$

4   Surface Evolution and Representation

Here our aims are threefold:

1. We would like to be able to represent a surface by piecewise spherical or planar patches – this would be useful in a variety of contexts.
2. We would like to be able to evolve a surface – here the idea is that a surface may have complicated features such as cusps or catastrophes, and we want to find a differentiable representation. We can then generate such a surface by evolution from some simple smooth starting point.
3. Wavefront propagation and reflection – this turns out to be linked with (2), and of course is useful in its own right.

These three are linked overall by the way geometric algebra help with thems, and particularly the conformal representation.

4.1   Surface representation

Here the machinery of the conformal representation turns out to be very useful. We start by triangulating the surface in the sense of putting down a grid of points where we identify groups of 3 points together with one interior point.

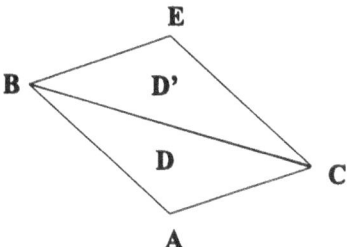

Fig. 1. Example of points for triangulation

If the interior point is to be taken as in the same plane as the other three points we will have a planar representation; if not, then we have a spherical representation. The conformal representation of the sphere is $S_1^* = A \wedge B \wedge C \wedge D$ say, so that S_1 is a vector and S_1^* is a 4-vector – note that if we compare with the notation used in previous sections, $\Sigma = S^*$. Now take another sphere, represented by the 4-vector S_2^* and the vector S_2, where $S_2^* = B \wedge C \wedge D' \wedge E$ where D and E are as shown in Figure 1. Now we find the equation of the intersection of these spheres. A point X on the intersection must simultaneously satisfy

$$X \cdot S_1 = 0 \quad \text{and} \quad X \cdot S_2 = 0$$

Now $X \cdot (S_1 \wedge S_2) = (X \cdot S_1) S_2 - (X \cdot S_2) S_1$ and we know $S_2 \neq \lambda S_1$ for any scalar λ (i.e. the spheres are distinct). Thus X lies on the intersection iff

$$X \cdot (S_1 \wedge S_2) = 0 \tag{51}$$
$$\Longleftrightarrow X \wedge [(S_1 \wedge S_2) I] = 0 \tag{52}$$

The intersection is thus the circle $C = (S_1 \wedge S_2)I$. This is a very neat way of finding the intersection in the case where the two spheres were originally specified by 4 points each (if the spheres were specified by the radius and centre, then we could do this fairly easily by conventional means). Indeed, given 4 points it is also not too hard to recover the conventional equation of a sphere in terms of its radius and centre, but things become slightly more complex when we are considering general 3d circles.

Given a general circle C (a trivector), how do we find its radius? (the radius of curvature of the intersection line in the above case). To do this we start with the unit circle in the $x - y$ plane and take as three points on it those shown in Figure 2. Now for any unit length vector, $\hat{x}$ say, we know that

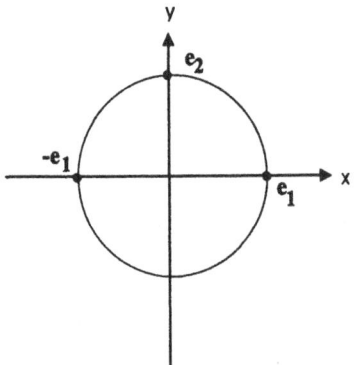

Fig. 2. Unit circle with three key points marked

$F(\hat{x}) = n + 2\hat{x} - \bar{n} = 2(\hat{x} + \bar{e})$, recalling the definition of $n, \bar{n}$ in equation 5. In particular we have

$$F(e_1) \wedge F(e_2) \wedge F(-e_1) = 16ie_3 \wedge \bar{e} = 16ie_3\bar{e}$$

where $i = e_1 e_2 e_3$ and we have used the facts that $e_1 \wedge e_2 = ie_3$ and $(ie_3) \cdot \bar{e} = 0$. Normalising, we can write $\hat{C} = ie_3\bar{e}$ ($\hat{C}$ meaning the unit circle, in this case, in the $x - y$-plane). Next, we note that

$$\frac{-\hat{C}^2}{(n \wedge \hat{C})^2} = 1$$

since $(ie_3\bar{e})(ie_3\bar{e}) = -1$ and $n \wedge \hat{C} = ie_3 e\bar{e}$. This therefore gives us a way of finding the radius of **any** circle centred on the origin. Let $C = D_\alpha \hat{C} \tilde{D}_\alpha$ where D_α is the dilatation rotor $D_\alpha = e^{\frac{\alpha}{2}e\bar{e}}$ introduced earlier, which satisfied

$$D_\alpha F(x) \tilde{D}_\alpha = e^\alpha F(e^{-\alpha}x)$$

Now, $D_\alpha n \tilde{D}_\alpha = e^{-\alpha}n$ (recall equations 22) and thus

$$n \wedge C = e^\alpha D_\alpha (n \wedge C) \tilde{D}_\alpha$$

since $D_\alpha(n \wedge C)\tilde{D}_\alpha = D_\alpha n \tilde{D}_\alpha \wedge D_\alpha C \tilde{D}_\alpha$. If we dilate the unit circle to a circle of radius ρ (so that $\rho = e^\alpha$) we thus find that

$$\frac{-C^2}{(n \wedge C)^2} = \rho^2 \tag{53}$$

Therefore, we can now position this circle anywhere we wish in 3d by applying the rotors for spatial rotations and translations. Neither of these involve further scale factors, so we deduce that $\frac{-C^2}{(n \wedge C)^2} = \rho^2$ is an identity for **any** circle C in 3d space. Note that $n \wedge C$ is the *plane* in which the circle lies (recall Section 1.1).

We now have the spheres and their intersections which can be used to represent the surface. The GA approach is again very useful here, since it enables us to retain full information for any point in space as to whether it is **inside** or **outside** any given spherical patch. This is because all orientation information is preserved and the representation for a hole differs from that of a sphere. Also for lines in space which we may wish to intersect with the surface, the fact that we can still use the $\lambda A + (1 - \lambda)B$ type construct (even though working in the conformal geometry) means that we can tell where an intersection occurs relative to an ordered list of points on the line. For example, for the above construct

$$\begin{aligned}
\lambda < 0 &\implies \text{intersection occurs past } B \\
0 < \lambda < 1 &\implies \text{intersection occurs between } A \text{ and } B \\
\lambda > 1 &\implies \text{intersection occurs before } A
\end{aligned}$$

Overall the GA conformal framework allows a systematic approach to be taken, based directly upon measured points on a surface and with fast lookup procedures to determine where a given space point or line lies with respect to the surface. This methodology is currently being implemented in a case of simulated surfaces coupled with ray-tracing to simulate radar reflection.

4.2   Surface evolution and wavefront propagation

We treat these two topics together since the particular technique we are investigating works in similar ways for each. Note a fuller account of the methods used here is in preparation – our main emphasis here will be on examples.

In [12] the problem of collisional avoidance in robotics was considered as an example of 'propagation' of surfaces. Suppose we have a surface $\mathcal{B}$ (we will describe shortly how we represent the surface in practice) which is the surface of a fixed obstacle. Let the surface of the robot interacting with this fixed surface be $\mathcal{A}$. In [12] it is shown that the collision avoidance problem for this robot is the same as the Huygens propagation of a wavefront $(-\mathcal{A})$ from the surface $\mathcal{B}$ (the minus just shows that we have to reverse the normal direction in moving between the two cases). If the wavefront $\mathcal{A}$ in fact just corresponds to reradiation of spherical wavelets ($\mathcal{A}$ is a sphere) then we have actual spherical wavefront propagation off $\mathcal{B}$, and the caustics (envelopes) of the wavefronts at successive times will be surfaces of constant phase in the geometric optics approximation. By using $\mathcal{A}$ (or $-\mathcal{A}$) to propagate $\mathcal{B}$, we 'evolve' $\mathcal{B}$ to successive new surfaces, which may be much more complicated than $\mathcal{B}$, and in particular by evolving an initially well-behaved surface we may be able to achieve one which exhibits the typical 'catastrophes' which occur for caustics in wavefront propagation. The method is still able to account for these in a fully differentiable manner however, since they are linked via a deterministic differentiable process to the initial differentiable surface.

In [12] a conformal geometry approach is employed to represent the surface. The key point of the method in [12] is to write the position p on the surface as a function of the set of normal directions m. In other words we regard the surface as 'indexed' by m and represent the position at a given value of m as $p[m]$ – the square brackets reflect the fact that this is not a single-valued function, since the same m may have several associated positions p if the object is not convex. Thus far, we do not need conformal geometry, but the next step is to write the representation as

$$\mathcal{R}(m) = m - n(p \cdot m) \tag{54}$$

Here, n is the null vector $e + \bar{e}$ introduced earlier. Thus, we adjoin to the normal m a multiple of a null vector given by the projection of p onto the normal. The neat feature of this is that we can now write

$$\mathcal{R}(m) = T_{-p} m \tilde{T}_{-p} \tag{55}$$

where $T_a = 1 + \frac{1}{2} na$, as earlier, is the conformal translation rotor. The result derived in [12] is that if we propagate a surface $\mathcal{B}(m)$ using the propagation function (wavefront) $\mathcal{A}(m)$ then the resulting surface, written as $\mathcal{A} \oplus \mathcal{B}(m)$ is described by the **composition** of the rotors corresponding to each surface individually:

$$T_{-p_{(\mathcal{A} \oplus \mathcal{B})}} = T_{-p_\mathcal{A}} T_{-p_\mathcal{B}}$$

This then suggests a spectral theory of surfaces with the indexing quantity as the normal direction and propagation as multiplication of 'direction spectra' (the translation rotors).

In any practical application of this formalism we have at some stage got to invert a given $\mathcal{R}(m)$ to find explicitly the current set of positions in the new surface. Since p enters into $\mathcal{R}(m)$ only via its projection on m, there is not sufficient information in $\mathcal{R}(m)$ itself to do this. In [12] this is solved by introducing derivatives of $\mathcal{R}(m)$. This enables $p[m]$ to be recovered, via quite a complicated inversion formula. Here we suggest a different technique, which achieves the required propagation much more simply and quickly, at least for the case of spherical wavefront propagation. We note that the problem we are addressing here is essentially the same as that considered in the *level set* method (see, e.g. [13]) applied in this instance to propagation with a constant velocity.

We explain the technique via two examples. First consider an initial surface consisting of a parabola in 2d as shown in Figure 3. We wish to carry

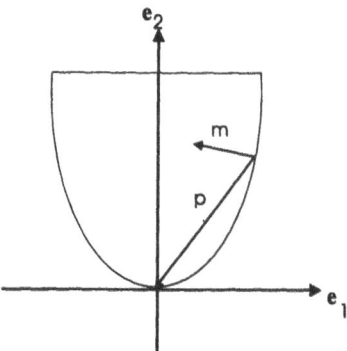

Fig. 3. Simple initial parabolic surface in 2d

out spherical Huygens propagation starting from this surface. It is easy to see that for a sphere, radius ρ and centre c, the representation is:

$$\mathcal{R}(m) = m - n(c \cdot m - \rho)$$

(this is because the inward pointing normal, the one we are using by convention, is $m = \frac{(c-p)}{\rho}$ for a point on a sphere).

The Huygens construction effectively recentres each sphere on the points of the surface we wish to propagate and the construction in terms of composing rotors effectively tells us to add the ps at the same m (composition of rotors just corresponds to the sum of the translations). Thus the propagated $\mathcal{R}(m)$, starting from $\mathcal{R}(m) = m - (p \cdot m)n$ is

$$\mathcal{R}^{prop}(m) = m - [(p \cdot m) - \rho]n$$

For the specific example of propagating a parabola, we can write

$$p = te_1 + \frac{1}{2}t^2 e_2 \tag{56}$$

$$m = \frac{te_1 - e_2}{\sqrt{1 + t^2}} \tag{57}$$

where t is a parameter (actually equal to the x-coordinate of a position). We find we can express $p \cdot m$ as $-\frac{1}{2}\frac{(m \cdot e_1)^2}{m \cdot e_2}$ and thus obtain

$$\mathcal{R}^{prop}(m) = m + n\left(\rho + \frac{1}{2}\frac{(m \cdot e_1)^2}{m \cdot e_2}\right) \tag{58}$$

It is an expression like this which needs to be inverted to find $p[m]$ using the differential $\mathcal{R}^{prop}(m)$ as in [12].

But this is actually much more complicated than we need. Instead one can note that propagating spherical wavefronts corresponds to geometric optics, and therefore just ray tracing, with rays normal to the wavefronts. All we need is to move out from any given position along the normal at that position by a distance ρ, in order to achieve the same effect as above. For example, in the parabola case, let us write

$$m = \cos\theta e_1 + \sin\theta e_2$$

and specify this as the inward normal by taking θ in the range $-\pi < \theta < 0$. (The link to the parameter t is $\cos\theta = t/\sqrt{1+t^2}, \quad \sin\theta = -1/\sqrt{1+t^2}$). Then we have that

$$p = -\cot\theta e_1 + \frac{1}{2}\cot^2\theta e_2$$

and the propagated position is just

$$p - \rho m = (-\cot\theta - \rho\cos\theta)e_1 + (\frac{1}{2}\cot^2\theta - \rho\sin\theta)e_2$$

The surfaces found this way are plotted in Figure 4 for $\rho = 0, 1, 2$. We see

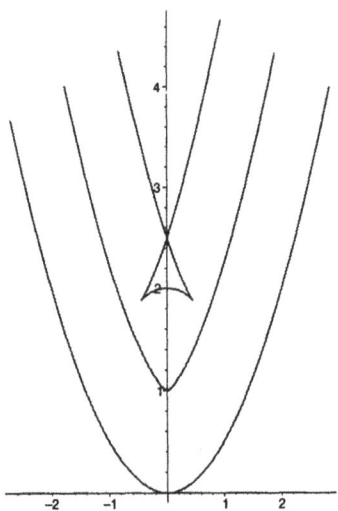

Fig. 4. Circular wavefront propagation of an initial parabola. Note the development of a 'swalowtail catastrophe'.

immediately that we have an initially smooth differentiable surface ($\rho = 0$), developing a cusp ($\rho = 1$) and then a swallowtail catastrophe ($\rho = 2$).

As a second example, also considered in [12], let us consider an initially cardioid-type surface, and propagate this. The initial cardioid is shown in Figure 5. The equation we use is

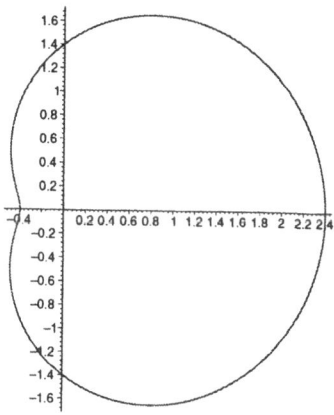

Fig. 5. Initial configuration of the cardioid-like shape used for circular wavefront propagation

$$p_x = a \left(\frac{1}{2} + b \cos \theta + \frac{1}{2} \cos 2\theta \right) \tag{59}$$

$$p_y = a \left(b \sin \theta + \frac{1}{2} \sin 2\theta \right) \tag{60}$$

i.e. an expression in terms of circular harmonics. The corresponding inward pointing normal is found to be

$$m = \frac{-1}{\sqrt{1 + 2b \cos \theta + b^2}} \{ (b \cos \theta + \cos 2\theta) e_1 + (b \sin \theta + \sin 2\theta) e_2 \} \tag{61}$$

Forming $p - \rho n$ as before, we get a succession of surfaces which initially collapse inwards, pass through each other and then eventually propagate outwards in a more or less circular fashion except for a swallowtail catastrophe. This matches what was found in [12] using the inversion approach.

One can also, of course, work in 3d. The above examples have been in 2d, but all of the formulae are 3-dimensional and may be applied to any initial surface for which we can find parametric representations of p and m. Figure 6 shows an example of a 3d cardioid surface which has been propagated in this fashion. Since the apparatus developed in [12] seems to be unnecessary at least for this spherical wavefront propagation, it might be wondered where the conformal representation, or indeed geometric algebra, enters into this problem. The answer lies in what happens if we want to **reflect** a developing wavefront/surface off another surface. Figure 7 shows what happens when

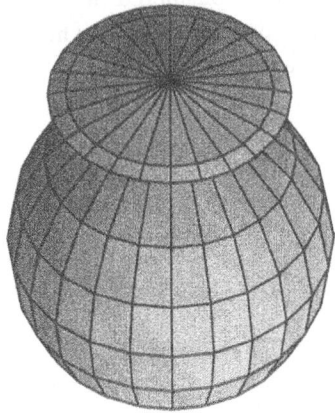

Fig. 6. A frame from the 3d development of cardioid propagation

the developing cardioid figure reflects off a plane surface. This is in 2d, but we also have a 3d demonstration where an initially cylindrically symmetric propagating surface is reflected off an offset spherical surface, resulting in a quite general 3d shape, see Figure 8 The way in which these examples were computed used the GA formula for reflection in 3d;

$$a' = -mam$$

for the reflection of a vector a to a vector a' in the local normal m, together with the above results for the intersection of lines and planes and lines and spheres etc in the conformal geometry. The 'lines' in this case are the 'rays' coresponding to normals to the wavefront. The conformal approach enables one to have a straightforward synthetic algorithm for all these computations, which is basically very simple, but generates quite impressively sophisticated results. Extensions to non-uniform propagation velocity are currently being considered.

References

1. Hestenes, D. and Sobczyk, G. (1984) Clifford Algebra to Geometric Calculus: A unified language for mathematics and physics. D. Reidel, Dordrecht.
2. Hestenes, D. (1999). New Foundations for Classical Mechanics, Kluwer Academic Publishers, 2nd edition.
3. Doran, C.J.L. and Lasenby, A.N. (1999). Physical applications of geometric algebra, Available at http://www.mrao.cam.ac.uk/~clifford/ptIIIcourse/.
4. Lasenby, J., Fitzgerald, W., Lasenby, A. and Doran, C., (1998). New geometric methods for computer vision: An application to structure and motion estimation, International Journal of Computer Vision, **26**, 3, 191–213.

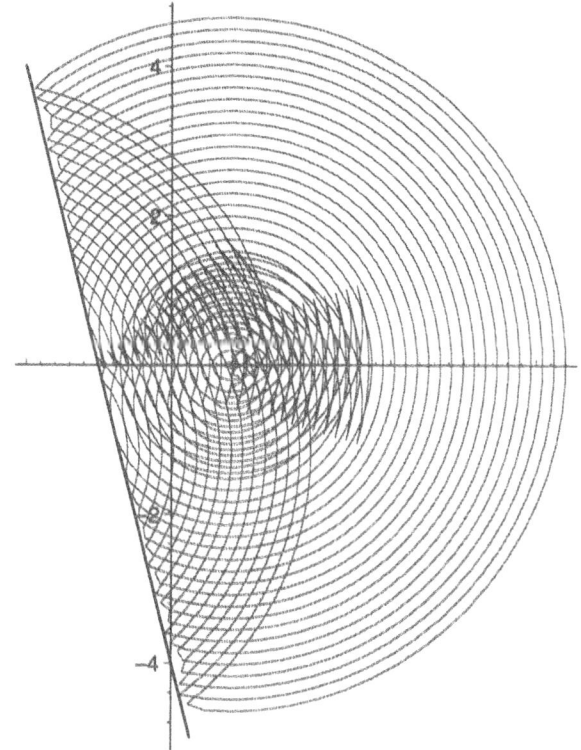

Fig. 7. Reflection of 'cardioid' wavefront from a plane sheet

5. Dorst, L., Mann, S. and Bouma, T. "GABLE: A MAT-LAB tutorial for geometric algebra," (2000). Available at `http://www.carol.wins.uva,nl/~leo/clifford/gablebeta.html`.

6. Lasenby, J. and Bayro-Corrochano, E.D. (1999) Analysis and Computation of Projective Invariants from Multiple Views in the Geometric Algebra Framework. International Journal of Pattern Recognition and Artificial Intelligence – Special Issue on Invariants for Pattern Recognition and Classification, **13**, No.8, p1105, ed. M.A. Rodrigues.

7. Hestenes, D. and Ziegler, R. (1991) Projective Geometry with Clifford Algebra. Acta Applicandae Mathematicae, **23**: 25–63.

8. Li, H., Hestenes, D. and Rockwood, A. (Summer 2000) Generalized Homogeneous Coordinates for Computational Geometry. in G. Sommer, editor, Geometric Computing with Clifford Algebras. Springer.

9. Hestenes, D. (Summer 2000) Old wine in new bottles: a new algebraic framework for computational geometry. To appear in Advances in Geometric Algebra with Applications in Science and Engineering, eds. Bayro-Corrochano, E.D. and Sobcyzk, G. Birkhauser Boston.

10. Dress, A. and Havel, T. (1993) Distance geometry and geometric algebra. Foundations of Physics, **23**, 1357-1374.

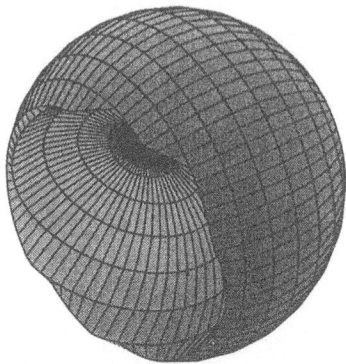

Fig. 8. Stage in propagation of an initially cylindrically symmetric cardioid-shaped 3d wavefront reflecting off an offset sphere

11. Brannan, D.A, Esplen, M.F. and Gray, J.J. (1999) *Geometry* Cambridge University Press.
12. Dorst, L. (Summer 2000) Objects in contact: boundary collisions as geometric wave propagation. To appear in Advances in Geometric Algebra with Applications in Science and Engineering, eds. Bayro-Corrochano, E.D. and Sobcyzk, G. Birkhauser.
13. Gomes, J. and Faugeras, O. (1999) Reconciling distance functions and level sets. Scale Space'99. LNCS **1682**, 70-81, eds. M. Nielsen *et al.*. Springer-Verlag, Berlin Heidelberg.

Interactive Design of Complex Mechanical Parts using a Parametric Representation

Hassan Ugail, Michael Robinson, Malcolm I. G. Bloor, and
Michael J. Wilson

Department of Applied Mathematics, The University of Leeds Leeds LS2 9JT
United Kingdom

Summary. In CAD, when considering the question of new designs of complex
mechanical parts, such as engine pistons, a parametric representation of the design
is usually defined. However, in general there is a lack of efficient tools to create and
manipulate such parametrically defined shapes.

In this paper, we show how the geometry of complex mechanical parts can be
parameterised efficiently enabling a designer to create and manipulate such geometries within an interactive environment. For surface generation we use the PDE
method which allows surfaces to be defined in terms of a relatively small number
of design parameters. The PDE method effectively creates surfaces by using the
information contained at the boundaries (edges) of the surface patch. An interactively defined parameterisation can then be introduced on the boundaries (which
are defined by means of space curves) of the surface. Thus, we show how complex
geometries of mechanical parts, such as engine pistons, can be efficiently parameterised for geometry manipulation allowing a designer to create alternative designs.

1 Introduction

There can be of no doubt that in the design and manufacture of complex
mechanical parts the trend nowadays is towards the extensive use of computer aided techniques. There exist many Computer-Aided Design (CAD)
systems, which use a variety of geometry representation schemes, e.g. B-
Rep [8] , CSG [8], feature based modelling [[9] and [11]] and variational methods [[10], [5] and [7]]. Examples of existing commercial CAD systems include
CATIA from Dessault Systemme, Pro/Engineer from Parametric Technology
and PowerSHAPE from Delcam International. Using standard commercial
CAD packages, complex mechanical parts can be built up from an intersecting series of geometric solids which form 'primary' surfaces and rolling
ball blend surfaces between the primary surfaces [[16] and [17]]. Thus, existing commercial CAD systems can create and manipulate the geometry of
complex mechanical parts, although the design process may not always be
straightforward.

As Imam [6] notes, when considering the design of engineering surfaces, it
is essential that CAD systems, used to create the design, be able to param-
eterise the geometry of the design efficiently. In parametric design, the basic
approach is to develop a generic description of an object or class of objects,

in which the shape is controlled by values of a set of design variables or parameters. A new design, created for a particular application is obtained from this generic template by selecting particular values for design parameters so that the item has properties suited to that application. As noted above, it is often the case that commercial systems often fail to generate a parametric model of the complex mechanical part in question. This is often due to the lack efficient tools which enable to define complex geometries in the form of a generic parametric model.

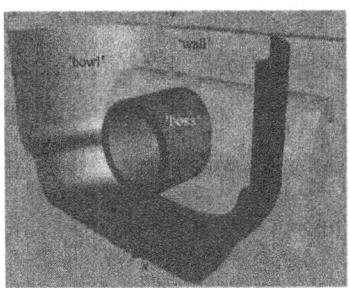

Fig. 1. The basic components of a traditional model of a piston

Consider a typical design for the inside of a piston of an internal combustion engine. The traditional design is constructed from a series of relatively simple geometric parts. Fig. 1 shows a typical design of the piston where the internal 'bowl' and the 'boss' form the major components of the piston. Using a standard CAD package, the geometry of the piston can be generated as follows.

The internal bowl of the piston is constructed by rotating a given profile through 2π about z-axis. This is intersected by two surfaces, formed by translating a second profile in the direction of the x-axis and is referred as the piston wall. The walls are intersected by two cylinders which form the boss. The result of this is shown in Fig. 1, where the other half of the (symmetrical) part is removed for the sake of clarity. The basic primary surfaces then have blends added at all sharp edges and various cuts made into them. The blends are traditionally formed as simple 'rolling ball' blends [12] .

An important point to note in the design process outlined above is the lack of consistency in creating different parts of the design. More importantly, the design process lacks the ability to easily change the design in order to take alternative designs into consideration. In other words, there is a lack of a flexible enough parametric representation to easily create alternative designs.

The aim of this paper is to show how a PDE parametric model [15] can be used to design and manipulate complex mechanical parts. The design parameters of the PDE parametric model are those introduced via the boundary conditions which specify the shape of the surface. Adjustments of the design parameters allow the user to select from a whole range of possible designs once an initial generic parameterisation has been specified. Often the initial design will be 'close' to the desired final design, so that adjustments of the design parameters will allow user to fine tune the surface shape once an initial approximation of the desired surface has been generated.

2   The PDE Method and Geometry Generation

In geometric design, it is common practice to define curves and surfaces using some form which represents the surface parametrically. Thus, surfaces are defined in terms of two parameters u and v so that any point on the surface $\underline{X}$ is given by an expression of the form:

$$\underline{X} = \underline{X}(u, v). \tag{1}$$

Equation (1) can be viewed as a mapping from a domain Ω in the (u, v) parameter space to Euclidean 3-space. In the case of the PDE method this mapping is defined as a partial differential operator:

$$L^m_{uv}(\underline{X}) = \underline{F}(u, v), \tag{2}$$

where the partial differential operator L is of degree m. Thus, effectively, surface design is treated as an appropriately posed boundary-value problem with appropriate boundary conditions imposed on $\partial\Omega$, the boundary of Ω. The partial differential operator L is usually taken to be that of elliptic type and the degree m of this operator depends on the level of surface control and continuity required on the shape of the surface. The function $\underline{F}(u, v)$ is included for completeness and is generally taken to be zero.

The PDE method has been discussed before by a number of different references, e.g. [16], [1] and [2]]. It has been shown how surfaces satisfying a wide range of functional requirements can be created by a suitable choice of the boundary conditions and appropriate values for the various design parameters associated with the method [[15], [13] and [4]].

2.1   Interactive Design

For the work described here, and for the majority of work carried out on the PDE method described elsewhere, the PDE chosen is of the form:

$$\left(\frac{\partial^2}{\partial u^2} + a^2 \frac{\partial^2}{\partial v^2} \right)^2 \underline{X}(u, v) = 0, \tag{3}$$

where the condition on the function $\underline{X}(u, v)$ and its normal derivatives $\frac{\partial X}{\partial n}$ can be imposed at the edges of the surface patch. The parameter a is a special design parameter which controls the relative smoothing of the surface in the u and v directions [2]. For periodic boundary conditions (e.g. $0 \le u \le 1$,

$0 \leq v \leq 2\pi$), a pseudo-spectral method has been developed for the solution of equation (3) which allows $\underline{X}(u,v)$ to be expressed in closed form [3].

As far as interactive design is concerned the boundary conditions are often defined in terms of curves in 3-space. For example, Fig. 2 shows a typical set of boundary curves and the corresponding PDE surface showing the port of a bifurcated transfer port of a 2-stroke engine. Here the value of a was taken to be 1.0100. Note that the curves marked p_1 and p_2 correspond to the boundary conditions on the function $\underline{X}(u,v)$. A vector field corresponding to the difference between the points on the curves marked p_1 and p_2 and those marked d_1 and d_2 respectively, corresponds to the conditions on the function $\frac{\partial X}{\partial n}$ such that

$$\frac{\partial X}{\partial n} = \left[\underline{p}(v) - \underline{d}(v)\right] s, \tag{4}$$

where s is a scalar. The conditions defined by p_1, p_2 and d_1, d_2 are known as the 'positional boundary conditions' and 'derivative boundary conditions' respectively [13]. Note that the surface patch will not necessarily pass through the curves which define the derivative boundary conditions.

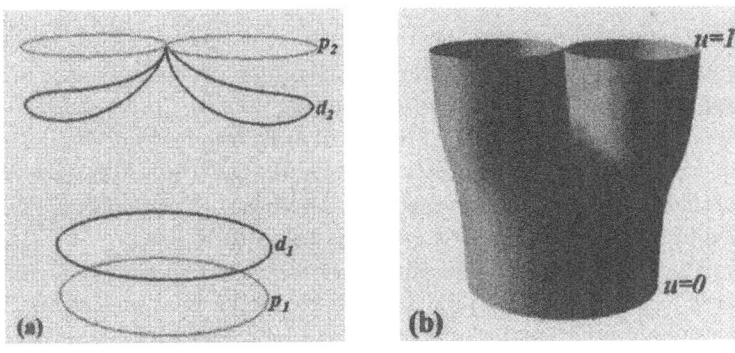

Fig. 2. Typical PDE surface. (a) The boundary curves. (b) The corresponding PDE surface patch

2.2   Interactive Design using the PDE Parametric Model

In this work the definition of the shape geometry is carried out using the PDE parameter model discussed in [15], where the parameterised boundary curves are used to define the shape of the surface. Essentially, this parameterisation is defined in such a way that linear transformations, such as translation, rotation and dilation, of the boundary curves can be carried out interactively which in turn result in a change in the shape of the surface. The result of this is that the designer is presented with tools which enable him/her to create and modify the geometry an intuitive manner. For convenience, the parameterisation on the boundary curves are denoted using the notation c_{kP_i} $(k = 1, 2)$, $(i = x, y, z)$. Here c defines the curve, with the letter p denoting the position curves and the letter d denoting the derivative curves. The index k ranges from 1 to 2 denoting the $u = 0$ and $u = 1$ boundary edges (respectively) of the surface. The letter P denotes the type of parameter, T for a translation, R for a rotation and D for a dilation. Finally the letter i denotes the coordinate directions relevant to a particular type of parameter. Adjustments to the values of these parameters along with the value of a in equation (3) can be used to create and manipulate complex geometries.

As mentioned earlier, the effect of these parameters on the surface shape is very intuitive. For example, Table 1 shows the values of the chosen parameters for $d = 1$ for the surface shown in Fig. 2.

Table 1. Values for the design parameters for the boundary $d = 1$ of the surface shown in Fig. 2

parameter	value
d_{1T_x}	0.000
d_{1T_y}	-0.850
d_{1T_z}	0.000
d_{1D_x}	0.771
d_{1D_y}	0.792
d_{1D_z}	0.000
d_{1R_x}	3.138
d_{1R_y}	0.000
d_{1R_z}	0.000

In order to show the effect of the design parameters we now choose a different set of values for the parameters for the boundary $d = 1$ of the surface shown in Fig. 2. The new values chosen for the parameters are shown in Table 2 and the resulting surface is shown in Fig. 3. Note the value of the a for the surface shown in Fig. 3 is the same as that shown in Fig. 2, i.e. $a=1.0100$. Essentially, the new values of the parameters produced a dilation followed by a translation which is followed by a rotation of the boundary curve $d = 1$. The parameters introduced on the boundary curves are varied using a graphical interface where the corresponding surface is visualised simultaneously. The spectral approximation method to the solution of the PDE, mentioned earlier, is fast enough for the surfaces to be created and manipulated in real time.

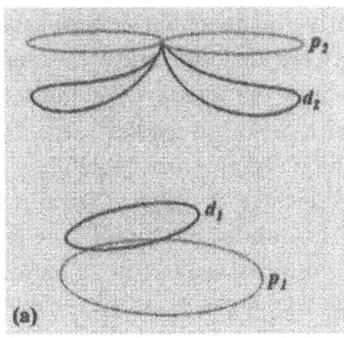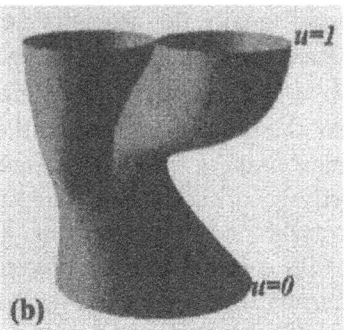

Fig. 3. The effect on the shape of the surface by changing the design parameters corresponding to the boundary $d = 1$. (a) The boundary curves. (b) The corresponding PDE surface patch

Table 2. Values for the design parameters for the boundary $d = 1$ of the surface shown in Fig. 3

parameter	value
d_{1T_x}	-0.310
d_{1T_y}	-0.850
d_{1T_z}	0.000
d_{1D_x}	0.371
d_{1D_y}	0.792
d_{1D_z}	0.000
d_{1R_x}	3.138
d_{1R_y}	0.000
d_{1R_z}	0.220

In order to build shapes corresponding to complex mechanical parts, more than one surface patch often needs to be joined together with common boundaries enabling to form a composite surface [[14] and [13]]. The parametric model discussed above has been extended to cater for such composite bodies. For example, Fig. 4 shows a bifurcated port for a modern internal combustion engine. This composite surface has been created using three surface patches, where the final shape resulted in the manipulation of the design parameters introduced onto the boundary curves defining the composite body.

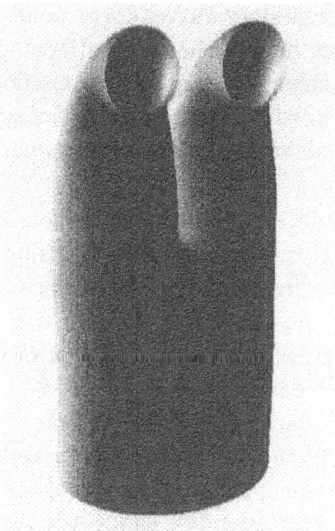

Fig. 4. A bifurcated port for a modern internal combustion engine created interactively.

3 Interactive Design and Manipulation of an Engine Piston

In this section we show how a generic shape of a typical piston of an internal combustion engine can be created, i.e. we show how a typical design can be created from 'scratch' and show how such a design can then be modified using the PDE parametric model discussed earlier.

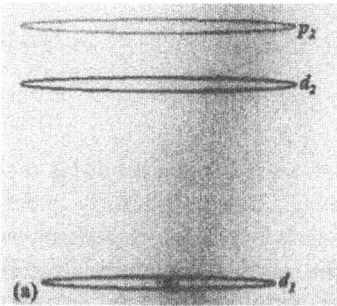
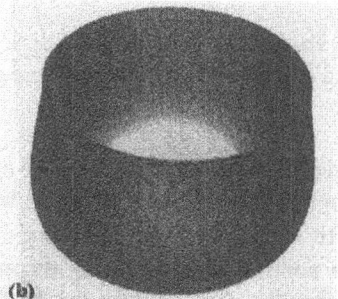

Fig. 5. Generic geometry of the bowl of the piston. (a) The boundary curves. (b) The PDE surface patch corresponding to the bowl of the piston

Fig. 5(a) shows the boundary curves corresponding to the surface shown in Fig. 5(b). This surface defines the internal bowl of the piston. Note that the boundary curve p_1 corresponding to the positional boundary condition at $u = 0$ of the internal bowl is scaled down to a single point. This is carried out by choosing very small values for the dilation parameters p_{1D_x}, p_{1D_y} and p_{1D_z}.

In order to accommodate the boss, part of the internal bowl of the piston needs to be removed. This is carried out by using a curve drawn on the (u, v) parameter space corresponding the surface. Since the points in $\Re^3$ corresponding the (u, v) points of any curve drawn on the parameter space is guaranteed to lie on the surface, the shape of the portion to be removed from the surface can be determined by choosing an appropriate shape of the curve on the (u, v) parameter space. In the case of the shape shown in Fig. 6(b), the shape of the corresponding curve on the (u, v) parameter space is an ellipse. Once the appropriate shape of the curve on the parameter space is determined, the corresponding shape can be removed. This is carried out by discarding the points which belong to interior of the curve in the (u, v) parameter space and solving the original PDE with a reparameterisation accounting for the curve drawn on the (u, v) parameter space. Further details of how this can be carried out is discussed in [14].

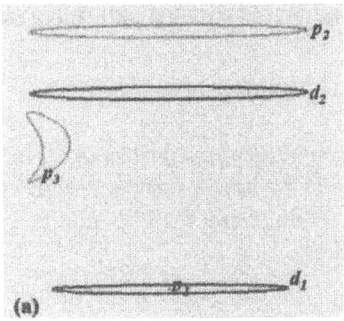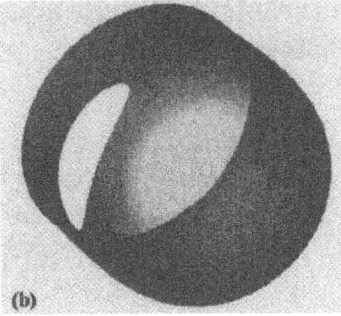

Fig. 6. Part of the bowl removed to accommodate boss. (a) The boundary curves. (b) The corresponding PDE surface patch

Once the desired portion of the surface from the internal bowl is removed, a second surface corresponding to the boss can be created in the usual manner with the curve in $\Re^3$ corresponding to that in the (u, v) parameter space for the portion removed being as one of the positional boundary conditions. Fig. 7(b) shows a generic design for the piston where the corresponding boundary curves are shown in Fig. 7(a).

Once the initial geometry of the piston is created in the manner described above, the PDE parametric model can be used to manipulate the geometry.

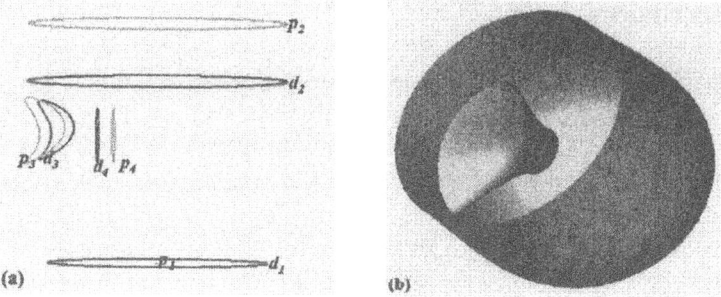

Fig. 7. Main components of piston created using two surface patches. (a) The PDE boundary curves. (b) The PDE surface patches corresponding to piston (the bowl and boss)

This enables one to create alternative designs starting with an initial design. For example, consider the shape of the pistons shown in Fig. 8. Here the geometry of the boss has been changed. This alteration from the shape shown in Fig. 7(b) is carried out using the dilation parameters p_{kD_z} and d_{kD_z} with $k = 1, 2$, i.e. the boundary curves corresponding to the surface patch have been scaled along z-direction. Note that the curve on the (u, v) parameter space since the boundary p_3 corresponds to the hole in the surface patch corresponding to the bowl of the piston, the dilation in this case is carried out on the (u, v) parameter space. Thus, during the manipulation of the boundary condition p_3, of the piston boss, the surface corresponding to the bowl need to be recalculated.

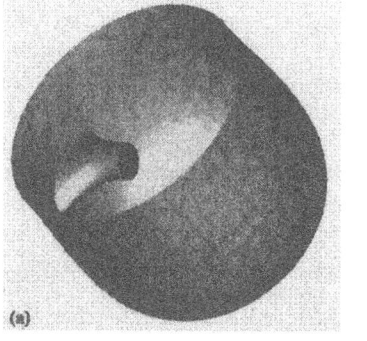 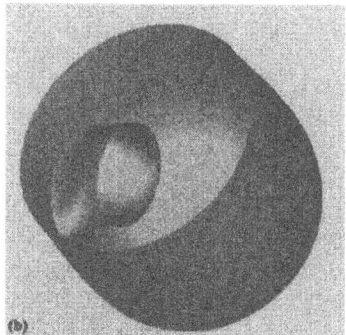

Fig. 8. Alternative designs of the boss of the piston

4 Conclusion

In this paper we have shown how a parameterisation of a given geometry can effectively make the design and manipulation of complex mechanical parts very intuitive. In particular, we have shown how the PDE parametric model allows complex designs to be created and manipulated in an interactive environment. Since the boundary conditions play a vital role in determining the shape of the surface, the design parameters have been introduced on the boundary curves which define the shape. These parameters are chosen in such a way that simple transformations of the boundary curves can be carried out. The reason for choosing such a model is that the design parameters introduced in this manner allows a designer to easily create and manipulate complex geometries without having to know the mathematical details of the solutions of PDEs.

The PDE parameteric model discussed in this paper mainly allow global maninpulations of the shapes in terms of the boundary conditions which define them. It is an intention to extend the model to cater for local manipulation of the surfaces too.

5 Acknowledgements

The authors would like to acknowledge the support of EPSRC Grant GR/L05730 and thank Michael Hildyard of AEG Automative for his interest in the work.

References

1. Bloor, M. I. G., and Wilson, M. J. (1989) Generating blend surfaces using partial differential equations, Computer-Aided Design, **21**, 165–171
2. Bloor, M. I. G., and Wilson, M. J. (1990) Using Partial Differential Equations to Generate Freeform Surfaces, Computer Aided Design **22**, 202–212
3. Bloor, M.I.G., and Wilson, M.J. (1996) Spectral Approximations to PDE Surfaces, Computer-Aided Design, **28**, 145–152
4. Dekanski, C.W., Bloor, M.I.G., and Wilson, M.J. (1995) The Generation of Propeller Blades Using the PDE Method, Journal of Ship Research, **39**, 108–116
5. Hagen, H., Schulze, G. (1990) Variational Principles in Curve and Surface Design. In Hagen, H., and Roller, D., (Eds.), Geometric Modelling, Springer, Berlin, pp. 161–184
6. Imam, M.H. (1982) Three-Dimensional Shape Optimisation, International Journal for Numerical Methods in Engineering, **18**, 661–673
7. Kallay, M. (1993) Constrained Optimisation in Surface Design. In Falcidieno, B., and Kunii, T.C., (Eds.), Modelling in Computer Graphics, Springer, Berlin, pp. 85–93
8. Mortenson, M.E. (1985) Geometric Modelling. Wiley, New York
9. Nakajima, N., and Gossard, D. (1982) Basic Study in Feature Descriptor, MIT CAD Technical Report, Massachusetts Institute of Technology, Cambridge, MA

10. Nowacki, H., and Rees, D. (1983) Design and Fairing of Ship Surfaces. In Barnhill, R.E., and Boehm, W., (Eds.) Surfaces in CAGD, North-Holland, Amsterdam, pp. 121–134.

11. Roller, D. (1989) Design Features: An approach to High Level Shape Manipulations. Computers in Engineering, **12**, 185–191

12. Rossignac, J.R., and Requicha, A.A.G. (1984) Constant Radius Blending in Solid Modelling. Computers in Mechanical Engineering, **3**, 65–73

13. Ugail, H., Bloor, M. I. G. and Wilson, M. J. (1998) On Interactive Design Using the PDE method, Mathematical Methods for Curves and Surfaces II, (eds) M. Dæhlen, T. Lyche, and L. L. Schumaker, Vanderbilt University Press Nashville TN, 493–500

14. Ugail, H., Bloor, M.I.G., and Wilson, M.J. (1999a) Techniques for Interactive Design Using the PDE Method, ACM Transactions on Graphics, **18**, (2), 195–212

15. Ugail, H., Bloor, M.I.G., and Wilson, M.J. (1999b) Manipulations of PDE Surfaces Using an Interactively Defined Parameterisation, Computers and Graphics, **24**, (3), 525–534

16. Vida, J., Martin, R. R., and Varady, T. (1994) A survey of blending methods that use parametric surfaces, Computer-Aided Design, **26**, (5), 341–365

17. Woodward, C.D. (1987) Blends in Geometric Modelling. In Martin, R.R. (Ed.), Mathematical Methods of Surfaces II, Oxford University Press, UK, pp. 255–297

Surfaces in the Mind's Eye

Jan Koenderink, Andrea van Doorn, and Astrid Kappers

Universiteit Utrecht, Buys Ballot Laboratorium,
PO Box 80000, 3508TA Utrecht, The Netherlands

Summary. "Pictorial reliefs" are surfaces of objects in pictorial space, the mental entity that happens to human observers when looking at (or rather: into) pictures. We discuss the structure of pictorial relief and pictorial space. It is determined by the nature of various "depth cues" which are intrinsically ambiguous. General arguments suggest that the structure of pictorial relief is invariant under a certain group of affinities that preserve planarity in general and the picture plane in particular. The subgroup of depth scalings has already been identified as of crucial importance in vision by sculptors. We have also found evidence of the occurrence of shears involving both the picture plane dimensions and depth. Observers apparently use these to "adjust their mental perspective".

1 Introduction

Most people think nothing of it when they look at a photograph and perceive a "physical scene". Yet no other species (except maybe some other primates) has this ability. Moreover, photographs are mere pieces of paper covered with pigments in a certain simultaneous order. Perhaps the paper was painted. Then you don't look at some physical record, but at the expression of somebodies figment of imagination. Perhaps the paper was simply overgrown with darkish fungus in a peculiar pattern. Then the scene exists only in the mind of the observer. You can't blame the fungus.

Pictures are an important aspect of our culture. We look *at* a picture and see a piece of paper, but we also look *into* "pictorial space". Pictorial space contains pictorial objects whose pictorial surfaces arrest the movement of our vision into pictorial space. It is the layout of pictorial objects and the attitude and shape (often called "pictorial relief") of their surfaces that structure our perception.

2 Pictorial Relief

2.1 Cues

"Pictorial relief" is a *mental* entity. One wonders how it relates to the world and whether one may "measure" it in some objective sense. These are important questions that have occupied philosophers and psychologists for ages.

A major dichotomy in vision is that between the "visual field" and the "visual world". Both are mental entities. The visual field is a two dimensional simultaneous order of colors. A painting is a good model when not simply understood as a canvas covered with paints. A painting is a statement that has to be read, thus the painting *as I see* it is also a mental entity. The painting as a physical object and a painting as I see it roughly stand in the relation as a peculiar structure of wood pulp covered with carbon particles to my seeing written language and even my understanding of a poem. A monkey doesn't see written language and I don't understand a Chinese poem.

My "visual world" is a three dimensional scene as it is seen by me. Its parts are *visual things*, not colors. Apart from entities being simultaneously present (as in the visual field) I also have a radial, or depth, order. Things have spatial attitudes, they may partly occlude each other, and so forth. Various people (perhaps most cogently the remarkable bishop Berkeley) have argued that visual worlds are figments of the imagination because visual fields cannot possibly specify visual worlds. Berkeley[2] argued that elements of the visual world can only be due to the interpretation of essentially arbitrary signs ("depth cues"). For instance, when there is a small man, next to a large one in my visual field, I see the small one as farther away. Such cues are (both phylogenetically and ontogenetically) learned by arbitrary association. Some cues became actually part of the hardware of many species.

Nowadays many of these cues are reasonably well understood. They are increasingly applied in computer vision. Examples are the binocular disparity cue, the shading cue, the contour cues, the occlusion cues, and many others. The fact that we have good theories doesn't undermine Berkeley's conclusion that the cues as our brain applies them became part of my mental makeup through arbitrary association though.

2.2   Operationalization

Is it possible to measure pictorial relief in any objective sense? This is an important question of methodology in experimental psychology. In fact, there exists a large corpus of methods[6]. Classically one measured discrimination thresholds (show two surfaces, ask "which surface is flattest, answer either this or that"), used scaling methods ("how curved is this surface on a scale from one to seven?"), or used methods of replication ("place this board in the same attitude with respect to your body as you perceive yonder surface"). Only very recently methods have been developed that yield actual geometrical entities as psychophysical responses.

A generic method[11] is to use some type of "gauge object" (see figure 1). This is what the carpenter does: Hold a yardstick next to an object, apply a protractor, take an object in compasses, and so forth. In all cases the carpenter judges the "fit". In a similar way one may place a picture of a gauge object in a picture of a scene and ask whether it "fits". Here is an example: Superimpose the wire frame rendering of an ellipse upon the picture of an

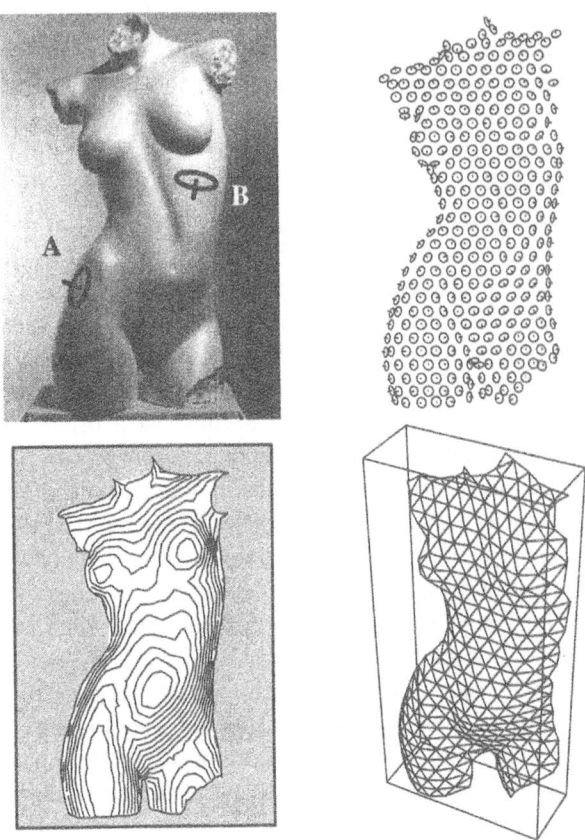

Fig. 1. *A simple operationalization of "pictorial relief", introduced in reference [11]. On the upper left we show a picture, superimposed on it are two "gauge figures". You are supposed to see at first blush that the one at A "fits", the one at B miserably fails to fit. After probing many locations in random order we obtain the results shown on top right. Integration yields the depth field at bottom left, also shown as a surface in "pictorial space" at bottom right. One of the dimensions in the latter figure is "depth".*

object. Ask the observer to adjust orientation and eccentricity of the ellipse such that it "looks like a circle painted upon the pictorial surface". Most people find this task easy. Interpret the ellipse as the projection of a circle and thus *operationally define* the "spatial attitude of the local pictorial surface element". This yields a local sample. You may repeat the process to sample a large number of locations. Thus you obtain a field of empirical "pictorial surface normals". You may check this field for integrability. Remarkably, we find that it is always integrable within the scatter revealed by repeated trials: Apparently the observers sample some kind of "surface" that exists *in their minds.* Integrate the field and you have one particular operationalization of the "pictorial relief" for the given observer, for the given picture. It is as easy as that.

This is one example. Many others are possible. We have gained experience with about half a dozen such methods in recent years[12,4].

3   Ambiguity and the Notion of Veridicality

3.1   The Veridicality Issue

Consider a physical scene including a photographer taking a picture. The picture will be *of* that "fiducial" scene. Do a psychophysical experiment using the picture as a stimulus. The response is another scene, say the surface of some object. Is the response like the real scene? (See figures 2 and 3.) This is the "veridicality" question. If the response is like the real scene one says that the "perception is veridical". Most of us don't doubt that our perceptions are typically veridical. Many people find it disturbing when the response turns out *not* to be veridical. They suspect the observer of having fallen victim to some "illusion".

Yet the veridicality issue is a moot one. Who is to decide? Let the photographer (only guy who ever saw the fiducial scene say) meet with a fatal accident: Then who will be the judge? Suppose the "photograph" is fake, actually a painting. Then the "fiducial scene" exists only in the artist's mind. Or suppose the "photograph" is a blank piece of paper overgrown with darkish fungus. Then there isn't any "fiducial scene" to begin with! Yet all this makes no difference whatsoever in the psychophysical experiment, the response will be the same. It is the "veridicality" question that makes no sense.

3.2   "Shape from X" and Ambiguity

In computer vision one implements algorithms for "Shape from X" (SFX). For instance, in "shape from shading" one computes a surface from a field of gray levels[8]. The idea is simple enough: First one has a model of the relevant physics. This allows "computer graphics", that is to say, given the scene one can predict what the photograph will be like. Then one *inverts* the process: Given the photograph one computes the scene. The problem is that the inversion is never uniquely possible. (For instance, see Belhumeur *et al.*[1] for an analysis of the shape from shading case, and Koenderink *et al.*[17] for a general analysis.) The various solutions are related by some group of "ambiguity transformations".

What this implies is that the orbit of the fiducial scene under the group of ambiguity transformations is a "metamer" of the fiducial scene. Each scene metameric with the fiducial scene would *have yielded the same photograph.* Thus all metameric scenes have an equal claim on the epithet "veridical"! Usually there exist infinities of them.

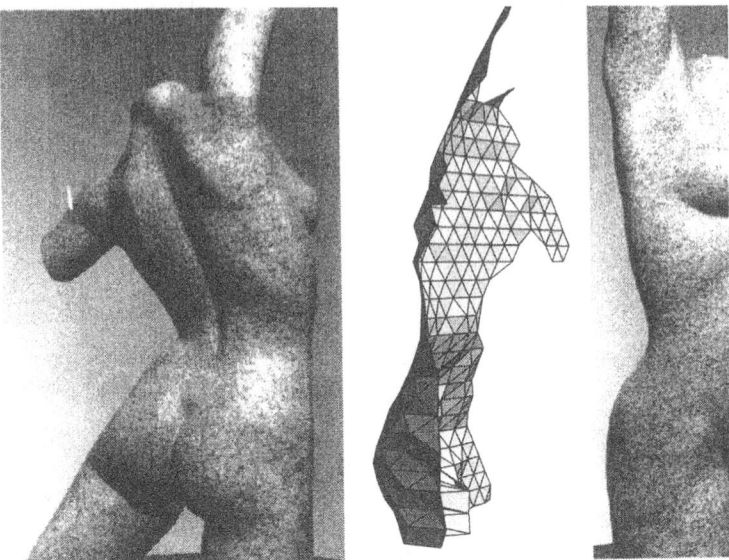

Fig. 2. *Pictorial reliefs are rarely "veridical" in the naïve sense. The picture in the middle shows the relief (depth towards the right) for the stimulus on the left. We rotated the (physical) object over 90° about the vertical and made another photograph, the picture on the right. Notice that the contours of the pictorial relief and the latter figure are different. The relief is not like the real thing, at least not in a naïve, Euclidian way.*

4   The Beholder's Share[5]

Since pictures underdetermine scenes, the pictorial relief must be due to the picture *as well* as to the "creative imagination" of the beholder. The picture acts as a "constraint" on the beholder's creativity. Of course the creative imagination is not altogether free. It is the result of eons of evolution and of a lifetime's experience. The beholder is an expert in "ecological optics", both in the realm of generic properties (causal connections, laws of ecological optics) and particular properties (the shapes, materials and events that recur in the biotope).

That this is a true *creative* force and not simply a bag of tricks (SFX algorithms) is clear because you really have to stick your neck out in order to be into a position to apply such tricks in the first place. For instance, you can't simply run a "shape from shading" algorithm on a picture. You first have to segment it into regions of uniform luminous atmospheres, and so forth. It is much as with the "Gestalt Laws" of "early vision"[19]: They are spontaneously acting (creative) forces rather than simple "filters".

The creative imagination is seen at work when you undergo a "Rohrschach test"[21] or (emulating Leonardo[18]) look at random patterns and "see" var-

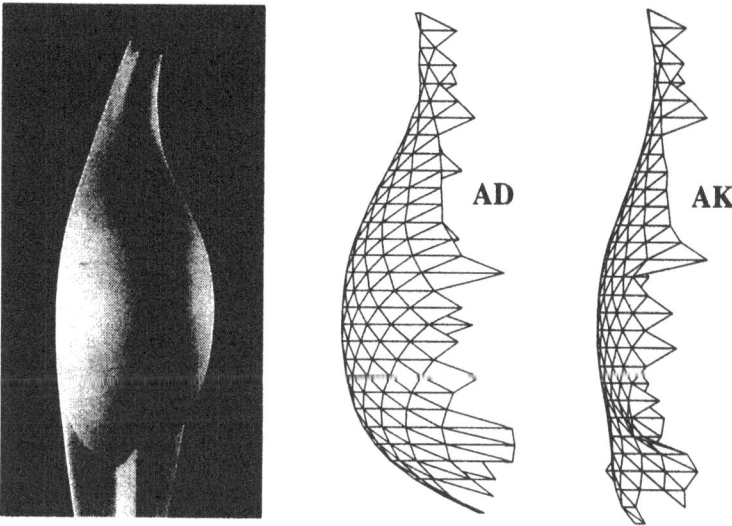

Fig. 3. *On the left the stimulus (one of Brancusi's "Birds"). On the right the pictorial reliefs for observers AD and AK (reference [11]). Notice that these are not the same, thus at least one of the observers can't be "veridical". Probably neither is.*

ious scenes. When you look long enough at a Necker cube[20] you notice spontaneous depth reversals: They *happen to you*. You can't control the creative imagination consciously.

5   The Nature of Ambiguity in Pictorial Relief

It is hopeless to try to figure out the group of ambiguities for generic photographs of scenes. Science just isn't up to the task. Only for a few—irrealistically simplified—SFX problems has the task been done. Nor is it to be expected that things will clear up in the near future.

Fortunately it is possible to guess at the nature of the ambiguity group from simple *a priori arguments*. For simplicity I assume orthographic projection here. Notice that the ambiguities should conserve the picture plane, for the pictorial space should "fit the picture". Next notice the fact that virtually all cues allow you to detect deviations from planarity. For instance, the shading cue will reveal local bumps and dents through shading and shadow. This means that the ambiguities conserve planes. But this limits the ambiguity group to certain affinities, namely depth scalings and shears involving depth but conserving the frontoparallel dimensions.

5.1   Hildebrand's "Depth Flow"

The German sculptor Adolf Hildebrand[7] wrote a book (in the 1890's) in which the group of depth scalings plays a major role. He noticed that observers typically don't discriminate between bas–relief and sculpture in the round when seen from the right vantage points. He thought of vision as of a "depth flow" and the sculptor's task as modulating the depth flow through ridges and ruts[13] of the pictorial relief.

We have been able to show that this idea has much to recommend it. First of all we noticed that pictorial reliefs for different observers are generally different (thus not veridical, see figures 2 and 3), but that the local depths of different observers tend to show very high linear correlations (coefficients of determination typically in the 0.9–1.0 range). We also find that slight changes in viewing conditions lead to different pictorial reliefs even for a single observer, but that—again—the local depths in one condition are typically highly correlated with the local depths in another condition[11,14,4]. (See figure 4.)

In the latter cases there is only a single picture, thus all pictorial cues are the same. The difference is in the (conflicting) cues that apply to the perception of the picture plane (picture as a "thing"), rather than the pictorial space. Examples of such cues are binocular disparity (monocular versus binocular viewing), accommodation (free viewing or via a lens at focal distance from the picture), or perspective (viewing the picture obliquely or frontoparallel). In all cases the differences can be described in terms of a "depth scaling" which may assume (empirically) values of up to a factor of five[14].

Another, somewhat surprising finding suggests that Hildebrand's idea that observers need nice, continuous depth flows has something to it. When you put two dots on a picture and ask "which one is closest" (in pictorial space of course) you find that observers find the task a natural one but may not always be able to perform it well. It turns out to be the case that they do well when both points happen to be on a single slope (continuous flow) but are bad at it when the points are on either side of a ridge or rut[15]. (See figure 5.) This is surprising, because we can find the pictorial relief with the gauge figure method and use these results to solve the task almost perfectly. Thus all the necessary information is in the observer's head, but the observers can't *use* that. It is like they have a picture of the flow field without the magnitudes of the flow vectors.

5.2   Changes in "Mental Perspective"

Recently we found cases where the pictorial reliefs (for the same picture, same observer) as operationalized through two slightly different tasks turned out to be quite different[16]. The local depth values didn't correlate at all (coefficients of determination not significantly different from zero). These are cases where Hildebrand's ideas apparently don't apply. We thought that

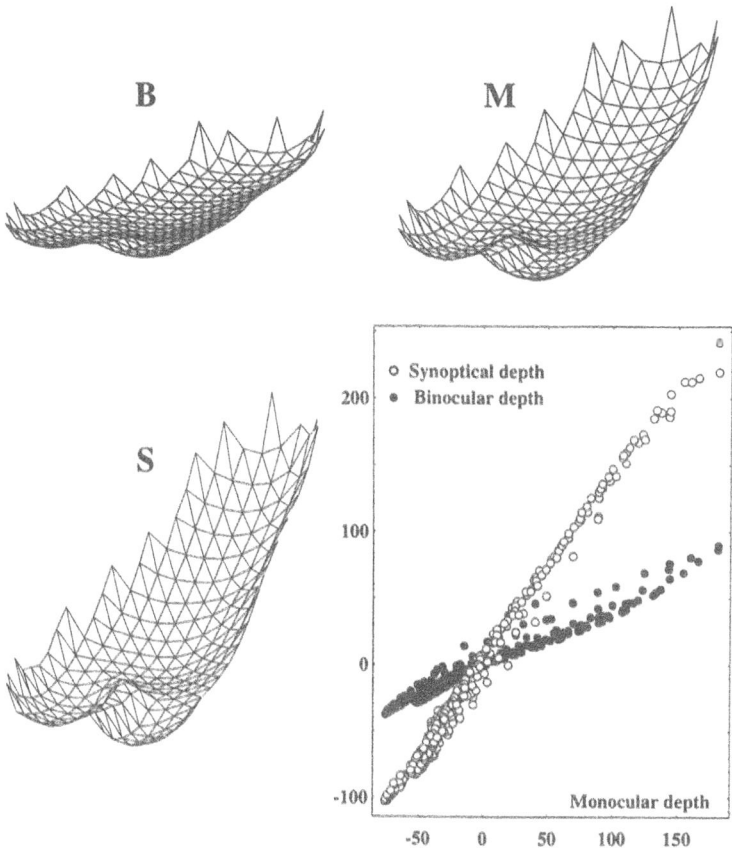

Fig. 4. Pictorial reliefs for one observer using different viewing methods, binocular viewing (B), monocular viewing (M), and "synoptical" viewing (S). (See reference [14].) In synoptical viewing both eyes obtain exactly the same optical pattern. (This is different from binocular viewing, because in that case the two eyes obtain different perspectives of the picture as a thing.) In these reliefs the depth dimension is the vertical. In the scatter plot we have monocular depth against binocular and synoptical depth, for all vertices of the reliefs. Notice the high correlations: Change of viewing conditions results in dilations or contractions of the depth domain. This makes sense, because the pictorial cues are the same, only the (conflicting) cues that reveal the (planar) picture surface differ.

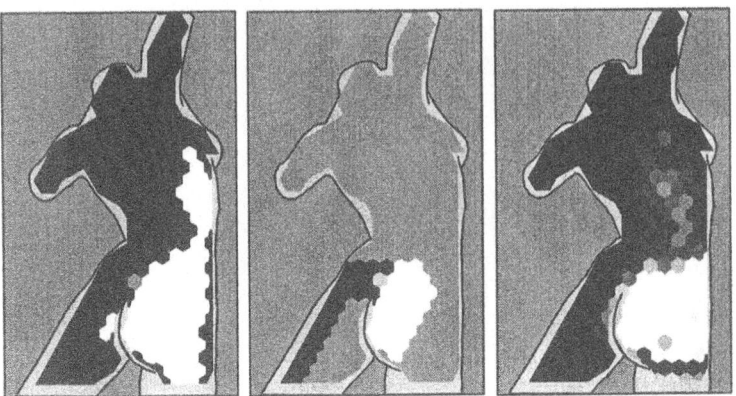

Fig. 5. *In these figures we consider the depth relation of any point within the contour to a fiducial point (it is the gray point in the middle figure). On the left all pixels nearer are drawn white, all pixels further away black. In the middle all pixels that can be reached by travelling towards the observer (in pictorial space) are white, those that can be reached by travelling away black, all other pixels gray (for these the depth relation would be "undecided"; they can't be reached by either going away or towards the observer starting from the fiducial point). In the righthand figure we show an empirical result: Pixel values indicate the probability of being judged "nearer" than the fiducial point (white 100%, black 0%). Notice that the result is perhaps more like the middle figure, thus observers may follow the slope but can't judge absolute depth values very well. (See reference [15].)*

this might be an instance of shear ambiguities and this turned out to be the case. When we performed multiple correlations of local depth values *and local image coordinate values* coefficients of determination went up in the 0.9–1.0 range. We needed shears that turned the frontoparallel plane into up to 50° oblique attitudes. (See figures 6 through 8.)

What does this mean? One likely interpretation is that the observers "changed their mental perspective". Notice that you may take *any point* on the pictorial relief and turn it into a (locally) frontoparallel plane through application of a shear. This is also what would happen when you were able to change your vantage point and put your eye on the normal direction of that point. Of course you can't do that in pictorial space because rotations would reveal parts of the scene that are not in the picture! But the shears enable you to do something very similar to a change of vantage point. Apparently the observers were exploiting the freedom left by the ambiguities to adjust their mental perspective.

In somewhat complicated scenes we find that observers often select different "mental viewpoints" for different parts of the scene[22]. Apparently the pictorial reliefs have some of the properties of Picasso's drawings where the artist simultaneously shows various aspects of a sleeping nude[3].

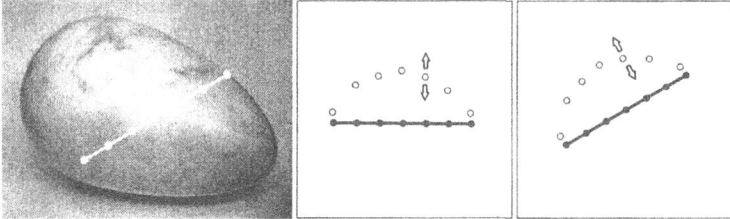

Fig. 6. *Two methods that operationalize pictorial relief. (Introduced in [16].) On the left a stimulus with superimposed line with points. In the middle and on the right two "response windows". The observer is instructed to drag the points in the response window (using a mouse or trackball; everything happens on a computer screen) so as to reproduce the cross section of the pictorial object with the plane defined by the fiducial line and the depth dimension. The two methods differ only in the orientation of the line in the response window: Either parallel to the fiducial line over the stimulus, or horizontal. Repeating this for many fiducial lines of various orientations allows one to construct a pictorial surface.*

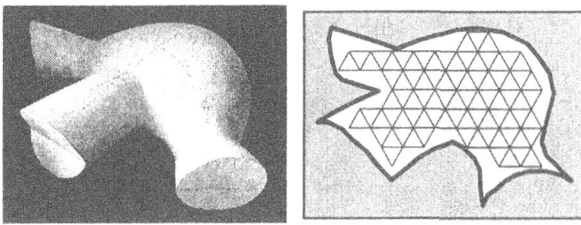

Fig. 7. *A stimulus picture ("Turtle" by Brancusi) and a triangulation of its interior. The fiducial lines in the methods discussed in the legend of figure 6 are taken from this triangulation.*

5.3   Multiple Images

In forensic investigations one records a scene through a number of photographs from various vantage points, rather than a single overview. Indeed, human observers appear to be able to use such a documentation to obtain a much more detailed "mental view" of the scene than would be possible from a single wide shot. In studying the records the observer must integrate multiple views of the same object that may differ in many respects: perspective, visible parts, light and shadow, textural details, and many more. This ability is by no means understood. One is in no position to replace the human observer with a software implementation, nor is this likely to be possible in the near future.

Consider a simple case: Two photographs of a matte white statue in the same illumination, taken from the same vantage point. The only change between the two shots is that the statue was rotated about the vertical over an

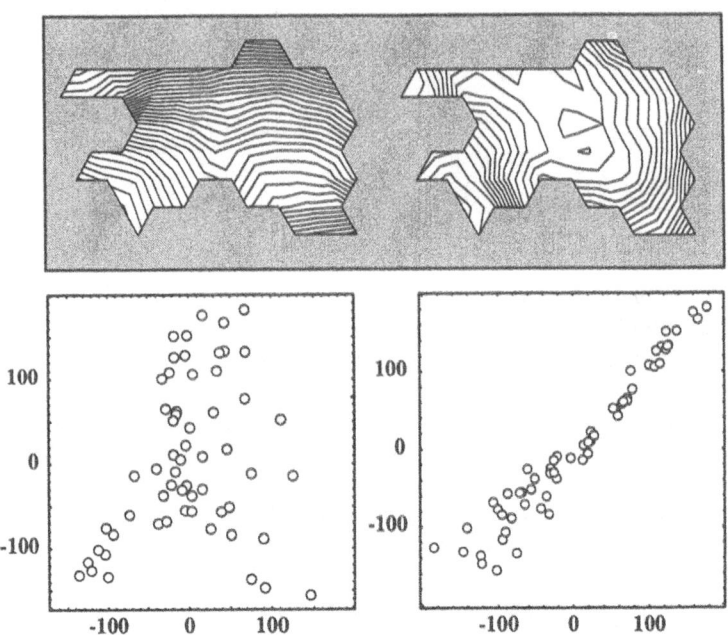

Fig. 8. *On the top two pictorial reliefs (curves of equal depth) for the Brancusi turtle stimulus obtained with the two methods (see reference [16]). Notice the enormous difference! In the bottom row on the left a scatter plot of the depths obtained from the two methods: Correlation is nonsignificant. Bottom right the scatter plot after taking the ambiguity into account. The correlations become very significant indeed. The two "different" results are actually identical modulo an ambiguity transform!*

angle of several tens of degrees. The outlines will be different, at corresponding locations the shading will be different, some parts will be visible in both shots, some in only one. When there are no obvious surface markings there is no way to locate corresponding points on the basis of image structure as such. The only way to locate corresponding points is to find the pictorial reliefs for both shots and compare these. There is obviously a chance that one might be able to find correspondences between localized differential invariants of the pictorial surfaces.

As expected, human observers find the task an easy one. We present the two photographs side by side on a CRT, in monochrome say. We put a dot in a contrasting color on one photograph and let the observer (using a mouse or trackball device) put a corresponding dot on the other. Here "correspondence" is understood *in the pictorial space common to both pictures*. Observers find this easy although they are lousy at judging the rotation of the object between the shots. (Errors of up to 30° are common.) They will easily do a thousand settings per hour.

Fig. 9. *Two pictures of the same object in the same illumination taken from the same vantage point. Between shots the object was turned over 45° about the vertical.*

We did the experiment for various rotations up to 90°. (See figures 9 and 10.) In the latter case the depth dimension in one picture is the horizontal picture plane dimension in the other picture! From the results we can reconstruct the object quite precisely, and also find the actual amounts of rotation with great (about 1°) precision[9].

Observers will even do the task in pairs of photographs that show no common parts. In one case we prepared photographs for a fixed setting, only changing the illumination of the scene. It is possible to find illuminations where one (left or right) or even both contours merge into the background (either in the light or in the dark) and thus literally don't appear on the photographs. We presented cases where one picture lacked the lefthand contour, the other the righthand contour. The observers have a hard time to notice that the contours aren't actually there, they complete then "in their mind's eye". From the results it is obvious that they often commit serious errors even though they find the task easy and sincerely believe to be dead–on[10].

6 Conclusion

"Pictorial reliefs" are entities that exist in the minds of human observers when they look at pictures, be they photographs, paintings or drawings. We have shown that these entities have many of the properties of true surfaces in a three dimensional space. The space is polarized with respect to the "direction of view", which is a dimension ("depth") that is a mental entity and that somehow augments the two dimensions that span the picture plane.

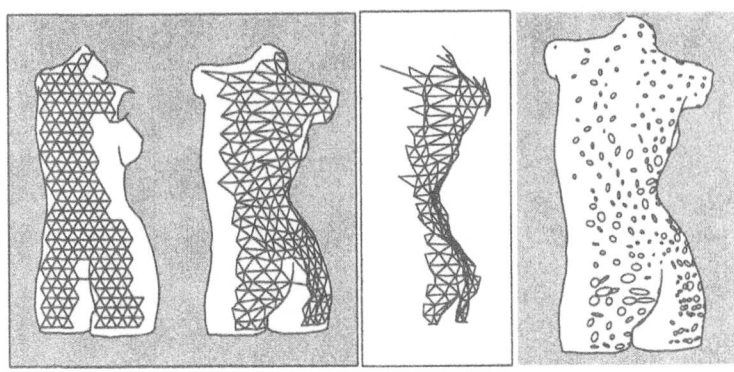

Fig. 10. *Result of correspondence judgments on the pictures of figure 9 (see reference [9]). On the left the lefthand figure has a fiducial triangulation, the righthand figure shows the corresponding triangulation. In the rightmost figure covariance ellipses for the correspondences are shown. In the middle we show a depth reconstruction from the correspondences, depth in the horizontal direction. It turns out that such reconstructions virtually don't depend on the physical amount of rotation: Results more closely approximate the physical object in the case of multiple images.*

The depth dimension is more volatile than the frontoparallel dimensions, but the two types do "mix" when the observer adjusts the perspective of the "mind's eye".

Pictorial reliefs can be "measured" in various ways. Of course these methods of measurement actually are "operational definitions" of pictorial relief, which otherwise remains a figment of the mind of the observer. It is to be expected *a priori* that different operationalizations need not be mutually coherent.

One way to characterize pictorial relief is to delineate the groups of "ambiguity transformations" that relate results of subclasses of operationalizations. Examples are the class of depth scalings which apparently apply to the observation of pieces of sculpture and "sculptural" paintings. We have found that a slightly wider class, transformations conserving plane (as a family) and the picture plane in particular, may apply to a very wide class of cases. It would be appropriate to develop the differential invariants of pictorial reliefs in this setting.

References

1. Belhumeur P. N., Kriegman D. J., Yuille A. L. (1999) The bas–relief ambiguity. International Journal of Computer Vision **35**, 33–44
2. Berkeley G. (1975) Philosophical Works, including the Works on Vision. Dent, London

3. Bouret J. (1957) Picasso, dessins. Georges Lang, Paris, 73 (Femme couchée, encre, 1941)
4. Doorn A. J. van, Koenderink J. J. (1996) How to probe different aspects of surface relief. In: Kappers A. M. L., Overbeeke C. J., Smets G. J. F., Stappers P. J. (Eds.) Studies in ecological psychology, EWEP 4. Delft University Press, Delft, 115–130
5. Gombridge E. H. (1959) Art and Illusion. Part III The Beholder's Share. Phaidon Press, London
6. Graham C. H. (Ed.) (1965) Vision and Visual Perception. John Wiley & Sons, New York
7. Hildebrand A. (1945) (original 1893) The Problem of Form in Painting and Sculpture. (Transl. Meyer M. and Ogden R. M.) G. E. Stechert & Co, New York
8. Horn B. K. P., Brooks M. J. (Eds.) (1989) Shape from Shading. The M.I.T. Press, Cambridge, Massachusetts
9. Koenderink J. J., Kappers A. M. L., Pollick F. E., Kawato M. (1997) Correspondence in pictorial space. Perception and Psychophysics 59, 813–827
10. Koenderink J.J., Doorn A. J. van, Arend L., Hecht H. (in press) Ecological optics and the creative eye. In: Mausfeld R. and Heyer D. (Eds.) Perception and the physical world. Wiley & Sons, New York
11. Koenderink J. J., Doorn A. J. van, Kappers A. M. L. (1992) Surface perception in pictures. Perception and Psychophysics 52, 487–496
12. Koenderink J. J., Doorn A. J. van, Kappers A. M. L. (1996) Pictorial surface attitude and local depth comparisons. Perception and Psychophysics 58, 163–173
13. Koenderink J. J., Doorn A. J. van (1998) The structure of relief. Advances in Imaging and Electron Physics 103, 65–150
14. Koenderink J. J., Doorn A. J. van, Kappers A. M. L. (1994) On so-called paradoxical monocular stereoscopy. Perception 23, 583–594
15. Koenderink J. J., Doorn A. J. van (1995) Relief: Pictorial and otherwise. Image and Vision Computing 13, 321–334
16. Koenderink J. J., Doorn A. J. van, Kappers A. M. L, Todd J. T. (submitted) Ambiguity and the "mental eye" in pictorial relief. Perception
17. Koenderink J. J., Doorn A. J. van (1997) The generic bilinear calibration-estimation problem. International Journal of Computer Vision 23, 217–234
18. Leonardo da Vinci (1989, first published 1651) In: Kemp M. (Ed.) Leonardo on painting. Yale University Press, New Haven
19. Metzger W. (1975) Gesetze des Sehens. Verlag Waldemar Kramer, Frankfurt a. M.
20. Necker L. A. (1832) Observations on some remarkable phænomena seen in Switzerland; and an optical phænomenon which occurs on viewing of a crystal or geometrical solid. Phil. Mag. (Ser. I) 3, 329–337
21. Rohrschach H. (1832) Psychodiagnostik. Huber, Bern and Berlin
22. Todd J. T., Koenderink J. J., Doorn A. J. van, Kappers A. M. L. (1996) Effects of changing viewing conditions on the perceived structure of smoothly curved surfaces. Journal of Experimental Psychology, Human Perception and Performance 22, 695–706

Shape-from-Texture from Eigenvectors of Spectral Distortion

Eraldo Ribeiro and Edwin R Hancock

Department of Computer Science,
University of York, York Y01 5DD, UK

Summary. This paper presents a simple approach to the recovery of dense orientation estimates for curved textured surfaces. We make two contributions. Firstly, we show how pairs of spectral peaks can be used to make direct estimates of the slant and tilt angles for local tangent planes to the textured surface. We commence by computing the affine distortion matrices for pairs of corresponding spectral peaks. The key theoretical contribution is to show that the directions of the eigenvectors of the affine distortion matrices can be used to estimate local slant and tilt angles. In particular, the leading eigenvector points in the tilt direction. Although not as geometrically transparent, the direction of the second eigenvector can be used to estimate the slant direction. The main practical benefit furnished by our analysis is that it allows us to estimate the orientation angles in closed form without recourse to numerical optimisation. Based on these theoretical properties we present an algorithm for the analysis of curved regularly textured surfaces. The second contribution of the paper is to show how initial orientation estimates delivered by the eigen-analysis can be refined using a process of robust smoothing. We apply the method to a variety of real-world and synthetic imagery. We show that the new shape-from-texture method can reliably estimate surface topography.

1 Introduction

The topic of shape-from-texture has been studied in the computer vision literature for almost three decades. Early work hinged on the use of texture-gradients [1,2]. Bajcsy and Lieberman were among the first to demonstrate the use of texture gradient as a depth cue [2]. Drawing on psychophysics, Stevens has provided an analysis of the information content of texture-gradient [3] and has shown how it can be used to recover surface orientation through the estimation of slant and tilt angles [4]. Witkin [5] has developed a statistical method for recovering local orientation and hence surface shape from natural two dimensional imagery. The method commences by assuming a uniform distribution of edge orientation, i.e. an isotropic texture, and models how this distribution transforms under perspective geometry. The shape of the edge orientation distribution can be used to estimate slant and tilt angles. Several authors have developed methods which assume that a fronto-parallel view of a planar sample of the texture is available. For instance, Ikeuchi [6] has shown how the process of estimating surface shape from regular texture patterns can be simplified using spherical projection. Aloimonos and Swain [7] have computed a texture analogue of the reflectance map from the area gradient of the fronto-parallel texture elements. The regularised variational method of Ikeuchi and Horn [8] for shape-from-shading is adapted to recover surface orientation. In this way poor initial surface orientation estimates can be improved using iterative relaxation operations [6,7]. However, the smoothing of the field of surface normals is based on a simple neighbourhood averaging method.

Several authors have attempted to elucidate more general frameworks for shape-from-texture. For instance Blake and Marinos [9] have critically assessed Witkin's [5] edge isotropy assumption. They have developed an optimisation-based method for recovering surface orientation from second-order texture moments. Kanatami and Chou [10] have further extended Witkin's work by modelling the statistical transformation of texture density. They consider the effects of perspective geometry when isotropy and homogeneity assumptions apply. Garding [11,12] has developed an elegant framework based on differential geometry. This simplifies the analysis of perspective geometry and provides a mathematical setting in which curvature can be directly recovered from weakly isotropic textures.

More recently, the use of spectral information has been explored as an alternative to texture-gradient information. The use of frequency domain or spectral measurements represents a way of overcoming some of the restrictive requirements imposed by the need to work with accurately determined texture gradients. Moreover, it increases the range of natural textures that can be accommodated. Frequency properties can be captured using a number of different representations including the Fourier transform, the wavelet function, Gabor wavelets and the Wigner distribution [13–17]. The frequency domain method has been used extensively in the recovery of the perspective pose of texture planes. In an important series of papers, Krumm and Shafer [18–20] recover slant and tilt angles for periodic surface textures. They commence by introducing the idea of using measurements of the affine distortion to recover the parameters of perspective projection from spectral peaks [18]. Parameter estimation is based on exhaustive numerical search. The method is later extended to provide a means of segmenting multiple planar patches from textured images [19]. Super and Bovik [14,21] have an equivalent method which focuses on linearising the Jacobian of the perspective transformation in the frequency domain. The method shares with that of Krumm and Shafer the feature of using exhaustive numerical search.

Turning our attention to curved surfaces, Krumm and Shafer [20] have shown how to relate the frequency distortion between local planar patches under an affine transform. Malik and Rosenholtz [15,22] use Garding's framework [11] to estimate local orientation and curvature. Whereas Garding uses departures from isotropy as the texture measurement, Malik and Rosenholtz use Krumm and Shafer's ideas to construct affine distortion measures. The texture measurements are based on frequency domain derivatives. The method uses numerical minimisation to recover the five parameters needed to estimate orientation and curvature. The method requires a good initial estimate of the magnitude of the curvature. It can be viewed as minimising a measure of spectral back-projection error.

2   Motivation

The observations underpinning this paper are twofold. First, although considerable effort has gone into formalising the problem of recovering shape from texture gradient, the demonstration of the method on real-world imagery is limited. The reason for this is that the estimation of texture gradient requires texture primitives to be extracted as a prerequisite. Secondly, although frequency-based methods are not dependant on the accurate segmentation of texture primitives, the underlying theory is less developed. For instance, the literature described above can be criticised on a number of grounds. First, in the methods of both Krumm and Shafer [20] and Malik and Rosenholtz [15,22] the recovery of local surface orientation is based on numerical optimisation. In addition, the method of Malik and Rosenholtz [15,22] has difficulty in distinguishing between curved and planar surfaces, is sensitive to initial parameter values and needs curvature to be specified as a parameter.

Based in these observations, our aim in this paper is to present an improved method for recovering shape from curved or planar textured surfaces. We choose to work in the frequency domain. The reasons for this are that frequency domain methods avoid difficulties associated with pre-segmenting structural texture primitives. Moreover, they are more amenable to local adaptation of scale to accommodate variations in texture density across highly inclined surfaces [23]. We follow Krumm and Shafer [20] by assuming that the local texture variations due to the perspectivity of the surface can be approximated in an affine manner. In the Fourier domain, this means that the local frequency content of the texture also undergoes affine distortion. We represent the local texture variations by estimating the frequency domain affine distortions between local spectral peaks. The affine distortion matrices constitute our representation of local surface structure.

We make three contributions. First, we show how local orientation can be estimated in closed form from an affine distortion matrix. The method is based on an eigenvector analysis of the distortion matrix. Our main theoretical contribution is to show that the directions of the eigenvectors can be used to directly estimate the slant and tilt angles. In other words, the local distortions provide a direct route to surface orientation. This result applies under the assumption that the underlying surface is painted with a uniform texture and is viewed under perspective projection onto the image plane. In other words, we assume the texture is homogeneous but not isotropic. Our main contribution is therefore to develop a direct and simple method for estimating slant and tilt angles.

Having established this simple property we proceed to develop a practical algorithm for shape from texture. The recovery of a dense field of tangent plane orientations is a two-step process. The first step is to make an initial estimate of the local surface orientation using the spectral distortions between corresponding spectral peaks. Once the surface normal estimates are to hand, the second step is to improve the consistency of the orientation field through the use of local contextual information. Conventionally, this second step is realised using the Horn and Brooks [24] method which involves smoothing the estimated directions of the surface normals through a process of local averaging. Here we adopt a more elaborate smoothing method which has proved successful in the shape-from-shading domain [25,26]. We use robust error kernels rather than quadratic smoothness penalties to improve the organisation of the needle map. This allows us to preserve fine surface detail whilst removing the effects of local noise.

Our final contribution is to show that the new method leads to reliable curvature estimates. Based on synthetic data we show that the robust smoothing does not unduly bias the estimation of curvature. Moreover, we experiment with the new shape-from-texture on a variety of demanding real world images of curved texture surfaces. Here the method produces qualitatively good results.

3   Geometric Modelling

This paper is concerned with recovering a dense map of surface orientations for curved surfaces which are uniformly painted with periodic textures. Although these may seem to be a restrictive condition, it is worth noting that the built environment is rich in surfaces which satisfy it. Our approach is a spectral one which is couched in the Fourier domain. However, before we proceed to develop the spectral representation, we must first review the underlying geometry. In particular we are interested in modelling the transformation between the curved texture surface and the image plane under perspective geometry. There are two processes at play in determining this projection. Firstly, there is the variation in the local surface orientation due to the underlying curvature of the surface. Here we assume that the surface can be locally approximated by planar patches. The second geometric factor is the perspective distortion of the locally planar surface patches when viewed from the image-plane. This latter process is parameterised in terms of the local slant and tilt angles of the tangent planes to the texture surface. Locally, the perspective foreshortening of the planar patches is approximated by an affine distortion.

We therefore commence by reviewing the projective geometry for the perspective transformation of points on a plane. Specifically, we are interested in the perspective transformation between the object-centred co-ordinates of the points on a local tangent plane to the texture surface and the viewer-centred co-ordinates of the corresponding points on the image plane. To be more formal, suppose that the camera has focal length $f < 0$. Consider two corresponding points. The point with co-ordinates $\mathbf{X_t} = (x_t, y_t, z_t)^T$ lies on the texture surface while the corresponding point on the image plane has co-ordinates $\mathbf{X_i} = (x_i, y_i, f)^T$. The curved texture surface is represented by the height-function $h = F(x_t, y_t)$. We represent the local orientation of the viewed texture surface at the point $\mathbf{X_t}$ using the slant σ and tilt τ angles. For a local tangent plane to the surface, the slant is the angle between viewer line of sight and the normal vector of the plane. The tilt is the angle of rotation of the normal vector to the texture plane around the line of sight axis. The local surface geometry is illustrated in Figure 1. Furthermore, since we regard the

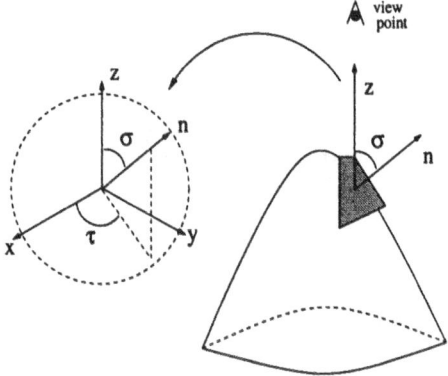

Fig. 1. Slant and tilt geometry of a local plane on the surface

texture as being homogeneously "painted" on the texture surface, the texture height z_t is always equal to zero. With these ingredients the perspective transformation between the texture-surface and image co-ordinate systems is given in matrix form by

$$\begin{bmatrix} x_i \\ y_i \\ z_i \end{bmatrix} = \frac{f}{h - x_t \sin \sigma} \times \left\{ \begin{bmatrix} \cos \sigma \cos \tau & -\sin \tau & \sin \sigma \cos \tau \\ \cos \sigma \sin \tau & \cos \tau & \sin \sigma \sin \tau \\ -\sin \sigma & 0 & \cos \sigma \end{bmatrix} \begin{bmatrix} x_t \\ y_t \\ 0 \end{bmatrix} + \begin{bmatrix} 0 \\ 0 \\ h \end{bmatrix} \right\}$$

The first term inside the curly braces represents the rotation of the local tangent plane to the texture surface in slant and tilt. The second term represents the displacement of the rotated plane along the optic axis. Finally, the multiplying term outside the braces represents the non-linear foreshortening in the slant direction. When expressed in this way, z_i is always equal to f since the image is formed at the focal plane of the camera. As a result we can confine our attention to the following simplified transformation of the x

and y co-ordinates

$$\begin{bmatrix} x_i \\ y_i \end{bmatrix} = \frac{f}{h - x_t \sin \sigma} \begin{bmatrix} \cos \tau & -\sin \tau \\ \sin \tau & \cos \tau \end{bmatrix} \begin{bmatrix} \cos \sigma & 0 \\ 0 & 1 \end{bmatrix} \begin{bmatrix} x_t \\ y_t \end{bmatrix} \tag{1}$$

This transformation can be represented using the shorthand $(x_i, y_i)^T = T_p(x_t, y_t)^T$, where T_p is the 2×2 transformation matrix. As written above, the transformation T_p can be considered as a composition of two transformations. The first of these is a non-uniform scaling proportional to the displacement in the slant direction. The second transformation is a counterclockwise rotation by an amount equal to the tilt angle.

Unfortunately, the non-linear nature of the perspective transformation makes the Fourier domain analysis of the texture somewhat intractable. To overcome this difficulty it is usual to use a linear approximation of the perspective projection [19,14].

To proceed, let $X_{oi} = (xo_t, yo_t, h)^T$ be the location of the origin or expansion point for the local co-ordinate system of the affine transformation. This origin projects to the point (xo_i, yo_i, f) on the image plane. We denote the local coordinate system on the image plane by $X_i' = (x_i', y_i', f)$ where $x_i = x_i' + xo_i$ and $y_i = y_i' + yo_i$. The first-order Taylor expansion is obtained by computing the Jacobian $J(.)$ of X_i. The required partial derivatives are calculated at the expansion point X_{oi}. After the necessary algebra, the resulting linear Taylor approximation is

$$T_A(X_{oi}) = J(X_{oi}) = J(T_p X_t) \mid_{(X_i'=0)} \tag{2}$$

where

$$J(X_i) = \begin{bmatrix} \frac{\partial}{\partial x_i'} x_t(x_i', y_i') & \frac{\partial}{\partial y_i'} x_t(x_i', y_i') \\ \frac{\partial}{\partial x_i'} y_t(x_i', y_i') & \frac{\partial}{\partial y_i'} y_t(x_i', y_i') \end{bmatrix} \tag{3}$$

Rewriting T_A in terms of the slant and tilt angles we have

$$T_A(X_{oi}) = \frac{\Omega}{h f \cos \sigma} \begin{bmatrix} xo_i \sin \sigma + f \cos \tau \cos \sigma & -f \sin \tau \\ yo_i \sin \sigma + f \sin \tau \cos \sigma & f \cos \tau \end{bmatrix} \tag{4}$$

where $\Omega = f \cos \sigma + \sin \sigma (xo_i \cos \tau + yo_i \sin \tau)$. Hence, the affine transformation matrix T_A depends only on the expansion point (xo_i, yo_i), which is a constant, together with the slant and tilt angles, which are the goal of our analysis.

4   Affine Distortion of the Power Spectrum

In Section 3 we developed an affine approximation of the perspective projection using a first-order Taylor series. In this Section, we will develop these ideas one step further by showing how the frequency content of the local texture plane transforms under local affine geometry.

Our starting point is a recent result due to Bracewell et al [27] which relates the effect of an affine transformation in the spatial domain to the Fourier-domain representation of a signal. Suppose that $G(.)$ represents the Fourier transform of a signal. Furthermore, let $\mathbf{X}$ be a vector of spatial co-ordinates and let $\mathbf{U}$ be the corresponding vector of frequencies. We are interested in determining the distribution of image-plane frequency vectors $\mathbf{U}_i$ when the distribution of texture-surface frequency vectors $\mathbf{U}_t$ is known. According to Bracewell et al, the distribution of image-plane frequencies $\mathbf{U}_i$ resulting from the Fourier transform of the affine transformation $\mathbf{X}_i = A\mathbf{X}_t + B$ is given by

$$G(\mathbf{U_i}) = \frac{1}{|det(A)|} e^{2\pi j \mathbf{U_t}^T A^{-1} \mathbf{B}} G[(A^T)^{-1} \mathbf{U_t}] \tag{5}$$

In our case, the affine transformation matrix is T_A defined in Equation 4 and there are no translation coefficients, i.e., $\mathbf{B} = \mathbf{0}$. As a result Equation 5 simplifies to:

$$G(\mathbf{U_i}) = \frac{1}{|det[T_A(X)]|} G[(T_A(X)^T)^{-1} \mathbf{U_t}] \tag{6}$$

The effect of the affine transformation of co-ordinates $T_A(X)$ is to induce an affine transformation T_A^{-T} on the local texture-plane frequency distribution. The spatial domain transformation matrix and the frequency domain transformation matrix are simply the inverse transpose on one-another. Applying the Fourier property of Equation 6 to the linearised version of the perspective transformation, the relationship between the texture plane and image spectra is $\mathbf{U}_i = T_A(X)^{-T}\mathbf{U}_t$. This spectral property has also been exploited by Malik and Rosenholtz [22] in their work on local shape-from-texture.

Here, we will consider only the affine distortion in the positions of frequency peaks. In other words we will not consider the distribution of the energy amplitude or phase in our analysis. For practical purposes we will use the local power spectrum to locate the positions of spectral peaks.

5   Spectral Distortion across the Image Plane

The main contribution is this paper is to show how the locally affine approximation to the perspective transformation of the texture spectra can be used to recover estimates of the local surface orientation, i.e. the slant and tilt parameters of curved surfaces. The idea underpinning our new shape-from-texture method is to measure local affine distortions between neighbouring spectral peaks. We demonstrate that the eigenvectors of the affine distortion matrix can be used to estimate the slant and tilt directions.

5.1 Spectral Distortion

The starting point for our analysis is the result presented in Section 4 where we showed how the spectral peaks on the texture surface are related to their counterparts on the image plane by the affine transformation T_A given in Equation 4. We aim to exploit this property to recover shape-from-texture. Provided that the texture distribution painted on a curved surface is homogeneous, observed changes in the image plane texture pattern can be attributed to variations in surface orientation. Our aim is to compute the slant and tilt angles of local tangent planes to a textured surface using the observed distortions of the texture spectrum across the image plane. To do this we measure the affine distortion between corresponding spectral peaks.

To commence, we consider the point S on the curved texture surface. Suppose that the neighbourhood of this point can be approximated by a local planar patch. This planar patch undergoes perspective projection onto the image plane. Using the result presented in Section 3 we make a locally affine approximation to this perspective projection. Further suppose that we sample the texture projection of the local planar patch at two neighbouring points A and B laying on the image plane. The co-ordinates of the two points are respectively $\mathbf{X}_A = (x, y)^T$ and $\mathbf{X}_B = (x + \Delta x, y + \Delta y)^T$ where Δx and Δy are the image-plane displacements between the two points.

Suppose that the local planar patch on the texture surface has a spectral peak with frequency vector $\mathbf{U}_S = (u_s, v_s)^T$. On the image plane, the corresponding frequency vectors for the spectral peaks at the points $\mathbf{X}_A$ and $\mathbf{X}_B$ are respectively $\mathbf{U}_A = (u_A, v_A)^T$ and $\mathbf{U}_B = (u_B, v_B)^T$. Using the Fourier domain affine projection property presented in Section 3, the texture-surface peak frequencies are related to the image plane peak frequencies via the equations $\mathbf{U}_A = (T_A(X_A)^{-1})^T \mathbf{U}_S$ and $\mathbf{U}_B = (T_A(X_B)^{-1})^T \mathbf{U}_S$ where $T_A(X_A)$ is the local affine approximation to the perspective projection of the planar surface patch at the point A and $T_A(X_B)$ is the corresponding affine projection matrix at the point B. As a result, the frequency vectors for the two corresponding spectral peaks on the image-plane are related to one-another via the local distortion $\mathbf{U}_B = \Phi \mathbf{U}_A$ where $\Phi = (T_A(X_A)T_A(X_B)^{-1})^T$. As a result, the texture-surface spectral distortion matrix Φ is a 2x2 matrix. This matrix relates the affine distortion of the image plane frequency vectors to the 3-D pose or orientation parameters of the local planar patch on the surface. Substituting for the affine approximation to the perspective transformation from Equation (4), the required matrix is given in terms of the slant and tilt angles as

$$\Phi = \frac{\Omega\,(\mathbf{A})}{\Omega^2\,(\mathbf{B})} \begin{bmatrix} \Omega\,(\mathbf{A}) + \Delta y \sin \sigma \sin \tau & -\Delta y \sin \sigma \cos \tau \\ -\Delta x \sin \sigma \sin \tau & \Omega\,(\mathbf{A}) + \Delta x \sin \sigma \cos \tau \end{bmatrix} \tag{7}$$

where $\Omega(A) = f \cos \sigma + \sin \sigma (x \cos \tau + y \sin \tau)$ and $\Omega(B) = f \cos \sigma + \sin \sigma \times ((x + \Delta x) \cos \tau + (y + \Delta y) \sin \tau)$ The above matrix represents the distortion of the spectrum sampled at the location B with respect to the sample at the location A. In the next Section we show how to solve directly for the parameters of surface orientation, i.e. the slant and tilt angles, using the eigen-structure of the transformation matrix Φ.

5.2   Eigenstructure of the Affine Distortion Matrix

Let us consider the eigenvector equation for the distortion matrix Φ, i.e. $\Phi \mathbf{w}_\lambda = \lambda \mathbf{w}_\lambda$ where $\lambda = (\lambda_1, \lambda_2)$ are the eigenvalues of the distortion matrix Φ and $\mathbf{w}_\lambda$ are the corresponding eigenvectors. Since Φ is a 2x2 matrix the two eigenvalues are found by solving the quadratic eigenvalue equation $det[\Phi - \lambda I] = 0$ where I is the 2x2 identity matrix. The explicit eigenvalue equation is

$$\lambda^2 - Trace(\Phi)\lambda + det(\Phi) = 0 \tag{8}$$

where $Trace(\Phi)$ and $det(\Phi)$ are the trace and determinant of Φ. Substituting for the elements of the transformation matrix Φ, we have

$$\lambda^2 - \left[\frac{\Omega(\mathbf{A})}{\Omega(\mathbf{B})} + \frac{\Omega^2(\mathbf{A})}{\Omega^2(\mathbf{B})}\right]\lambda + \left[\frac{\Omega(\mathbf{A})}{\Omega(\mathbf{B})} \times \frac{\Omega^2(\mathbf{A})}{\Omega^2(\mathbf{B})}\right] = 0 \tag{9}$$

The two eigenvalue solutions of the above quadratic equation are

$$\lambda_1 = \frac{\Omega^2(\mathbf{A})}{\Omega^2(\mathbf{B})} \qquad and \qquad \lambda_2 = \frac{\Omega(\mathbf{A})}{\Omega(\mathbf{B})} \tag{10}$$

The corresponding eigenvectors are $\mathbf{w}(\lambda_1) = [\mathbf{w}_x(\lambda_1), \mathbf{w}_y(\lambda_1)]^T$ and $\mathbf{w}(\lambda_2) = [\mathbf{w}_x(\lambda_2), \mathbf{w}_y(\lambda_2)]^T$. More explictly, the eigenvectors are:

$$\mathbf{w}(\lambda_1) = [1, \tan\tau]^T \qquad and \qquad \mathbf{w}(\lambda_2) = \left[1, -\frac{\Delta x}{\Delta y}\right]^T \tag{11}$$

As a result we can directly determine the tilt angle from the vector components of the eigenvector associated with the eigenvalue λ_1. The tilt angle is given by

$$\tau = \arctan\left[\frac{\mathbf{w}_y(\lambda_1)}{\mathbf{w}_x(\lambda_1)}\right] \tag{12}$$

Once the tilt angle has been obtained, we recover the slant angle by solving the equation

$$\lambda_2 = \frac{\Omega(\mathbf{A})}{\Omega(\mathbf{B})} = \frac{f\cos\sigma + \sin\sigma(x\cos\tau + y\sin\tau)}{f\cos\sigma + \sin\sigma(x + \Delta x)\cos\tau + (y + \Delta y)\sin\tau} \tag{13}$$

The solution is

$$\sigma = \arctan\left[\frac{f(\lambda_2 - 1)}{(y(1 - \lambda_2) - \lambda\Delta y)\sin\tau + (x(1 - \lambda_2) - \lambda_2\Delta x)\cos\tau}\right] \tag{14}$$

With the slant and tilt angles to hand the surface normal $\mathbf{n}$ may be computed.

6   Computing Local Planar Orientation

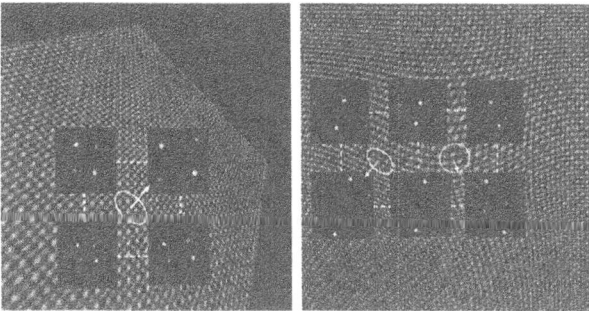

Fig. 2. Affine distortion of the local spectra across the image plane according to the local surface orientation. The arrows are the estimated normal vectors at the center of the local planes.

In this Section we explain how to recover local planar surface orientation using our local affine distortion method. Our method shares with Krumm and Shafer [19] and Super and Bovik [14] the feature of using the affine distortion of spectra to estimate surface orientation. However, these two methods recover surface orientation by exhaustive spectral back-projection and error enumeration for all slant and tilt angles. The orientation is selected so as to numerically minimise a back-projection error. This is clearly a highly time consuming process due to the amount of search and numerical minimisation required. Moreover, unless good initialisation values are to hand, the method is prone to convergence to local optima.

Instead, our method solves for the local surface orientation parameters in closed-form. To directly recover the planar orientation angles we use the eigenvectors of the spectral distortion matrix. We assume that the texture is homogeneous over the entire surface. The consequence of this assumption is that the spectral content of the texture does not change systematically over the curved texture surface. As a result of this assumption the local spectral distortions measured on the image plane are attributable solely to changes in perspectivity. Shape effects such as changes in local surface orientation, which are attributable to surface curvature, must be assessed at a more global level.

The first step in orientation recovery is to estimate the affine distortion matrix which represents the transformation between different local texture regions on the image plane. These image texture regions are assumed to belong to a single local planar patch on the curved texture surface. We do this by selecting pairs of neighbouring points on the image plane. At each point there may be several clear spectral peaks. Since the affine distortion matrix Φ has four elements that need to be estimated, we need to know the correspondences between at least two different spectral peaks at the different locations. Suppose that $U_1^{p_1} = (u_1^{p_1}, v_1^{p_1})^T$ and $U_1^{p_2} = (u_1^{p_2}, v_1^{p_2})^T$ represent the frequency vectors for two distinct spectral peaks located at the point with co-ordinates $\mathbf{X}_1 = (x_1, y_1)^T$ on the image plane. The frequency vectors are used to construct the columns of a 2x2 spectral measurement matrix $V_1 = (U_1^{p_1} | U_1^{p_2})$. Further, suppose that $U_2^{p_1} = (u_2^{p_1}, v_2^{p_1})^T$ and $U_2^{p_2} = (u_2^{p_2}, v_2^{p_2})^T$ represent the frequency vectors for the corresponding spectral peaks at the point $\mathbf{X}_2 = (x_2, y_2)^T$. The corresponding spectral measurement matrix is $V_2 = (U_2^{p_1} | U_2^{p_2})$. Under the affine model presented in Section 4, the spectral measurement matrices are related via the equation $V_2 = \Phi V_1$. As a result the local estimate of the spectral distortion matrix is $\Phi = (V_1^T)^{-1} V_2$.

In practice, we only make use of the most energetic peaks appearing in the power spectrum. That is to say we do not consider the detailed distribution of frequencies. Our method requires that we supply correspondences between spectral peaks so that the distortion matrices can be estimated. We use the energy amplitude of the peaks to establish the required correspondences. This is done by ordering the peaks according to their energy amplitude. The ordering of the amplitudes of peaks at different image locations establishes the required pattern of spectral correspondences.

After estimating the affine transform between two local spectral peaks we can directly apply the eigenvector analysis described in Section 5 to estimate the tilt and the slant angles by using Equations 12 and 14. However, there are some obstacles to the direct estimation of the local spectral distortion matrix. The first of these is related to the choice of spectral scale. If the window size used in the spectral estimator is mismatched then the power spectrum becomes de-focused and poor peak localisation results. The second problem arrises if there is no significant affine distortion between the corresponding spectral peaks used in the analysis. In other words, in choosing the locations of the spectral samples we must strike a compromise. If the locations are too close to one-another then we risk poor orientation estimation since the local affine distortions due to global perspectivity are too small to detect. If, on the other hand, the points at too far apart or span a high curvature feature, then the local distortions in the power spectrum are likely to be due to significant changes in local surface orientation. We can overcome this latter effect through smoothing the field of local orientation estimates.

Finally, in Figure 2 we provide an illustration of the affine distortion of the local spectra across a plane viewed under perspective geometry and across a curved surface. At each location we show the computed spectra. The brightness of the peaks is proportional to their energy. Notice that the energy ordering is preserved from location to location. It is also important to note that the distortions due to curvature are greater than those due to perspectivity.

7 Robust Smoothing of the Needle-map

The orientation estimates returned by the new shape-from-texture method are likely to be noisy and inconsistent when viewed from the perspective of local smoothness. In order to improve the consistency of our needle map, and hence the surface shape description, we employ an iterative smoothing process to update the estimated normal vectors. Aloimonos and Swain [7] have used Horn and Brooks [24] local averaging method for this purpose. However, in order to avoid the over-smoothing of local surface detail associated with high curvature features, we use a robust smoothing method. Rather than using a quadratic penalty of the sort which underpins the method of Aloimonos and Swain [7], this method uses robust error kernels. to gauge the effect of the smoothness error. The reason for this is that the quadratic penalty grows indefinitely with increasing smoothness error. This can have the undesirable effect of over-smoothing genuine surface detail. Examples of such surface structures include ridges and ravines. By using robust error kernels we can moderate the effects of smoothing over regions of genuine surface detail and allow a more faithful topographic representation to be recovered [26].

This robust smoothing technique has been successfully exploited in the recovery of shape-from-shading by Worthington and Hancock [26]. Here it is used as a regulariser that is applied in conjunction with data-closeness constraints furnished by the image irradiance equation. In fact, the most effective error kernel was found to be the log-cosh function, which can be viewed as a continuous counterpart of Huber's kernel. In the present work the robust smoothing process is applied to the initial surface normal estimates provided by our shape-from-texture method. According to Worthington and Hancock [26], the smoothness error or consistency of the field of surface normals is measured using the derivatives of the needle-map in the x and y directions. The robust smoothness penalty is

$$I = \int \int \left\{ \left(\rho_\sigma \left(\left\| \frac{\partial \mathbf{n}}{\partial x} \right\| \right) + \rho_\sigma \left(\left\| \frac{\partial \mathbf{n}}{\partial y} \right\| \right) \right) \right\} dx dy \qquad (15)$$

In the above measure, $\rho_\sigma(\eta)$ is the robust error kernel used to gauge the local consistency of the needle-map or field of surface normals. The argument of the kernel η is the measured error and the parameter σ controls the width of the kernel. It is important to note the robust-error kernels are applied separately to the magnitudes of the derivatives of the needle-map in the x and y directions. Applying variational calculus to the smoothness penalty function yields the following equation for updating the surface normals where $\mathbf{n}_{i,j}^{(k)}$ is the estimated surface normal at the pixel with row index i and column index j at iteration k of the smoothing process

$$
\begin{aligned}
\mathbf{n}_{i,j}^{(k+1)} &= \left\| \frac{\partial \mathbf{n}_{i,j}^{(k)}}{\partial x} \right\|^{-1} \left[\frac{\partial}{\partial x} \left(\rho'_\sigma \left(\left\| \frac{\partial \mathbf{n}_{i,j}^{(k)}}{\partial x} \right\| \right) \right) + \rho'_\sigma \left(\left\| \frac{\partial \mathbf{n}_{i,j}^{(k)}}{\partial x} \right\| \right) \times \right. \\
&\quad \left. \left(\mathbf{n}_{i+1,j}^{(k)} + \mathbf{n}_{i-1,j}^{(k)} - \left\| \frac{\partial \mathbf{n}_{i,j}^{(k)}}{\partial x} \right\|^{-2} \left(\frac{\partial \mathbf{n}_{i,j}^{(k)}}{\partial x} \cdot \frac{\partial^2 \mathbf{n}_{i,j}^{(k)}}{\partial x^2} \right) \frac{\partial \mathbf{n}_{i,j}^{(k)}}{\partial x} \right) \right] \\
&\quad + \left\| \frac{\partial \mathbf{n}_{i,j}^{(k)}}{\partial y} \right\|^{-1} \left[\frac{\partial}{\partial y} \left(\rho'_\sigma \left(\left\| \frac{\partial \mathbf{n}_{i,j}^{(k)}}{\partial y} \right\| \right) \right) + \rho'_\sigma \left(\left\| \frac{\partial \mathbf{n}_{i,j}^{(k)}}{\partial y} \right\| \right) \times \right. \\
&\quad \left. \left(\mathbf{n}_{i,j+1}^{(k)} + \mathbf{n}_{i,j-1}^{(k)} - \left\| \frac{\partial \mathbf{n}_{i,j}^{(k)}}{\partial y} \right\|^{-2} \left(\frac{\partial \mathbf{n}_{i,j}^{(k)}}{\partial y} \cdot \frac{\partial^2 \mathbf{n}_{i,j}^{(k)}}{\partial y^2} \right) \frac{\partial \mathbf{n}_{i,j}^{(k)}}{\partial y} \right) \right]
\end{aligned}
\tag{16}
$$

As stated in Equation 16, the smoothing process is entirely general. Any robust error kernel $\rho_\sigma(\eta)$ can be inserted into the above result to yield a needle-map smoothing process. However, it must be stressed that performance is critically determined by the choice of error-kernel. Worthington and Hancock [26] found the most effective error kernel was the log-cosh sigmoidal-derivative M-estimator. The kernel has the functional form

$$
\rho_\sigma(\eta) = \frac{\sigma}{\pi} \log \cosh \left(\frac{\pi \eta}{\sigma} \right)
\tag{17}
$$

8   Curvature Estimation

Once the smoothed needle-map is to hand, then we can use the surface normals to estimate curvature. In our experiments, we have investigated the quality of the shape-index of Koenderink and Van Doorn as a scale-invariant measure of surface topography.

The differential structure of a surface is captured by the Hessian matrix, which may be written in terms of surface normals as

$$
\mathcal{H} = \begin{pmatrix} \left(\frac{\partial \mathbf{n}}{\partial x} \right)_x & \left(\frac{\partial \mathbf{n}}{\partial x} \right)_y \\ \left(\frac{\partial \mathbf{n}}{\partial y} \right)_x & \left(\frac{\partial \mathbf{n}}{\partial y} \right)_y \end{pmatrix}
\tag{18}
$$

where $(\cdots)_x$ and $(\cdots)_y$ denote the x and y components of the parenthesised vector respectively. The eigenvalues of the Hessian matrix, found by solving the equation $|\mathcal{H} - \kappa \mathbf{I}| = 0$, are the principal curvatures of the surface, denoted $\kappa_{1,2}$. The shape index is defined in terms of the principal curvatures

$$
\phi = \frac{2}{\pi} \arctan \frac{\kappa_2 + \kappa_1}{\kappa_2 - \kappa_1} \qquad \kappa_1 \geq \kappa_2
\tag{19}
$$

and the overall magnitude of curvature is measured by the curvedness $c = \sqrt{k_1^2 + k_2^2}$. The shape index may be expressed in terms of surface normals thus

$$\phi = \frac{2}{\pi} \arctan \frac{\left(\frac{\partial \mathbf{n}}{\partial x}\right)_x + \left(\frac{\partial \mathbf{n}}{\partial y}\right)_y}{\sqrt{\left(\left(\frac{\partial \mathbf{n}}{\partial x}\right)_x - \left(\frac{\partial \mathbf{n}}{\partial y}\right)_y\right)^2 + 4\left(\frac{\partial \mathbf{n}}{\partial x}\right)_y \left(\frac{\partial \mathbf{n}}{\partial y}\right)_x}} \qquad (20)$$

The relationship between the shape-index, the mean and Gaussian curvatures, and the topographic class of the underlying surface are summarised in Table 1. The table lists the topographic classes (i.e. dome, ridge, saddle ridge etc.) and the corresponding shape-index interval.

Class	Symbol	H	K	Region-type	Shape index interval
Dome	D	-	+	Elliptic	$[\frac{5}{8}, 1)$
Ridge	R	-	0	Parabolic	$(\frac{3}{8}, \frac{5}{8})$
Saddle ridge	SR	-	-	Hyperbolic	$(\frac{1}{8}, \frac{3}{8})$
Plane	P	0	0	Hyperbolic	Undefined
Saddle-point	S	0	-	Hyperbolic	$[-\frac{1}{8}, \frac{1}{8})$
Saddle-rut	SV	+	-	Hyperbolic	$[-\frac{3}{8}, -\frac{1}{8})$
Rut	V	+	0	Parabolic	$(-\frac{5}{8}, -\frac{3}{8})$
Cup	C	+	+	Elliptic	$(-\frac{5}{8}, -1)$

Table 1. Topographic classes.

9   Experiments with curved surfaces

We have experimented with both synthetic surfaces with known ground truth and real-world images. The former are used to assess the accuracy of the method, while we use the latter to demonstrate the practical utility of the method.

9.1   Artificial Textures

We commence by assessing the ability of the method to recover reliable slant, tilt and shape-index information. We use the ground truth surface normals to investigate the systematics of the errors in the estimated needle-map. In Figure 3 we show scatter plots of the estimated slant, tilt and shape index versus their ground truth values. In each case there is a clear regression line. Not surprisingly, the most dispersed quantity is the shape-index. The tilt angle has the tightest regression line.

The parameter of our spectral distortion method is the distance between the points used to estimate the affine distortion matrix on the image plane.

As pointed out earlier, if this distance is too small then the affine distortion becomes undetectable. If, on the other hand, the distance is too large then we sample changes in surface orientation rather than perspective foreshortening.

In Figure 4 for the smoothed (a) and unsmoothed (b) needle-maps we plot the linear regression coefficients extracted from the scatter plots of ground truth versus estimated shape-index as a function of the inter-point distances. If the shape-index measurements are unbiased then the linear regression co-efficient should be unity. The main feature to note is that there is a critical value of the distance which results in a maximum value of the regression co-efficient. For the smoothed needle-maps, the linear regression coefficient is closest to unity (0.97) when the inter-point distance is $r = 16$ pixels; this represents an improvement over the initial unsmoothed value of $\mu = 0.51$. For the unsmoothed needle-maps, the best regression coefficient (0.84) is obtained when $r = 48$ pixels; here the corresponding smoothed value is $\mu = 0.93$.

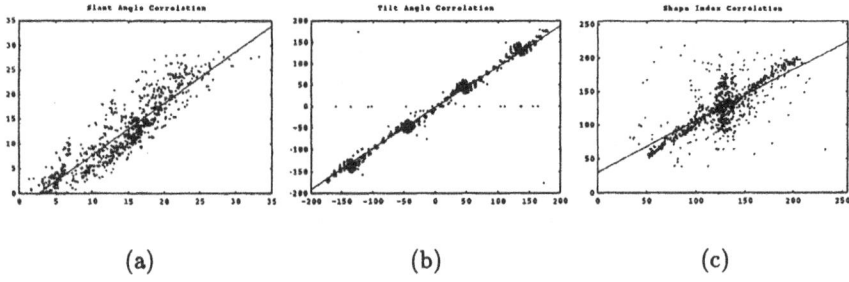

(a)   (b)   (c)

Fig. 3. Scatter Correlation Plots (Robust Kernel). (a) Slant Angle Correlation($\mu = 0.91$); (b) Tilt Angle Correlation($\mu = 0.96$); (c) Shape Index Correlation($\mu = 0.78$). Where μ is the linear correlation coeficient

9.2   Real world textures

In this section we experiment with real world textured surfaces. We have generated the images used in this study by moulding regularly textured sheets into curved surfaces. The images used in this study are shown in the first column of Figure 5. There are two sets of images. The first four have been created by placing a table-cloth with a rectangular texture pattern on top of surfaces of various shapes. From top-to-bottom, the surface shapes are a ridge, a bulge, a series of ripples and a sphere (a balloon). The second group of images which appear in the fifth, sixth and seventh rows have been created by bending a regularly textured sheet of wrapping paper into various tubular shapes. The first of these is a cylinder, the second is a "wave" while the final

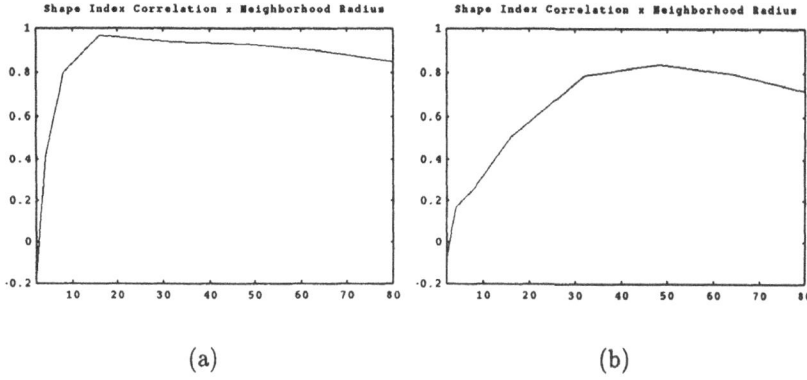

Fig. 4. Plot of the shape index correlation in terms of the neighbourhood radius. (a) Smoothed needle map; (b) Non-Smoothed needle map.

example is an irregular tube. The textures in all seven images show strong perspective effects.

The remaining columns of Figure 5, from left to right, show the initial needle-map, the final smoothed needle-map, the estimated shape-index and the estimated curvedness. In the case of this real world data, the initial needle maps are more noisy and disorganised than their synthetic counterparts. However, there are clear regions of needle-map consistency. When robust smoothing is applied to the initial needle maps, then there is a significant improvement in the directional consistency of the needle directions. In the case of the ridge in the first image, the defining planes are uniform and the ridge-line is cleanly segmented. In the case of the bulge in the second image, the ridge or fold structure at the bottom of the picture is well identified. The ripples stand our clearly in the third image. The radial needle-map pattern emerges clearly in the case of the sphere. This radial pattern is also clear for the three "tubular" objects. The shape-index information in the fourth column and the curvedness in the fifth column reveal the changes in surface topography. In particular, the bright patches in the curvedness plot represent highly curved regions and these correspond to ridge and ravine structures on the surfaces.

Finally, we illustrate how our method can be used to identify multiple texture planes. Here we use the Gustavson-Kessel [28] fuzzy clustering method to locate clusters in the distribution of needle-map directions after robust smoothing has been applied. Each cluster is taken to represent a distinct plane. The mean surface normal direction for the cluster represents the orientation of a distinct plane. Figure 6 shows results obtained for real world images for a textured box and a textured ridge. In each figure panel (a) is the original image, panel (b) shows the initial distribution of needle-map direc-

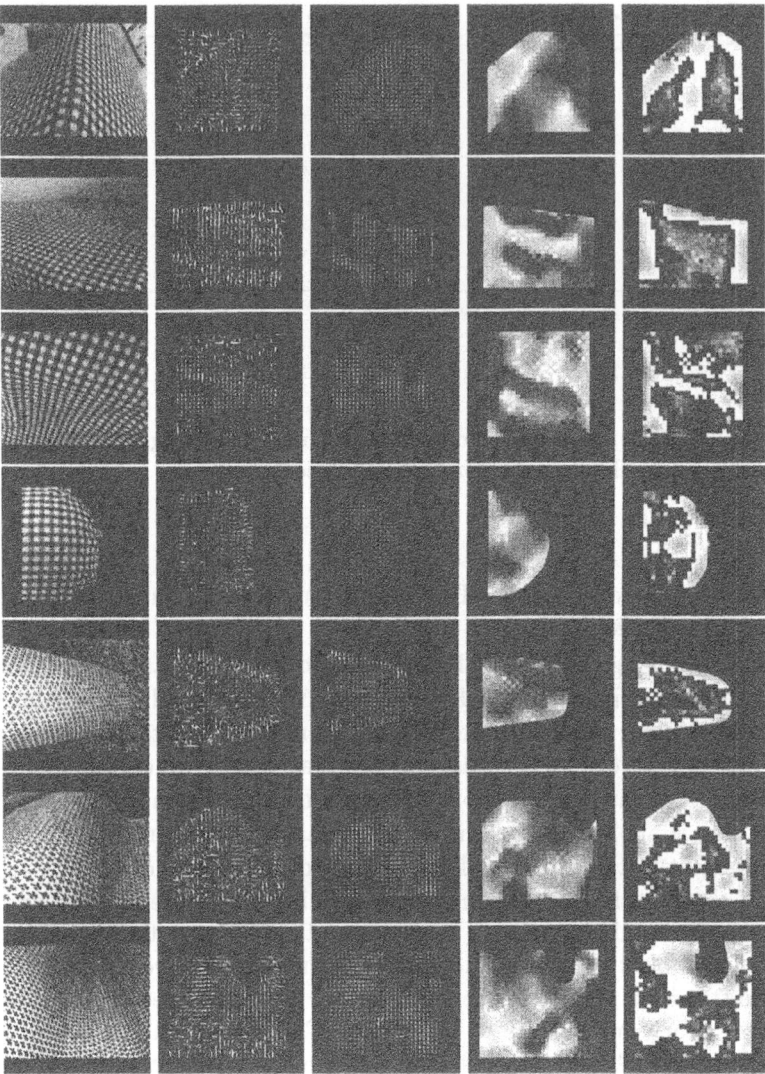

Fig. 5. Real curved surfaces. (a) original image; (b) recovered needle map; (c) Smoothed needle map; (d) Shape index map; (e) Curvedness map.

tions, panel (c) is the final needle-map distribution after robust smoothing and panel (d) is the segmentation obtained by clustering the needle-map directions. The surface normal data is visualised on a unit sphere. The points indicate the positions where the surface normals intercept the sphere. In both cases the initial distribution of needle-map direction is dispersed. After robust smoothing clear clusters develop corresponding to distinct planes. The cluster

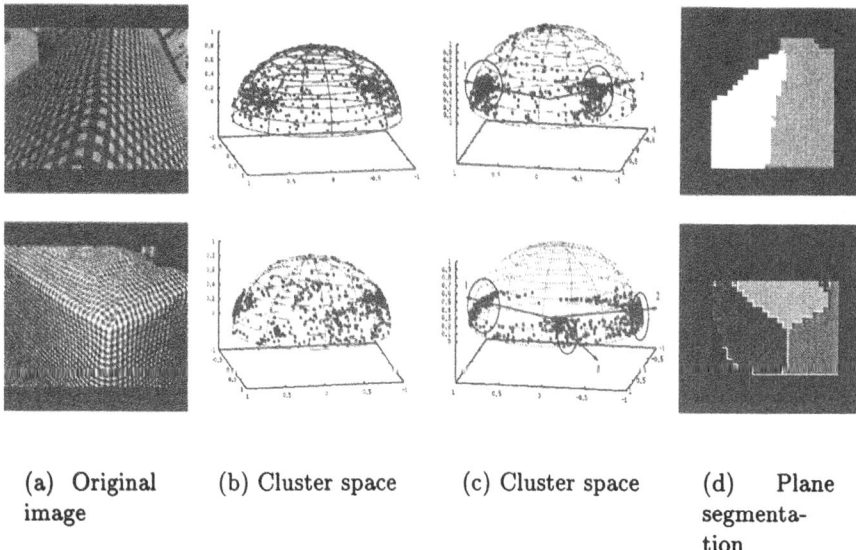

| (a) Original image | (b) Cluster space | (c) Cluster space | (d)   Plane segmentation |

Fig. 6. Experiments on 3D-plane segmentation.

centres detected by Gustavson-Kessel method are indicated by circles, whose radius indicates the cluster-variance. The segmentations shown are obtained by labelling the smoothed surface normals according to the nearest cluster. In both cases the planar segmentation is in good subjective agreement with the image contents.

10   Conclusions

We have presented a new method for estimating the local orientation of tangent planes to curved textured surfaces. The method commences by finding affine spectral distortions between neighbouring points on the image plane. The directions of the eigenvalues of the local distortion matrices can be used to make closed form estimates of the slant and tilt directions. The initial orientation estimates returned by the new method are iteratively refined using a robust smoothing technique to produce a needle map of improved consistency.

The method is demonstrated on both synthetic imagery with known ground truth and on real-world images of man-made textured surfaces. The method proves useful in the analysis of both planar and curved surfaces. Moreover, the extracted needle maps can be used to make reliable estimates of surface curvature information.

There are a number of ways in which the ideas presented in this paper could be extended. Firstly, our texture measures are relatively crude and could be refined to allow us to analyse textures that are regular but not necessarily periodic. Secondly, there is scope for improving the quality of the needle map through measuring the back-projection error associated with the smoothed surface normal directions. Finally, we are investigating ways of utilising the extracted needle maps for recognising curved textured objects.

References

1. A. Rosenfeld. A note on automatic detection of texture gradients. *TC*, 24:988–991, 1975.
2. R. Bajcsy and L. Lieberman. Texture gradient as a depth cue. *Computer Graphics and Image Processing*, 5:52–67, 1976.
3. K.A. Stevens. The information content of texture gradients. *Biological Cybernetics*, 42:95–105, 1981.
4. K.A. Stevens. Slant-tilt: the visual encoding of surface orientation. *Biological Cybernetics*, 46:183–195, 1983.
5. A. P. Witkin. Recovering surface shape and orientation from texture. *Artificial Intelligence*, 17:17–45, 1981.
6. K. Ikeuchi. Shape from regular patterns. *Artificial Intelligence*, 22:49–75, 1984.
7. J. Aloimonos and M.J. Swain. Shape from texture. *Biological Cybernetics*, 58(5):345–360, 1988.
8. K. Ikeuchi and B.K.P. Horn. Numerical shape from shading and occluding boundaries. *Artificial Intelligence*, 17:141–184, 1981.
9. A. Blake and C. Marinos. Shape from texture: estimation, isotropy and moments. *Artificial Intelligence*, 45(3):323–380, 1990.
10. K. Kanatani and T. Chou. Shape from texture: General principle. *Artificial Intelligence*, 38:1–48, 1989.
11. J. Garding. Shape from texture for smooth curved surfaces. In *European Conference on Computer Vision*, pages 630–638, 1992.
12. J. Garding. Shape from texture for smooth curved surfaces in perspective projection. *J. of Mathematical Imaging and Vision*, 2:329–352, 1992.
13. L.G. Brown and H. Shvaytser. Surface orientation from projective foreshortening of isotopic texture autocorrelation. *IEEE Trans. on Pattern Analysis and Machine Intelligence*, 12(6):584–588, 1990.
14. B.J. Super and A.C. Bovik. Planar surface orientation from texture spatial frequencies. *Pattern Recognition*, 28(5):729–743, 1995.
15. J. Malik and R. Rosenholtz. A differential method for computing local shape-from-texture for planar and curved surfaces. In *IEEE Conference on Vision and Pattern Recognition*, pages 267–273, 1993.
16. Ko Sakai and L.H. Finkel. A shape-from-texture algorithm based on human visual psychophysics. In *IEEE Conference on Vision and Pattern Recognition*, pages 527–532, 1994.
17. J.Y. Jau and R.T. Chin. Shape from texture using wigner distribution. *Computer Vision, Graphics and Image Processing*, 52:248–263, 1990.
18. J. Krumm and S. A. Shafer. Local spatial frequency analysis of image texture. In *IEEE International Conference on Computer Vision*, pages 354–358, 1990.
19. J. Krumm and S. A. Shafer. Shape from periodic texture using spectrogram. In *IEEE Conference on Computer Vision and Pattern Recognition*, pages 284–289, 1992.

20. J. Krumm and S. A. Shafer. Texture segmentation and shape in the same image. In *IEEE International Conference on Computer Vision*, pages 121–127, 1995.
21. B.J. Super and A.C. Bovik. Filters for directly detecting surface orientation in an image. In *SPIE Conference on Visual Communications and Image Processing*, pages 144–155, 1992.
22. J. Malik and R. Rosenholtz. Recovering surface curvature and orientation from texture distortion: a least squares algorithm and sensitive analysis. *Lectures Notes in Computer Science - ECCV'94*, 800:353–364, 1994.
23. J.V. Stone and S.D. Isard. Adaptive scale filtering: A general method for obtaining shape from texture. *IEEE Trans. on Pattern Analysis and Machine Intelligence*, 17(7):713–718, 1995.
24. B.K.P. Horn. *Robot Vision*. MIT Press, Massachusetts, 1986.
25. P. L. Worthington and E.R. Hancock. Needle map recovery using robust regularizers. *Image and Vision Computing*, 17(8):545–559, 1998.
26. P.L. Worthington and E.R. Hancock. New constraints on data-closeness and needle map consistency for shape-from-shading. *IEEE Trans. on Pattern Analysis and Machine Intelligence*, 21(12):1250–1267, December 1999.
27. R.N. Bracewell, K.-Y. Chang, A.K. Jha, and Y.-H. Wang. Affine theorem for two-dimensional fourier transform. *Electronics Letters*, 29(3):304, 1993.
28. J. C. Bezdek. *Pattern Recognition with Fuzzy Objective Algorithms*. Plenum Press, 1981.

Camera Calibration from Symmetry

Kwan-Yee K. Wong, Paulo R. S. Mendonça, and Roberto Cipolla

University of Cambridge, Department of Engineering,
Trumpington Street, Cambridge, CB2 1PZ, UK

Summary. This paper addresses the problem of calibrating a pinhole camera from images of a surface of revolution. Camera calibration is the process of determining the intrinsic or internal parameters (i.e. aspect ratio, focal length and principal point) of a camera, and is important for both motion estimation and metric reconstruction of 3D models. In this paper, a novel and simple calibration technique has been introduced which is based on the symmetry of images of surfaces of revolution. Traditional techniques for camera calibration involve taking images of some precisely machined calibration pattern (such as a calibration grid). The use of surfaces of revolution, which are commonly found in daily life (e.g. bowls and vases), makes the process easier as a result of the reduced cost and increased accessibility of the calibration objects. In this paper, it is shown that 2 images of surface of revolution will provide enough information for determining the aspect ratio, focal length and principal point of a camera. An analytical error model is developed, providing variances and confidence intervals of the parameters estimated. The techniques presented in this paper have been implemented and tested with both synthetic and real data. Experiment results show that the camera calibration method presented here is both practical and accurate.

1 Introduction

An essential step for motion estimation and 3D Euclidean reconstruction, two important tasks in computer vision, is the determination of the intrinsic parameters of cameras. This process, known as *camera calibration*, usually involves taking images of some special patterns with known geometry (see [15,10,8] and [6, Chapter 3]). Such methods do not require direct mechanical measurements on the cameras, and often produce very good results. Nevertheless, they involve the design and use of highly accurate tailor-made calibration patterns, which are both difficult and expensive to manufacture.

In this paper a novel technique for camera calibration is introduced. It relates the ideas from [2,11] for calibration from vanishing points to symmetry properties of images of surfaces of revolution [16,13,17,5,12]. The method presented here allows the camera to be calibrated from two or more images of surfaces of revolution, which are commonly found in daily life (bowls, vases etc.). The use of such objects has the advantage of easy accessibility and low cost, in contrast to the traditional calibration pattern.

Section 2 shows how the symmetry that appears in the images of surfaces of revolution provides information about vanishing points related to a set of three mutually orthogonal directions. By extending the techniques for calibration from vanishing points, such information can be used in the development of practical algorithms for camera calibration. These algorithms, detailed in Section 3, are capable of dealing with both known and unknown aspect ratio, whereas previous techniques for calibration based on vanishing points can only handle the former case. The error model is described in Section 4. Section 5 first presents results of experiments conducted on synthetic data, which are used to perform an evaluation of the robustness of the algorithm in the presence of noise. Experiments on real data show the usefulness of the proposed method. Finally, conclusions are presented in Section 6.

2   Theoretical Background

The major contributions of [2] and [17] are briefly reviewed in this section, which provides the mathematical background for the algorithms developed in this paper.

In [17] it has been shown that the perspective image of a surface of revolution exhibits a special symmetry which can be expressed in terms of a transformation known as a *harmonic homology* (see details in [14, Chapter IX]). Consider a surface of revolution S. The image of S taken by a pinhole camera $\mathbf{P}$ is a curve ε. Let $\mathbf{l}_s$ be the image of the axis of revolution of S_r in the camera $\mathbf{P}$. The optical center of $\mathbf{P}$ and the axis of revolution define a plane Π, whose normal direction is $\mathbf{n}_x$. The image of the point at infinity in the direction $\mathbf{n}_x$ is the vanishing point $\mathbf{v}_x$.

If $\mathbf{v}_x$ and $\mathbf{l}_s$ are represented in homogeneous coordinates as $\mathbf{v}_x = [u \ v \ 1]^\mathrm{T}$ and $\mathbf{l}_s = [\cos\theta \ \sin\theta \ -d]^\mathrm{T}$, the 2D collineation $\mathbf{W}$ given by

$$\mathbf{W} = \mathbb{I} - 2\frac{\mathbf{v}_x \mathbf{l}_s^\mathrm{T}}{\mathbf{v}_x^\mathrm{T} \mathbf{l}_s} \tag{1}$$

is a harmonic homology. The profile ε will be invariant to this transformation, which simply maps one side of the profile (with respect to the image of the axis of rotation) to the other

Consider now any two vectors $\mathbf{n}_y$ and $\mathbf{n}_z$ parallel to Π and orthogonal to each other, which together with $\mathbf{n}_x$ form a set of three mutually orthogonal directions. By construction, the vanishing points corresponding to the directions of $\mathbf{n}_y$ and $\mathbf{n}_z$ will lie on $\mathbf{l}_s$ (hereafter, referred to as *axis of revolution*).

These three vanishing points can be used to determine the focal length and the principal point of the camera $\mathbf{P}$ assuming that it has zero skew and aspect ratio 1, as shown in [2,3]. In that paper it is proved that the principal point will coincide with the orthocenter of the triangle with vertices given by the vanishing points, and it follows that the square root of the product of the distances from the orthocenter to any vertex and to the opposite side will give the focal length (see Fig. 1(a)). As a result, given a harmonic homology $\mathbf{W}$ defined by the vanishing point $\mathbf{v}_x$ and the axis of revolution $\mathbf{l}_s$, the principal point of the camera $\mathbf{P}$ will lie on a line $\mathbf{l}_x$ passing through $\mathbf{v}_x$ and perpendicular to $\mathbf{l}_s$. The product of the distances from the principal point to $\mathbf{v}_x$ and to $\mathbf{l}_s$ will give the square of the focal length (see Fig. 1(b)).

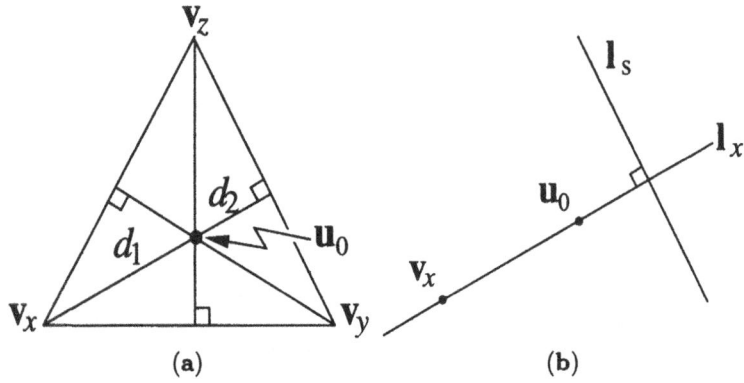

(a)   (b)

Fig. 1. (a) The principal point $\mathbf{u}_0$ of the camera coincides with the orthocenter of the triangle with vertices given by the vanishing points $\mathbf{v}_x$, $\mathbf{v}_y$ and $\mathbf{v}_z$, and the product of the distances d_1 and d_2 is equal to the square of the focal length. (b) The vanishing point $\mathbf{v}_x$ and the axis of revolution $\mathbf{l}_s$ define a line $\mathbf{l}_x$ along which the principal point must lie

3   Algorithms and Implementation

3.1   Estimation of the Harmonic Homology $\mathbf{W}$

The profile ε of each surface of revolution is extracted from the image by applying a Canny edge detector (see Fig. 2). The harmonic homology $\mathbf{W}$ that maps each side of the profile ε to its symmetrical counterpart is then estimated by minimizing the geometrical distances between the original profile and its transformed version. This can be done by sampling N evenly spaced points $\mathbf{x}_i$ along the profile ε and optimizing the cost function

$$Cost_{\mathbf{W}}(\mathbf{v}_x, \mathbf{l}_s) = \sum_{i=1}^{N} \text{dist}(\varepsilon, \mathbf{W}(\mathbf{v}_x, \mathbf{l}_s)\mathbf{x}_i)^2, \tag{2}$$

where $\text{dist}(\varepsilon, \mathbf{W}(\mathbf{v}_x, \mathbf{l}_s)\mathbf{x}_i)$ is the distance between the original profile ε and the transformed sample point $\mathbf{W}(\mathbf{v}_x, \mathbf{l}_s)\mathbf{x}_i$. The 4 parameters for the optimization are θ and d, which define the axis of revolution $\mathbf{l}_s$, and ρ and r, which define the vanishing point $\mathbf{v}_x$ in the x-direction (see Section 2).

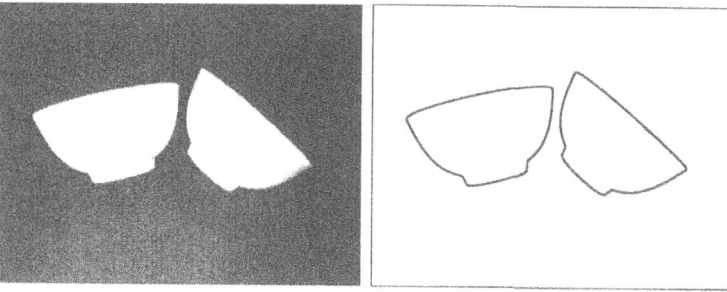

Fig. 2. The profiles of the surfaces of revolution (bowls) are extracted by Canny edge detector

The success of most non-linear optimization problems requires a good initialization so as to avoid convergence to a local minimum. This is achieved here by using bitangents of the profile [16]. Two points near a bitangent are selected and a polynomial is fitted to the profile in the neighbourhood of each point. The bitangent and the bitangent points can then be obtained analytically from the two polynomials. Given two bitangents $l(p_1, p_2)$ and $l(q_1, q_2)$ on the two sides of the profile ε with bitangent points p_1, p_2 and q_1, q_2 respectively (see Fig. 3), the intersection of the two bitangents $(l(p_1, p_2), l(q_1, q_2))$ and the intersection of the diagonals $(l(p_1, q_2), l(q_1, p_2))$ give two points which define a line for an estimate of l_s. An estimate for the vanishing point v_x is given by the point of intersection of the lines $l(p_1, q_1)$ and $l(p_2, q_2)$. The initialization of l_s and v_x from bitangents often provides an excellent initial guess for the optimization problem. This is generally good enough to avoid any local minimum and allows convergence to the global minimum in a small number of iterations. The estimation of the harmonic homology W is summarized in Algorithm 1.

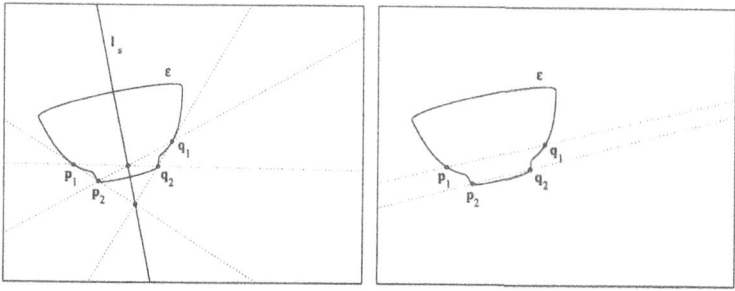

Fig. 3. Initialization of the optimization parameters l_s and v_x from the bitangents and lines formed by bitangent points

---

Algorithm 1 Estimation of the harmonic homology $\mathbf{W}$

---

extract the profile ε of the surface of revolution by using a Canny edge detector;
sample N evenly spaced points $\mathbf{x}_i$ along ε;
initialize the axis of revolution $\mathbf{l}_s$ and the vanishing point $\mathbf{v}_x$ by identifying bitangents in ε;
while not converged **do**
    transform each point $\mathbf{x}_i$ by $\mathbf{W}$;
    compute the distances between ε and the transformed points $\mathbf{W}\mathbf{x}_i$;
    update $\mathbf{l}_s$ and $\mathbf{v}_x$ to minimize the function in (2);
end while

3.2 Estimation of Intrinsic Parameters

Known Aspect Ratio When the aspect ratio of the camera is 1, the line $\mathbf{l}_x$ passing through the principal point (u_0, v_0) and the vanishing point $\mathbf{v}_x$ will be perpendicular to the axis of revolution $\mathbf{l}_s$ (see Section 2). As a result, $\mathbf{l}_x$ can be expressed in terms of $\mathbf{v}_x$ and $\mathbf{l}_s$, and is given by $\mathbf{l}_x = [l_2 \ -l_1 \ l_1v_2 - l_2v_1]$, where $\mathbf{v}_x$ and $\mathbf{l}_s$ are represented by $[v_1 \ v_2 \ 1]^{\mathrm{T}}$ and $[l_1 \ l_2 \ l_3]^{\mathrm{T}}$ respectively. Given two such lines $\mathbf{l}_{x1}$ and $\mathbf{l}_{x2}$, the principal point (u_0, v_0) will be given by the point of intersection of the two lines. When more than two lines are available, the principal point (u_0, v_0) can be estimated by linear least-squares method from

$$\begin{pmatrix} \mathbf{l}_{x1}^{\mathrm{T}} \\ \mathbf{l}_{x2}^{\mathrm{T}} \\ \vdots \\ \mathbf{l}_{xN}^{\mathrm{T}} \end{pmatrix} \begin{pmatrix} su_0 \\ sv_0 \\ s \end{pmatrix} = \mathbf{0}, \tag{3}$$

where $N \geq 2$ is the total number of lines (i.e. number of profiles) and s is a scale factor. The estimated principal point (u_0, v_0) is then projected on each line $\mathbf{l}_{xi}$ orthogonally as $\mathbf{u}_{0i}$, and the focal length f will be given by

$$f = \frac{1}{N} \sum_{i=1}^{N} \sqrt{\mathrm{dist}(\mathbf{u}_{0i}, \mathbf{v}_{xi}) \times \mathrm{dist}(\mathbf{u}_{0i}, \mathbf{l}_{si})}, \tag{4}$$

where $\mathrm{dist}(\mathbf{u}_{0i}, \mathbf{v}_{xi})$ is the distance between the vanishing point $\mathbf{v}_{xi}$ and the projected principal point $\mathbf{u}_{0i}$, and $\mathrm{dist}(\mathbf{u}_{0i}, \mathbf{l}_{si})$ is the distance between the axis of revolution $\mathbf{l}_{si}$ and the projected principal point $\mathbf{u}_{0i}$. Note that the terms for summation are actually the focal lengths estimated from each pair of $\mathbf{v}_{xi}$ and $\mathbf{l}_{si}$ with the estimated principal point projected onto the corresponding $\mathbf{l}_{xi}$ (see Section 2), and the focal length f is taken to be the mean of these estimated values. When the aspect ratio a is known but not 1, there exists a homography $\mathbf{T}(a)$ given by

$$\mathbf{T}(a) = \begin{pmatrix} \frac{1}{a} & 0 & 0 \\ 0 & 1 & 0 \\ 0 & 0 & 1 \end{pmatrix} \tag{5}$$

that transforms the image coordinate system such that the resulting camera would have aspect ratio 1 with the original focal length preserved. After transforming the camera to one with aspect ratio 1, the algorithm presented above can be applied in the same way and the principle point will be given by (au_0, v_0), where (u_0, v_0) are obtained from (3).

Unknown Aspect Ratio When the aspect ratio a of the camera is unknown, the principal point (u_0, v_0) and focal length f cannot be obtained directly from the vanishing points $\mathbf{v}_{zi}$ and axes of revolution $\mathbf{l}_{si}$. However, a search for a can be performed by optimizing the cost function

$$\begin{aligned} Cost_a(a) = \mathrm{VAR}(\{&\mathrm{dist}(\mathbf{u}'_{0i}, \mathbf{T}(a)\mathbf{v}_{zi}) \\ \times\ &\mathrm{dist}(\mathbf{u}'_{0i}, \mathbf{T}^{-\mathrm{T}}(a)\mathbf{l}_{si})\}_{i=1}^{N}), \end{aligned} \tag{6}$$

where VAR is the variance of the data set. $\mathbf{u}'_{0i}$ is the projection of the principal point $\mathbf{u}'_0$ onto the line passing through $\mathbf{T}(a)\mathbf{v}_{zi}$ and orthogonal to $\mathbf{T}^{-1}(a)\mathbf{l}_{si}$. $\mathrm{dist}(\mathbf{u}'_{0i}, \mathbf{T}(a)\mathbf{v}_{zi})$ is the distance between the transformed vanishing point $\mathbf{T}(a)\mathbf{v}_{zi}$ and the projected principal point $\mathbf{u}'_{0i}$, and $\mathrm{dist}(\mathbf{p}'_{0i}, \mathbf{T}^{-1}(a)\mathbf{l}_{si})$ is the distance between the transformed axis of revolution $\mathbf{T}^{-1}(a)\mathbf{l}_{si}$ and the projected principal point $\mathbf{u}'_{0i}$.

Since the search space is only one-dimensional, instead of performing an extensive search, one can just plot the cost against typical range of a and obtain the result directly (see Fig. 4).

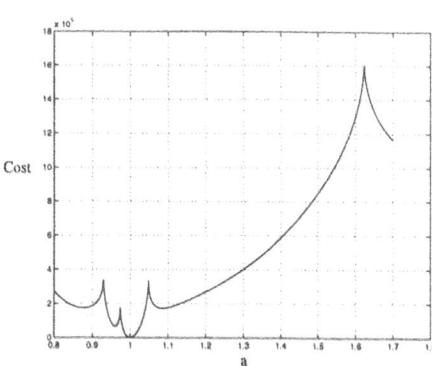

Fig. 4. Plot of the cost function in (6) against a from 0.8 to 1.7 with step size 0.001 for a camera with aspect ratio 1

After convergence, the aspect ratio a obtained can then be used to transform the vanishing points $\mathbf{v}_{xi}$ and axes of revolution $\mathbf{l}_{si}$ using the homography $\mathbf{T}(a)$, and the principal point (u_0, v_0) and focal length f can be recovered as in the case of known aspect ratio.

4   Error Model

An important part of any estimation problem is the determination of the uncertainties of the parameters estimated. It is not enough to state the value of a focal length after calibrating a camera: a confidence interval for the value estimated must also be provided. A comprehensive introduction to this and related problems can be found in [9].

4.1   Error Propagation

Consider the cost function

$$\xi^2(\boldsymbol{\theta}) = \sum_{i=1}^{N} (\mathbf{y}_i - \mathbf{f}(\mathbf{x}_i; \boldsymbol{\theta}))^{\mathrm{T}} (\mathbf{y}_i - \mathbf{f}(\mathbf{x}_i; \boldsymbol{\theta})). \tag{7}$$

The value $\boldsymbol{\theta} = \hat{\boldsymbol{\theta}}$ that minimizes (7) is a *maximum likelihood estimator* (MLE) for the parameter $\boldsymbol{\theta}$ of the parametric model $\mathbf{y} = \mathbf{f}(\mathbf{x}; \boldsymbol{\theta})$. For noise free-data, i.e., $(\mathbf{x}_i, \mathbf{y}_i)$ exactly satisfying the model, the *residual* $\xi^2(\hat{\boldsymbol{\theta}})$ in (7) will be zero. However, due to the presence of noise, the data will not perfectly fit the model, and $\xi^2(\hat{\boldsymbol{\theta}}) \geq 0$. It is possible to propagate this residual and obtain the corresponding uncertainty for the parameter $\hat{\boldsymbol{\theta}}$ by projecting the points $(2\xi^2(\hat{\boldsymbol{\theta}}), \boldsymbol{\theta}) \cap (\xi^2(\boldsymbol{\theta}), \boldsymbol{\theta})$ onto the plane $(0, \boldsymbol{\theta})$ (see Fig. 5). If $\mathbf{H}|_{\hat{\boldsymbol{\theta}}}$ is the Hessian of $\xi^2(\boldsymbol{\theta})$ computed in $\hat{\boldsymbol{\theta}}$, it can be shown that the covariance matrix $\mathbf{C}(\hat{\boldsymbol{\theta}})$ of $\hat{\boldsymbol{\theta}}$ will be given by $\mathbf{C}(\hat{\boldsymbol{\theta}}) = 2\xi^2 \mathbf{H}|_{\hat{\boldsymbol{\theta}}}^{-1}$.

Analysis of several covariance matrices for the parameters of $\mathbf{l}_s$ and $\mathbf{v}_x$ showed that the uncertainty is essentially in $\mathbf{v}_x$. This was expected, since the vanishing point $\mathbf{v}_x$ corresponds to the intersection of lines which are nearly parallel. Thus it is justified, in order to simplify the analysis, to ignore the uncertainty in the parameters of $\mathbf{l}_s$. The covariance matrix of u and v in $\mathbf{v}_x$ will be henceforth denoted by $\mathbf{C}(\mathbf{v}_x)$.

4.2   Covariance Matrix of the Principal Point

This error propagation technique can be directly used to determine the covariance matrix $\mathbf{C}(\mathbf{u}_0)$ of the principal point $\mathbf{u}_0$. The computation of the hessian of the function (2) is carried out during the optimization process.

After the computation of the homologies, $\mathbf{u}_0$ is recovered from (3), which can be rewritten as

$$\mathbf{L}_x \begin{bmatrix} \mathbf{u}_0 \\ 1 \end{bmatrix} = \mathbf{0}. \tag{8}$$

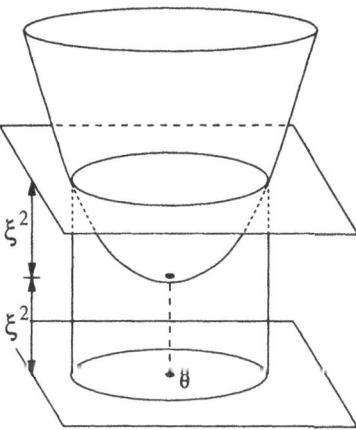

Fig. 5. The covariance matrix of the parameter $\hat{\theta}$ can be found by projecting points of the manifold $(2\xi^2(\hat{\theta}), \theta) \cap (\xi^2(\theta), \theta)$ onto the plane $(0, \theta)$

If $\mathbf{L}_x = [\mathbf{N}_x - \mathbf{d}]$, then the least squares solution of (3) will be given by $\hat{\mathbf{u}}_0 = \mathbf{N}_x^+ \mathbf{d}$, where $\mathbf{N}_x^+$ is the Moore-Penrose inverse of $\mathbf{N}_x$. Therefore,

$$\mathbf{C}(\hat{\mathbf{u}}_0) = \mathbf{N}_x^+ \text{diag}(\sigma^2(\mathbf{d})) \mathbf{N}_x^{+T}, \tag{9}$$

where $\text{diag}(\sigma^2(\mathbf{d}))$ corresponds to the diagonal matrix whose entries are $\sigma^2(d_i)$.

4.3 Variance of the Focal Length

The computation of the variance of the focal length as estimated by (4) is more elaborated. Each term in the summation in (4) is a separate (but not independent) estimate f_i of f, given by $f_i = \sqrt{d_{i1}}\sqrt{d_{i2}}$. In order to simplify the calculations some hypothesis will be assumed: (a) $\sqrt{d_{i2}}$ is constant; since $d_{i2} = \mathbf{v}_x^T \mathbf{l}_s$, d_{i2} is not constant; nevertheless, the vanishing point $\mathbf{v}_x$ is, in general, tens of thousands of pixels away from $\mathbf{l}_s$; the covariance matrix of $\mathbf{v}_x$, in all the experiments made, indicates an uncertainty of less than 50 pixels for this parameter, justifying the approximation; (b) f_i is independent of f_j if $i \neq j$; in fact, they all depend on $\mathbf{u}_0$, but the influence of $\mathbf{l}_s$ and $\mathbf{v}_x$ is dominant. Under this hypothesis it can be shown that

$$\text{VAR}(f) \approx \frac{1}{N^2} \sum_{i=1}^{N} \frac{1}{4 d_{i1}} \text{VAR}(d_{i1}) d_{i2}. \tag{10}$$

Despite all the approximations, (10) and (9) produce good results, as shown by the results of the experiments.

5   Experimental Results

Experiments on both synthetic and real data have been performed, and the results will be presented in the following subsections.

5.1   Synthetic Data

Generation of Data The experimental setup consists of a surface of revolution which is made up of two spheres intersecting each other. The intrinsic parameters of the synthetic camera are given by the calibration matrix $\mathbf{K}$ as in [6], where focal length $f = 700$, aspect ratio $a = 1$ and with principal point at $(320, 240)$, and the images have a dimension of 640×480. Each sphere $\mathbf{Q}$ will be projected as a conic $\mathbf{C}$ on the image plane and the projection is given by [4] $\mathbf{C} = (\mathbf{PQ}^{-1}\mathbf{P}^T)^{-1}$, where $\mathbf{P}$ is the 3×4 projection matrix of the camera, $\mathbf{Q}$ is a 4×4 matrix representing the sphere and $\mathbf{C}$ is a 3×3 matrix representing the conic. The profile in each image is found by projecting each sphere into the image as a conic and finding points on each conic that lie outside the other conic. The profiles formed by the three cameras are shown in Fig. 6.

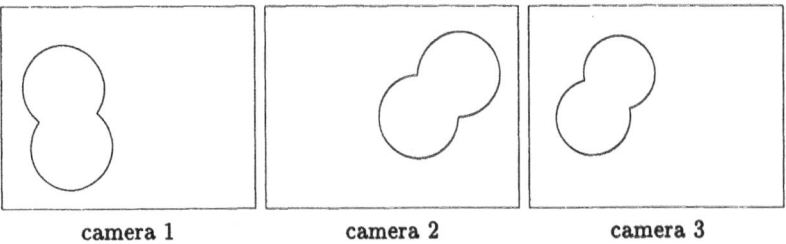

camera 1   camera 2   camera 3

Fig. 6. Profiles of the surface of revolution in the images taken by the synthetic camera at the three different positions

In order to evaluate the robustness of the algorithms described in Section 3, different levels of noise have been added along the normal direction of each pixel on the profile. Nevertheless, whereas for corner features the assumption of uncorrelated noise is acceptable, the noise affecting adjacent edgels of profiles exhibits a strong correlation. This is due to the fact that one of the steps of edge detection is a smoothing of the image, which introduces correlation between adjacent pixels. To simulate this effect, uncorrelated uniform noise was convolved with the same Gaussian kernel used to smooth the image, and the output of this filtering was used as to disturb the edgels of the profile.

Experiments and Results Experiments on noise-free data (see Fig. 6) and data with six different noise levels have been performed. The six noise levels are 0.5, 0.7, 1.0, 1.2 and 1.5 pixel respectively. For each noise level, 10 experiments have been conducted using the algorithms described in Section 3 for unknown aspect ratio. One hundred evenly spaced points have been sampled from each profile for the estimation of the harmonic homology $\mathbf{W}$.

noise level	mean(f)	std(f)	error(f)
ground truth	700.00	0.00	0.00%
0.0	698.15	0.00	-0.20%
0.5	686.34	13.21	-1.95%
0.7	674.65	23.50	-3.62%
1.0	676.05	28.15	-3.42%
1.2	685.69	7.70	-2.04%
1.5	638.67	30.02	-8.76%

Table 1. Results of calibration of focal length from profiles of surface of revolution using synthetic data with different noise levels

noise level	mean(a)	std(a)	error(a)
ground truth	1.0000	0.0000	0.00%
0.0	1.0000	0.0000	0.00%
0.5	1.0031	0.0063	0.31%
0.7	1.0059	0.0078	0.59%
1.0	1.0014	0.0086	0.14%
1.2	0.9950	0.0027	-0.50%
1.5	1.0080	0.0069	0.80%

Table 2. Results of calibration of the aspect ratio from profiles of surface of revolution using synthetic data with different noise levels

The results of the experiments are shown in Table 1 , Table 2 and Table 3. The estimated parameters being shown for each noise level are the mean values over the 10 experiments. The errors being listed are the percentage errors of the mean value of each parameter relative to the ground truth focal length. As the noise level increases, both the relative errors and standard deviations increase. For a noise level of 1.0 pixel, the error for the focal length and principal point is less than 5%.

noise level	mean(u_0, v_0)	std(u_0, v_0)	error(u_0, v_0)
ground truth	(320.00,240.00)	(0.00,0.00)	(0.00%,0.00%)
0.0	(321.05,240.06)	(0.00,0.00)	(0.15%,0.01%)
0.5	(318.54,247.18)	(12.73,8.19)	(-0.21,%1.03%)
0.7	(311.87,258.64)	(18.00,13.06)	(-1.16%,2.66%)
1.0	(322.20,256.65)	(22.06,16.96)	(0.31%,2.38%)
1.2	(339.01,247.17)	(7.70,4.09)	(2.72%,1.02%)
1.5	(316.64,253.99)	(17.62,14.94)	(-0.48%,2.00%)

Table 3. Results of calibration of the principal point. Errors are relative to the focal length

5.2   Real Data

Ground Truth The camera used in the experiments for real data is a FUJI MX-700 digital camera and the image size is chosen to be 640×480. The ground truth for the camera's intrinsic parameters is obtained by using a calibration grid.

11 images of the calibration grid have been taken with the camera at different orientations. Corners are extracted from each image using a Canny edge detector and line fitting techniques. For each image, a linear least-squares method [1] is first used to obtain estimates for the 11 parameters of the projection matrix. These estimates are then used to initialize the optimization for the projection matrix elements which minimize the re-projection errors [7] while enforcing aspect ratio 1 and zero skew.

Experiments and Results A set of real images have been used for the calibration of the digital camera, consisting of 3 images of two bowls, providing 4 profiles of surface of revolution (see Fig. 7). The results of calibration from the images using the algorithm presented in Section 3 for unknown aspect ratio are shown in Table 4. The errors shown are the percentage error of the each parameter relative to the ground truth focal length $f = 685.00$ estimated from the calibration grid.

parameter	f	a	u_0	v_0
MLE	697.53	1.0100	318.36	244.40
ground truth	685.00	1.0008	322.60	232.15
error	1.83%	0.9245%	-0.62%	1.79%
predicted std	10.37	-	5.19	3.41

Table 4. Results of calibration using the profiles extracted from the images of the bowls

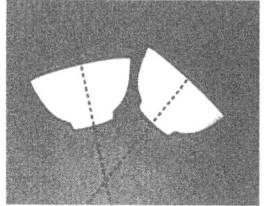 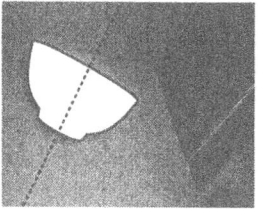 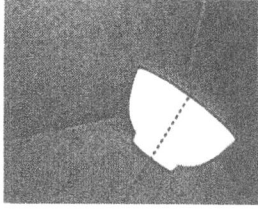

Fig. 7. 3 images of the bowls with the extracted profiles and estimated axes of revolution represented by solid and dash lines respectively

6   Conclusions

By exploiting the symmetry of surfaces of revolution and properties of vanishing points, a practical technique for camera calibration has been developed. The use of a surface of revolution makes the calibration process easier, in not requiring the use of any precisely adjusted device such as a calibration grid. Note that a surface of revolution can be generated by rotating any arbitrarily shaped object on a turntable [12] and thus the method can be seamlessly incorporated into reconstruction systems based on turntable sequences. The proposed method is promising, as demonstrated by experimental results on synthetic and real data. The focal lengths are estimated with high accuracy, with an error smaller than 4% for realistic levels of noise.

The experiments with synthetic data have demonstrated that the calibration algorithm is remarkably robust, performing well for any practical level of noise. The estimation of the focal length seems to be biased, and the focal length is consistently underestimated. Nevertheless, this trend was not confirmed by the real data experiments, which produced a nearly 2% overestimation of f. This suggests that the error model for the synthetic data experiments must be improved.

The error analysis developed in Section 4 is accurate, except in the prediction of the range of v_0. This parameter, however, presented instability even for the calibration obtained from the grid, which produced results in the interval 225.69 – 239.7245. It also has a standard deviation larger then all the other parameters, computed from the 11 values found by using the calibration grid. A natural extension of this ideas is to use the error analysis not only to predict uncertainty bounds, but also to improve the quality of the estimation.

Acknowledgements

Paulo R. S. Mendonça gratefully acknowledges the financial support of CAPES, Brazilian Ministry of Education, grant BEX1165/96-8.

References

1. Y. I. Abdel-Aziz and H. M. Karara. Direct linear transformation into object space coordinates in close-range photogrammetry. In *Proc. Symp. Close-Range Photogrammetry*, pages 1–18, University of Illinois, Urbana, 1971.

2. B. Caprile and V. Torre. Vanishing points for camera calibration. *Int. Journal of Computer Vision*, 4:127–139, 1990.

3. R. Cipolla, T. Drummond, and D. Robertson. Camera calibration from vanishing points in images of architectural scenes. In *Proc. British Machine Vision Conference*, pages 382–391, Nottingham, UK, 1999.

4. G. Cross and A. Zisserman. Quadric reconstruction from dual-space geometry. In *Proc. 6th Int. Conf. on Computer Vision*, pages 25–31, 1998.

5. R.W. Curwen, C.V. Stewart, and J.L. Mundy. Recognition of plane projective symmetry. In *Proc. 6th Int. Conf. on Computer Vision*, pages 1115–1122, 1998.

6. O. Faugeras. *Three-Dimensional Computer Vision: a Geometric Viewpoint*. MIT Press, 1993.

7. O.D. Faugeras and G. Toscani. The calibration problem for stereo. In *Proc. Conf. Computer Vision and Pattern Recognition*, pages 15–20, 1986.

8. O.D. Faugeras and G. Toscani. Camera calibration for 3D computer vision. In *Proc. Inter. Workshop Mach. Vision Mach. Intell.*, pages 25–34, 1987.

9. K. Kanatani. *Statistical Optimization for Geometric Computation: Theory and Practice*. Elsevier Science, Amsterdam, 1996.

10. R. K. Lenz and R. Y. Tsai. Techniques for calibration of the scale factor and image center for high accuracy 3D machine vision metrology. *IEEE Trans. Pattern Analysis and Machine Intell.*, 10(5):713–720, 1988.

11. D. Liebowitz and A. Zisserman. Combining scene and auto-calibration constraints. In *Proc. 7th Int. Conf. on Computer Vision*, pages 293–300, 1999.

12. P. R. S. Mendonça, K-Y. K. Wong, and R. Cipolla. Circular motion recovery from image profiles. In B. Triggs, R. Szeliski, and A. Zisserman, editors, *ICCV Vision and Algorithms Workshop: Theory and Practice*, Corfu, Greece, 21–22 September 1999. Springer–Verlag.

13. J. L. Mundy and A. Zisserman. Repeated structures: Image correspondence constraints and 3d structure recovery. In J. L. Mundy, A. Zisserman, and D. Forsyth, editors, *Applications of Invariance in Computer Vision*, volume 825 of *Lecture Notes in Computer Science*, pages 89–106. Springer–Verlag, 1994.

14. J. G. Semple and G. T. Kneebone. *Algebraic Projective Geometry*. Oxford University Press, 1952.

15. R.Y. Tsai. A versatile camera calibration technique for high-accuracy 3D machine vision metrology using off-the-shelf tv cameras and lenses. *IEEE Journal of Robotics and Automation*, RA-3(4):323–344, 1987.

16. A. Zisserman, D. Forsyth, J. Mundy, and C. A. Rothwell. Recognizing general curved objects efficiently. In J.L Mundy and A. Zisserman, editors, *Geometric Invariance in Computer Vision*, chapter 11, pages 228–251. MIT Press, Cambridge, Mass., 1992.

17. A. Zisserman, J. L. Mundy, D. A. Forsyth, J. Liu, N. Pillow, C. Rothwell, and S. Utcke. Class-based grouping in perspective images. In *Proc. 5th Int. Conf. on Computer Vision*, pages 183–188, 1995.

Dynamic Shapes of Arbitrary Dimension: The Vector Distance Functions

Olivier Faugeras[1,2] and Jose Gomes[1]

[1] INRIA
2004 route des Lucioles
B.P. 93, 06902 Sophia-Antipolis Cedex
France
[2] MIT
AI-Lab
545 Technology Square
Cambridge, MA 02139
USA

Summary. We present a novel method for representing and evolving objects of arbitrary dimension. The method, called the Vector Distance Function (VDF) method, uses the vector that connects any point in space to its closest point on the object. It can deal with smooth manifolds with and without boundaries and with shapes of different dimensions. It can be used to evolve such objects according to a variety of motions, including mean curvature. If discontinuous velocity fields are allowed the dimension of the objects can change. The evolution method that we propose guarantees that we stay in the class of VDFs and therefore that the intrinsic properties of the underlying shapes such as their dimension, curvatures can be read off easily from the VDF and its spatial derivatives at each time instant. The main disadvantage of the method is its redundancy: the size of the representation is always that of the ambient space even though the object we are representing may be of a much lower dimension. This disadvantage is also one of its strengths since it buys us flexibility.

1 Introduction and history

In this paper we present a general method for representing objects of arbitrary dimension embedded in spaces of arbitrary dimension. The representation method is also the basis for evolving such objects according to a variety of motions, including mean curvature. We are not limited to objects of constant dimension, for example we can cope with open curves or surfaces, or even with objects such as the one shown in figure 1 which is the union of an open curve, an open surface, and a volume. The history of this program is quite long and can be traced back in Computer Vision to the early work on snakes by Kass, Witkins and Terzopoulos [29]. But the motivation is of course much older, see [31]. Their idea was to find a practical method for evolving planar curves, considered as embedded in an image, when submitted to force fields created by elements in the image, e.g. edges, corners, or interactively,

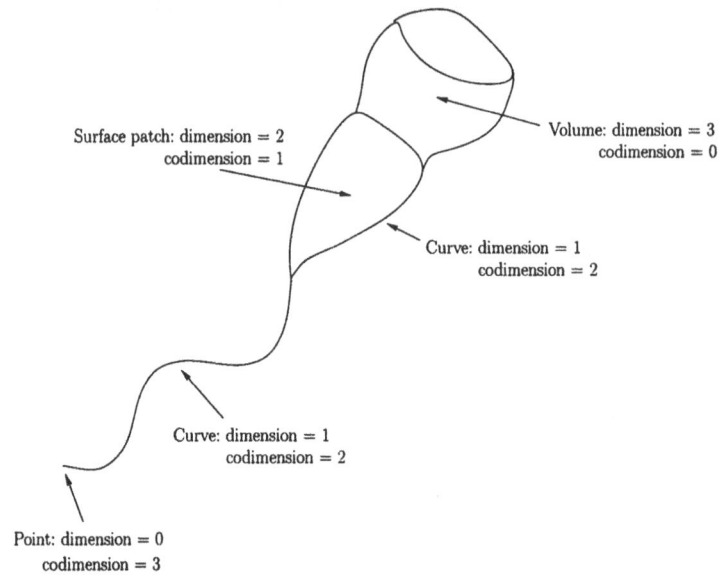

Surface patch: dimension = 2
codimension = 1

Volume: dimension = 3
codimension = 0

Curve: dimension = 1
codimension = 2

Curve: dimension = 1
codimension = 2

Point: dimension = 0
codimension = 3

Fig. 1. An object difficult to model with existing techniques.

by the user. Mathematically, these fields were forcing the curve to evolve according to a second order ordinary differential equation (ODE). Evolving the curve was then equivalent to solving this ODE starting from an initial curve provided by the user. Computationally, the curve was represented by a set of splines and its evolution reduced to the evolution of their control points. The method was quite efficient and could cope with open and closed curves. Improvements of the computational side allowed the Oxford and Cambridge groups around Andrew Blake and Roberto Cipolla to come up with real-time implementations on sequences of images for tracking silhouettes of objects ([8,15]).

It was not too difficult to generalize the idea to 3-D in order to cope with surfaces and this was done by a number of people after the original paper [43]. See [33] for a nice survey of these methods. One of the problems that were met in both the 3-D and the 2-D cases is the problem that the control points of the approximating splines, even though they were chosen somewhat uniformly distributed along the initial curve or surface, had a tendency not to remain so during the course of the evolution, creating numerical problems. These numerical problems were not well understood in particular because the general problem of curve and surface evolution was not well understood to start with.

As it is often the case in science and technology, it turned out that this problem was receiving at the same time a great deal of attention in the community of mathematicians. They were looking at a special case, namely that of the evolution of a closed simple smooth planar curve when submitted to some velocity field. In a series of papers, Epstein, Gage, Grayson, Hamilton, and others proved three important facts. First they proved that under some mild hypotheses it is only the normal part of the velocity field, i.e. its component along the normal to the curve that is relevant for the evolution of the curve [17]. Second, in the case where the velocity is the curvature of the curve, a type of motion known as mean curvature motion, they showed that even if the initial curve is not convex, it becomes so in finite time [27]; and third they showed that after becoming convex, the curve goes through a singularity in finite time, disappearing as a circle with zero radius [21,22]. The importance of these results cannot be underestimated because in many applications where one evolves planar curves, one of the terms in the velocity is very often a function of curvature. It is in effect always the case when the velocity of the curve is derived from a variational principle written intrinsically, i.e. as a function of its arclength.

The extension to higher dimensions is both straightforward and difficult. It is in effect straightforward to see that the mean curvature motion equation

$$C_t = \kappa \mathbf{N}$$

of a planar closed smooth curve C becomes

$$\mathcal{M}_t = \mathcal{H}\mathbf{N} \tag{1}$$

in the case of a smooth closed hypersurface $\mathcal{M}$ of $\mathbb{R}^n$ ($n¿2$) (its dimension is $n - 1$, therefore its codimension is 1, meaning that it can be described locally by one equation in the space coordinates, or that its tangent space $T_{\mathbf{x}}\mathcal{M}$ is of dimension $n - 1$ at every point $\mathbf{x}$, or that its normal space $N_{\mathbf{x}}\mathcal{M}$ is of dimension 1 at every point), where $\mathcal{H}$ is the mean curvature of the manifold, i.e. the mean of its $n - 1$ principal curvatures, and $\mathbf{N}$ is one of its unit normal. In the case of a smooth closed manifold of dimension strictly less than $n - 1$, i.e. of codimension strictly greater than 1, the evolution equation simply becomes

$$\mathcal{M}_t = \mathcal{H}$$

where $\mathcal{H}$ is the mean curvature vector of the manifold, a vector that belongs to the normal space of the manifold (of dimension $k > 1$) and whose definition is given in [41] or in [24].

But the simplicity of these equations hides the complexity of the corresponding motion. There are no theorems analog to the ones by Epstein, Gage, Grayson and Hamilton for the case of hypersurfaces, and practically nothing is known in the case of arbitrary codimension. In fact there exist counterexamples to the straightforward generalization of their theorems: the case of the dumbell which breaks into two pieces before becoming convex is one such counterexample [28,38].

Beside this work in differential geometry, another community was busy at the same time with inventing numerical methods and algorithms for evolving interfaces between different materials. Since these interfaces were often closed planar curves or surfaces that could develop discontinuities (shocks) and change topology, the method of level sets was introduced by Osher and Sethian [34] to deal exactly with these issues with which the standard Lagrangian techniques could not deal. Their Eulerian approach consists in replacing the evolution of the manifold of interest (a planar curve or a 3D surface in their case) with that of a function defined in a space of one dimension higher than the embedding space of the manifold and whose zero-crossings are precisely the evolving manifold. This function satisfies a Partial Differential Equation (PDE) which is very closely related to that satisfied by the evolving manifold:

$$u_t = \frac{1}{n-1} div(\frac{\nabla u}{\|\nabla u\|})\|\nabla u\| \qquad (2)$$

While Sethian went ahead to develop efficient algorithms for implementing this idea such as the narrow-banding and the fast marching methods [39], other mathematicians, such as Evans and Spruck [19,14] characterized the set of functions for which the method was indeed valid. One such function, and an obvious candidate, is the signed distance function to the manifold: by definition, its zero-crossings are the manifold under consideration and it is smooth in its vicinity. This function, being easy to compute, is often used in the applications of the level set method. One major problem it suffers from is the fact that the signed distance function is *not* a solution of the above PDE, as shown for example in [6].

On the other hand, as shown by the authors [23], the solution of the above PDE causes quickly numerical problems because its level sets tend to get very close to each other. The solution which has been adopted by the vast majority of authors is to reinitialize the function u after a fixed number of iterations (sometimes at each iteration!) so that it remains the distance function to its 0 level set. This is of course incorrect unless one proves that one is solving the initial PDE. As shown by Zhao, Chan, Meriman and Osher [44] and by the authors [26], it is possible to preserve the distance function, but the usual level-set PDE has to be modified in a non-trivial manner.

As an interesting application of these ideas Steve Zucker and his collaborators developed a nice theory of two-dimensional shape representation [30] based on the deformation of the boundary C_0 of the shape by an equation of the type

$$C_t = (1 + \alpha\kappa)\mathbf{N} \quad C(.,0) = C_0(.).$$

They use the singularities of the evolving curve to represent the original shape; the resulting representation can be used for matching [40].

Another interesting contribution to the world of snakes was made by Caselles and his collaborators who observed that since the mean curvature motion is the gradient flow that minimizes the $n-1$-dimensional area of a hypersurface (a curve in the plane, a surface in space, etc ...), if one changed the metric of the hypersurface by introducing e.g. a function of the intensity of the n-dimensional image in which it evolves, one obtained a new evolution equation that is similar to (1) and can be implemented by the level set technique using an equation similar to (2) [10–12,9]. This was the starting point of a large body of work using these ideas in such areas as image segmentation [42,13], motion analysis [35], and stereo [20]

All this was done in the framework of the evolution of manifolds of codimension 1. In 1996, Ambrosio and Soner, inspired by ideas of De Giorgi, published a paper [3] in which they showed that the level-set method could be extended to the case of arbitrary codimension. Their idea is to replace the evolution of the smooth manifold under mean curvature motion by that of a tubular neighbourhood of the manifold, in effect a hypersurface. They show that the evolution of this tube is related to that of the manifold in a simple way and that it is not the mean curvature motion of the hypersurface. This theory has been applied by Lorigo et al. [32] to the problem of detecting blood vessels in volumetric medical images, the blood vessels being considered as the tubular neighborhoods of 3D curves. A different approach, closer to what we propose in the present paper, has been proposed by Sapiro and collaborators in a preliminary paper on tracking curves on a surface [7].

There are a number of problems with Ambrosio and Soner's approach, the main one being that it sweeps in some sense the dust under the rug: even though it does evolve correctly the manifold of interest, it turns out that recovering this manifold is in itself a major problem since it is not explicitly represented.

It is therefore natural to turn to a different approach and to attempt to represent an arbitrary smooth manifold $\mathcal{M}$ of dimension k as the intersection of $n - k$ hypersurfaces; the evolution of the hypersurfaces is computed in order to guarantee that their intersection evolves as required for $\mathcal{M}$ and that they remain transverse. This approach is natural since it is based on the definition of the dimension (or the codimension). It was suggested by Ambrosio and Soner in [3] but not pursued because it was thought to be too difficult. The corresponding program can nonetheless be achieved, as described in [25]. We will not pursue it here because it is not easy to deal with such objects as the one represented in figure 1 and to deal with changes in the dimension of the manifold during the evolution. Instead we will follow a slightly counterintuitive idea that was proposed by Ruuth, Merriman, Xin and Osher in a discrete setting, [37]. We were inspired by the last section of this technical report and our paper elaborates on some of the suggestions of these authors and generalizes them in a variety of directions. The idea is to introduce redundancy in the representation of the manifold $\mathcal{M}$: instead of representing it as the intersection of k hypersurfaces, we propose to represent it as the intersection of n hypersurfaces. These hypersurfaces are related in a natural manner to the distance of the points of $\mathbb{R}^n$ to $\mathcal{M}$ and evolve in such a way as to guarantee that their intersection evolves according to the desired evolution for $\mathcal{M}$. Introducing this redundancy allows for more flexibility in the representation: manifolds with non constant dimensions (in space) and boundaries such as the one in figure 1 can now be represented and evolved. Their dimension can even change in time, i.e. increase or decrease.

The plan of the paper is as follows. In section 2 we introduce our redundant representation, called the Vector Distance Function (VDF), for arbitrary smooth manifolds of dimension k. In section 3 we study some of its differential properties. In section 4 we start looking at the problem of evolving a manifold by evolving its VDF instead: we show that this problem has a very simple solution that guarantees that the VDF remains a VDF at all times. In section 5 we illustrate some of the theoretical results with the deformation of an helix under mean curvature motion. Section 6 is devoted to an important generalization of our results to the case of smooth manifolds with a boundary which we show to be closely related to the problem of changing dimension. In section 7 we illustrate some of the new results with the example of a spherical sector. We conclude in section 8.

2   The vector distance function (VDF) to a smooth manifold

Let $\mathcal{M}$ be a closed subset of $\mathbb{R}^n$. For every point x we note $\delta(x)$ the distance $dist(\mathbf{x}, \mathcal{M})$ of $\mathbf{x}$ to $\mathcal{M}$. This function is Lipschitz continuous and therefore almost everywhere differentiable [18]. The same holds for the function $\eta(\mathbf{x}) = \frac{1}{2}\delta^2(\mathbf{x})$. We note $\mathbf{u}(\mathbf{x})$ its derivative, defined almost everywhere:

$$\mathbf{u}(\mathbf{x}) = D\eta(\mathbf{x}) = \delta(\mathbf{x})D\delta(\mathbf{x})$$

This equation shows that, since $\delta(\mathbf{x})$ satisfies a.e. the eikonal equation

$$\|D\delta\| = 1, \tag{3}$$

$\mathbf{u}(\mathbf{x})$ is a vector of length $\delta(\mathbf{x})$. Moreover, at a point $\mathbf{x}$ where δ is differentiable, let $\mathbf{y} = P_{\mathcal{M}}(\mathbf{x})$ be the unique projection of $\mathbf{x}$ onto $\mathcal{M}$. This point is such that

$$\delta(\mathbf{x}) = \|\mathbf{x} - \mathbf{y}\|.$$

Moreover, if $\mathcal{M}$ is smooth at $\mathbf{y}$, the vector $\mathbf{x} - \mathbf{y}$ is normal to $\mathcal{M}$ at $\mathbf{y}$ and parallel to $D\delta(\mathbf{x})$, see figure 2:

$$\mathbf{u}(\mathbf{x}) = \mathbf{x} - \mathbf{y} \equiv \mathbf{x} - P_{\mathcal{M}}(\mathbf{x})$$

The vectors such as $\mathbf{x} - \mathbf{y}$, normal at $\mathbf{y}$ to $\mathcal{M}$, define the *characteristics* of the distance function δ. Starting from a point $\mathbf{y}$ of $\mathcal{M}$ and following a characteristic, i.e. a direction in the normal space $N_{\mathbf{y}}\mathcal{M}$, we either go to infinity or reach a point $\mathbf{z}$ at finite distance where δ is not differentiable and therefore $\mathbf{u}$ is not defined. Such a point belongs to the skeleton of $\mathcal{M}$. Because of the previous properties of the function $\mathbf{u}$, we have the following proposition

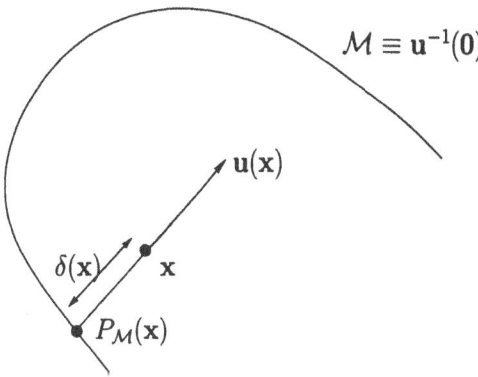

$$\mathcal{M} \equiv \mathbf{u}^{-1}(0)$$

Fig. 2. The projection of the point $\mathbf{x}$ on the smooth manifold $\mathcal{M}$ is noted $P_{\mathcal{M}}(\mathbf{x})$. The VDF to $\mathcal{M}$ at $\mathbf{x}$ is equal to $\mathbf{x} - P_{\mathcal{M}}(\mathbf{x})$.

Proposition 1 *Let $\mathbf{x}$ be a point of $\mathbb{R}^n$ where $\mathbf{u}$ is defined. The following relation*

$$\mathbf{u}(\mathbf{x} + \alpha\mathbf{u}(\mathbf{x})) = (1 + \alpha)\mathbf{u}(\mathbf{x}) \tag{4}$$

holds true for all values of $+\infty \geq \alpha_m > \alpha \geq -1$ such that $\mathbf{x} + \alpha_m\mathbf{u}(\mathbf{x})$ is the first point on the characteristic where $\mathbf{u}$ is not defined.

Proof. Use the equation $\mathbf{u}(\mathbf{x}) = \mathbf{x} - P_{\mathcal{M}}(\mathbf{x})$ and the fact that $P_{\mathcal{M}}(\mathbf{x} + \alpha\mathbf{u}(\mathbf{x})) = P_{\mathcal{M}}(\mathbf{x})$ for all α's such that $\alpha_m > \alpha \geq -1$, see figure 3.

It may be interesting to pause here and make the remark that the VDF to a smooth manifold $\mathcal{M}$ is an implicit representation of this manifold:

Proposition 2 *Let $\mathcal{M}$ be a smooth closed manifold and $\mathbf{u}$ its VDF, defined a.e.. Then*

$$\mathcal{M} = u^{-1}(0) \tag{5}$$

In effect, $\mathcal{M}$ is the intersection of the n hypersurfaces of equations $u_i(\mathbf{x}) = 0$, $i = 1, \cdots, n$. The differential $D\mathbf{u}$ of $\mathbf{u}$ which is defined a.e. if $\mathcal{M}$ is smooth provides some interesting information about the dimension of $\mathcal{M}$ at the points of $\mathcal{M}$ and, as we will see in the next section, at all points of $\mathbb{R}^n$ where it is defined. At a point $\mathbf{x}$ of $\mathcal{M}$ this differential $D\mathbf{u}(\mathbf{x})$ is a mapping from $\mathbb{R}^n$ into $\mathbb{R}^n$ whose rank is $n - k$, the codimension of $\mathcal{M}$ at $\mathbf{x}$ and whose nullspace is of dimension k, the dimension of $\mathcal{M}$.

Because we are interested in evolving the manifold $\mathcal{M}$ through the evolution of $\mathbf{u}$, while keeping $\mathbf{u}$ a VDF, we are interested in finding a characterization of the VDFs analog to the one for distance functions, (3). This will be our first step in the exploration of the differential properties of the function $\mathbf{u}$ that will be pursued in the next section.

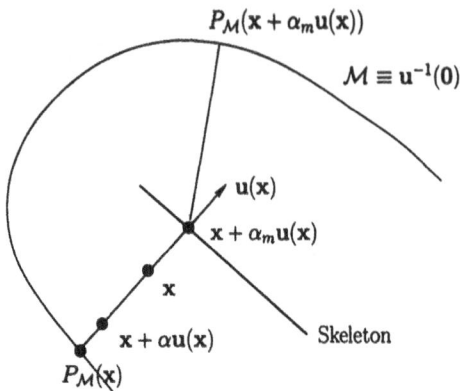

Fig. 3. The VDF is defined a.e.. In particular it is not defined on the skeleton of the manifold $\mathcal{M}$: the point $\mathbf{x} + \alpha_m \mathbf{u}(\mathbf{x})$ is at equal distances from the two points $P_{\mathcal{M}}(\mathbf{x})$ and $P_{\mathcal{M}}(\mathbf{x} + \alpha_m \mathbf{u}(\mathbf{x})$.

Proposition 3 *Let $\mathbf{u} : \mathbb{R}^n \longrightarrow \mathbb{R}^n$ be such that*

$$(D\mathbf{u})^T \mathbf{u} = \mathbf{u} \quad a.e. \tag{6}$$

and $\mathbf{u}$ is continuous at all points of the set $\mathcal{M} = \mathbf{u}^{-1}(0)$. Then $\mathbf{u}(\mathbf{x}) = D\eta(\mathbf{x})$ a.e., where $\eta(\mathbf{x})$ is the function $\frac{1}{2} dist^2(\mathbf{x}, \mathcal{M})$ to the set $\mathcal{M}$.

Proof. Define the function $\phi : \mathbb{R}^n \to \mathbb{R}$ by

$$\forall \mathbf{x} \in \mathbb{R}^n, \quad \phi(\mathbf{x}) = \|\mathbf{u}(\mathbf{x})\| \tag{7}$$

and compute its first order derivative with respect to $\mathbf{x}$

$$\nabla \phi = \frac{D\mathbf{u}^T \mathbf{u}}{\|\mathbf{u}\|} = \frac{\mathbf{u}}{\|\mathbf{u}\|}$$

Hence

$$\|\nabla \phi\| = 1 \quad a.e.$$

which means that ϕ is equal to the distance function to the set $\mathcal{M}$ plus a constant

$$\phi = \delta + C$$

In addition, the combination of $\phi = \delta + C$ and (7) shows that

$$\mathbf{u} = (\delta + C)\nabla \delta$$

The continuity of $\mathbf{u}$ on $\mathbf{u}^{-1}(0)$ implies that $C = 0$. Indeed, let x_0 be a point of $\mathcal{M}$ and $\mathbf{n}$ be a unit vector of $N_{x_0}\mathcal{M}$. We consider the line

$\lambda : \mathbf{x}_0 + \lambda \mathbf{n}$ and the variations of δ and $\nabla \delta$ along this line. Figure 4 shows that the product $\delta \nabla \delta$ is continuous on $\mathcal{M}$ but $\nabla \delta$ is not.

Finally

$$\mathbf{u} = \phi \nabla \phi = \nabla \left(\frac{\delta^2}{2} \right)$$

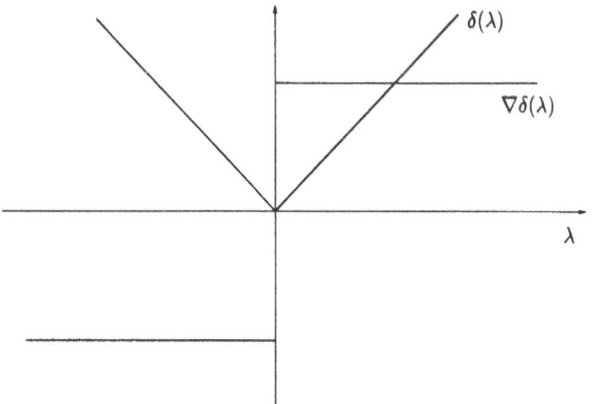

Fig. 4. The function $\nabla \delta$ is discontinuous at $\lambda = 0$ along the line $\mathbf{x}_0 + \lambda \mathbf{n}$. On the other hand, the product $\delta \nabla \delta$ is continuous at $\lambda = 0$.

Equation (6) is the characteristic equation of the class of Vector Distance Functions. Unfortunately, since the notion of viscosity solutions is not readily available for systems of PDEs we cannot state and prove a result analog to the one caracterizing the distance function to $\mathcal{M}$ as the unique viscosity solution of the eikonal equation (see for example [5,18]).

The previous proposition says nothing about the regularity of the set $\mathcal{M}$. In fact, the proof of proposition 3 assumes that $\mathcal{M}$ is smooth enough to have a normal space at every point. This is explained later in the paper. If we want that set to be a smooth manifold, then it is likely that $\mathbf{u}$ must satisfy some extra regularity conditions. We have not pursued this direction but a clue can be found in the paper [2] where it is shown that, given a smooth manifold $\mathcal{M}$, the first and second fundamental forms of $\mathcal{M}$ can be recovered from the *third* order derivatives of the distance function δ to $\mathcal{M}$ at all points where it is differentiable.

We can also make good use of some standard results in Differential Geometry but we need first some properties of VDFs that will be proved in section 3.

Let us make two additional remarks that are useful in the application of the theory of VDFs.

Corollary 1 *A VDF* **u** *satisfies a.e. the following two properties:*

$$\mathbf{u} = \frac{1}{2} D\|\mathbf{u}\|^2$$

and

$$D\mathbf{u} = (D\mathbf{u})^T$$

Proof. Both properties follow from the fact that we have proved in proposition 3 that the magnitude of **u** is a distance function: the first one is the last equation in the proof, the second one is a consequence of the fact that $D\mathbf{u}$ is the second order derivative of $1/2\delta^2$, hence a Hessian, hence symmetric.

To provide the reader with some intuition, we show in figure 5 the VDF of a smooth manifold of dimension 0, a point, embedded in $\mathbb{R}^2$. Similarly,

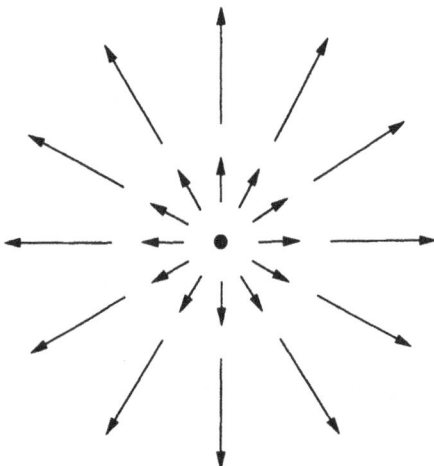

Fig. 5. The VDF of a point: a radial field equal to **0** at the point.

figure 6 shows the VDF of a smooth manifold of dimension 1, an infinite line, embedded in $\mathbb{R}^2$, and figure 7 shows the VDF of another smooth manifold of dimension 1, a circle, also embedded in $\mathbb{R}^2$.

3 Properties of VDFs

We now study some differential properties of the VDF **u**. Most of them can be found in [3]. For those which cannot be found there, see [24]. Equation (4) yields, for $\alpha = -1$ the (almost obvious) equation:

$$\mathbf{u}(\mathbf{x} - \mathbf{u}(\mathbf{x})) = \mathbf{0} \tag{8}$$

We use this equation to prove the following proposition:

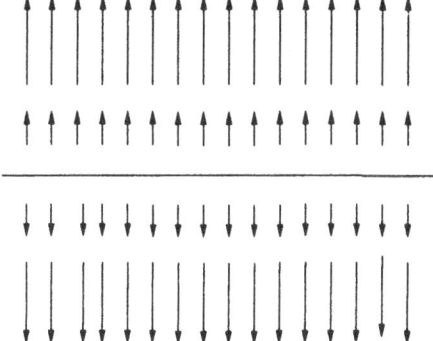

Fig. 6. The VDF of a line: a translationally invariant field equal to **0** on the line.

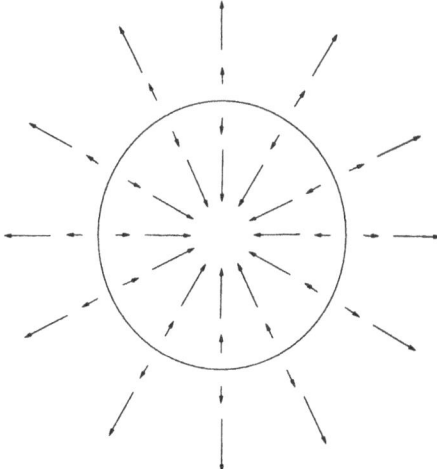

Fig. 7. The VDF of a circle: a radial field equal to **0** on the circle and undefined at the center.

Proposition 4 *The derivative of a VDF satisfies the following relation at each point of $\mathcal{M}$:*

$$Du = (Du)^2. \tag{9}$$

Therefore it is a projector on a vector subspace.

Proof. Take the derivative of (8) with respect to $\mathbf{x}$ to obtain

$$Du(\mathbf{x} - \mathbf{u}(\mathbf{x}))(\mathbf{I} - Du(\mathbf{x})) = \mathbf{0} \tag{10}$$

which is true a.e.. Let us specialize this relation to the case where $\mathbf{u}(\mathbf{x}) = \mathbf{0}$, i.e. for $\mathbf{x}$ on $\mathcal{M}$:

$$Du = (Du)^2.$$

This equation says that $Du(\mathbf{x})$ is a projector for all points of $\mathcal{M}$.

It is not difficult to find out which vector subspace it is a projector on by using some other properties of the distance vector functions (see below).

In the sequel we denote by u_i, $i = 1, \cdots, n$ the coordinates of u. The equation $D\mathbf{u}\mathbf{u} = \mathbf{u}$ can be rewritten in terms of the coordinates u_i of u and their gradients:

$$\mathbf{u} = \sum_{i=1}^{n} u_i \nabla u_i \tag{11}$$

We group in the next proposition a number of useful results on $D\mathbf{u}$ and $\mathbf{D}^2\mathbf{u}$.

Proposition 5 *The square $(D\mathbf{u})^2$ of the derivative $D\mathbf{u}$ of u is given by*

$$(D\mathbf{u})^2 = \sum_{i=1}^{n} \nabla u_i \nabla u_i^T$$

The second order derivative $\mathbf{D}^2\mathbf{u}$ of u is related to $D\mathbf{u}$ by the following matrix equation

$$\mathbf{D}^2\mathbf{u}\,\mathbf{u} + (D\mathbf{u})^2 = D\mathbf{u} \tag{12}$$

where the matrix $\mathbf{D}^2\mathbf{u}\,\mathbf{u}$ is given by

$$\mathbf{D}^2\mathbf{u}\,\mathbf{u} = \sum_{i=1}^{n} u_i \mathbf{D}^2 u_i \, ,$$

and $\mathbf{D}^2 u_i$ is the Hessian matrix of the function $u_i : \mathbb{R}^n \longrightarrow \mathbb{R}$.

Proof. Take the derivative of both sides of equation (11) to obtain an expression for $D\mathbf{u}$:

$$D\mathbf{u} = \sum_{i=1}^{n} \left(\nabla u_i \nabla u_i^T + u_i \mathbf{D}^2 u_i \right) \tag{13}$$

We compare this equation with the one obtained by deriving $D\mathbf{u}\mathbf{u} = \mathbf{u}$, i.e.

$$\mathbf{D}^2\mathbf{u}\,\mathbf{u} + (D\mathbf{u})^2 = D\mathbf{u}$$

Indeed, we show that

$$(D\mathbf{u})^2 = \sum_{i=1}^{n} \nabla u_i \nabla u_i^T$$

and hence that the matrix $\mathbf{D}^2\mathbf{u}\,\mathbf{u}$ is equal to $\sum_{i=1}^{n} u_i \mathbf{D}^2 u_i$. In general we have

$$\mathbf{D}^2\mathbf{u}\,\mathbf{x} = \sum_{i=1}^{n} x_i \mathbf{D}^2 u_i \tag{14}$$

In order to see this, write $\left(\sum_{i=1}^{n} \nabla u_i \nabla u_i^T \right) \mathbf{x} = \sum_{i=1}^{n} (\nabla u_i \cdot \mathbf{x}) \nabla u_i$ and note that this is equal to $D\mathbf{u}\,\mathbf{y}$ where the ith coordinate of y is $\nabla u_i \cdot \mathbf{x}$. Hence $\mathbf{y} = D\mathbf{u}\,\mathbf{x}$ and the result follows.

The previous proof allows to show the following important proposition

Proposition 6 *When evaluated at* $\mathbf{x}$ *on* $\mathcal{M}$, $D\mathbf{u}(\mathbf{x})$ *is the projector on the normal space* $N_{\mathbf{x}}\mathcal{M}$. *This implies that* $D\mathbf{u}(\mathbf{x})$ *has* $n - k$ *(the dimension of* $N_{\mathbf{x}}\mathcal{M}$*) eigenvalues equal to 1 and* k *(the dimension of* $T_{\mathbf{x}}\mathcal{M}$*) eigenvalues equal to 0. In particular, the rank of* $D\mathbf{u}$ *is equal to* k *on* $\mathcal{M}$.

Proof. In the special case where $\mathbf{u} = \mathbf{0}$, i.e on $\mathcal{M}$, equation (13) simplifies to

$$D\mathbf{u} - \sum_{i=1}^{n} \nabla u_i \nabla u_i^T, \tag{15}$$

which shows that the image of the linear operator $D\mathbf{u}(\mathbf{x})$, i.e. $D\mathbf{u}(\mathbf{x})(\mathbb{R}^n)$, at every point $\mathbf{x}$ of $\mathcal{M}$, is the vector space of dimension $n - k$ generated by the vectors ∇u_i, $i = 1, \cdots, n$, i.e. the normal space $N_{\mathbf{x}}\mathcal{M}$ to $\mathcal{M}$ at point $\mathbf{x}$. Combining this with our observation that $D\mathbf{u}$ is a projector at every point of $\mathcal{M}$ allows us to state that $D\mathbf{u}(\mathbf{x})$ is the projector on the normal space to $\mathcal{M}$ at $\mathbf{x}$. In particular we have

$$D\mathbf{u}\nabla u_i = \nabla u_i \quad i = 1, \cdots, n \tag{16}$$

on $\mathcal{M}$, which is not obvious from (15).

This property allows us to say something about the regularity of $\mathcal{M}$. Let $\mathbf{u}$ be a solution of the problem described in proposition 3. If $\mathbf{u}$ is such that the rank of $D\mathbf{u}$ is constant and equal to k on $\mathcal{M} = \mathbf{u}^{-1}(0)$, then because the set where the rank of $D\mathbf{u}$ is equal to k is open (the rank is a lower semicontinuous function), this open set contains $\mathcal{M}$ and theorem 12 of chapter 2 in volume I of [41] says that $\mathcal{M}$ is a smooth manifold of dimension $n - k$ (we do not make more precise what we mean by smooth here).

We now come to the study of the eigenvalues and eigendirections of $D\mathbf{u}$ outside $\mathcal{M}$. The results are quite simple if we consider the line defined by the two points $\mathbf{x}$ and $P_{\mathcal{M}}(\mathbf{x})$: let $\mathbf{n}$ be the unit vector parallel to $\mathbf{x} - P_{\mathcal{M}}(\mathbf{x})$ and consider the line $s \longrightarrow P_{\mathcal{M}}(\mathbf{x}) + s\mathbf{n} \equiv \mathbf{x}(s)$; such a line is called a *characteristic* line. Consider further the values $s_{min} < 0$ and $s_{max} > 0$ (possibly infinite) such that $D\mathbf{u}(\mathbf{x}(s))$ is defined on the open interval $I =]s_{min}, s_{max}[$; the eigendirections of $D\mathbf{u}(\mathbf{x}(s))$ are constant and $n - k$ eigenvalues are equal to 1 for all s in I. Such an open segment is called a *characteristic* segment. More precisely, we have the following proposition

Proposition 7 *The eigendirections of the symetric matrix* $D\mathbf{u}$ *are constant along each characteristic segment.*
Moreover, if a ray is parameterized by its arclength s, *starting at* $P_{\mathcal{M}}(\mathbf{x})$,

then each one of the eigenvalues of $D\mathbf{u}$ has one of the three following forms:

$$\lambda(s) = 1, \quad or$$
$$\lambda(s) = 0, \quad or$$
$$\lambda(s) = \frac{s}{s \pm c}, \quad c > 0 \tag{17}$$
$$\forall s \in I$$

where c depends on the particular eigendirection. The first form corresponds to the eigenvectors of $D\mathbf{u}$ which are elements of the normal space to $\mathcal{M}$ at $P_\mathcal{M}(\mathbf{x})$; there are $n - k$ such eigenvalues. The second and third forms (the second form is obtained from the third form by taking $c = \infty$) correspond to eigenvectors of $D\mathbf{u}$ which are elements of the tangent space to $\mathcal{M}$ at $P_\mathcal{M}(\mathbf{x})$. There are k such eigenvalues.

The proof can be found in [3] or in [24].
As a final property of the VDFs, we state without proof (the proof is in [24]) the following proposition

Proposition 8 *The mean-curvature vector of $\mathcal{M}$ at $\mathbf{x}$ is equal, up to a scale factor, to the Laplacian $\Delta\mathbf{u}(\mathbf{x})$ of $\mathbf{u}$ at the same point:*

$$\mathcal{H}(\mathbf{x}) = -\frac{1}{k}(\Delta\mathbf{u}(\mathbf{x})) \quad \forall\mathbf{x} \quad \in \mathcal{M}$$

4   How to evolve a smooth manifold by evolving its VDF

Let us consider a family $\mathcal{M}(\mathbf{p}, t)$ of smooth manifolds of dimension k, where $\mathbf{p}$ is a k-dimensional vector parameterizing $\mathcal{M}$ at each time instant t. We assume the initial conditions

$$\mathcal{M}(\cdot, 0) = \mathcal{M}_0(\cdot),$$

where $\mathcal{M}_0$ is a smooth manifold of dimension k. Furthermore the evolution of the family $\mathcal{M}$ is governed by the following PDE

$$\mathcal{M}_t(\mathbf{p}, t) = \mathcal{H}(\mathcal{M}(\mathbf{p}, t), t) + \Pi^N_{\mathcal{M}(\mathbf{p},t)}(\mathcal{D}(\mathcal{M}(\mathbf{p}, t), t)) \overset{def}{=} \mathbf{V}(\mathcal{M}(\mathbf{p}, t), t) \tag{18}$$

where $\mathcal{H}(\mathcal{M}(\mathbf{p}, t), t)$ is the mean curvature vector at the point $\mathcal{M}(\mathbf{p}, t)$ of the manifold $\mathcal{M}$ (see [41] or [24] for definitions) and $\mathcal{D}(\mathbf{x}, t)$ is a vector field defined on $\mathbb{R}^n \times \mathbb{R}^+$ representing a velocity induced on $\mathcal{M}$ by the data. $\Pi^N_{\mathcal{M}(\mathbf{p},t)}$ is the projection operator on the normal space $N\mathcal{M}_{\mathcal{M}(\mathbf{p},t)}$ to $\mathcal{M}$ at the point $\mathcal{M}(\mathbf{p}, t)$, see figure 8. The goal of this section is to explore ways of evolving $\mathbf{u}$, the VDF to $\mathcal{M}$, instead of $\mathcal{M}$ while guaranteeing three conditions:

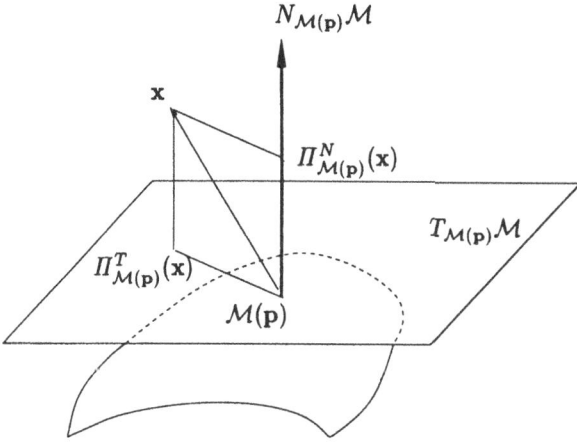

Fig. 8. Any point $\mathbf{x}$ of $\mathbb{R}^n$ can be projected orthogonally onto the normal, $N_{\mathcal{M}(\mathbf{p})}\mathcal{M}$, and tangent, $T_{\mathcal{M}(\mathbf{p})}\mathcal{M}$, spaces of the smooth manifold $\mathcal{M}$ at the point $\mathcal{M}(\mathbf{p})$ by the linear operators $\Pi^N_{\mathcal{M}(\mathbf{p})}(\mathbf{x})$ and $\Pi^T_{\mathcal{M}(\mathbf{p})}(\mathbf{x})$, respectively.

1. That $\mathbf{u}$ remains a VDF at all time instants.
2. That $\mathbf{u}^{-1}(0) = \mathcal{M}$ at all times where $\mathcal{M}$ is defined.
3. That the manifold $\mathcal{M}$ evolve according to (18).

In other words, we look for a vector field $\mathbf{b}(\mathbf{x}, t)$ defined a.e. on $\mathbb{R}^n$ and for some interval $[0, T[$ of $\mathbb{R}^+$ such that

$$
\begin{aligned}
(D\mathbf{u})^T \mathbf{u} &= \mathbf{u} \\
\mathbf{u}(\mathcal{M}(\mathbf{p}, t), t) &= 0 \\
\mathbf{u}_t(\mathbf{x}, t) &= \mathbf{b}(\mathbf{x}, t) \\
(\mathbf{u}^{-1}(0))_t &= \mathcal{H} + \Pi^N_{\mathcal{M}}(\mathcal{D}(\mathcal{M}, t))
\end{aligned}
\tag{19}
$$

It is not too difficult to find a characterization of $\mathbf{b}$.

Proposition 9 *The velocity field $\mathbf{b}$ of the VDF $\mathbf{u}$ is characterized by the first order, quasilinear Partial Differential Equation:*

$$
D\mathbf{b}\,\mathbf{u} = (\mathbf{I} - D\mathbf{u})\mathbf{b},
\tag{20}
$$

with initial conditions

$$
\mathbf{b}(\mathcal{M}(\mathbf{p}, t), t) = -\mathbf{V}(\mathcal{M}(\mathbf{p}, t), t),
\tag{21}
$$

wherever $\mathbf{u}$ is defined and differentiable.

Proof. We start with the equation (6) that characterizes the VDFs a.e. and take its time derivative:

$$
(D\mathbf{u}_t)^T \mathbf{u} + (D\mathbf{u})^T \mathbf{u}_t = \mathbf{u}_t
$$

We then replace u_t with its value b and obtain:

$$(Db)^T u = (I - (Du)^T)b$$

Since Du is a symmetric operator (corollary 1) we can force Db to be symmetric and we obtain equation (20).

Conversely, if the field b satisfies (20) wherever u is defined and differentiable, and its derivative Db is symmetric as is the derivative Du, we have:

$$(Db)^T u + (Du)^T b = b$$

Using the fact that $b = u_t$, we rewrite this equation as

$$(Du)_t^T u + (Du)^T u_t = u_t$$

The lefthand side appears as the time derivative of $(Du)^T u$ hence

$$(Du)^T u = u + c$$

where $c(x,t)$ is a vector field such that $c_t = 0$. If we choose $c(.,0) = 0$ we then have

$$(Du)^T u = u \quad \text{a.e.} \quad \forall t \geq 0,$$

which, according to proposition 3, implies that u is a VDF for all times $t \geq 0$ where u is defined a.e..

The next problem we want to address is that of relating in a simple way the field b to the velocity field of the manifold $\mathcal{M} = u^{-1}(0)$. The result is extremely simple, see figure 9.

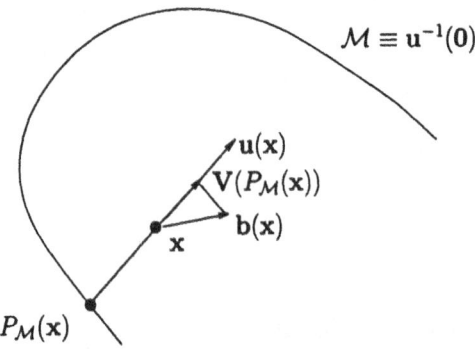

Fig. 9. The velocity field $b(x)$ of the VDF u at point x where u is defined is equal to the velocity of the projection $P_{\mathcal{M}}(x)$ of x on the smooth manifold $\mathcal{M}$.

Proposition 10 *At all points* **x** *where* **u** *is defined, its component in the normal space* $N_{P_{\mathcal{M}}(\mathbf{x})}\mathcal{M}$ *to* $\mathcal{M}$ *at the point* $P_{\mathcal{M}}(\mathbf{x})$ *is equal to minus the normal velocity* $\mathbf{V}(P_{\mathcal{M}}(\mathbf{x}), t)$ *of the point* $P_{\mathcal{M}}(\mathbf{x})$:

$$\mathbf{b}(\mathbf{x}, t)_{|N_{P_{\mathcal{M}}(\mathbf{x})}\mathcal{M}} = \mathbf{b}(\mathbf{x} - \mathbf{u}(\mathbf{x}, t))_{|N_{P_{\mathcal{M}}(\mathbf{x})}\mathcal{M}} \equiv \mathbf{b}(P_{\mathcal{M}}(\mathbf{x}))_{|N_{P_{\mathcal{M}}(\mathbf{x})}\mathcal{M}}$$
$$\equiv -\mathbf{V}(P_{\mathcal{M}}(\mathbf{x}), t) \qquad (22)$$

Proof. We use the method of characteristics (see e.g. [4]) and rewrite this equation along the characteristics of **b** so that a set of ODEs are obtained.

In detail, we build the $2n$-dimensional characteristic vectors (see figure 10)

$$\mathbf{a} = \begin{pmatrix} \mathbf{u} \\ (\mathbf{I} - (D\mathbf{u})^T)\mathbf{b} \end{pmatrix} \qquad (23)$$

We fix a point $\begin{pmatrix} \mathbf{x}_0 \\ \mathbf{b}_0 \end{pmatrix} \in \mathbb{R}^{2n}$ such that $\mathbf{x}_0$ is not on $\mathcal{M}$, i.e. $\mathbf{u}(\mathbf{x}_0) \neq 0$. The characteristic curve of the vector field **b** passing through this point is tangent to **a** so that it can be described by the following ODE's

$$\begin{pmatrix} \dot{\mathbf{x}} \\ \dot{\mathbf{b}} \end{pmatrix} = \mathbf{a} \qquad (24)$$

The first three equations of this set of six yield

$$\dot{\mathbf{x}}(q) = \mathbf{u}(\mathbf{x}) = \mathbf{x}(q) - P_{\mathcal{M}}(\mathbf{x}_0),$$

since the characteristic going through $\mathbf{x}_0$ is the line defined by $\mathbf{x}_0$ and $P_{\mathcal{M}}(\mathbf{x}_0)$. This ODE is readily integrated:

$$\mathbf{x}(q) - P_{\mathcal{M}}(\mathbf{x}_0) = (\mathbf{x}_0 - P_{\mathcal{M}}(\mathbf{x}_0))e^q$$

The parameter q is such that for $q = 0$ the point $\mathbf{x}(q)$ is at $\mathbf{x}_0$, and when q goes to $-\infty$ it tends to $P_{\mathcal{M}}(\mathbf{x}_0)$. Let us rewrite the last equation as

$$\mathbf{x}(q) = P_{\mathcal{M}}(\mathbf{x}_0) + \mathbf{u}(\mathbf{x}_0)e^q,$$

and let us do the change of variable $s = d_0 e^q$ ($d_0 = \|\mathbf{u}(\mathbf{x}_0)\|$ is the distance of x_0 to $P_{\mathcal{M}}(\mathbf{x}_0)$) which corresponds to a change of origin on the characteristic (the origin being now $P_{\mathcal{M}}(\mathbf{x}_0)$ instead of $\mathbf{x}_0$), and a change of velocity. Let us now look at the last three equations of (23) and (24):

$$\dot{\mathbf{b}}(\mathbf{x}(q)) = (\mathbf{I} - D\mathbf{u}^T(\mathbf{x}(q)))\,\mathbf{b}(\mathbf{x}(q))$$

This equation is readily integrated as

$$\mathbf{b}(p) = \exp\left(\int_0^p (\mathbf{I} - (D\mathbf{u})^T(q))dq\right)\mathbf{b}(0)$$

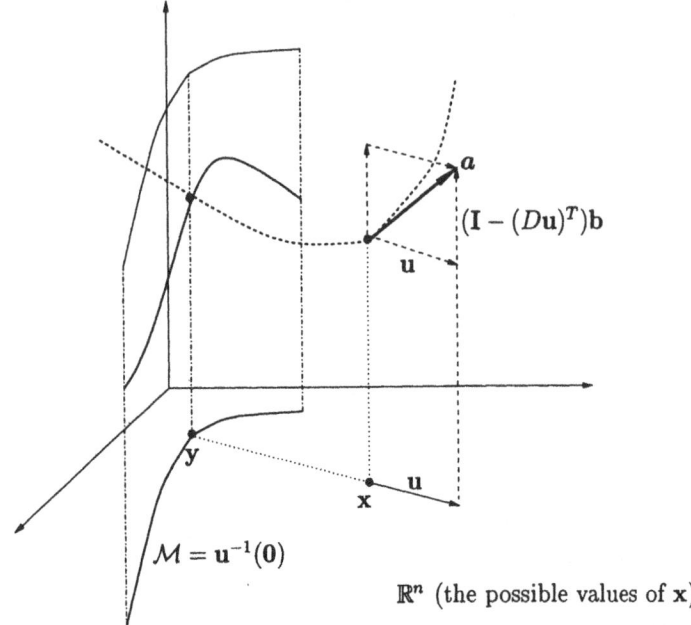

Fig. 10. The horizontal space represents the Euclidean space $\mathbb{R}^n$ where $\mathcal{M}$ "lives": it is the set of possible values of $\mathbf{x}$. The vertical space also represents $\mathbb{R}^n$, the set of possible values of $\mathbf{b}$. A "surface" over the horizontal space whose tangent plane is never vertical represents the graph of a function $\mathbb{R}^n \longrightarrow \mathbb{R}^n$.

We compute an analytical form of the righthand side which, by means of the previous change of variable $s = d_0 e^q$ and an abuse of notations, namely $\mathbf{b}(s) \stackrel{def}{=} \mathbf{b}(\mathbf{x}(s))$ and similarly $D\mathbf{u}(s) \stackrel{def}{=} D\mathbf{u}(\mathbf{x}(s)))$ is also integrated by quadrature with the help of proposition 7 which tells us that along a characteritic ray, all symmetric matrices $D\mathbf{u}$ can be diagonalized in the same orthonormal basis, their eigenvalues being given by equations (17) in the same proposition. In this basis, the matrix $D\mathbf{u}$ can be written

$$D\mathbf{u}(q) = diag(1, \cdots, 1, \frac{d_0 e^q}{d_0 e^q \pm c_1}, \cdots, \frac{d_0 e^q}{d_0 e^q \pm c_k})$$

where there are $n - k$ 1's and k positive values $c_l, l = 1, \cdots, k$, some of them possibly infinite to take into account the fact that some of the eigenvalues not equal to 1 can be equal to 0.

Hence the integrand $\mathbf{I} - (D\mathbf{u})^T(q)$ is equal to

$$diag(0, \cdots, 0, 1 - \frac{d_0 e^q}{d_0 e^q \pm c_1}, \cdots, 1 - \frac{d_0 e^q}{d_0 e^q \pm c_k})$$

We then have to compute the values of the integrals

$$\int_0^p (1 - \frac{d_0 e^q}{d_0 e^q \pm c_l}) dq \quad l = 1, \cdots, k$$

Applying the previous change of variable $s = d_0 e^q$, these integrals can be rewritten as:

$$\int_{d_0}^{d_0 e^p} (\frac{1}{s} - \frac{1}{s \pm c_l}) ds,$$

which is found to be equal to

$$\log \frac{(d_0 \pm c_l) e^p}{d_0 e^p \pm c_l}$$

Taking the exponential of these values we obtain the relation between the lth components of the velocity field at the point $P_{\mathcal{M}}(\mathbf{x}_0) + \mathbf{u}(\mathbf{x}_0) e^p$, noted $b_l(p)$ and at the point $\mathbf{x}_0$, noted $b_l(0)$:

$$b_l(p) = \frac{(d_0 \pm c_l) e^p}{d_0 e^p \pm c_l} b_l(0)$$

Let us rewrite this expression as

$$b_l(0) = \frac{d_0 e^p \pm c_l}{(d_0 \pm c_l) e^p} b_l(p)$$

When $p \longrightarrow -\infty$, $b_l(p)$ goes to 0 since it is a component of the velocity field of $\mathcal{M}$ in its tangent space $T\mathcal{M}$. Since $b_l(0)$ is well defined the product $b_l(p) e^{-p}$ must have a limit, noted α_l when $p \longrightarrow -\infty$:

$$b_l(0) = \pm \frac{c_l}{d_0 \pm c_l} \alpha_l \tag{25}$$

This shows that the components of $\mathbf{b}(\mathbf{x}_0)$ in the tangent space to $\mathcal{M}$ at $P_{\mathcal{M}}(\mathbf{x}_0)$ are in general non zero. The constants α_l depend upon the geometry of $\mathcal{M}$ and of the various derivatives of the velocity field on $\mathcal{M}$. The exact values for these parameters can be found in [24] but are not important in practice, see section 8. To summarize, in an orthonormal basis where $D\mathbf{u}(\mathbf{x}_0)$ and $D\mathbf{u}(P_{\mathcal{M}}(x_0))$ are diagonal, the first $n - k$ coordinates of $\mathbf{b}(\mathbf{x}_0)$ and $\mathbf{b}(P_{\mathcal{M}}(x_0))$ are equal (they correspond to components of the velocity in the normal space to $\mathcal{M}$). The remaining k components, corresponding to the tangent space to $\mathcal{M}$ are in general non zero and are given by (25).

5   Example I: the helix

We consider an helix $\mathcal{M}$ embedded in $\mathbb{R}^3$ by

$$\sigma \to \mathbf{x}(\sigma) = \begin{pmatrix} r e^{i\sigma} \\ \sigma \end{pmatrix}, \quad \sigma \mathbb{R}.$$

We adopt the cylindrical coordinates (r, θ, z). This helix is a smooth manifold of dimension 1, codimension 2, without a boundary. Its Frenet frame is made of the tangent τ, the normal ν and the binormal β, respectively

$$\tau(\sigma) = \frac{1}{\sqrt{r^2 + 1}} \begin{pmatrix} re^{i(\sigma + \frac{\pi}{2})} \\ 1 \end{pmatrix}, \quad \nu(\sigma) = \begin{pmatrix} -e^{i\sigma} \\ 0 \end{pmatrix}$$

$$\text{and } \beta(\sigma) = \frac{1}{\sqrt{r^2 + 1}} \begin{pmatrix} -e^{i(\sigma + \frac{\pi}{2})} \\ r \end{pmatrix}$$

Based on symmetry considerations, it is shown in [24] that it is only necessary to compute $\mathbf{u}$ for points of cylindrical coordinates $(\rho, 0, z)$ and that in the Frenet frame, as stated in proposition 7, the matrix $D\mathbf{u}$ is diagonal and equal to

$$D\mathbf{u}(\rho, 0, z) = \begin{pmatrix} \dfrac{r(\rho \cos \sigma - r)}{r\rho \cos \sigma + 1} & 0 & 0 \\ 0 & 1 & 0 \\ 0 & 0 & 1 \end{pmatrix}$$

The first eigenvalue (corresponding to the tangential vector τ) is better understood by introducing explicitely the distance to the helix, i.e. $s = \|\mathbf{u}\|$. To do so, we remark that $\mathbf{u} = s\mathbf{n}$ where $\mathbf{n}$ is a unit normal vector, thus of the form $\mathbf{n} = \nu \cos \alpha + \beta \sin \alpha$. Now, noticing the following

$$\mathbf{a} \cdot \nu = r - \rho \cos \sigma$$

we see that the first eigenvalue is

$$\frac{r(\rho \cos \sigma - r)}{r\rho \cos \sigma + 1} = \frac{s}{\frac{1}{\kappa \cos \alpha} + s}$$

where $\kappa = \frac{r}{r^2 + 1}$ is the curvature of the helix, which confirms again the result of proposition 7.

We shall now make this helix evolve in time according to mean curvature motion and verify that its (time dependent) distance vector satisfies the equation (22). First of all, we show the helix remains an helix with decreasing radius until it becomes (in infinite time) the y axis. The fact, the mean-curvature vector of the helix is $\kappa \nu$ suggests that the trace of the evolving helix in the $(\mathbf{i}, \mathbf{j})$ plane (a circle of initial radius r) evolves with a radius satisfying the following ODE

$$\dot{R}(t) = -\frac{R(t)}{R^2(t) + 1} = -\kappa(t)$$

with initial condition $R(0) = r$. This integrates as

$$\frac{R^2(t) - r^2}{2} + \ln \frac{R(t)}{r} = t$$

which expresses implicitely the function $t \to R(t)$.

We then verify that the time dependant helix embedded by

$$\sigma \to \begin{pmatrix} R(t)e^{i\sigma} \\ \sigma \end{pmatrix}, \quad \forall \sigma \in \mathbb{R}$$

satisfies the following ODE

$$\mathcal{M}_t = \begin{pmatrix} \dot{R}(t)e^{i\sigma} \\ 0 \end{pmatrix} = -\kappa(t)\boldsymbol{\nu}$$

i.e. it is *the* solution to the mean curvature motion and has the correct initial value. As a corollary, the distance vector to this evolving helix is

$$\mathbf{u}(\rho, 0, z, t) = \begin{pmatrix} \rho i \\ z \end{pmatrix} - \begin{pmatrix} R(t)e^{i\sigma(t)} \\ \sigma(t) \end{pmatrix}$$

where $\sigma(t)$ is defined by

$$\rho R(t) \sin \sigma(t) + \sigma(t) - z = 0$$

We compute

$$\mathbf{u}_t(\rho, 0, z, t) = -\begin{pmatrix} (\dot{R} + iR\dot{\sigma})e^{i\sigma} \\ \dot{\sigma} \end{pmatrix} = -\kappa(t)\boldsymbol{\nu}(\sigma(t)) - \dot{\sigma}\sqrt{R^2 + 1}\boldsymbol{\tau}(\sigma(t))$$

We find, as proved in proposition 10, that the normal component of $\mathbf{b}(\rho, 0, z, t)$ is equal to minus the velocity of the point $P_{\mathcal{M}}(\rho, 0, z)$.

6 Smooth manifolds with a boundary and changes in dimension

A simple example will introduce the new issues of this section. In the plane $\mathbb{R}^2$, we consider the time dependent segment $[A(t), B(t)]$ whose endpoints have respectively the velocities $v_A(t) = \alpha i$ and $v_B(t) = \beta i$ with $\alpha < 0 < \beta$ and with initial condition $A(0) = B(0) = O$. This situation describes a point transforming into a segment with increasing length: it is a prototype of a change of dimension followed by the evolution of a smooth manifold with boundary. We note $\mathbf{x} = [x, y]^T$ the coordinates of a point of $\mathbb{R}^2$. The VDF to this object is easily shown to satisfy

$$\mathbf{u}_t(x, y, t) + D\mathbf{u}\mathbf{V} = 0 \quad \text{with} \quad \mathbf{V}(x, y, t) = \begin{cases} v_A & \text{if } x < \alpha t, \\ 0 & \text{if } \alpha t < x < \beta t, \quad (26) \\ v_B & \text{if } x > \beta t. \end{cases}$$

This equation is a variant of a well-known class of PDEs called the *transport equations* [18]. One of its solution is readily shown to be

$$\mathbf{u}_0(\mathbf{x} - \mathbf{A}) \quad \text{if} \quad x < \alpha t$$
$$\mathbf{u}_0(x, 0) \quad \text{if} \quad \alpha t < x < \beta t$$
$$\mathbf{u}_0(\mathbf{x} - \mathbf{B}) \quad \text{if} \quad x > \beta t$$

At first glance (see figure 11), the vector field $\mathbf{b} = -D\mathbf{u}\mathbf{V}$ is not of the form presented in the previous section because it is only piecewise-smooth, $\mathbf{b}$ being discontinuous on the lines of equations $x = \alpha t$ and $x = \beta t$. But, as shown in the figure, the evolution of the point O is quite remarkable: it is a smooth manifold of dimension 0 which is turned into a smooth manifold with boundary (the segment AB). The velocity field $\mathbf{b}$ is discontinuous on the vertical axis at time $t = 0$ which has the effect of allowing the point to "spread" to a line segment. At time $t > 0$ the velocity field $\mathbf{b}(., t)$ is of the form $\mathbf{V}(P_{AB}(.))$ everywhere except on the previous two lines. We note that the *normal* component of $\mathbf{V}$ is continuous along AB, being equal to 0 everywhere, while its *tangential* component is discontinuous at A and B, i.e. at the boundary of the curve AB. This last point is the reason why AB can grow in time. We will see more on this in section 6.1.

The situation we have just described is archetypal of all cases in higher dimensions and codimensions and reveals an undesirable lack of generality in the analysis of the previous section and suggests that *non-continuous* $\mathbf{b}$'s may also be interesting since they can account for changes of dimension and tangent velocities at the boundary of a manifold (both are intimately related as it can be learned from the segment example).

In order to deal with these new issues, it is necessary in the first place to provide ourselves with a model for the intuitive but vague notion of "changing dimension". This model is valid regardless of the dimensions but for the sake of simplicity, it will be presented in the case of manifolds embedded in $\mathbb{R}^3$. In this model, the target manifold $\mathcal{M}$ has a boundary $\partial\mathcal{M}$ of dimension $k - 1$. Thus, the study of the VDFs to manifolds with boundary cannot be dissociated from the content of this section.

6.1   VDFs of manifolds with a boundary and their singularities

The previous section has motivated the introduction of manifolds with boundaries. Well-known examples of such manifolds are open curves in the plane or in 3D, surface patches in 3D, regions of the plane or volumes in 3D. Such manifolds are easily modelled by noticing that, if their dimension is k each of their points is contained in a neighborhood which is either homeomorphic to $\mathbb{R}^k$ or to the closed half-space $H^k \equiv \{(x_1, \cdots, x_k) \in \mathbb{R}^k : x^k \geq 0\}$. We note $\mathcal{M}$ the manifold and $\partial\mathcal{M} \subset \mathcal{M}$ its boundary which we assume to be a smooth manifold of dimension $k - 1$. $\mathcal{M}$ is also a function of the time t.

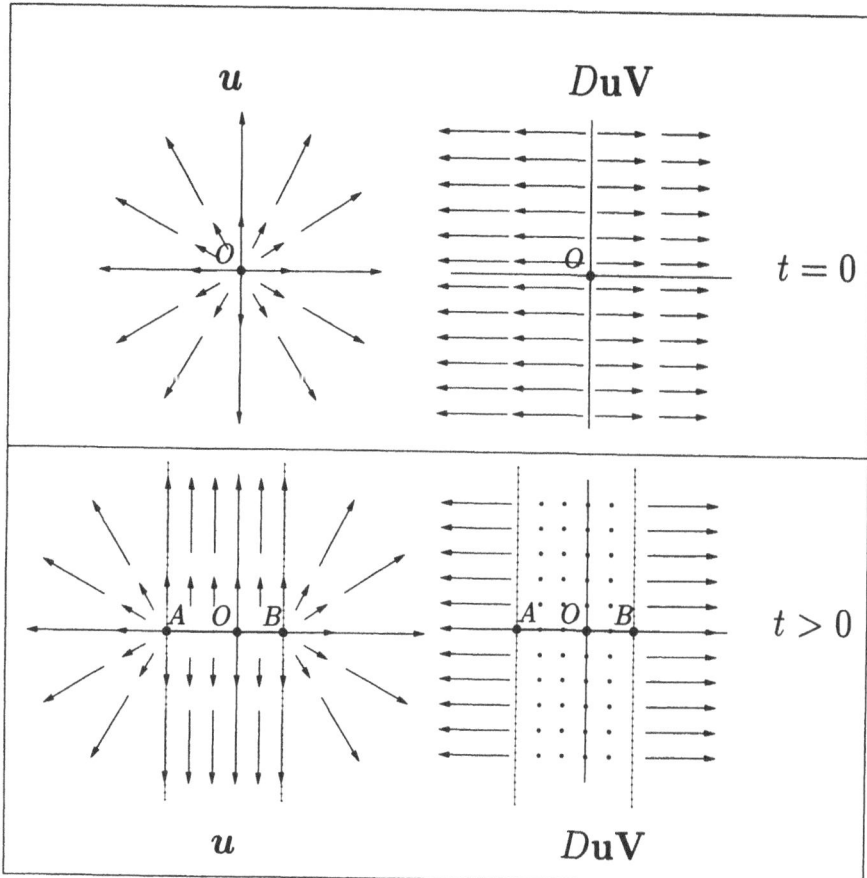

Fig. 11. Changing dimension: If we allow the velocity field **b** to be discontinuous, we can induce changes in the dimension of the manifold. The example shows the simplest of all manifolds, a single point O in the plane and its VDF (upper lefthand corner), the velocity field $\mathbf{b}(.,0) = -D\mathbf{u}V$ is discontinuous on the vertical axis (upper righthand corner). At a later time $t > 0$ the point O has become the line segment AB (its VDF is shown in the lower lefthand corner); the new velocity $\mathbf{V}(.,t)$ is shown in the lower righthand corner. The initial point, a smooth manifold without boundary of dimension 0 is turned into a closed (in the topological sense) open (in the usual sense) curve, a smooth manifold with boundary (the two endpoints A and B) of dimension 1.

We show that the boundary $\partial\mathcal{M}$ introduces new singularities for the VDF to $\mathcal{M}$, in effect singularities of order 1: both $D\mathbf{u}$ and $\mathbf{u}_t$ are discontinuous on some hypersurfaces defined by $\partial\mathcal{M}$ and the normal space of $\mathcal{M}$ at points of $\partial\mathcal{M}$.

This is shown in the next proposition

Proposition 11 *The spatial derivative $D\mathbf{u}$, of the VDF to the smooth manifold $\mathcal{M}$ with boundary $\partial\mathcal{M}$ is discontinuous on a special hypersurface $R_{\partial\mathcal{M}}$,* *defined in the proof, generated by ∂M and the normal space to M at points of ∂M.*

Proof. To see this we consider the set, noted $\mathcal{M}_a$ (a for above), of points **x** such that $P_\mathcal{M}(\mathbf{x}) \in \mathcal{M}\backslash\{\partial\mathcal{M}\}$, and the set, noted $\mathcal{M}_b$ (b for beside), of points **x** such that $P_\mathcal{M}(\mathbf{x}) \in \partial\mathcal{M}$. The boundary between these two sets is the hypersurface, noted $R_{\partial\mathcal{M}}$, of $\mathbb{R}^n$ generated by $\partial\mathcal{M}$ and all the vectors normal to $\mathcal{M}$ at a point of $\partial\mathcal{M}$. For example, in the case of an open curve in the plane, this set is equal to the two lines orthogonal to the curve at its two endpoints, see figure 12. As further examples, in the case of an open 3D curve,

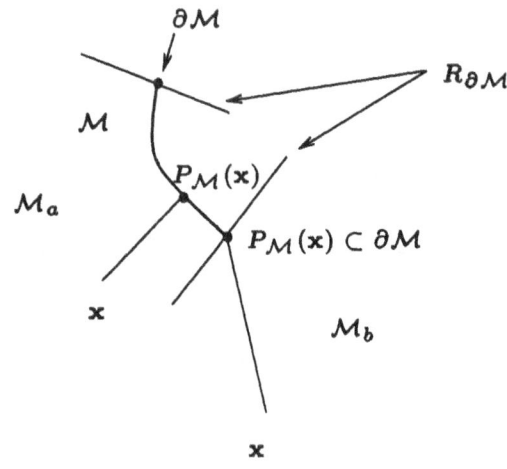

Fig. 12. The manifold $\mathcal{M}$ is an open curve of $\mathbb{R}^2$; the two endpoints are the boundary $\partial\mathcal{M}$.

$R_{\partial\mathcal{M}}$ is made of the two planes orthogonal to the curve at its two endpoints, in the case of an open surface patch $R_{\partial\mathcal{M}}$ is the ruled surface generated by its boundary and the normals to the surface, see figure 13.

In region $\mathcal{M}_a$, the VDF is that of a manifold of dimension k whereas in region $\mathcal{M}_b$, the VDF "sees" only $\partial\mathcal{M}$, a manifold of dimension $k-1$.

Because of proposition 7, in region $\mathcal{M}_b$, $D\mathbf{u}$ has $k-1$ eigenvalues equal to 1 whereas it has k such eigenvalues in region $\mathcal{M}_a$, therefore $D\mathbf{u}$ is discontinuous on $R_{\partial\mathcal{M}}$.

Let us now look at the discontinuities of $\mathbf{u}_t$. In each of the two regions $\mathcal{M}_a$ and $\mathcal{M}_b$ we are actually evolving the VDFs of $\mathcal{M}\backslash\{\partial\mathcal{M}\}$ and $\partial\mathcal{M}$, respectively, since this is what the VDF "sees". Therefore, in each of these two regions, $\mathbf{u}_t$ satisfies the properties determined in section 4. But since the normal spaces at a point of $\partial\mathcal{M}$ considered as a point of $\partial\mathcal{M}$ and as a point of $\mathcal{M}$ differ (the first one contains the second and is of dimension higher by one)

$$N_\mathbf{x}\mathcal{M} \subset N_\mathbf{x}\partial\mathcal{M} \quad \forall\mathbf{x} \in \partial\mathcal{M} \quad \text{and} \quad dim(N_\mathbf{x}\partial\mathcal{M}) = dim(N_\mathbf{x}\mathcal{M}) + 1,$$

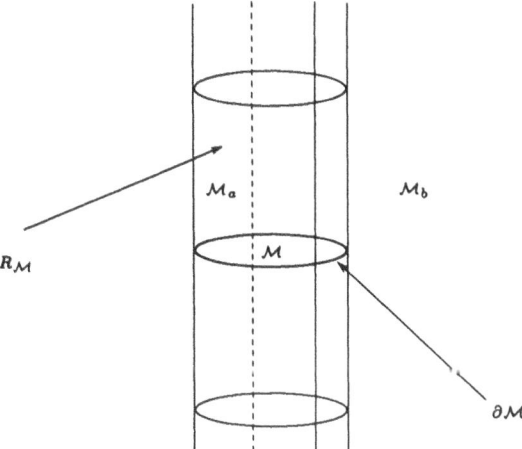

Fig. 13. The manifold $\mathcal{M}$ is a surface patch of $\mathbb{R}^3$; its boundary $\partial\mathcal{M}$ is a smooth closed curve which generates a ruled surface: the rulings are the normals to $\mathcal{M}$.

we may expect some problems. We have the following proposition for the discontinuities of $\mathbf{u}_t$

Proposition 12 *The tangential component of the time derivative $\mathbf{u}_t$, of the VDF to the smooth manifold $\mathcal{M}$ with boundary $\partial\mathcal{M}$ can be discontinuous on the hypersurface $R_{\partial\mathcal{M}}$ generated by $\partial\mathcal{M}$ and the normal space to $\mathcal{M}$ at points of $\partial\mathcal{M}$. The normal component is continuous.*

Proof. Let us consider a point $\mathbf{x}$ of $\mathcal{M}_b$. We note $\mathbf{y} = P_{\partial\mathcal{M}}(\mathbf{x}) = P_{\mathcal{M}}(\mathbf{x})$. $\mathbf{u}_t(\mathbf{x}, t)$ has two components, one in $N_{\mathbf{y}}\partial\mathcal{M}$ and one in $T_{\mathbf{y}}\partial\mathcal{M}$

$$\mathbf{u}_t(\mathbf{x}, t) = \mathbf{u}_t(\mathbf{x}, t)_{| N_{\mathbf{y}}\partial\mathcal{M}} + \mathbf{u}_t(\mathbf{x}, t)_{| T_{\mathbf{y}}\partial\mathcal{M}}$$

The first component, noted $\mathbf{N}(\mathbf{y}, t)$ depends only on $\mathbf{y}$ and t as long as $P_{\mathcal{M}}(\mathbf{x}) = \mathbf{y}$. It can be uniquely decomposed as the sum of a vector $\mathbf{N}_1(\mathbf{y}, t)$ in $N_{\mathbf{y}}\mathcal{M}$ and a vector $\mathbf{N}_2(\mathbf{y}, t)$ in $N_{\mathbf{y}}\partial\mathcal{M} \cap T_{\mathbf{y}}\mathcal{M}$. This comes from the fact that $N_{\mathbf{y}}\mathcal{M} + T_{\mathbf{y}}\mathcal{M} = \mathbb{R}^n$ and $N_{\mathbf{y}}\mathcal{M} \subset N_{\mathbf{y}}\partial\mathcal{M}$ which imply that

$$N_{\mathbf{y}}\mathcal{M} + (N_{\mathbf{y}}\partial\mathcal{M} \cap T_{\mathbf{y}}\mathcal{M}) = N_{\mathbf{y}}\partial\mathcal{M}$$

Let us now consider a point $\mathbf{x}'$ of $R_{\partial\mathcal{M}}$ such that $P_{\mathcal{M}}(\mathbf{x}') = \mathbf{y}$. Since $\mathbf{x}'$ is in $\mathcal{M}_a$, the velocity $\mathbf{u}_t(\mathbf{x}', t)$ has two components, one, noted $\mathbf{N}'(\mathbf{y}, t)$ is in the normal space $N_{\mathbf{y}}\mathcal{M}$ at $\mathbf{y}$ and the other one is in $T_{\mathbf{y}}\mathcal{M}$:

$$\mathbf{u}_t(\mathbf{x}', t) = \mathbf{N}'(\mathbf{y}, t) + \mathbf{u}_t(\mathbf{x}', t)_{| T_{\mathbf{y}}\mathcal{M}}$$

Therefore we must have $\mathbf{N}_1(\mathbf{y}, t) = \mathbf{N}'(\mathbf{y}, t)$ since otherwise the manifold $\mathcal{M}$ will spread out in its normal space and create a discontinuity at $\mathbf{y}$, a contradiction with the smoothness hypothesis.

The tangential components are equal to $N_2(y) + u_t(x,t)_{|T_y \partial M}$ for $u(x,t)$ and, using the fact that $T_y \mathcal{M} = T_y \partial \mathcal{M} + (N_y \partial \mathcal{M} \cap T_y \mathcal{M})$, $u_t(x',t)_{|N_y \partial \mathcal{M} \cap T_y \mathcal{M}} + u_t(x',t)_{|T_y \partial \mathcal{M}}$. If we let x tend to x', the tangential components may be discontinuous without introducing a discontinuity on $\mathcal{M}$.

An example of this is shown in figure 14 in the case of an open curve of $\mathbb{R}^2$. Another example is shown in in figure 15 in the case of a surface patch of $\mathbb{R}^3$.

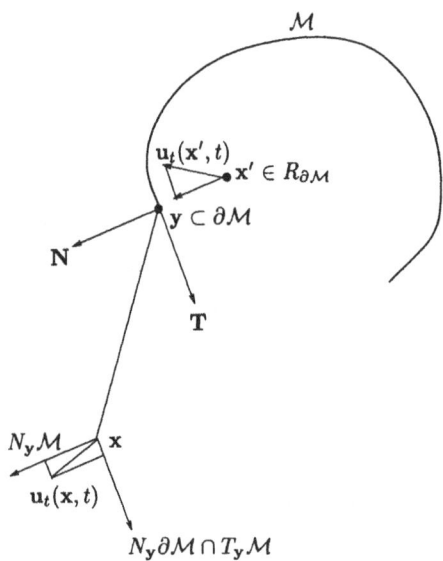

Fig. 14. The time derivative $u_t(x,t)$ of the VDF of the open planar curve $\mathcal{M}$ at point $x \in \mathcal{M}_b$ can be uniquely decomposed as the sum of a component in $N_y \mathcal{M}$ and a component in $N_y \partial \mathcal{M} \cap T_y \mathcal{M} = T_y \mathcal{M}$. The component in $N_y \mathcal{M}$ must be equal to that of $u_t(x',t)$ at any point $x' \in R_{\partial \mathcal{M}}$ such that $P_{\mathcal{M}}(x') = P_{\mathcal{M}}(x)$. See proof of proposition 12.

6.2   Modelling a change of dimension

An initial smooth manifold W of dimension $m < n$ at $t = 0$ increases its dimension if it "spreads" out in the direction of some priviledged normal directions and becomes another manifold $M(t)$, $t > 0$, of higher dimension $k \in \{m+1, \cdots, n\}$. The increment in dimension is equal to the number of linearly independent orthogonal normal directions where this filling occurs. The modelling of a decrease in dimension is obtained by reversing the direction of time.

Let us make this clearer in some simple cases in $\mathbb{R}^3$. For instance (see figure 16), the normal space to a point (a smooth manifold of dimension 0)

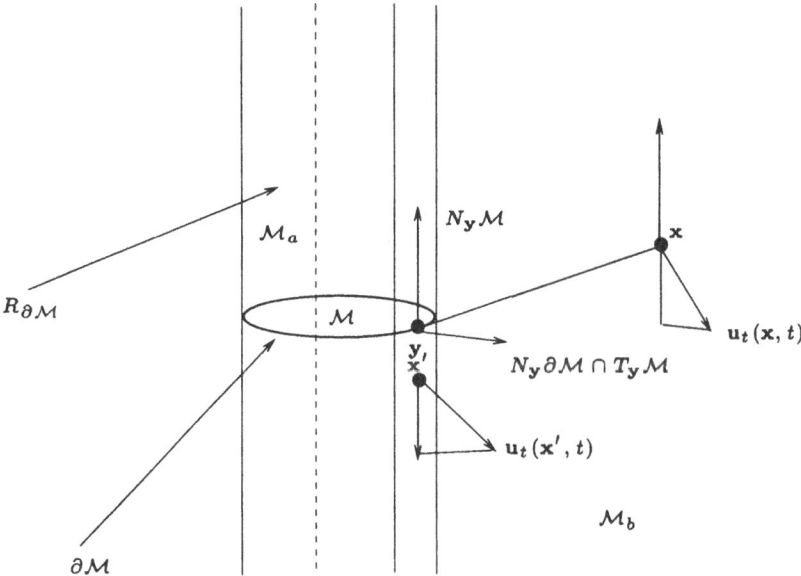

Fig. 15. The time derivative $\mathbf{u}_t$ of the VDF of the surface patch $\mathcal{M}$ at point $\mathbf{x} \in \mathcal{M}_b$ can be uniquely decomposed in the sum of a component in $N_\mathbf{y}\mathcal{M}$ and a component in $N_\mathbf{y}\partial\mathcal{M} \cap T_\mathbf{y}\mathcal{M} \subset T_\mathbf{y}\mathcal{M}$. The component in $N_\mathbf{y}\mathcal{M}$ must be equal to that of $\mathbf{u}_t(\mathbf{x}', t)$ at any point $\mathbf{x}' \in R_{\partial\mathcal{M}}$ such that $P_\mathcal{M}(\mathbf{x}') = P_\mathcal{M}(\mathbf{x})$. See proof of proposition 12.

is $\mathbb{R}^3$ itself, i.e. it is three-dimensional. In order to change its dimension, the point may "choose" a certain normal direction and spread out in that direction to become an arc of a curve (case a). Alternatively, it may "choose" to spread out over a two-dimensional subspace of its normal space and transform smoothly into a small surface patch (case b). It can even fill the entire embedding space by spreading out in all of its normal directions (case c). In all directions of spread, the velocity field is like the one described in the previous section, equation (26) with the values of α and β varying with the normal. Similarly, an embedded curve in $\mathbb{R}^3$ (a smooth manifold of dimension 1) has a two-dimensional normal space at each of its points and thus, its dimension can increase by 1 or 2 according to the number of orthogonal normal directions that are given a priviledged role. By choosing a particular smooth normal vector field, the curve can spread in these nornal directions to become a ribbon (case d) which is a smooth manifold with boundary of dimension 2. Alternatively, the curve may spread in all directions of its normal spaces, thereby becoming a smooth manifold with boundary of dimension 3, a volume ((see figure 17). To be complete, a surface patch (a smooth manifold of dimension 2 with or without boundary) has a one-dimensional normal space. By allowing the velocity field $\mathbf{b}$ to be discontinuous on the patch with jumps that may depend on the normal at each point, the patch can become a smooth manifold of dimension 3 with boundary, a volume (see figure 18).

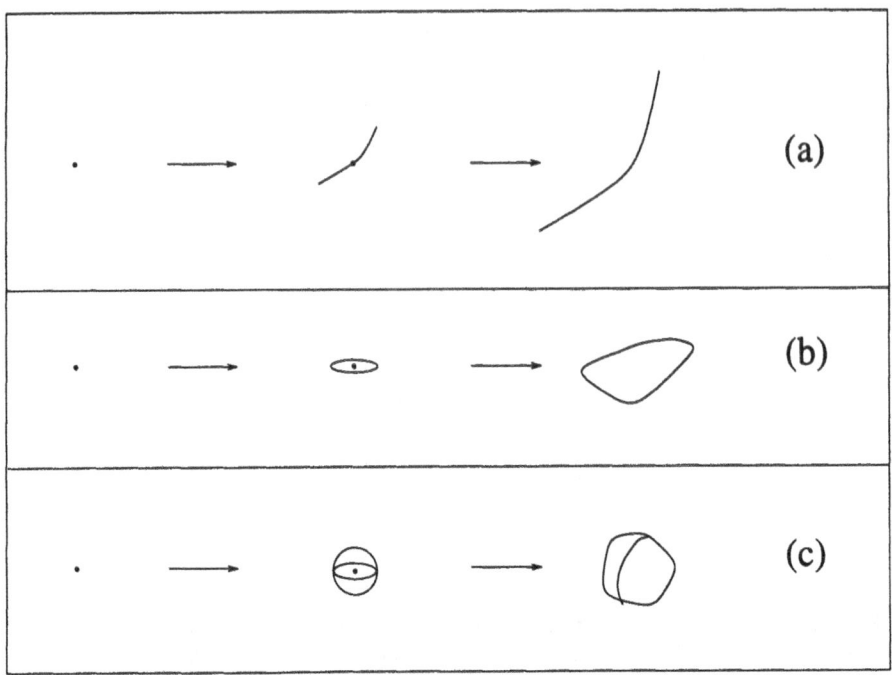

Fig. 16. A point ($m = 0$) spreads out to become a curve ($k = 1$): (a), a patch of a surface ($k = 2$): (b) and a volume ($k = 3$): (c). See text.

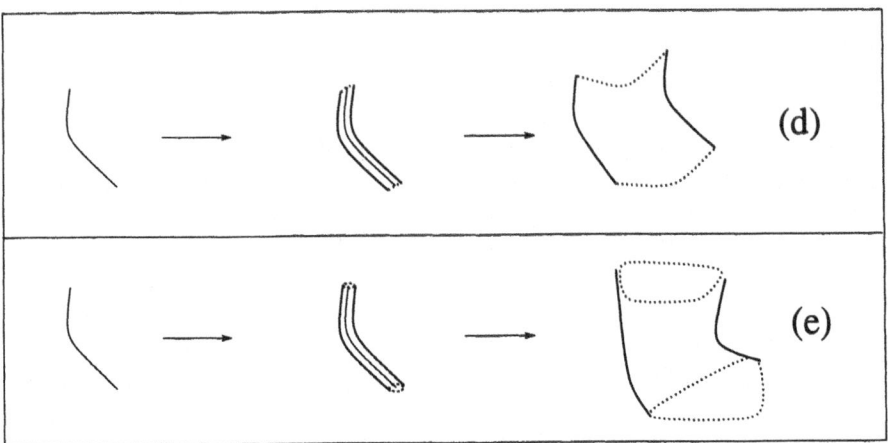

Fig. 17. A curve ($m = 1$) spreads out to become a surface patch ($k = 2$): (d), and a volume ($k = 3$): (e). See text.

Palpably, the important missing ingredient for the formalization of this "spreading" is that of a k-dimensional neighbourhood of a smooth manifold of dimension $m < n$, $k > m$ $\mathcal{W}$. Now, the key tool in Riemannian geometry

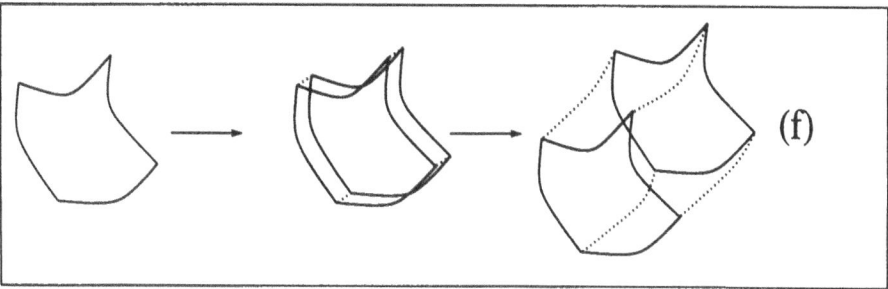

(f)

Fig. 18. A surface patch ($m = 2$) spreads out to become a volume ($k = 3$): (f). See text.

for discussing neighbourhoods of embedded manifolds is the *exponential map*: a parameterization by arc-length of the geodesic curves.

Definition 1. Let g be a Riemannian metric[1] and

$$\mathbb{R} \to \mathbb{R}^3 \tag{27}$$

$$\lambda \to \exp_g(\boldsymbol{x}, \boldsymbol{v}, \lambda) \tag{28}$$

be an embedding of the corresponding geodesic curve passing through $\boldsymbol{x}$, tangent to the unit vector $\boldsymbol{v}$ such that λ is the arc length.

See [41,16] for important results concerning the exponential map and its use in the study of neighbourhoods. As a simple example, if $g(\mathbf{x}) = \mathbf{I}_3 \forall \mathbf{x}$, $\exp_g(\boldsymbol{x}, \boldsymbol{v}, \lambda) = \mathbf{x} + \lambda \boldsymbol{v}$ is the line going through x and parallel to $\boldsymbol{v}$. By choosing different metrics g, in particular non-constant ones, one obtains curved geodesics which can be used to model smooth manifolds $\mathcal{M}$ containing $\mathcal{W}$. The appeal of this notion is due to the fact it allows to define what is a neighbourhood of radius λ of a manifold. We are going to use it right now for this purpose with $\lambda = t$, the time, in order to clarify the way a manifold can smoothly spread out over a neighbourhood of itself. To begin with, let us revisit some of the examples of figure 17.

First, we return to the example of a curve transforming into a ribbon (17 d). Let $\mathcal{W}$ be this curve, $\boldsymbol{n}$ a differentiable unit normal vector field along $\mathcal{W}$ and g a Riemannian metric of $\mathbb{R}^3$ (see figure 17 d). Consider a point $\boldsymbol{x} \in \mathcal{W}$ and the geodesic curve passing through x and tangent to $\boldsymbol{n}(\mathbf{x})$ i.e. the curve $\mathcal{G}_\mathbf{x}(s)$ embedded by $s \to \exp_g(\mathbf{x}, \pm \boldsymbol{n}, s). 0 \leq s \leq t$. The open curve $\mathcal{G}_\mathbf{x}$ is normal to $\mathcal{W}$ at x and models the change of dimension of the point x into a curve whose boundary is the two points $A_\mathbf{x} \equiv \left\{ \exp_g(\boldsymbol{x}, \boldsymbol{n}, t), B_\mathbf{x} \equiv \exp_g(\boldsymbol{x}, -\boldsymbol{n}, t) \right\}$.

For $t > 0$ small enough, consider the union of all these geodesic curves

$$\mathcal{M}(t) = \cup_{\mathbf{x} \in \mathcal{W}} \mathcal{G}_\mathbf{x}$$

---

[1] $g(\mathbf{x})$ is at every point x of $\mathbb{R}^3$ a positive quadratic form of $\mathbb{R}^3$; the function $\mathbf{x} \longrightarrow g(\mathbf{x})$ is assumed to be smooth

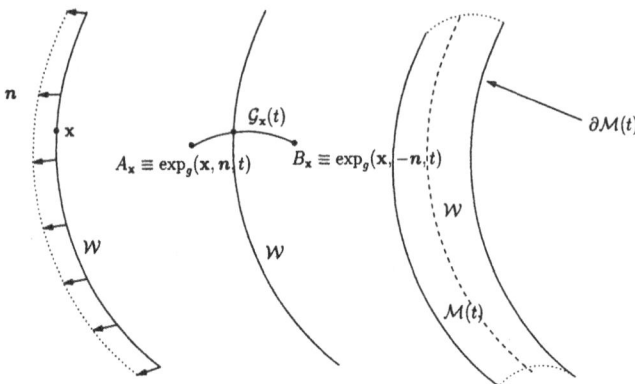

Fig. 19. Given a smooth field of normals $n(\mathbf{x})$ along the curve $\mathcal{W}$, and a riemannian metric g on $\mathbb{R}^3$, the set $\mathcal{M}(t)$ for t small enough, union of all the geodesic curves $\mathcal{G}_\mathbf{x}$ is a smooth surface patch whose boundary $\partial\mathcal{M}(t)$ is the union of all the points $A_\mathbf{x}$ and $B_\mathbf{x}$, see text.

It is a smooth surface of $\mathbb{R}^3$ which has for boundary a smooth curve. Indeed, $\mathcal{M}(t)$ can be embedded by

$$\mathcal{W} \times [0, t[\to \mathbb{R}^3$$
$$(\mathbf{x}, s) \to \exp_g(\mathbf{x}, \pm n(\mathbf{x}), s)$$

and its border $\partial\mathcal{M}(t)$ by

$$\mathcal{W} \to \mathbb{R}^3$$
$$\mathbf{x} \to \exp_g(\mathbf{x}, \pm n(\mathbf{x}), t)$$

The proof these two maps are embeddings is a straightforward consequence of the important known result that, for short geodesics, the exponential map is a diffeomorphism.

Next, we show how these ideas apply in the case of a curve transforming into a tubular volume (see figure 18 e). The increase of dimension should be two and since the normal space at each point $N_\mathbf{x}\mathcal{W}$ of the curve is twodimensional, it means that the spread must occur in all the normal directions. We note $P\mathcal{W}$ the subset of the normal bundle $N\mathcal{W}$ corresponding to unit normal vectors (we have dimension($P\mathcal{W}$) = dimension($N\mathcal{W}$) − 1). Hence, the target manifold $\mathcal{M}(t)$ can be embedded by

$$P\mathcal{W} \times [0, t[\to \mathbb{R}^3 \tag{29}$$
$$((\mathbf{x}, n), s) \to \exp_g(\mathbf{x}, n, s) \tag{30}$$

and its border $\partial\mathcal{M}(t)$ by

$$P\mathcal{W} \to \mathbb{R}^3 \tag{31}$$
$$(\mathbf{x}, n) \to \exp_g(\mathbf{x}, n, t) \tag{32}$$

which shows that the dimension of $\mathcal{M}(t)$ is 3 and that of $\partial\mathcal{M}(t)$ is 2.

To conclude the presentation of the model with a general note, the change of dimension is entirely defined by the choice of a submanifold of PW and a metric g of the embedding space. The submanifold of PW describes the set of directions where W is supposed to spread out and its dimension is equal to the increment of dimension minus one. As far as the metric g is concerned, its purpose is to describe the shape of $\mathcal{M}(t)$.

In terms of PDEs, this spreading can be achieved with a straightforward generalization of the PDE (26). Let us note v_{A_x} and v_{B_x} the tangent vectors to $\mathcal{G}_x$ at A_x and B_x, respectively. The VDF to $\mathcal{M}(t)$ satisfies the following PDE

$$\mathbf{u}_t + D\mathbf{u}\mathbf{V} = 0 \text{ with } \mathbf{V}(\mathbf{x}, t) = \begin{cases} v_{A_x} & \text{if } \mathbf{x} \in \mathcal{M}(t)_b \text{ and } P_{\mathcal{M}}(\mathbf{x}) = A_\mathbf{x}, \\ v_{B_x} & \text{if } \mathbf{x} \in \mathcal{M}(t)_b \text{ and } P_{\mathcal{M}}(\mathbf{x}) = B_\mathbf{x}, \\ 0 & \text{if } \mathbf{x} \in \mathcal{M}(t)_a, \end{cases}$$

At $t = 0$, this reduces to

$$\mathbf{u}_t + D\mathbf{u}\mathbf{V} = 0 \text{ with } \mathbf{V}(\mathbf{x}, t) = \begin{cases} n(P_W(\mathbf{x})) & \text{if } (\mathbf{x} - P_W(\mathbf{x})) \cdot n(P_W(\mathbf{x})) > 0, \\ -n(P_W(\mathbf{x})) & \text{if } (\mathbf{x} - P_W(\mathbf{x})) \cdot n(P_W(\mathbf{x})) < 0 \end{cases}$$

7 Example II: the sphere

Consider the spherical sector, part of the sphere of radius 1, centered at the origin O from which we take away the part above the plane of equation $x_3 = \cos\alpha$, see figure 20. This is a smooth manifold $\mathcal{M}$ with a boundary $\partial\mathcal{M}$ equal to the circle of radius $\sin\alpha$ centered at the point C of coordinates $[0, 0, \cos\alpha]^T$. The surface $R_{\partial\mathcal{M}}$ is the half cone of vertex O containing $\partial\mathcal{M}$. For a point $\mathbf{x}$ in $\mathcal{M}_a$, the VDF is equal to

$$\mathbf{u}(\mathbf{x}) = \mathbf{x} - \frac{\mathbf{x}}{\|\mathbf{x}\|}$$

For a point $\mathbf{x}'$ in $\mathcal{M}_b$, the VDF is equal to

$$\mathbf{u}(\mathbf{x}) = \mathbf{x} - \frac{\mathbf{z}}{\|\mathbf{z}\|}\sin\alpha - \mathbf{C}$$

where $\mathbf{z} = [x_1, x_2, 0]^T$ is the projection of $\mathbf{x}$ on the plane of equation $x_3 = 0$. It is therefore easy to compute $D\mathbf{u}$

$$D\mathbf{u}(\mathbf{x}) = \begin{cases} (1 - \frac{1}{\|\mathbf{x}\|})\mathbf{I}_3 + \frac{\mathbf{x}\mathbf{x}^T}{\|\mathbf{x}\|^3} & \text{if } \mathbf{x} \in \mathcal{M}_a \\ \mathbf{I}_3 - \frac{\sin\alpha}{\|\mathbf{z}\|}\mathbf{I}_3^2 + \frac{\mathbf{z}\mathbf{z}^T}{\|\mathbf{z}\|^3}\sin\alpha & \text{if } \mathbf{x} \in \mathcal{M}_b \end{cases}$$

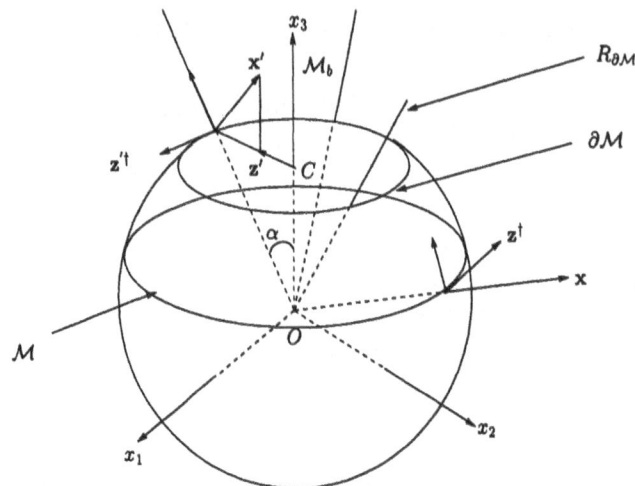

Fig. 20. A spherical sector.

whre $\mathbf{I}_3^2 = diag(1,1,0)$. We verify that in an orthonormal basis built from the two vectors $(\mathbf{x} - \mathbf{u}(\mathbf{x}), \mathbf{z}^\dagger)$ $(\mathbf{z}^\dagger = [-x_2, x_1, 0]^T)$ and their cross-product, the matrix $D\mathbf{u}$ becomes diagonal

$$D\mathbf{u}(\mathbf{x}) = \begin{cases} diag(1, 1 - \frac{1}{\|\mathbf{x}\|}, 1 - \frac{1}{\|\mathbf{x}\|}) & \text{if } \mathbf{x} \in \mathcal{M}_a \\ diag(1, 1, 1 - \frac{\sin\alpha}{\|\mathbf{z}\|}) & \text{if } \mathbf{x} \in \mathcal{M}_b \end{cases}$$

In region $\mathcal{M}_a$, the VDF "sees" a smooth manifold of dimension 2 (a sphere) and, according to proposition 7, there is one eigenvalue equal to 1 and the other two are equal to $\frac{\|\mathbf{x}\|-1}{\|\mathbf{x}\|} \equiv \frac{s}{s+1}$ if we let $s = \|\mathbf{x}\| - 1$. As expected, these two eigenvalues tend to 0 when s does, i.e. when the point $\mathbf{x}$ approaches the sphere. They blow up at the center ($s = -1$) which is both a center of curvature and on the skeleton.

In region $\mathcal{M}_b$, the VDF "sees" a smooth manifold of dimension 1 (a circle) and, according to the same proposition, there are two eigenvalues equal to 1 and one equal to $\frac{\|\mathbf{z}\|-\sin\alpha}{\|\mathbf{z}\|}$. This eigenvalue tends to zero when $\|\mathbf{z}\|$ tends to $\sin\alpha$, i.e. when $\mathbf{x}$ approaches the circle.It blows up when $\|\mathbf{z}\| = 0$, i.e. on the x_3-axis, the skeleton of the circle.

Moreover, when $\mathbf{x}$ is on the half cone $R_{\partial\mathcal{M}}$ of equation $\|\mathbf{z}\|^2 - x_3^2 \tan^2\alpha = 0$ one verifies that $\|\mathbf{x}\|^2 \sin^2\alpha = \|\mathbf{z}\|^2$. Hence when $\mathbf{x}$ tends to $R_{\partial\mathcal{M}}$ in $\mathcal{M}_b$ $D\mathbf{u}(\mathbf{x})$ tends to $diag(1, 1, 1 - \frac{1}{\|\mathbf{x}\|})$ which is different from $diag(1, 1 - \frac{1}{\|\mathbf{x}\|}, 1 - \frac{1}{\|\mathbf{x}\|})$ outside the sphere. Hence we have an example of the result in proposition 11.

8   Some remarks and conclusion

The method of the VDFs for representing arbitrary smooth manifolds with or without a boundary finds its roots in the work of Ambrosio and Soner [3] where we found inspiration and some of the technical results that we needed, and in the work of Ruuth, Merriman, Xin and Osher [37] that inspired us the idea of the VDF representation. Our contributions are the development of a) a method for evolving a VDF instead of a smooth closed manifold while guaranteeing that it stays a VDF over time and that the manifold evolves correctly, b) a theory that describes changes of dimension by a generalized transport equation and c) a theory that extends a) to deal seamlessly with smooth manifolds with boundaries. Moreover, in our approach, the function we evolve is regular on the manifold of interest unlike for example the method presented in [36].

These theoretical developments are being implemented. Let us make two remarks concerning this implementation. The first remark is related to the computation of the velocity field **b**. Even though proposition 10 provides a characterization of **b(x)** in terms of $\mathbf{V}(\mathbf{x} - \mathbf{u}(\mathbf{x}))$, and the tangential terms given in the proof, our favorite method is to compute **b** by applying proposition 9 and solving the quasilinear PDE (20) with initial conditions (21).

The second remark is that because of the accumulation of numerical errors, the function **u** may drift away from the class of VDFs. In order to correct for this drift, we suggest that it may be a good idea to combine the solution of (19) with that of

$$\mathbf{u}_t = ((D\mathbf{u})^T + (\alpha - 1)\mathbf{I})\mathbf{u}, \tag{33}$$

where

$$\alpha = Trace(D\mathbf{u}(D\mathbf{u})^T) + \mathbf{u} \cdot \Delta\mathbf{u} - div\mathbf{u}.$$

Equation (33) is the Euler-Lagrange equation of the following functional

$$\frac{1}{2}\int_{\Omega} \|D\mathbf{u})^T\mathbf{u} - \mathbf{u}\|^2 \ d\mathbf{x}$$

where Ω is some neighbourhood of $\mathcal{M}$. This functional arises naturally from proposition 3.

To conclude, we think that the VDF method for representing and evolving shapes has the following advantages: it can deal with smooth manifolds with and without boundaries, with shapes of different dimensions; if discontinuous velocity fields are allowed dimension can change; the evolution method that we propose guarantees that we stay in the class of VDFs and therefore that the intrinsic properties of the underlying shapes such as their dimension, curvatures can be read off easily from the VDF and its spatial derivatives. The main disadvantage is its redundancy: the size of the representation is always that of the ambient space even though the object we are representing is of a much lower dimension. This disadvantage is also one of its strengths since it buys us flexibility.

References

1. Boston, MA, June 1995. IEEE Computer Society Press.
2. L. Ambrosio and C. Mantegazza. Curvature and distance function from a manifold. *J. Geom. Anal.*, 1996. To appear.
3. Luigi Ambrosio and Halil M. Soner. Level set approach to mean curvature flow in arbitrary codimension. *J. of Diff. Geom.*, 43:693–737, 1996.
4. V. I. Arnold. *Geometrical Methods in the Theory of Ordinary Differential Equations*. Springer-Verlag New York Inc., 1983.
5. G. Barles. *Solutions de viscosité des équations de Hamilton-Jacobi*. Springer-Verlag, 1994.
6. G. Barles, H.M. Soner, and P.E. Souganidis. Front propagation and phase field theory. *SIAM J. Control and Optimization*, 31(2):439–469, March 1993.
7. M. Bertalmio, G. Sapiro, and G. Randall. Region Tacking on Surfaces Deforming via Level-Sets Methods. In Mads Nielsen, P. Johansen, O.F. Olsen, and J. Weickert, editors, *Scale-Space Theories in Computer Vision*, volume 1682 of *Lecture Notes in Computer Science*, pages 58–69. Springer, September 1999.
8. A. Blake and M. Isard. *Active Contours*. Springer-Verlag, 1998.
9. V. Caselles and B. Coll. Snakes in Movement. *SIAM Journal on Numerical Analysis*, 33:2445–2456, December 1996.
10. V. Caselles, R. Kimmel, and G. Sapiro. Geodesic active contours. In *Proceedings of the 5th International Conference on Computer Vision* [1], pages 694–699.
11. V. Caselles, R. Kimmel, and G. Sapiro. Geodesic active contours. *The International Journal of Computer Vision*, 22(1):61–79, 1997.
12. V. Caselles, R. Kimmel, G. Sapiro, and C. Sbert. 3d active contours. In M-O. Berger, R. Deriche, I. Herlin, J. Jaffre, and J-M. Morel, editors, *Images, Wavelets and PDEs*, volume 219 of *Lecture Notes in Control and Information Sciences*, pages 43–49. Springer, June 1996.
13. V. Caselles, R. Kimmel, G. Sapiro, and C. Sbert. Minimal surfaces based object segmentation. *IEEE Transactions on Pattern Analysis and Machine Intelligence*, 9(4):394–398, 1997.
14. Y.G. Chen, Y. Giga, and S. Goto. Uniqueness and existence of viscosity solutions of generalized mean curvature flow equations. *J. Differential Geometry*, 33:749–786, 1991.
15. R. Cipolla and P. Giblin. *Visual Motion of Curves and Surfaces*. Cambridge University Press, 2000.
16. M. P. DoCarmo. *Riemannian Geometry*. Birkhäuser, 1992.
17. C.L. Epstein and Michael Gage. The curve shortening flow. In A.J. Chorin, A.J. Majda, and P.D. Lax, editors, *Wave motion: theory, modelling and computation*. Springer-Verlag, 1987.
18. L.C. Evans. *Partial Differential Equations*, volume 19 of *Graduate Studies in Mathematics*. American Mathematical Society, 1998.
19. L.C. Evans and J. Spruck. Motion of level sets by mean curvature: I. *Journal of Differential Geometry*, 33:635–681, 1991.
20. Olivier Faugeras and Renaud Keriven. Variational principles, surface evolution, pde's, level set methods and the stereo problem. *IEEE Trans. on Image Processing*, 7(3):336–344, March 1998.
21. M. Gage. Curve shortening makes convex curves circular. *Invent. Math.*, 76:357–364, 1984.

22. M. Gage and R.S. Hamilton. The heat equation shrinking convex plane curves. *J. of Differential Geometry*, 23:69–96, 1986.

23. J. Gomes and O. Faugeras. Reconciling Distance Functions and Level Sets. In Mads Nielsen, P. Johansen, O.F. Olsen, and J. Weickert, editors, *Scale-Space Theories in Computer Vision*, volume 1682. Springer, 1999.

24. J. Gomes and O. Faugeras. Representing and evolving smooth manifolds of arbitrary dimension embedded in R^n as the intersection of n hypersurfaces: The vector distance functions. Technical report, INRIA, 2000.

25. J. Gomes and O. Faugeras. Shape representation as the intersection of k hypersurfaces. Technical report, INRIA, 2000.

26. J. Gomes and O.D. Faugeras. Reconciling Distance Functions and Level Sets. *Journal of Visual Communication and Image Representation*, 11:209–223, 2000.

27. M. Grayson. The heat equation shrinks embedded plane curves to round points. *J. of Differential Geometry*, 26:285–314, 1987.

28. M. Grayson. A short note on the evolution of surfaces via mean curvature. *Duke Math J.*, pages 555–558, 1989. Proof that the dumbell can split in 2 under mean curvature motion.

29. M. Kass, A. Witkin, and D. Terzopoulos. SNAKES: Active contour models. *The International Journal of Computer Vision*, 1:321–332, January 1988.

30. B. Kimia, A. R. Tannenbaum, and S. W. Zucker. Shapes, schoks and deformations i: The components of two-dimensional shape and the reaction-diffusion space. *ijcv*, 15:189–224, 1995.

31. Jan J. Koenderink. *Solid Shape*. MIT Press, 1990.

32. L. Lorigo, O. Faugeras, W.E.L. Grimson, R. Keriven, R. Kikinis, and C-F. Westin. Co-dimension 2 geodesic active contours for mra segmentation. In *Proc. Int'l Conf. Information Processing in Medical Imaging*, pages 126–139, June 1999.

33. J. Montagnat, H. Delingette, N. Scapel, and N. Ayache. Representation, shape, topology and evolution of deformable surfaces. Application to 3D medical image segmentation. Technical report, INRIA, 2000.

34. S. Osher and J. Sethian. Fronts propagating with curvature dependent speed : algorithms based on the Hamilton-Jacobi formulation. *Journal of Computational Physics*, 79:12–49, 1988.

35. Nikos Paragios and Rachid Deriche. Geodesic active contours and level sets for the detection and tracking of moving objects. *IEEE PAMI*, 22(3):266–280, March 2000.

36. S. J. Ruuth, B. Merriman, J. Xin, and S. Osher. Diffusion-Generated Motion by Mean Curvature for filaments. Technical Report 98-47, UCLA Computational and Applied Mathematics Reports, November 1998.

37. S.J. Ruuth, B. Merriman, and S. Osher. A fixed grid method for capturing the motion of self-intersecting interfaces and related pdes. Technical Report 99-22, UCLA Computational and Applied Mathematics Reports, July 1999.

38. J.A. Sethian. Recent numerical algorithms for hypersurfaces moving with curvature-dependent speed:Hamilton-Jacobi equations and conservation laws. *J. Differential Geometry*, 31:131–136, 1990.

39. J.A. Sethian. Theory, algorithms, and applications of level set methods for propagating interfaces. Technical Report PAM-651, Center for Pure and Applied Mathematics, University of California, Berkeley, August 1995. To appear Acta Numerica.

40. K. Siddiqi, A. Shokoufandeh, S. J. Dickinson, and S. W. Zucker. Shock Graphs and Shape Matching. *The International Journal of Computer Vision*, 35(1):13–32, November 1999.

41. Michael Spivak. *A Comprehensive Introduction to Differential Geometry*, volume I-V. Publish or Perish, Berkeley, CA, 1979. Second edition.

42. H. Tek and B. B. Kimia. Image Segmentation by Reaction-Diffusion Bubbles. In *Proceedings of the 5th International Conference on Computer Vision* [1].

43. D. Terzopoulos, A. Witkin, and M. Kass. Constraints on Deformable Models: Recovering 3D shape and Nonrigid Motion. *Artificial Intelligence*, 36(1):91–123, 1988.

44. Hong-Kai Zhao, T. Chan, B. Merriman, and S. Osher. A variational level set approach to multiphase motion. *Journal of Computational Physics*, 127(0167):179–195, 1996.

Least–Squares Fitting of Algebraic Spline Curves via Normal Vector Estimation

Bert Jüttler

Darmstadt University of Technology, Dept. of Mathematics,
Schlossgartenstr. 7, 64289 Darmstadt, Germany

Summary. We describe an algorithm for fitting implicitly defined algebraic spline curves to given planar data. By simultaneously approximating points and associated normal vectors, we obtain a method which is both computationally simple, as the result is obtained by solving a system of linear equations, and geometrically invariant. The initial result of the curve fitting procedure is improved by iteratively adjusting the associated field of normal vectors. It is planned to generalize the approach to algebraic spline surfaces.

1 Introduction

This paper is devoted to the reconstruction of planar curves from scattered data. The (possibly noisy) data are assumed to be generated be taking sample points of a certain planar object. We construct a planar curve that approximately matches the shape of the data. Our approach is based on implicitly defined algebraic spline curves. More precisely, the planar curve is obtained as the zero contour of a bivariate real spline function.

The present study is intended to serve as a prototype for planned research on methods for reconstructing algebraic spline surfaces from scattered data in 3–space. However, the reconstruction of planar shapes from measurement data is also an interesting subject in its own right (see [13] for references), with various applications, e.g. in medical imaging.

Compared to the parametric representations, such as NURB (Non–Uniform Rational B-spline, see [7,9]) curves and surfaces, implicitly defined curves and surfaces offer several advantages:

- The curve and surface fitting procedures do not need the estimation of auxiliary parameter values which are associated with the given data. In the parametric case, by contrast, these parameters have a strong influence to the resulting shape, and it may be difficult to generate appropriate values, in particular for more complex shapes (see e.g. [9]).
- It is possible to bypass the initial polygonalization resp. triangulation step that is required in many parametric curve and surface fitting procedures.
- They can be used to define planar domains resp. solids; the point membership can easily be decided by evaluating the sign of the generating real function.

- Relatively simple algorithms are available for computing the intersection(s) with straight lines, as this reduces to a one–dimensional root-finding problem. This is particularly advantageous for the visualization with the help of ray–tracing methods.
- Additional shape constraints can easily be added to the curve resp. surface fitting procedure. For instance, the convexity of the resulting curve resp. surface can be guaranteed by the convexity of the underlying real function; see [11] for suitable linear convexity criteria. This leads to relatively simple constrained optimization problems, as the feasible domain is a convex set. Convexity criteria for truly parametric representations, by contrast, are far more complicated, cf. [12].

On the other hand, implicitly defined curves and surfaces cause some extraneous difficulties which need to be taken care of.

- The visualization and evaluation of the surface needs special contour–finding algorithms, such as 'marching cubes', see [9].
- In order to exchange data with commercial CAD systems, the implicitly defined curves and surfaces have to be converted into the industrial NURBS standard, cf. [3].
- Approximation or interpolation schemes may produce algebraic curves or surfaces which consist of several disconnected components. This needs special treatment in order to avoid the resulting problems.

Methods for curve and surface fitting with implicitly defined algebraic curves and surfaces have been discussed in an enormous number of publications, and it is virtually impossible to give a complete survey. We list only a few references which had a major influence to the present research.

Pratt [14], Taubin [15], and Bajaj et al. [2] describe methods for implicit curve and surface fitting. The methods are based on the algebraic distance, combined with suitable normalizations of the unknown coefficients. For instance, Pratt's 'simple fit' method [14] keeps the value of one of the coefficients, leading to a linear normalization constraint. The results, however, are not geometrically invariant; they depend on the choice of the coordinates. Taubin's method [15] constrains the sum of the squared gradients at the data sites. This leads to a geometrically invariant quadratic normalization.

The approximants are computed by solving constrained quadratic programming problems (minimization of a quadratic objective function subject to linear, resp. quadratic, constraints). In the case of quadratic constraints (which are required in order to get geometrically invariant results), the solutions are found by numerically solving generalized eigenvalue problems, i.e., by computing the eigenvector which is associated with the smallest eigenvalue of a certain matrix. The dimension of the matrix is equal to the number of unknown coefficients. Based on experimental evidence, Pratt's and Taubin's methods have been compared by Umasuthan and Wallace [16].

In a number of publications, Bajaj and various co–authors [5,1] have developed implicit algebraic surfaces into a powerful tool for reconstructing curves and surfaces from measurement data ('reverse engineering'). Their approach focuses on the use of low–degree patches whose coefficients satisfy certain sign conditions, in order to guarantee the desired topology of the result.

Recently, Werghi et al. [17] have developed an incremental framework, incorporating geometric constraints (such as orthogonality), for fitting implicitly defined geometric primitives, such as planes and quadrics.

The present paper describes a novel approximation technique for implicitly defined algebraic spline curves. By simultaneously approximating points and associated normal vectors, we obtain a method which is both computationally simple (as the result is obtained by solving a system of linear equations) and geometrically invariant. This will help to overcome some of the limitations of the normalization–based algebraic curve fitting procedures.

In the remainder of this paper we describe an algorithm that fits an implicitly defined algebraic spline curve to given planar data. More precisely, consider a set of points

$$p_i = (p_{i,1} \ p_{i,2}) \in \mathbb{R}^2; \quad i = 1, \dots, N; \tag{1}$$

in the plane. The approximating curve is to be described as the zero contour of a bivariate real function $z = f(x_1, x_2)$. The function f is chosen as a piecewise polynomial function, leading to a approximating algebraic spline curve.

The approximating curve is constructed in several steps, as follows. Firstly we estimate normal vectors n_i which are associated with the given data. In the second step we fit an algebraic spline curve, matching both the data and the associated normals. Using the additional normal vector information, we are able to avoid the various normalizations that are used for fitting algebraic curves and surfaces. Finally, in order to improve the results, one may update the associated normals and iterate the curve fitting procedure.

2 Estimating associated normal vectors

As the first step of the curve fitting procedure, we generate unit normal vectors

$$n_i = (n_{i,1} \ n_{i,2}) \in \mathbb{S}^1; \quad i = 1, \dots, N; \tag{2}$$

from the unit circle $\mathbb{S}^1 = \{ z \in \mathbb{R}^2 \mid \|z\| = 1 \}$ which are associated with the given data (1). The estimation of normal vectors from measurement data is a standard problem in scattered data approximation. In many applications, the normal vectors can be generated directly from additional information accompanying the data. For instance, if the points p_i are generated from a certain image, such as in Computer Tomography, then the normal vectors could be chosen as the (normalized) gradients of the color function. If no additional information is available, however, then the normal vectors have to be estimated directly from the data. We summarize the basic idea of the method. More details can be found in [13], see also [8] for the 3D case.

In order to associate a unit normal vector n_i with one the points p_i, we fit a simple curve to the neighborhood of that point, see Figure 1 for a schematic illustration.

Firstly we compute the associated local line of regression L_i. It is found by minimizing the weighted sum of squared distances of the points $(p_j)_{j=1,...,N}$ from the line L_i. The weight function $w = w(r)$ is chosen such that influence of a point p_j to the line of regression L_i decreases with its distance $r = \|p_i - p_j\|$ from p_i. Possible choices include the characteristic function $w(r) = \chi_{[0,h]}$ (i.e., taking only points p_j with maximum distance h from p_i into account), or the exponential weight function $w(r) = \exp(-r^2/H^2)$, with certain suitable constants h, H. The resulting optimization problem has a quadratic objective function which minimized is subject to the quadratic equality constraint $\|n^*\|^2 = 1$, where n^* is the unit normal vector of L_i. It can be solved by computing the eigenvectors of a certain 2×2 matrix.

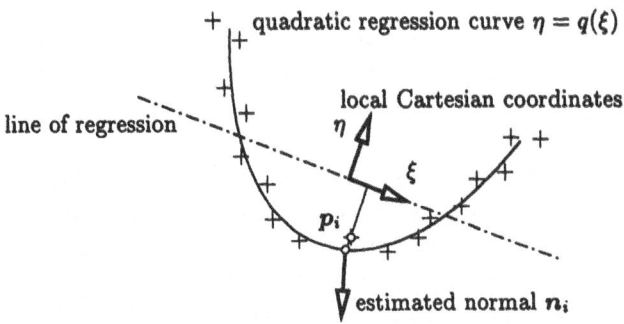

Fig. 1. Estimating the normal vector from the given data (shown as crosses).

In the second step we choose a new Cartesian coordinate system whose axis of abscissae is parallel to L_i, and fit a local quadratic regression curve to the data. This curve is the graph of a quadratic polynomial with respect to the new Cartesian system. Its coefficients are computed minimizing the weighted sum of squared residuals, leading to a 3×3 system of linear equations.

The direction $\pm n_i$ of the unit normal vector which is associated with p_i is then chosen such that it is parallel to the normal of the quadratic polynomial, evaluated at the abscissa of p_i. In order to get useful results, however, we have to guarantee that neighboring normal vectors have the same orientation. That is, if two points p_i, p_j are relatively close together, then the inner product $n_i \cdot n_j$ of the associated normal vectors is expected to be positive. In order to choose the orientation of the normals, one could compute the minimum spanning tree of the data and use the resulting neighborhood information, see [13]. As a cheaper alternative, one may simply choose an appropriate rectangular grid, and select – for each of the resulting quadrangular

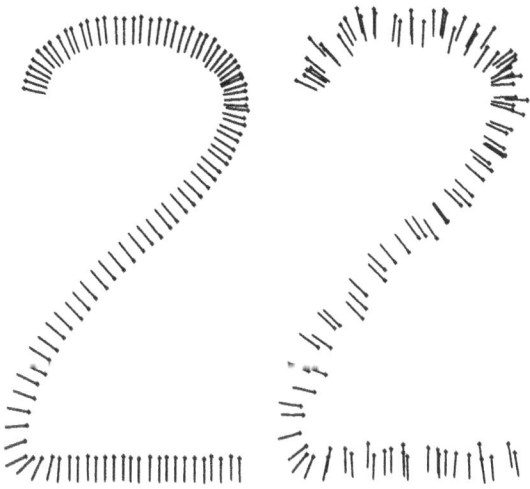

Fig. 2. Estimation of associated normal vectors from scattered data (examples with different levels of noise).

cells – a representative of the points which are contained in it (if such points exist). The orientation of the normals that are associated with the representatives is chosen according to the neighborhood structure of the grid. Then, the orientation of the remaining normals is chosen according to that of the representatives. Clearly, the success of this simple method depends on the appropriate specification of the grid size, requiring some user interaction.

Two examples are shown in Figure 2. We have sampled 100 points from a planar shape and added some noise to it, with two different levels. The plots show the data $(p_i)_{i=1,\ldots,N}$, along with the estimated normal vectors $(n_i)_{i=1,\ldots,N}$. The normal vectors have been estimated by considering – for each point – the 10 nearest neighbors, and fitting the line of regression and a quadratic regression curve to them.

3   Implicit algebraic tensor–product spline curves

The implicitly defined algebraic spline curve is described by a tensor–product spline function of (bi–) degree d ($d \geq 2$). The segments of the resulting spline are algebraic curves of order $2d$. The use of tensor–product spline offers several advantages, including simple implementation, simple conditions for global smoothness, simple evaluation, sufficient flexibility and refinability (e.g. using hierarchical B-spline representations, see [4])

The approximating curve is described as the zero contour $f(x, y) = 0$ of the tensor–product spline function

$$f(x, y) = \sum_{(i,j)\in\mathcal{I}} M_i(x)\, N_j(y)\, c_{i,j} \tag{3}$$

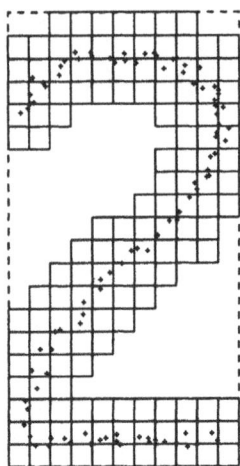

Fig. 3. The domain $\mathcal{D}$ of the tensor–product spline function (3) consists of all cells that contain data, and the neighbouring cells (within the enlarged bounding box of the data, shown as dashed lines).

with the real coefficients (control points) $c_{i,j}$, where $\mathcal{I}$ is a certain index set, see below. The basis functions $(M_i(x))_{i=1..m}$ and $(N_j(y))_{j=1..n}$ are B-splines of degree d with respect to the knot sequences $\mathcal{X} = (\xi_i)_{i=1..m+1}$ and $\mathcal{Y} = (\eta_j)_{j=1..n+1}$, see [6,9]. We choose the first and last $d+1$ knots of $\mathcal{X}$ resp. $\mathcal{Y}$ as the abscissas resp. ordinates of the slightly enlarged (by blowing it up) bounding box of the data $(p_k)_{k=1,...,N}$. The remaining inner knots are chosen equidistant.

The two knot sequences define a partition of the (slightly enlarged) bounding box $[\xi_1, \xi_{m+d+1}] \times [\eta_1, \eta_{n+d+1}]$ into rectangular cells. Clearly, not all of these cells necessarily contain data points. We choose the domain $\mathcal{D}$ of the tensor–product spline function (3) to be the union of all cells that contain data, and of the neighboring cells, cf. Figure 3. The index set $\mathcal{I}$ in (3) is chosen such that the summation includes all products $M_i(x)\,N_j(y)$ that do not vanish on $\mathcal{D}$,

$$\mathcal{I} = \{\,(i,j)\mid \exists\,k \in \{1,\ldots,N\}, r \in \{1,\ldots,m\}, s \in \{1,\ldots,n\}: \atop M_r(p_{k,1})\,N_s(p_{k,2}) \neq 0 \,\wedge\, \max\{|i-r|,|j-s|\} \leq 1\,\} \tag{4}$$

Clearly, the domain $\mathcal{D}$ of the tensor–product spline function (3) depends on the choice of the Cartesian coordinate system. An geometrically invariant choice could be obtained by fitting a line of regression (which serves as one of the coordinate axes) to all data.

In order to keep the algebraic order as low as possible (if this is desired), it would be more appropriate to chose a bivariate spline function of total degree d (e.g., defined with respect to a union of triangles, such as Powell–Sabin spline or simplex spline, see [9]), rather than a tensor–product one. Within the tensor–product setting, a lower algebraic degree could be guaranteed by introducing additional side–conditions. For instance, a biquadratic tensor-product spline, along with the linear constraints $f_{xxy} = 0$ and $f_{xyy} = 0$, gives C^1 spline curves of algebraic order 2, i.e., C^1 conic splines. Adding these constraints, however, leads to a highly redundant representation.

The approximation method which is described in the next section can be applied to any implicit algebraic spline curve, not only to curves in tensor-product representation. Using tensor–products we obtain a method which is computationally simple, but whose results depend on the choice of the system of coordinates (unless we choose the coordinates with the help of the line of regression, as outlined earlier). This dependency, however, is caused only by the choice of the space of functions, not by the approximation method. For instance, applying the approximation method to bivariate polynomials of total degree d gives geometrically invariant results.

4 Fitting curves to points with associated normals

Let $c = (c_{i,j})_{(i,j) \in \mathcal{I}}$ be the vector obtained by gathering all B-spline coefficients (control points) of the approximating algebraic spline curve, in a suitable ordering. Its components will be computed by minimizing a quadratic objective function $F = F(c)$, which is formed as a certain linear combination of four terms.

4.1 Approximating the data

The first two terms deal with the data $(p_i)_{i=1,\dots,N}$ and with the associated normal vectors $(n_i)_{i=1,\dots,N}$. The given data are approximated by minimizing the sum of the squared 'algebraic distances' (see [14,15]),

$$L(c) = \sum_{i=1}^{N} [\, f(p_{i,1}, p_{i,2})\,]^2. \tag{5}$$

The sum L is a homogeneous quadratic form of the unknown control points c. Hence, it is minimized by the null vector $c_{i,j} = 0$, leading to the tensor-product spline function $f(x, y) \equiv 0$. In order to get results which are more useful, one has to introduce a normalization. Various normalizations have been described in the literature [2,14,15], most of them based on a suitable norm in the coefficient space. Our approach is based on the simultaneous approximation of the data and the associated normal vectors, by minimizing – in addition to L – the sum

$$M(c) = \sum_{i=1}^{N} \| \nabla f(p_{i,1}, p_{i,2}) - n_i \|^2$$

$$= \sum_{i=1}^{N} (f_x(p_{i,1}, p_{i,2}) - n_{i,1})^2 + (f_y(p_{i,1}, p_{i,2}) - n_{i,2})^2. \tag{6}$$

That is, the gradients $\nabla f = (f_x, f_y)$ of the tensor–product spline function $f(x, y)$ at the given data p_i are to match the estimated normal vectors n_i. Consequently, as the given normals n_i are unit vectors, the algebraic distances in (5) can be expected to approximate the real distances.

4.2   Tension terms

Although the minimization of weighted linear combination of L and G leads to good results in many cases, it may produce approximating spline curves that split into several disconnected components. There are several possibilities to address this well–known problem of algebraic curve and surface fitting. For instance, based on the signs of the coefficients, one may derive criteria which guarantee the desired topology of the result, see [5].

As we wish to compute the solution by solving a system of linear equations, we use the simpler approach of adding suitable 'tension terms' that pull the approximating curve towards a simpler shape. If the the tension terms have a sufficiently strong influence (which is governed by certain weights), then the approximating curve has the desired topology.

A global tension term is given by the quadratic functional

$$G(c) = \iint_{\mathcal{D}} f_{xx}^2 + 2 f_{xy}^2 + f_{yy}^2 \, dx \, dy \tag{7}$$

which measures the deviation of $f(x, y)$ from a linear function. Hence, by increasing the influence of this tension term, the resulting spline curve gets closer to a straight line.

In addition to the global tension term we have tested a data–dependent tension term also. It is based on the following simple observation.

Lemma 1. *Consider a tensor–product polynomial $p(x, y)$. If the quadratic functional (with respect to the coefficients of the polynomial)*

$$Q = \iint_{[0,1]^2} \sum_{(i,j) \in \mathcal{L}} \left[\left(\frac{\partial}{\partial x} \right)^i \left(\frac{\partial}{\partial y} \right)^j p(x, y) \right]^2 \, dx \, dy \tag{8}$$

vanishes, with the index set

$$\mathcal{L} = \{(i, j) \in \mathbb{Z}_+^2 \mid j \geq 2 \ \vee \ (j = 1 \wedge i \geq 1)\}, \tag{9}$$

and $p_y(x, y) \neq 0$ holds, then the level curves $p(x, y) = d$ are graphs of polynomials $y = q_d(x)$.

Proof. If a tensor–product polynomial $p(x, y)$ satisfies

$$\left(\frac{\partial}{\partial x} \right)^i \left(\frac{\partial}{\partial y} \right)^j p(x, y) = 0 \quad \forall (i, j) \in \mathcal{L} \tag{10}$$

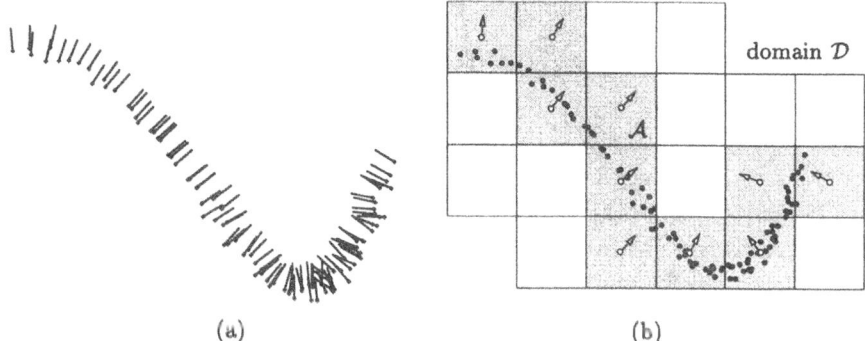

(a)   (b)

Fig. 4. Generating the data–dependent tension term $D(c)$. The data with the associated normal vectors (a) and the set $\mathcal{A}$ of cells with the normals n_c (b).

and $p_y(x, y) \neq 0$, then it has the form $p(x, y) = C\,y + q(x)$ with a real constant $C \neq 0$. Consequently, the level curves $p(x, y) = d$ are simply graphs of the polynomials $y = (d - q(x))/C$. The linear equations (10) are equivalent to the quadratic equation $Q = 0$.   □

By replacing the partial derivatives in (8) with *directional* derivatives with respect two perpendicular unit vectors, one obtains a condition for the level curves to be graphs of polynomials with respect to another system of coordinates. We obtain the data–dependent tension term L by collecting the left–hand sides of (8), and using suitable directional derivatives, whose directions depend on the data.

More precisely, let $\mathcal{A}$ be the set of all cells (belonging to the grid defined by the knot sequences $\mathcal{X}$ and $\mathcal{Y}$ of the spline function) which contain at least one of the data $(p_i)_{i=1,\dots,N}$. The data–dependent tension term is given by

$$D(c) = \sum_{C \in \mathcal{A}} \iint_C \sum_{(i,j) \in \mathcal{L}} \left[\left(\frac{\partial}{\partial n_c^\perp} \right)^i \left(\frac{\partial}{\partial n_c} \right)^j f(x, y) \right]^2 dx\, dy = 0, \quad (11)$$

where n_c is the normalized average of the associated normal vectors of the points in C, and $n_c^\perp$ obtained by a clockwise rotation of $\pi/2$.

An example is presented in Figure 4. Its left part (a) shows the data $(p_i)_{i=1,\dots,N}$ and the associated normals $(n_i)_{i=1,\dots,N}$. The right figure (b) visualizes the domain $\mathcal{D}$ of the tensor–product spline function, the cells $\mathcal{A}$ that contain data (marked in grey), and the associated normal vectors n_c (shown as arrows) which are used for generating D.

The tension term D measures – for each cell C – the deviation of the algebraic curves $f(x, y) = d$ from graphs of polynomials with respect to an adapted system of Cartesian coordinates. By minimizing it one may obtain the desired topology of the solution, without getting results which are 'too flat'. An example is given below.

4.3   Computing the solution

The approximating implicit spline curve is found by minimizing the weighted linear combination

$$F(c) = L(c) + w_1 \, M(c) + w_2 \, G(c) + w_3 \, D(c) \quad \rightarrow \quad \text{Min}, \qquad (12)$$

see (5), (6), (7) and (11), with certain non–negative weights w_1, w_2, and w_3. As this leads to a quadratic objective function of the unknown control points $c = (c_{i,j})_{(i,j) \in \mathcal{I}}$, the solution can be found by solving the sparse linear system of equations

$$\frac{\partial}{\partial c_{i,j}} F(c) = 0, \quad (i,j) \in \mathcal{I}, \qquad (13)$$

with the help of appropriate methods from numerical linear algebra. Alternatively, one may also compute the solution of (12) by generating an overconstrained system of linear equations, and computing a least–squares solution to it using QU factorization, see e.g. [6].

The weight w_1 controls the influence of the estimated normal vectors n_i to the resulting curve. The weights w_2 and w_3 act as tension parameters, see Section 4.5 for an example.

As the main benefit from the simultaneous approximation of points p_i and normal vectors n_i, the approximating spline curve is found by solving a system of linear equations. The required computations have the complexity $\mathcal{O}(h^2)$ (without taking the sparsity into account), where $h = |\mathcal{I}|$ is the number of coefficients.

The traditional, normalization–based methods [14,15], by contrast, generate the solution of the curve fitting problem by solving a generalized eigenvalue problem, requiring iterative numerical procedures for computing the result. According to [15], the complexity is $\mathcal{O}(h^3)$.

4.4   Existence and uniqueness

Under certain mild assumptions, the problem (12) can easily be shown to be uniquely solvable.

Proposition 1. *If the weights w_1 and w_2 are positive, and $w_3 \neq 0$, then the quadratic optimization problem (12) has a unique solution. Consequently, the rank of the square coefficient matrix of the linear system (13) equals the number of coefficients $|\mathcal{I}|$.*

Proof. It is easy to see that the quadratic functionals (5), (6), (7) and (11) are convex. That is,

$$Q(\lambda c^{(1)} + (1-\lambda) c^{(2)}) \leq \lambda \, Q(c^{(1)}) + (1-\lambda) \, Q(c^{(2)}), \quad 0 < \lambda < 1, \quad (14)$$

holds for all of them, $Q \in \{L, N, G, D\}$. However, they are not strictly convex, as both sides of (14) may be equal. This case is characterized by certain conditions to the difference vector of the coefficients, as follows.

Consider a tensor–product spline function $g(x, y)$ with the coefficients $c^{(1)} - c^{(2)} = (c_i^{(1)} - c_i^{(2)})_{i=1,...,N}$, where both sides of (14) are assumed to be equal. If $Q = L$ then these coefficients represent a function $f(x, y)$ such that the associated zero contour interpolates all data, i.e.

$$g(p_{i,1}, p_{i,2}) = 0 \quad \text{holds for} \quad i = 1, \ldots, N. \tag{15}$$

If $Q = N$ then the difference coefficients represent a function $f(x, y)$ whose gradient vanishes at all data, i.e.

$$\nabla g(p_{i,1}, p_{i,2}) = 0 \quad \text{holds for} \quad i = 1, \ldots, N. \tag{16}$$

Finally, if $Q = G$ then they simply describe a linear function $g(x, y)$, i.e.,

$$g_{xx}(x, y) = g_{xy}(x, y) = g_{yy}(x, y) = 0 \quad \text{holds for} \quad (x, y) \in \mathcal{D}. \tag{17}$$

Thus, the linear combination (12) with positive coefficients w_1, w_2 (and non-negative w_3) is even strictly convex, as only the zero function $(c^{(1)} - c^{(2)} = 0)$ simultaneously fulfills the three conditions (15), (16), (17). This proves the assertion. $\qquad\qquad\qquad\qquad\qquad\qquad\qquad\qquad\qquad\qquad\qquad\quad\square$

Note that the tension term G is the only part of the objective function (12) that acts on all cells of the domain $\mathcal{D}$. The other parts act only on cells which contain data. Consequently, the choice $w_2 = 0$ will always produce a singular system (13), if at least one of the products $M_i(x) N_j(y)$ vanishes at all data $(x, y) = (p_{i,1}, p_{i,2})$. This may happen very easily, as the domain of the spline function $f(x, y)$ consists of all cells containing data, and the neighboring cells.

4.5   Examples

As a first example, we approximate the two sets of data shown in Figure 2 by implicit algebraic tensor–product spline curves of degree $d = 2$. We choose knot vectors which generate a grid of 3×2 cells, leading to $|\mathcal{I}| = 20$ unknown B-spline coefficients $c_{i,j}$. The weights of the objective function are chosen as $w_1 = 1$, $w_2 = w_3 = 0.001$. The resulting curves are shown in Figure 5.

In order to compare these results with the ones described in the next section, we provide the ℓ_1 (total error), ℓ_2 (least–squares), and ℓ_∞ (maximum error) norm of the residual errors

$$\boldsymbol{R} = (R_i)_{i=1,...,N} \quad \text{with} \quad R_i = \inf\{ \|\boldsymbol{p}_i - \boldsymbol{z}\| \mid f(z_1, z_2) = 0 \}, \tag{18}$$

measuring the orthogonal distances from the data $\boldsymbol{p}_i$ to the approximating curve. The orthogonal distances have been generated by computing the foot-points of the data with the help of Newton–Raphson iterations. The curves

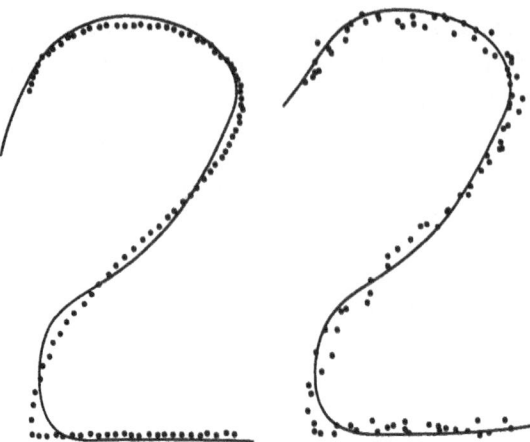

Fig. 5. Approximation of the data from Figure 2 by algebraic tensor–product spline curves of degree 2.

from Figure 5 give the following results, where the residuals R_l resp. R_r refer to the example on the left– resp. right–hand side:

$$\|R_l\|_1 = 5.286, \ \|R_l\|_2 = 0.623, \ \|R_l\|_\infty = 0.204,$$
$$\|R_r\|_1 = 5.959, \ \|R_r\|_2 = 0.732, \ \|R_r\|_\infty = 0.178. \tag{19}$$

The next figure visualizes the influence of the two different tension terms. Here, the knot vectors generate a grid of 6×4 cells, see Figure 4 for the resulting domain $\mathcal{D}$. This leads to $|\mathcal{I}| = 45$ unknown B–spline coefficients $c_{i,j}$. The curves in Figures 6a resp. b have been generated by adding global resp. data–dependent tension. We chose the weights of the objective function such that the ratios of the approximation and tension parts

$$\frac{L(c) + w_1 M(c)}{w_2 G(c) + w_3 D(c)} \tag{20}$$

have the same values (≈ 1.2) for both curves. Thus, both tension terms have approximately the same influence to the result. The data–dependent tension term gives the better result, as the global tension term tends to straighten out the resulting curve.

Clearly, as a slight modification of the method, one could also apply tension locally, to specific parts of the curve only. This could be achieved simply by using weight functions $w_2 = w_2(x,y)$ and $w_3 = w_3(x,y)$ which depend on the coordinates x and y.

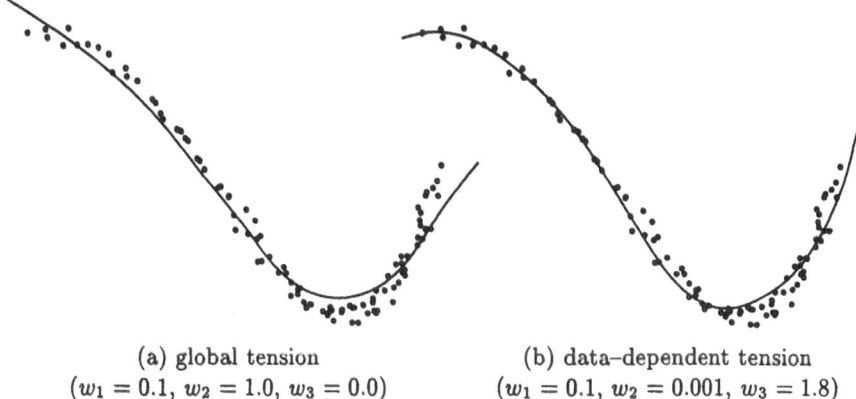

(a) global tension
$(w_1 = 0.1, \; w_2 = 1.0, \; w_3 = 0.0)$

(b) data–dependent tension
$(w_1 = 0.1, \; w_2 = 0.001, \; w_3 = 1.8)$

Fig. 6. Comparison of the tension terms. Approximating curve with global tension (a) and with data–dependent tension (b). Both tension terms have approximately the same influence.

5   Updating the objective function

By appropriately updating the objective function, the result of the curve fitting procedure can be improved. This leads to an iterative curve fitting procedure.

5.1   Weighted least–squares

Clearly, the algebraic distances (5) do not represent the real distances between the data and the approximating spline curve. In order to obtain a better approximation, it is a standard approach in implicit curve and surface fitting to use – instead of (5) – the weighted least–squares sum

$$L^*(c) = \sum_{i=1}^{N} [\, \omega_i \; f(p_{i,1}, p_{i,2}) \,]^2. \tag{21}$$

with certain positive weights ω_i, leading to the modified optimization problem

$$F(c) = L^*(c) + w_1 \, M(c) + w_2 \, G(c) + w_3 \, D(c) \quad \rightarrow \quad \text{Min.} \tag{22}$$

Ideally, the weights would have the values

$$\omega_i = \frac{1}{\|\nabla f(p_{i,1}, p_{i,1})\|}, \tag{23}$$

where f is the solution to (22). Then, the weighted least–squares sum (21) would be a good approximation of the squared Euclidean distances, see [15]. The following 'reweight procedure' is described in [15]; it is similar in spirit to various related methods of weighted least–squares.

(Iterative fitting procedure)

1. Compute an initial solution $f^{(1)}(x, y)$ by choosing the weights $\omega_i = 1$, $i = 1, \ldots, N$. Let $j = 1$.
2. Choose the weights $\omega_i = 1/\|\nabla f^{(j)}(p_{i,1}, p_{i,1})\|$ and compute the new solution $f^{(j+1)}$ of (22).
3. If a certain termination criterion is satisfied (e.g., the improvement of Euclidean distances and/or the change of weights stays below a certain limit) then stop, $f^{(j+1)}$ is the approximate solution. Otherwise increase j by 1 and continue with Step 2.

In the original algorithm [15], the computation of the next solution in Step 1 resp. 2 requires the solution of a generalized eigenvalue problem. By simultaneously approximating points and associated normal vectors, we have to solve a system of linear equations instead.

5.2   Adjusting the normal vectors

A similar iterative procedure can be applied to the normal vectors $(n_i)_{i=1,\ldots,N}$ which are associated with the given data. It is analogous to the method of 'parameter correction' for parametric curve and surface fitting, see [9, Section 4.4.3]. There, the new parameters of a point are chosen according to the location of its nearest neighbor on the approximating curve resp. surface; they are then used for computing an improved solution.

Similarly, we may use the gradients of the first approximation at the data p_i in order to adjust the normal vectors n_i. Clearly, the lengths of these vectors will be non-uniform in general. In order to avoid contraction to a sequence of null vectors, we scale them such that the sum of the squared lengths equals $\sqrt{N}$,

$$\sum_{i=0}^{N} \|n_i\|^2 = N. \tag{24}$$

This leads to the following modified Step 2 of the iterative fitting procedure from the previous section:

2′. Choose the normal vectors according to

$$n_i = \frac{\sqrt{N}}{\sqrt{\sum_{k=1}^{N} \|\nabla f^{(j)}(p_{k,1}, p_{k,2})\|^2}} \nabla f^{(j)}(p_{i,1}, p_{i,2}); \quad i = 1, \ldots, N; \tag{25}$$

and continue with the original Step 2.

Clearly, also the termination criterion in Step 3 has to be modified, by taking the change of the associated normal vectors into account.

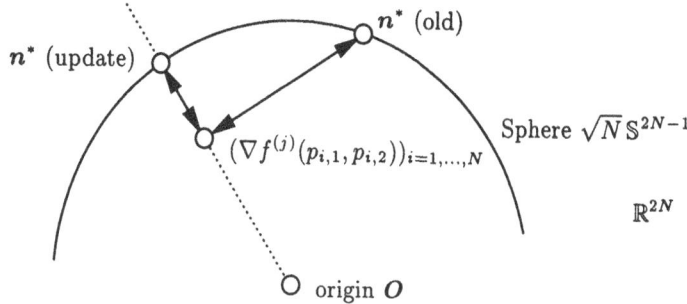

Fig. 7. Geometric interpretation of the normal vector adjustment. The arrows '$\leftrightarrow$' represent the value of the normal vector part $M(c)$ of the objective function.

One may identify the normal vectors

$$n^* = (n_i)_{i=1,\ldots,N} \tag{26}$$

with a single vector $n^* \in \mathbb{R}^{2N}$. The vectors satisfying (24) form the sphere $\sqrt{N}\,\mathbb{S}^{2N-1}$ with radius $\sqrt{N}$ and center at the origin. The adjustment of the normal vectors can be interpreted as follows. The new normals n^* are found by intersecting the ray

$$\lambda \left(\nabla f^{(j)}(p_{i,1}, p_{i,2}) \right)_{i=1,\ldots,N}, \quad \lambda \in \mathbb{R}_+, \tag{27}$$

with the sphere, see Figure 7. The normal vector part $M(c)$ of the objective function measures the distance of the gradients (which are also considered as a point in $\mathbb{R}^{2N}$) and the (previous) normals n^*. Consequently, the adjustment of the normal vectors (25) always leads to a smaller value of $M(c)$, as the new normals n^* have the minimum possible distance from the gradients.

5.3   Examples

The modified (by using Step 2′ instead of Step 2) iterative curve fitting procedure has been applied to the data of the previous examples, see Figures 2 and 5. This leads to the normal vectors $(n_i)_{i=1,\ldots,N}$ and the approximating curves which are shown in Figures 8 and 9. These result have been obtained after 12 iterations of the fitting procedure. In order to facilitate the comparison, the initial curves have been drawn in grey, along with the result after 12 steps.

Again we we provide the ℓ_1 (total error), ℓ_2 (least–squares), and ℓ_∞ (maximum error) norm of the residual errors, compare with the initial values (19):

$$\|R_l\|_1 = 2.807, \ \|R_l\|_2 = 0.381, \ \|R_l\|_\infty = 0.155,$$
$$\|R_r\|_1 = 4.613, \ \|R_r\|_2 = 0.555, \ \|R_r\|_\infty = 0.133. \tag{28}$$

According to our numerical experience, the iterative adjustment of the normal vectors leads only to relatively small changes in the shape of the

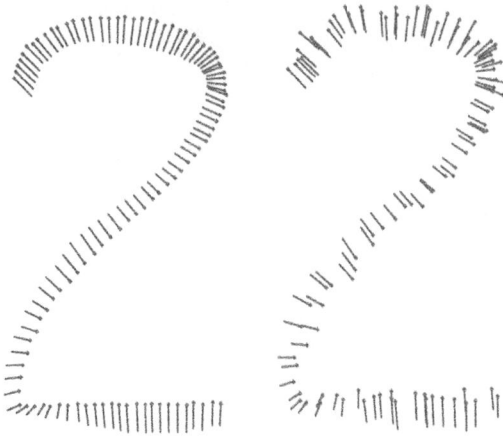

Fig. 8. Normal vectors obtained after 12 steps of the modified iterative curve fitting procedure, compare with the initial values in Figure 2.

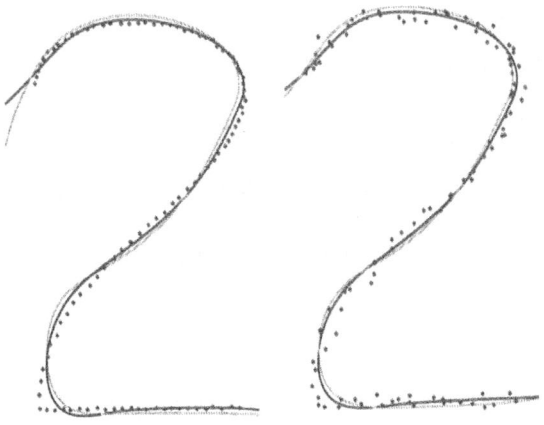

Fig. 9. Approximating curves obtained after 12 steps of the modified iterative curve fitting procedure. The initial curves (see also Figure 5) are shown in grey.

curves. Consequently, the initial estimates n_i seem to be fairly good. Also, the biggest improvement of the approximation errors is often achieved in the few first steps of the iterative fitting procedure, and the effects of the remaining steps are rather small. Figure 10 shows the ℓ_2 norms of the residual error, and the angle between new and old normal vector $n^* \in \mathbb{R}^{2N}$, for each of the 12 iteration steps of the previous example.

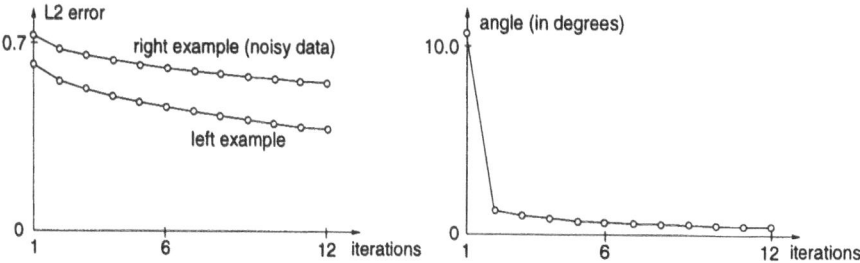

Fig. 10. The ℓ_2 norm of the residual errors (left) and the angle (in degrees) between new and old normal vector(s) $\boldsymbol{n}^* \in \mathbb{R}^{2N}$ for the 12 iteration steps.

Concluding remarks

We have described a novel technique for fitting implicitly defined algebraic spline curves to planar scattered data. By simultaneously approximating points and associated normal vectors, which are estimated from the data, one obtains a method which is both computationally simple, as the result is obtained by solving a system of linear equations, and geometrically invariant, as no auxiliary normalization of the spline coefficients is needed. Weighted least–squares and an iterative adjustment of the normal vectors have been used in order to improve the initial result.

As demonstrated in this paper, both the problems of implicit (algebraic) and parametric curve fitting can be dealt with by solving certain sequences of systems of linear equations. The fitting of parametric curves needs the estimation of auxiliary parameter values, which can then be iteratively improved via parameter correction. Analogously, curve fitting with implicit representations requires auxiliary normal vectors, which can then be adjusted, in order to obtain better results.

Also, the method presented in this paper can be seen as a contribution to methods for approximating dual data. That is, a curve resp. surface is to be generated by approximating tangents resp. tangent planes, rather than points, cf. [10]. Our curve fitting scheme deals with 'mixed' data, which are obtained by combining point and tangent information.

The success of the curve fitting scheme depends on the robustness of the method which is used for estimating the normal vectors from the data. The sensitivity of the result with respect to errors contained in the data should therefore be analyzed.

Future research will focus on implicitly defined algebraic spline surface. One the one hand, we plan to explore the use of low–degree spline surfaces, such as quadrics and cubicoids, for surface fitting applications. On the other hand, surfaces of higher order will be used in order for more complex objects, with potential applications in reverse engineering.

References

1. Bajaj, C.L. (2000) Publications, available at `www.ticam.utexas.edu/CCV/papers/fhighlights.html`.
2. Bajaj, C.L.; Ihm, I.; Warren, J. (1993) Higher–Order Interpolation and Least–Squares approximation Using Implicit Algebraic Surfaces. ACM Transactions on Grapaics **12**, 327–347.
3. Bajaj, C.L.; Xu, G. (1997) Spline approximations of real algebraic surfaces. J. Symb. Comput. **23**, 315-333
4. Forsey, D.; Bartels, R. (1995) Surface Fitting with Hierarchical Splines, ACM Transactions on Graphics **14**, 134–161.
5. Bernardini, F.; Bajaj, C.L.; Chen, J.; Schikore, D.R. (1999) Automatic Reconstruction of 3D CAD Models from Digital Scans. Int. J. Comp. Geom. Appl. **9**, 327–370.
6. Boehm, W.; Prautzsch, H (1993) Numerical methods. AK Peters, Wellesley (Mass.), and Vieweg, Braunschweig.
7. Farin, G. (1995) NURB curves and surfaces. AK Peters, Wellesley (Mass.).
8. Hoppe, H.; DeRose, T.; Duchamp, T.; McDonald, J.; Stuetzle, W. (1992) Surface reconstruction from unorganized points. Computer Graphics **26**, 71–78.
9. Hoschek, J.; Lasser, D. (1993) Fundamentals of Computer Aided Geometric Design. AK Peters, Wellesley (Mass.).
10. Hoschek, J.; Schwanecke, U. (1997) Interpolation and approximation with ruled surfaces. In: Cripps, R. (ed.) The Mathematics of Surfaces VIII, Information Geometers, Winchester, 213–232.
11. Jüttler, B. (1997) Surface fitting using convex tensor–product splines. J. Comp. Appl. Math. **84**, 23–44.
12. Koras, G.D.; Kaklis, P.D. (1999), Convexity conditions for parametric tensor-product B-spline surfaces. Adv. Comput. Math. **10**, 291–309.
13. Lee, I.-K. (2000), Curve reconstruction from unorganized points. Comput. Aided Geom. Des. **17**, 161–177.
14. Pratt, V. (1987) Direct Least–Squares Fitting of Algebraic Surfaces. ACM Computer Graphics **21** (Siggraph'87), 145–152.
15. Taubin, R. (1991) Estimation of Planar Curves, Surfaces, and Nonplanar Space Curves Defined by Implicit Equations with Applications to Edge and Range Image Segmentation. IEEE Trans. Pattern Analysis and Machine Intelligence **13**, 1115–1138.
16. Umasuthan, M.; Wallace, A.M. (1994) A comparative analysis of algorithms for fitting planar curves and surfaces defined by implicit polynomials. In: Fisher, R.B. (ed.) Design and applications of Curves and Surfaces (Mathematics of Surfaces V). Clarendon Press, Oxford, 495–514.
17. Werghi, N; Fisher, R.; Robertson, C.; Ashbrook, A. (1999) Object reconstruction by incorporating geometric constraints in reverse engineering. Comp. Aided Des. **31**, 363–399.

Use of Reverse Automatic Differentiation in Ship Hull Optimisation

R. Boudjemaa, M.I.G. Bloor and M.J. Wilson

Applied Mathematics Department, University of Leeds.

Summary.

In this paper we discuss the use of a gradient-based numerical optimization method for fairing ship hull surfaces. In order to improve the performance of the optimisation method, the derivatives of the objective function with respect to the design variables are computed using automatic differentiation. We will then compare the performance of the algorithm with results obtained using finite-difference to approximate the derivative information.

1 Introduction

Although the details of the design process differ from product to product, the broud outline of the process is often the same. For example, according to Raymer [1], describing aircraft design, the process consists of three different stages: 1) Conceptual Design, where the physical properties of the shape are defined and configured, e.g. size, weights and performance; 2) Preliminary Design, where the physical configuration is fixed and the geometric design of the object starts, and 3) Detailed Design, where each part of the object is designed and then the whole pieces are put together.

In the past, designers have had to rely on their own knowledge when it came to considering the physical properties of a new design. However, the arrival of computers and the increase in the accuracy numerical methods has opened the door for a new way of making design decisions. Simulation Based-Design (SDB), as explained by Kurtz [2], is the process in which a product can be studied, made, and tested without the need to build it physically. This, of course, relies on a computer being able to simulate the properties of any object using proper design parameters, as well as manipulate the geometry model describing the object's shape.

Shape optimisation has become a very important field due to its applications in many aspects of engineering and science, and much research has been done to minimize the time costs of the computation and to maximise the accuracy and efficiency of the results [10,21,22]. This work has involved both the surface generation methods and the optimisation methods.

It is often, though not by any means always, possible to quantify the desired physical property of a design by means of the value of an objective function or 'measure-of-merit'. Indeed, this is a necessary requirement if numerical optimisation is to be used in the design process. This can produce new designs with better performance. Direct numerical optimisation is often used since this allows constraints on the design to be straight forwadly applied in the optimisation process.

Most efficient methods for direct numerical optimisation require the calculation of a gradient of the objective function with respect to the design variables. The gradient information is used to construct a search direction for the function minimum. In the past this gradient was in most cases approximated using numerical techniques such as finite-difference. The only problem is that the approximated gradient information is likely to be innacurate once the function to be minimised is noisy, which can lead to a false result in the optimisation, or at least greatly slows the process down. Noise often results from inaccuracises in the calculation of the objective function, such as usually arise due to discrete nature of many computational methods.

The first attempt to improve the time consuming and sometimes inaccurate process of finding derivatives was performed by Francavilla *et al.* [10]. Their approach was to perform an analytical finite element differentiation obtaining an expression in terms of the derivatives with respect to each design variable. The derivatives were then approximated by finite differences. Kumar *et al.* [21] improved this method by using both finite differences and a semi-analytical approach.

Advances in computers and computer languages have allowed the development of a new method for obtaining accurate derivatives of any programmable function. In particular, the introduction of Automatic differentiation (AD) [11] allows the user to take any program and differentiate the code in that program with respect to the input variables, obtaining at the end the gradient of the program output with respect to the input variables. One of the earliest AD software's appeared in the early 1990's as described in [20], and was steadily developed [19,13] until the present. An immense amount of work has been done to improve the time and space cost of AD [12–20] and the process now comes in two 'flavours'; Reverse Automatic differentiation [13–16]. and Forward Automatic Differentiation [17,20].

In this paper we will illustrate the use of the RAD technique in the process of shape optimization. RAD will be used in obtaining the derivative information of the objective function with respect to the design variables, which will be then fed to the gradient-based optimization method (BFGS). The problem is to obtain a fair ship hull surface. This is done by using a set of bulk heads, from which we develop a surface between each pair of bulk heads using the PDE method [5–7], and then choose the boundary conditions on the basis of having surface continuity between each developed surface as well as the need to obtain a fair surface. To achieve this end we compute an objective function which represents a quantitative measure of surface quality related to surface curvature based on an integral of the *Gaussian* curvature.

After obtaining the objective function the next step is to minimise it. This is done using the BFGS optimisation method [8,9]. However there is a potential which is the inefficiency of the BFGS method when the objective function is noisy. This noise can be transfered through the derivative information, causing the BFGS method to converge to a local minimum or to a non existing solution. The approach to this problem adopted in this paper was to use Reverse Automatic Differentiation (RAD) method to obtain accurate derivatives.

The construction of ship hulls is one of the fields profoundly affected by CAD development and there is even a distinct field of research called Computer Aided Ship Hull Design (CASHD) [5]. This field involves developing computer techniques for creating ship hull surfaces.

The next section will show how the hull surface is generated using the PDE method. Section Three will deal with the way we are generating the objective function, while section Four deals with the BFGS optimisation method used. The fifth section introduces the process of AD and its two flavours. Section Six deals with the numerical result obtained by both the finite difference method and the RAD method, and a comparison of these results followed by the Conclusion.

2   Ship Hull Generation using PDE

The ship hull surface is generated using the PDE method introduced by Bloor and Wilson [6]. The method seeks a periodic solution $X(u, v)$ of the fourth-order elliptic PDE over the region $0 \leq u \leq 1$ and $0 \leq v \leq 2\pi$:

$$\left(\frac{\partial^2}{\partial u^2} + a^2 \frac{\partial^2}{\partial v^2} \right)^2 X = 0 \tag{1}$$

where a is a smoothing parameter. The general form for the periodic solution in v of (1) is given by

$$X(u, v) = A_0(u) + \sum_{n=1}^{\infty} (A_n(u)cos(nv) + B_n(u)sin(nv)) \tag{2}$$

where

$$A_0(u) = a_{00} + a_{01}u + a_{02}u^2 + a_{03}u^3 \tag{3}$$
$$A_n(u) = a_{n1}e^{anu} + a_{n2}ue^{anu} + a_{n3}e^{-anu} + a_{n4}ue^{anu} \tag{4}$$
$$B_n(u) = b_{n1}e^{anu} + b_{n2}ue^{anu} + b_{n3}e^{-anu} + b_{n4}ue^{anu} \tag{5}$$

$$X(0,v) = D_0(v) \tag{6}$$
$$X(1,v) = D_1(v) \tag{7}$$
$$X_u(0,v) = Q_0(v) \tag{8}$$
$$X_u(1,v) = Q_1(v) \tag{9}$$

$X(0,v)$ and $X(1,v)$ represent the function boundary conditions at $u = 0$ and $u = 1$ respectively; $X_u(0,v)$ and $X_u(1,v)$ represents the derivative boundary conditions at $u = 0$ and $u = 1$ respectively. $D_0(v)$, $D_1(v)$, $Q_0(v)$, and $Q_1(v)$ are specified vector functions which give the boundary conditions. a_{n1}, a_{n2}, a_{n3}, a_{n4}, b_{n1}, b_{n2}, b_{n3}, b_{n4} are vector constants computed from the boundary conditions imposed on the solution at $u = 1$ and $u = 0$. The growth and decay of the exponential functions contained in the A_n and B_n terms of (4) and (5) offers a smooth transition between the boundary conditions on $u = 0$ and $u = 1$. The rate of change in the exponential functions is affected by the an term which contains the smoothing parameter a, and the Fourier mode number n.

The ship hull surface is generated from a combination of N plane sections curves which are provided as discrete point data. The ship hull (shown in Figure(1)) was produced by generating a PDE surface between each pair of section curves. The original curves consists of 13 data points for the end curves and 15 data points for inside curves. The reparameterized curves consists of 161 data points. The boundary conditions were derived so that successive PDE patches meet at the common section curve with exact function value and without loosing continuity of surface normal. The derivative boundary conditions imposed at each section were computed using central differences of the function values at the two neighbouring section curves. The function boundary conditions and first derivative boundary conditions imposed on the solution are of the form

$$X(0,v) = S_0(v) \tag{10}$$
$$X(1,v) = S_1(v) \tag{11}$$
$$X_u(0,v) = G_0(v) \tag{12}$$
$$X_u(1,v) = G_1(v) \tag{13}$$

S_0 and S_1 are computed for each pair of section curves, where the curves are specified discretely, and then reparameterised in terms of arc-length. G_0 and G_1 are obtained using one-sided finite differences of the function boundary conditions.

$$G_j^k = \frac{1}{2}\left(S_j^{k+1} - S_j^{k-1}\right) \qquad (14)$$

where j implies the point computed at each section curve and k defines the section curve number. The derivative boundary conditions at the end sections of the ship were computed using one-sided differences with quadratic accuracy, thus

$$G_j^1 = \frac{4S_j^2 - 3S_j^1 - S_j^3}{2}$$

$$G_j^N = \frac{-4S_j^{N-1} + 3S_j^N - S_j^{N-2}}{2}. \qquad (15)$$

Figure (2) shows a PDE generated ship hull with $N = 18$ section curves. We will show below how the design variables are introduced through the derivative boundary conditions.

The boundary conditions although given as non-periodic curves, are converted to periodic curves using the method described by Bloor and Wilson in [5].

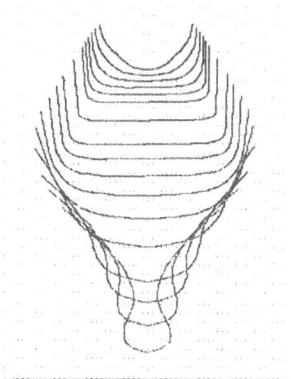

Fig. 1. The section curves of the ship hull.

3   Objective Function Computation

After generating the surface, an objective function was used to quantify the quality of the hull shape. The objective function depends on the shape of the

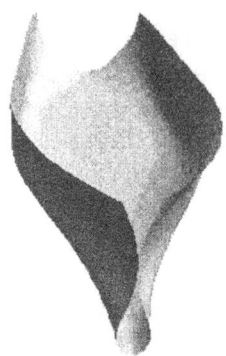

Fig. 2. A PDE Generated Ship Hull Surface.

object and hence is a function of the design variables. In order to optimise the surface shape, we will vary the design variables. The objective function that we used was the *area integral of the Gaussian curvature.*

$$I = \int_A |K| dA \tag{16}$$

where A is the surface area of the shape, in this case the ship hull, and K is the *gaussian* curvatures of the shape. This measure was used since in practice ship hulls are made as far as possible from developable surfaces, i.e. rolled steel plates, and a strictly developable surface has *mean* curvature $H = 0$.

The objective function was then obtained from the sum of integral(16) taken over all the surface patches.

After computing the objective function, a combination of design variables that minimised this function had to be found. This minimisation was carried out automatically using numerical techniques for optimisation.

4   Optimisation Method

Gradient methods for functional minimisation require the derivative of the function with respect to the design variables. Standard techniques, such as the steepest descent (SD) method, determine the search direction using the direction of maximum gradient and then minimise the function. The Broyden-Fletcher-Goldfarb-Shanno (BFGS) method is an efficient numerical method which defines the direction of search using an approximation to the Hessian matrix of the objective function. As described by Fletcher [8], the BFGS method iteratively constructs an approximation to the Hessian of the objective function. The method is often favoured due to its quadratic convergence, so that theoretically, for a quadratic function, it convergs in no more than n steps for a problem of n variables. However, the draw back is the fact that one needs to store the Hessian matrix and hence can be some memory problems if the number of variables is large. The Hessian is used through the *quasi-Newton's* method to define the search direction s^k, where $s^k = -H^k \nabla f(x^k)$. k is the number of iterations, $f(x)$ is the objective function, and H is a symmetric positive definite matrix which approximate the Hessian matrix. As defined in Dadone *et al.* [9], the updating formula from H is given by,

$$H^{k+1} = H + \left(1 + \frac{\gamma^{k^T} H^k \gamma^k}{\delta^{k^T} \gamma^k}\right) \frac{\delta^k \delta^{k^T}}{\delta^{k^T} \gamma^k} - \left(\frac{\delta^k \gamma^{k^T} H^k + H^k \gamma^k \delta^{k^T}}{\delta^{k^T} \gamma^k}\right) \quad (17)$$

where

$$\delta^k = \alpha^k s^k = x^{k+1} - x^k, \quad (18)$$
$$\gamma^k = \nabla f\left(x^{k+1}\right) - \nabla f\left(x^k\right). \quad (19)$$

and $\nabla f(x)$ is the gradient of the objective function.
This search direction is used inside a *line search* algorithm which has the following general structure:

1. Define x^0 as a starting initial value for the design parameters.
2. For $k = 0$ to convergence:
   compute the direction search s^k;
   find an α^k such that $f\left(x^k + \alpha^k s^k\right)$ is minimised;
   $x^{k+1} = x^k + \alpha^k s^k$.

In previous work, the gradient of the objective function was computed using finite-differences. The idea was to perturb each of the design variables at a time and then approximate the partial derivative of that design variable using the following (or similar) formula,

$$\frac{f(x_i + h) - f(x_i - h)}{2h} \quad (20)$$

where x_i is a design variable and h is the amplitude of the disturbance. In this paper, the gradient information $\nabla f(x)$ is computed both using *Reverse Automatic Differentiation* and using *Finite-Differences*, so that the effect of using both in BFGS optimisation may be observed.

5   Automatic Differentiation

Automatic Differentiation (AD) is a numerical process for obtaining accurate derivatives automatically. As explained by Bischof et al [20] , AD techniques rely on the fact that in a computer program every function is executed as a sequence of elementary operations such as additions, multiplications, and elementary functions like sin, cos, and exp. AD method applies in a repetitive way the basic rules of differentiation to the operations of a particular program until the derivatives are obtained. This process was made easy by the arrival of advanced computer languages, such as C and Fortran 90, where the programmer can derive his own data types and then define operations for that particular data type. In practice there are two 'flavours' to AD, Forward Automatic Differentiation and Reverse Automatic Differentiation. Beside having different process, the two flavours also differ in the memory space used and the computation time.

The cost AD algorithm, for both running time and memory space is a "one-off" cost, which means that AD requires to go through the program only once to generate the derivatives. This is not the case for FD since for each derivative it requires going through the program twice.

ADOL-C [13] is an example of a software package developed to differentiate both C/C++ program in the forward and reverse mode.

5.1   Forward Automatic Differentiation

This technique is explained in detail by [17,13,20]. Forward Automatic Differentiation (FAD) generates the gradient of a function by defining the *doublet* data type. This type consist of the function value and the gradient value. The program than defines the operation associated with this new type.

If for example $f(x)$ is a function of n variables, then we define the associated doublets as follows,

$$U_i = (x_i, \quad g_i)$$

where U_i is a doublet associated with the ith function variable, x_i is that variable and g_i is the corresponding gradient value. In this case x_i is an element of the vector x, so the corresponding g_i element is the i'th unary vector (see Figure (3)). If U_1 and U_2 are two doublets defined as follows,

$$U_1 = (x_1, \quad g_1) \quad U_2 = (x_2, \quad g_2)$$

then some of the operations associated with them are:

$$U_1 + U_2 = (x_1 + x_2, \quad g_1 + g_2)$$
$$U_1 * U_2 = (x_1 * x_2, \quad g_1 * x_2 + g_2 * x_1)$$
$$cosU_1 = (cosx_1, \quad -g_1 * sinx_1)$$
$$expU_1 = (expx_1, \quad g_1 * expx_1)$$

Softwares have been developed in both FORTRAN 77 and C languages, for example ADIFOR in [20] and ADIC in [19], which can forward differentiate any programmable function.

The process of computing the derivatives in FAD proceeds in the same direction as that of computing the function value. Figure(3) shows an example of how a simple function $f = x_1 x_2 e^{x_3}$ is computed and how the FAD method derives its gradient using doublets and the operations associated with them.

The only draw back in using FAD is that it tends to be slow if the function gets at all complicated, which in our case it is.

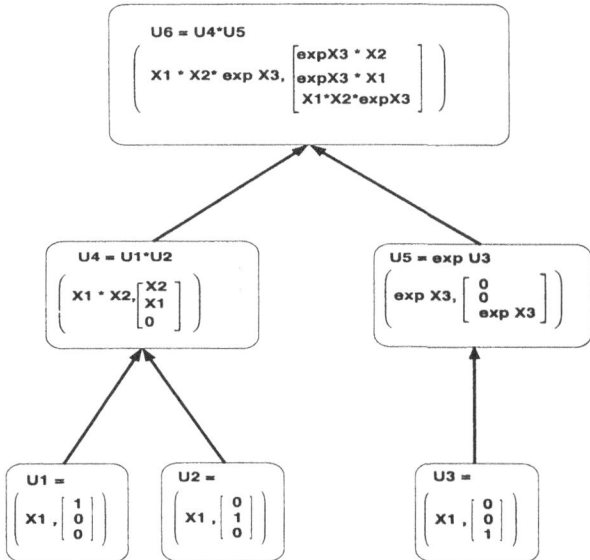

Fig. 3. The process of computing a simple function and obtaining its gradient using FAD method.

5.2   Reverse Automatic Differentiation

Reverse Automatic Differentiation (RAD) is the other flavour of AD. The method involves a different process to that of FAD in the direction of computing the gradient vector. The method starts first by building up the function and by defining it as a set of unary and binary operation such as, addition, subtraction, multiplication and so on, and then travelling backward in the program, i.e. starting from the end of the program and travelling back to the start of it, in order to obtain the derivative information. Just as in the case of FAD the function needs to be differentiable and capable of being computed by a computer program.

The function is then transformed into a binary tree where each new element depends on no more than two previous elements and no less than one. In particular, these elements are defined as nodes and each node could be an independent variable or a simple mathematical operation e.g. $+, -, *, /$ or an elementary function e.g. *sin, cos, exp*, $\cdots$ of no more than two and no less than one previous nodes. It is to be noted that in principle each node could depend on more than two previous nodes but for our case, and often in practice, it is reduced to maximum of two.

For example let $f(x)$ be a differentiable function of n independant variables and let $v_1, v_2, \cdots, v_m$ be a set of real variable defined as follows:

$$
\begin{aligned}
&do \quad i = 1, \cdots, n \\
&\qquad v_i = x_i; \\
&end \quad do; \\
&\quad do \quad j = n+1, \cdots, m \\
&\qquad v_j = h_j\left(v_1, \cdots, v_{j-1}\right) \\
&\quad end \quad do; \\
&\qquad f = v_m
\end{aligned} \tag{21}
$$

with the assumption that there are $m - n$ steps in computing the function f. The v_i represents the function value and h_i defines a unary operation, binary operation, or an elementary function. In order to compute the partial derivatives of the function with respect to the x_i we introduce new real variables $\bar{v}_i$ defined as *adjoints*. These adjoint variables will be assigned the derivatives values at the end of the process.

The adjoints are initialy all set to zero except for the last one i.e. $\bar{v}_m$ which is set to one. The adjoints are then computed by a process of *reverse accumulation* defined as follows,

$$
\begin{aligned}
&do \quad i = m, \cdots, n+1 \\
&\quad do \quad j = 1, \cdots, i-1 \\
&\qquad \bar{v}_j = \bar{v}_j + \frac{\partial h_i}{\partial v_j}\bar{v}_i \\
&\quad end \quad do; \\
&end \quad do; \\
&\quad do \quad k = 1, \cdots, n \\
&\qquad \frac{\partial f}{\partial x_k} = \bar{v}_k \\
&\quad end \quad do;
\end{aligned} \tag{22}
$$

In a sense RAD is the reverse process to that of building a function. This is shown by an example in Figures (5) and (6) where the function $(x1 + x2)(x3 + x4)$ is computed and differentiated.

In terms of programming each node is defined as follows,

type node;

integer :: operation
real :: adjoint, value
pointers:: arg1, arg2

end type node.

The integer variable will indicate the type of operation being carried out. This is done by assigning each operation a integer value at the beginning of the program. For example addition could be assigned the value 1, multiplication the value 2 and so on. In the case where the node is an independant variable, the integer part *operation* is assigned the value 0. The real part *value* represents the function value, while *adjoint* represents the derivative part. Pointers *arg1* and *arg2* link each node to its parent nodes, and in the case of a independant variable point to null. Figure(4) show the structure of a node and the use of the pointers between nodes.

In our case, the program that generates the objective function is taken as originally written, the only change made being in the type declaration part of it. A self developed RAD software package written in FORTRAN 90 language was used for differentiation. Each design variable is declared as type *node*. Once this is done the building of the function is done automatically using the predefined node operation and elementary functions. Once the function is constructed, we call upon a subroutine which travels backward in the program generating adjoint values using the RAD algorithm. Hence, when the subroutine reaches the beginning of the program, i.e. the design variables nodes, the gradient will be generated.

In our paper we use RAD instead of FAD for two reasons. First, the memory cost for our particular objective function is less in RAD when compared to the cost of FAD. Second, the computation time is far less for RAD than for FAD. The only draw backs of the RAD method is that it consumes a lot of memory.

6   Results

In this section, we present results for the optimisation of a PDE ship hull, where the gradient-based method used to select the optimum value of the

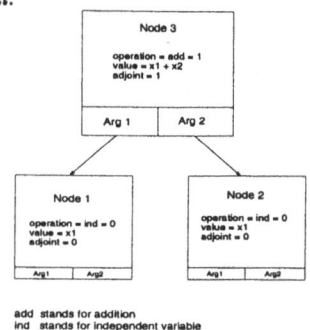

Fig. 4. The structure of a node.

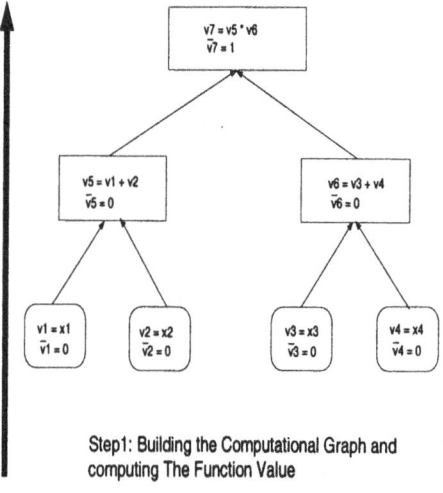

Fig. 5. The process of RAD function program building.

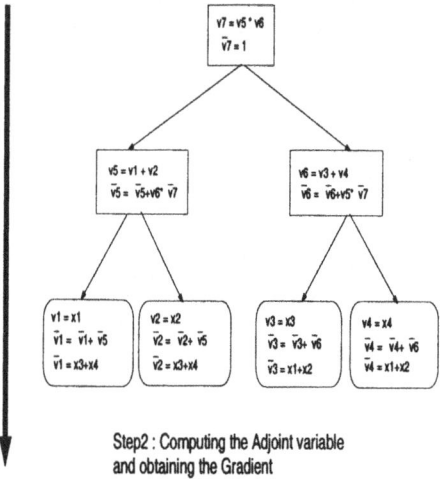

Fig. 6. The process of RAD program differentiation.

design variable (BFGS) computes its derivatives using RAD. For the sake of comparison, we also carry out surface optimisation using BFGS where the derivatives are computed by finite-differencing the objective function with respect . In the testing of the methods, only seven section curves were used to generate the hull and six PDE surfaces were studied because of limitation on time and memory. However, results could easily be generalized to all eighteen sections.

The test involved introducing the real variable ω_i into the derivative boundary condition of the ship hull, where the ω_i variables multiplied the y coordinates of the middle section curves as shown in Figure(7). Only 5 ωs where used since we only have 7 sections which two of them are the end sections where the boundary conditions we regard as being fixed. The new y coordinates were then used to compute the derivative boundary conditions in (12,13), thus

$$\hat{S}_y^i = S_y^i * \omega_i \tag{23}$$

$$\hat{G}_y^i = \frac{1}{2}\left(\hat{S}_y^{i+1} - \hat{S}_y^{i-1}\right) \tag{24}$$

where G_x^i and G_z^i are computed as before.

The new $\hat{G}_y^i$ derivative boundary conditions, where i represents the section curve number, are imposed on the solution of (1). The effect of this is to create a surface oscillation into the ship hull surface. This surface oscillation is shown in Figure(9) were only 7 sections of the ship hull were considered. This also gives an increase in the objective function as the Gaussian curvature is disturbed along with the surface area. This change is shown in Figure (8).

Our first test was to minimize the objective function of the PDE developed hull surface using the BFGS method with a gradient estimate of our objective function (16) computed using the Finite Difference method as defined in (20). The gradient of the function contains the partial derivative of the objective function with respect to each of the ω_i variables. We tested the optimisation procedures using different values of h. The results obtained are presented in Table(1), where $\underline{\omega}_0$ represents the vector of initial values, $f(\underline{\omega}_0)$ represents the initial value of the objective function, N represents the number of iterations of the BFGS method, and $\underline{\omega}_n$ the vector solutions to the minimal objective function $f(\underline{\omega}_n)$.

The value of h affects the accuracy of the derivative information fed to the BFGS method, which in turn affects the accuracy of the solution. In Table(1) we can see that for different values of h we obtain different values of $\underline{\omega}_n$ and

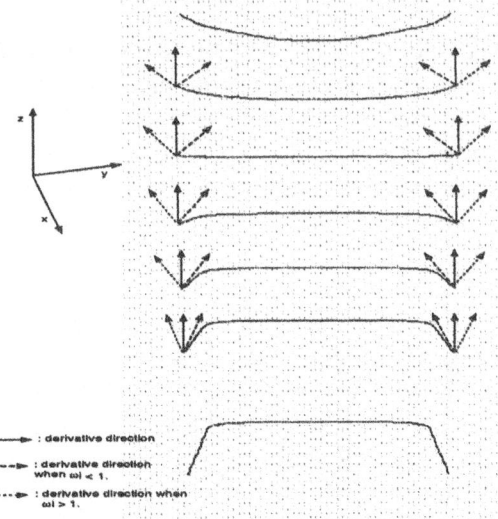

Fig. 7. The changes in the derivative boundary conditions as ω's change.

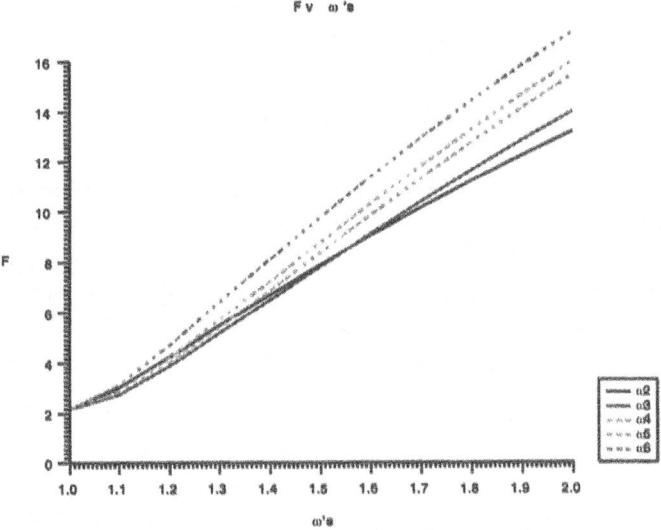

Fig. 8. The variation in the Objective Function as a function of ω.

$f(\underline{\omega}_n)$. Choosing the right h is an important step which the programmer has to take. This point will be discussed further in this section.

In each case, the surface starts off by having the shape shown in Figure(9a), and at the optimum has the same shape, more or less, as shown in Figure(9b).

$h = 0.01$				
$\underline{\omega}_0$	$f(\underline{\omega}_0)$	N	$\underline{\omega}_N$	$f(\underline{\omega}_N)$
2.			1.00006	
2.			1.01418	
2.	22.384746	15	1.0017	2.1423735
2.			1.005	
2.			1.00171	

$h = 0.1$				
$\underline{\omega}_0$	$f(\underline{\omega}_0)$	N	$\underline{\omega}_N$	$f(\underline{\omega}_N)$
2.			1.0072	
2.			1.0151	
2.	22.384746	14	1.0093	2.1705942
2.			1.0103	
2.			1.0077	

$h = 0.5$				
$\underline{\omega}_0$	$f(\underline{\omega}_0)$	N	$\underline{\omega}_N$	$f(\underline{\omega}_N)$
2.			1.0231	
2.			1.04157	
2.	22.384746	11	1.0157	2.34092481
2.			1.0403	
2.			1.01681	

Table 1. Optimisation using FD

The BFGS+RAD optimisation was then used to optimise the objective function (16) with respect to the ω_i variables. The results are shown in Table(2). Note that the final value of the objective function is close to that found for the case $h = 0.01$ shown in Table(1). Optimisation starting from initial states close to this optimium were also carried out,and in each case the optimisation returned to the same optimal value.

Figure(9) indicates the appearance of the trial surface. In (9a) we see the oscillation produced by disturbing the derivative boundary conditions and setting ω's to the value 2.0. After optimisation, we obtain a smooth shape as shown in (9b).

The running time for the optimization method BFGS+FD was between seven and ten minutes for a ship hull surface of seven curve sections. The

$\underline{\omega}_0$	$f(\underline{\omega}_0)$	N	$\underline{\omega}_N$	$f(\underline{\omega}_N)$
2.			0.99884	
2.			1.013645	
2.	22.384746	22	1.001189	2.1418227
2.			1.004465	
2.			1.001044	

Table 2. Optimisation using RAD

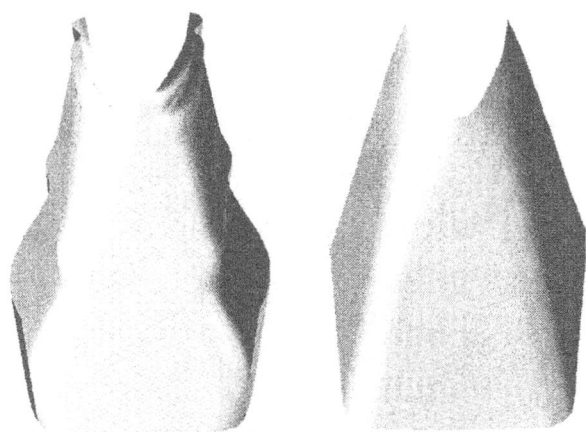

(a)Initial hull surface with oscillation   (b)Hull surface in optimum case

Fig. 9. The oscillation effect in the hull surface

running time for BFGS+RAD was nearly half an hour for also a surface of seven curve sections. However, these figures from the RAD calculation include both the cost of the optimisation and the cost of calculating the data structures for the derivative computation, which is a 'one-off' cost arried out only once at the start of the calculation.

As mentioned above, one of the problems of using a gradient-based method for optimisation is the presence of noise in the objective function. This noise can disturb the accuracy and efficiency of the BFGS method, if the gradient information is contaminated by this noise. In the present problem, noise is enhanced due to the fact that the objective function is computed discretely. The cost of the RAD data structure is significant because the costs of calculating the objective function are minimal in our case.

The reason for the increase runing time is that the BFGS take more iterations to converge. The likely reason for this is that where finite differences are used there is a greater likelihood of the algorithm converging to a local minimum closer to the original state than the case where AD is used. Also, in our case, unlike many cases where a physically-based objective function (such as hydrodynamics drag) is used, the cost of calculating the objective function is minimal, and therefore the overlead for the FD case in calculating derivatives is very small.

A second test was performed in which the level of noise was varied to compare the behaviour of the RAD method and the FD method when computing the gradient information. In order to control the noise level in the objective function, the numerical calculation of the objective function was modified. The original process of computing (10), involved using a discrete mesh (u, v), where $0 \le u \le 1$ and $0 \le v \le 2\pi$. Numerically, this is done by using an $m \times n$ mesh, where m and n are integer values choosen by the user. Algorithmically we may denote the process by,

$$i = 2, 3, \cdots, m$$
$$j = 2, 3, \cdots, n \qquad (25)$$
$$Sum = Sum + \left(\frac{I(i,j) + I(i-1, j-1) + I(i-1, j) + I(i, j-1)}{4} \right) H_u H_v$$

where

$$H_u = \frac{1.0}{n-1} \quad H_v = \frac{2\pi}{m-1}$$

and the evaluated value of Sum is the value of (16) for the entire u, v mesh. The value $I(i, j)$ denotes the value of integral(16) at the particular i, j point of the mesh. In the normal process of computing the integral, the values Hu and Hv are the same at all points in the mesh. This is illustrated in Figure (10). However, to artificially control the noise level in the calculation of Sum, these two distances Hu and Hv were modified by introducing the random integers α and β, where

$$\sum_{k=1}^{t_1} \alpha_k = n, \quad \sum_{k=1}^{t_2} \beta_k = m$$

where t_1 is the number of divisions of the u axis and t_2 is the number of divisions of the v axis. We introduce the 'noise' by randomly choosing the number of divisions of both u and v axis. Thus a computational mesh was used to calculate the value of the objective function from the set of integers $(\alpha_1, \alpha_2, \cdots, \alpha_{t_1}) \times (\beta_1, \beta_2, \cdots, \beta_{t_2})$, where the typical elements of that mesh had sides of length

$$H_u = \left(\frac{1.0}{n-1} \right) * \alpha_{k_1 + 1} \quad H_v = \left(\frac{2\pi}{m-1} \right) * \beta_{k_2 + 1}$$

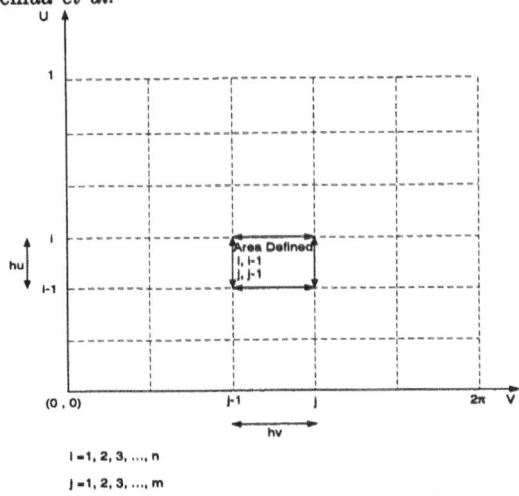

Fig. 10. The (u, v) patch of the objective function.

The variables α and β were allowed to vary between 1 and θ where $3 \leq \theta \leq min(m,n)$. The noise increases as θ gets bigger and decrease as it gets smaller. This is due to the fact that if θ get bigger, the variation in the mesh used to calculate the objective function gets larger. This is illustrated in Figure(11) as we see the function change when the noise magnitude is changed. The function was computed 100 times producing different value from one iteration to the next. Note that in each case the values of the design variables were constant, i.e. the shape was the same (i.e. initial shape Fig. 9(a)).

Figure(13) shows the derivatives calculation with $1 \leq \alpha, \beta \leq 5$. Figure(14) is the case where α and β are allowed to vary from 1 to 20, which gives more noise and therefore more inaccuracy in the derivatives information.

Figure(12) shows the variation in the (noisy) objective function as two of the ω_i's are varied ($i = 2, 3$) near the minimum of the objective function. A test was then performed to compare the derivatives calculated using finite difference and RAD with respect to variations in the value of ω_2. Note that the noise has a very short wavelength. These results are shown in Figures (13), (14), and (15).

Figure (13) shows the finite difference derivatives calculated for several values of h. Note that the derivatives estimates are poor for both large values of $h(= 0.5)$ and very small values of $h(= 0.01)$. Only for intermediate values of $h(= 0.1)$ does the gradient vary in approximately the right way.

In Figure (14) the level of the noise has been increased, and in this case the gradient variable is poorly estimated, even for $h = 0.1$.

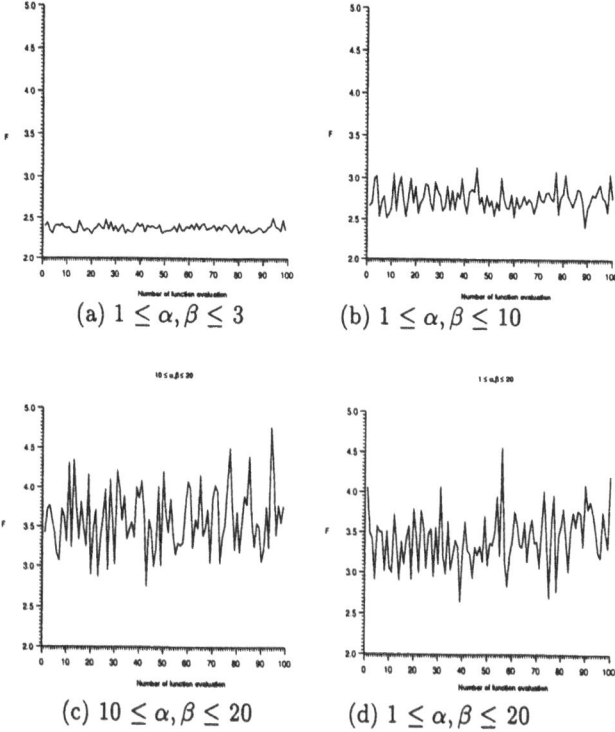

(a) $1 \leq \alpha, \beta \leq 3$ (b) $1 \leq \alpha, \beta \leq 10$

(c) $10 \leq \alpha, \beta \leq 20$ (d) $1 \leq \alpha, \beta \leq 20$

Fig. 11. The amplitude of the noise in the objective function as the as a function of the range over which α, β can vary.

The derivatives of the objective function with respect to the design variables were both computed using Finite Difference (FD) and RAD techniques.

Figure(15) shows that although noisy, the RAD method does manage to resolve the underlying trend in the gradient variable.

Some explanation of these results is needed. Obviously at each value of ω_i the computed value of the objective function will be subject to an error, i.e.

$$f_{calc}(\omega_i) = f_{exact}(\omega_i) + error_i$$

In the case where the derivatives are calculated using finite differences, both the function and error are differenced leading to an error in the derivatives:

$$\frac{\partial f}{\partial \omega_i} = \frac{(f_{exact}(\omega_i + h) + error_i + h) - (f_{exact}(\omega_i - h) - error_i - h)}{2h}$$

$$\frac{\partial f}{\partial \omega_i} = \frac{f_{exact}(\omega_i + h) - f_{exact}(\omega_i - h)}{2h} + \frac{2error_i + 2h}{2h}$$

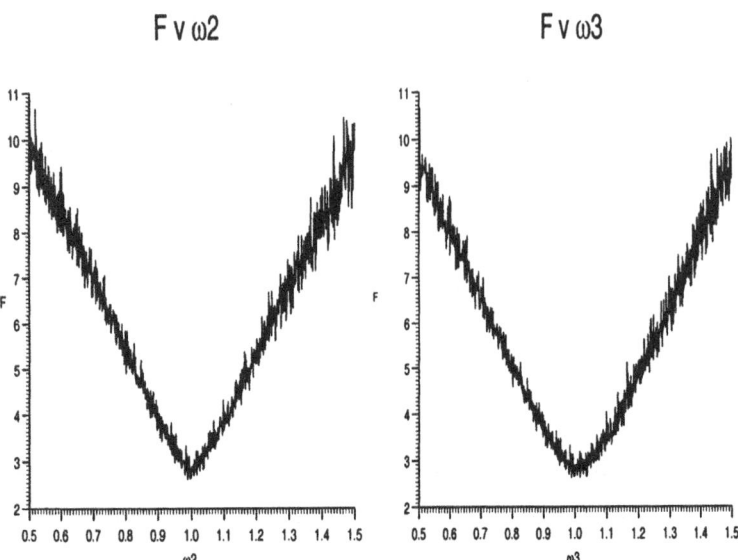

Fig. 12. Noisy objective function versus ω_1 and ω_2

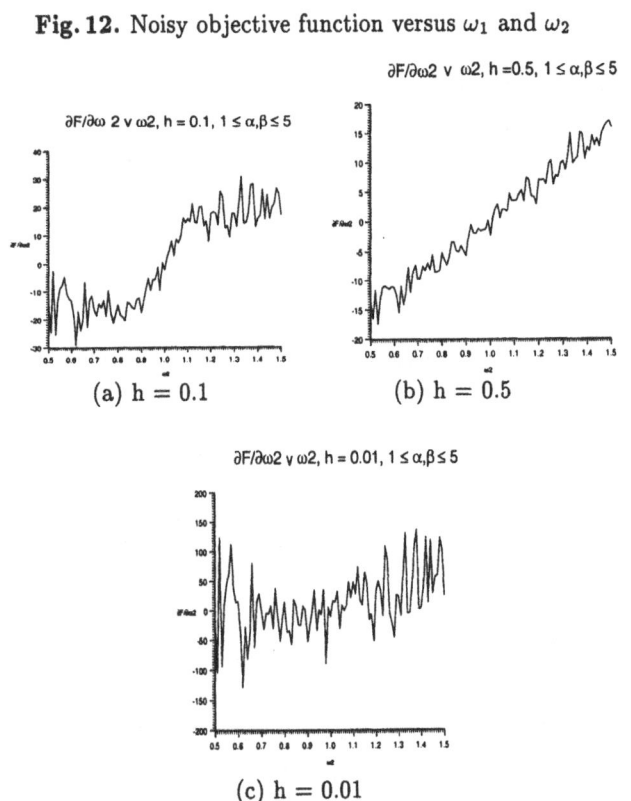

Fig. 13. Numerical derivatives of F with respect to ω_2 obtained by Finite Differences for various values of the step size h

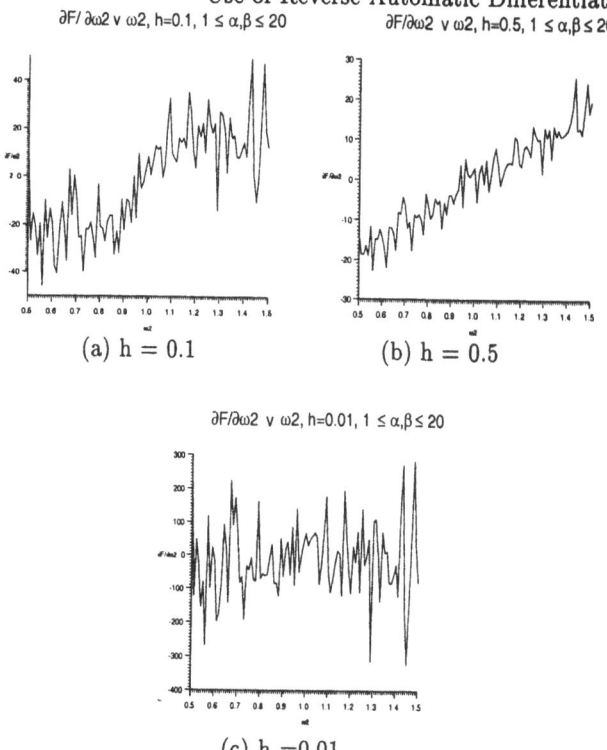

Fig. 14. Numerical derivatives of F with respect to ω_2 obtained by Finite Differences for various values of the step size h

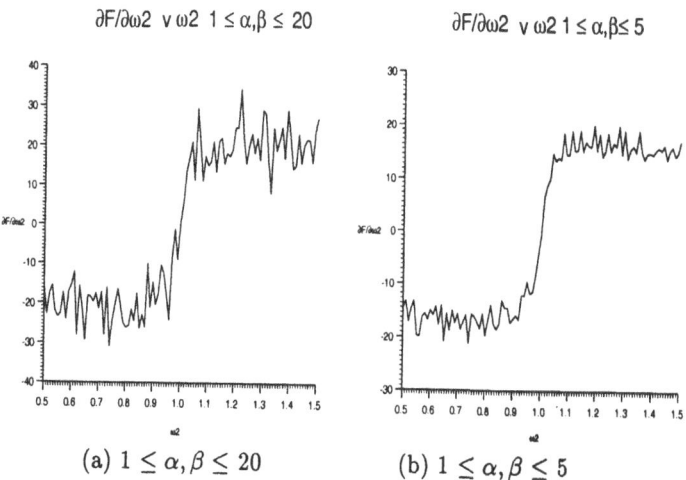

Fig. 15. Numerical derivatives of F with respect to ω_2 obtained by RAD for various values of the step size h

However, in the RAD case the noise itself is not differentiated in the calculation of the derivatives.

Finally, Table(3) shows the result of optimisation using FD method when $1 \leq \alpha, \beta \leq 5$ for different values of h, while Table(4) shows the results of optimisation using the RAD method when $1 \leq \alpha, \beta \leq 5$. Notice that in the FD case, the best results are obtained using $h = 0.1$, which previous results (Figure 13) show the best estimate of the gradient variation. However, the optimum obtained for $h = 0.1$ is significantly worse that the optimum obtained using RAD Table(4).

$h = 0.01$			
$\underline{\omega}_0$	$f(\underline{\omega}_0)$	N	$f(\underline{\omega}_N)$
1.5			
1.5			
1.5	14.35824	11	3.02381
1.5			
1.5			

$h = 0.1$			
$\underline{\omega}_0$	$f(\underline{\omega}_0)$	N	$f(\underline{\omega}_N)$
1.5			
1.5			
1.5	14.35824	11	2.8433
1.5			
1.5			

$h = 0.5$			
$\underline{\omega}_0$	$f(\underline{\omega}_0)$	N	$f(\underline{\omega}_N)$
1.5			
1.5			
1.5	14.35824	10	3.12713
1.5			
1.5			

Table 3. Optimisation using FD when $1 \leq \alpha, \beta \leq 5$

$\underline{\omega}_0$	$f(\underline{\omega}_0)$	N	$f(\underline{\omega}_N)$
1.5			
1.5			
1.5	14.35824	17	2.660905
1.5			
1.5			

Table 4. Optimisation using RAD when $1 \leq \alpha, \beta \leq 5$

7 Conclusion

From the results obtained, we can conclude that for this problem at least RAD is an efficient and accurate method of obtaining derivatives information. The RAD method when combined with a gradient optimisation method gives accurate results in the optimisation of the objective function. The numerical results shows that choosing the correct h for the FD method can make a significant change in the results.

The effect of noise from the calculation of the objective function on the estimated gradient was greatly reduced when using the RAD algorithm. In particular these results show that the RAD algorithm does not differentiate the noise in the original objective function. However in the case of the FD method, noise contaminates the derivative information since this method when approximates the derivatives, it differences the noise as well as the objective function.

Acknowledgements

The authors acknowledge financial support from the EPSRC (grant number GR/L11366), and the Algerian Ministry of Education. They would also like to thank Dr Ian Applegarth of the UK office of Kockums Computer Systems for providing the section data.

References

1. **D. P. Raymer**, *Aircraft Design: A conceptual Approach*, AIAA, AIAA Educational Series, Washington, DC, 1989 in **M. I. G. Bloor** and **M. J. Wilson**, *Efficient Parametrisation of Generic Aircraft Geometry*, Journal of Aircraft, Volume 32, Number 6, 1994, Pages 1269-1275.
2. **Paul Kurtz**, *Simulation Based Design*, Applied Research Laboratory, The Pennsylvania State University, from www.arl.psu.edu.
3. **T. W. Lowe**, *Functionality in Computer Aided Geometric Design*. PhD thesis, Leeds University, Leeds, Jannuary 1992.

4. **Günther Greiner**, *"Modelling of Curves and Surfaces Based on Optimization Techniques"* in Creating Fair and Shape-Preserving Curves and Surfaces edited by **H. Nowacki** and **P. D. Kaklis**, B. G. Teubner Stuttgart·Leipzig, 1998

5. **M. I. G. Bloor** and **M. J. Wilson**, *Parametric Geometry and Optimisation of Hull Forms*, Technical Report, Department of Applied Mathematics, University of Leeds, 1999.

6. **M. I. G. Bloor** and **M. J. Wilson**, *Using Partial Differential Equations to Generate Free-Form Surfaces*, Computer-Aided Design, volume 22, number 4, may 1990.

7. **M. I. G. Bloor** and **M. J. Wilson**, *Efficient Parametrisation of Generic Aircraft Geometry*, Journal of Aircraft, Volume 32,Number 6, 1994, Pages 1269-1275.

8. **R. Fletcher**, *Practical Methods of Optimization*, 2nd Edition, Jhon Wiley and Sons Ltd, Essex, 1987.

9. **A. Dadone,B. Mohammadi** and **N. Petruzzelli**, *Incomplete Sensitivities and BFGS Methods for 3D Aerodynamics Shape Design.*, Research Report No. 3633, INRIA, Domaine de Voluceau, Rocquencourt, BP 105, 78153 LE CHESNAY Cedex(France), March 1999.

10. **A. Francavilla, C.V. Ramakrishnan** and **O.C. Zienkiewicz**, *"optimization of Shape to minimize stress concentration"*, Journal of strain Analysis, 10, pp 63-70. 1975. in **T. W. Lowe**, *Functionality in Computer Aided Geometric Design*. PhD thesis, Leeds University, Leeds, Jannuary 1992.

11. **Andreas Griewank**. *" On Automatic Differentiation"* in Mathematical Programming: Recent Developments and Applications, edited by **M. Iri** and **K. Tanabe**, pages 83-108, Kluwer Academic Publishers, Amsterdam, 1989. CRPC version.

12. **James C. Newman**, **W. Kyle Anderson** and **David L. Whitfield**, *Multidisciplinary Sensitivity Derivatives Using Complex Variables*, Research Report MSSU-COE-ERC-98-08, Computational Fluid Dynamics Laboratory, NSF Engineering Research Center for Computational Field Simulation, Mississippi State University P.O.Box 9627, Mississippi State, MS 39762, July 1998.

13. **A. Griewank, D. Juedes, H. Mitev, J. Utke, O. Vogel**, and **A. Walther**, *ADOL-C: A Package for the Automatic Differentiation of Algorithms Written in C/C++*; this is the updated version of the paper published in ACM TOMS, vol. 22(2) June 1996, pp. 131-167, Algor. 755

14. **Bruce Christianson**, *Automatic Hessians by Reverse Accumulation*, IMA Journal of Numerical Analysis (1992)12, 135-150

15. **J. Grimm,L. Pottier** and **N. Rostaing-Schmidt**, *Optimal time and minimum space-time product for reversing a certain class of programs.*, Research Report No. 2794, INRIA, Domaine de Voluceau, Rocquencourt, BP 105, 78153 LE CHESNAY Cedex(France), Feb 1996.

16. **R. Giering** and **T. Kaminski** *Recipes for Adjoint Code Construction*, ACM Transactions on Mathematical Software, Vol. 24, No. 4, December 1998, Pages 437-474.

17. **M. Ulbrich** and **S. Ulbrich** *Automatic Differentiation: A Structure-Exploiting Forward Mode with Almost Optimal Complexity for Kantorovic Trees*,Technical Report No. IAMS1996.1TUM, Institut Fur Angewandtle Mathematik und Statistik, Technische Universitat Munchen, January 1996.

18. **Thomas F. Coleman** and **Arun Verma**, *The efficient computation of sparse Jacobian matrices using automatic differentiation*, Technical Report CTC95TR225, Cornell Theory Center, Cornell University, Ithaca NY 14850. October 17, 1996.

19. **C. Bischof, L. Roh**, and **A. Mauer**, *ADIC – An extensible automatic differentiation tool for ANSI-C*, Argonne Preprint ANL/MCS-P626-1196. To appear in Software: Practice and Experience.

20. **C. Bischof, A. Carle, G. Corliss, A. Griewank**, and **P. Hovland**, *ADIFOR - Generating Derivative Codes from Fortran Programs*, Argonne Preprint MCS-P263-0991, and CRPC Technical Report CRPC-TR91185. Published in Scientific Programming, 1(1), pp. 1-29, 1992.

21. **V. Kumar, M.D. German** and **S.J. Lee**, "A Geometry-Based 2-dimensional shape optimization methodology and a software system with Applications", Automation of Design, Analysis and Manufacturing, B. Prasad, **CAD/CAM Robotics and Factories of the Future volume 2**, Springer-Verlag, Berlin, 1989, pp. 5-10.

22. **Henry P. Moreton** and **Carlo H. Sequin**, *Functional Minimization for Fair Surface Design*, Computer Science Division, Technical Report,Department of Electrical Engineering and Computer Science,University of California.

Symmetry Sets and Medial Axes in Two and Three Dimensions

Peter Giblin

Department of Mathematical Sciences
The University of Liverpool
Liverpool L69 3BX, England

Summary. The medial axis of a plane curve (or surface) M is part of a larger set called the symmetry set. This is the closure of the locus of centres of all circles (spheres) tangent to M in at least two places. The structure of symmetry sets of surfaces is closely connected with that of *families* of plane curves. I describe a significant family of curves where the 'underground' activity of the symmetry set affects the evolution of the medial axis. Some relationships between the geometry (e.g. curvature) of the symmetry set and that of the generating curve (surface) are given, and the article ends with an examination of one of the most significant features of the symmetry set or medial axis of a surface, the so-called $A_1 A_3$ points, which are related to happenings encountered before in families of curves.

1 Introduction

In this paper I shall describe some recent work on symmetry sets and medial axes of surfaces. Some of this is joint work—very much work in progress—with Ben Kimia of Brown University. In fact it might be fair to describe this paper as *wishful thinking*: I shall describe some results, also fairly recent, on the corresponding constructions in the plane, to show the kind of detail which it would be good to have in 3 dimensions. Some progress has been made but there is a lot more to do.

Firstly, recall that the *symmetry set* of a smooth curve γ in the plane is the closure of the locus of centres of circles tangent to γ in at least two places [3,8]. The *medial axis* is the subset of the symmetry set where we restrict to circles whose radius equals the global minimum distance from the centre to γ. The medial axis represents a great simplification of the symmetry set and is more commonly used in shape recognition.

In §2 I shall describe an example involving a basic change of shape in a plane curve, and we can see the evolution of the medial axis, and, more importantly, the underlying 'unseen' evolution of the symmetry set which every now and then makes itself felt influencing the medial axis.

Given the symmetry set (or medial axis) of γ we can recover the curve from the extra information afforded by the radius function: the radius of the circle centred on the symmetry set and tangent twice to γ. Using this radius function in a systematic way gives a more 'intrinsic' approach to the reconstruction of γ from its symmetry set [9]. Generalisation of this idea to surfaces is just beginning; see §3.

The definition of symmetry set or medial axis for a curve is turned into the definition for a surface Γ by replacing 'circle' by 'sphere': we consider the locus of centres of spheres which are tangent to Γ at least twice. In the curve case, the symmetry set is singular when the circle is the osculating circle at one (or both) points of contact. When the circle is the osculating circle at a vertex of γ the symmetry set has an endpoint. In the surface case the symmetry set is a surface which is singular when the sphere is a principal sphere—equal in radius to one of the principal radii of curvature—at a point of contact with Γ. The symmetry set has a boundary edge when the sphere is a principal sphere at a *ridge point* of Γ. Ridge points have been investigated extensively; see for example [15,13].

The local structure of symmetry sets and medial axes of surfaces is known, and expounded for example in [3,11]. In this article I shall concentrate on one particularly interesting configuration (§4), which has quite strong connexions with events observed in the curve case in §2. Although the evolution of symmetry sets and medial axes of curves in generic families has been known for some time [2,10], and the evolution of ridges (crests) on surfaces is also known [4], the generic evolution of symmetry sets and medial axes in three dimensions is not known. This is quite a challenge. Likewise it would be very interesting to study global properties of symmetry sets, perhaps obtaining analogous results to those in [1] for the curve case.

It is a pleasure to thank Dirk Siersma for telling me about his result in §3.2 and Ben Kimia for inspiring discussions and also for some of the figures.

2   A dented ellipse

I shall describe a very instructive example of shape evolution which shows the symmetry set in action. The changes which occur on the medial axis during this evolution are only possible because of the underlying very subtle changes which are occurring on the symmetry set. (Motto: *the symmetry set matters.*) The full list of local changes on the symmetry set for a generic curve evolution was found in [2], and this example illustrates some of the more recondite—but vital!—of these changes in a vivid way.

For the purposes of illustration we shall take a symmetrical figure, in fact an ellipse, and introduce a symmetrical dent. Start with the ellipse parametrized by $(x, y) = (a \sin u, b(\cos u - 1))$; this has semi-axes a and b and is tangent to (and below) the x-axis at the origin. A dent is introduced by subtracting a term He^{-qu^2} from y, where $H > 0$ and $q > 0$. (For a bump, we take $H < 0$.) The values used in the illustration are $a = 1.5$, $b = 1$, $q = 20$, and H from 0 to 1.4. The sequence is shown in Figures 1, 2 and 3, and we now proceed to describe the crucial events which occur in these.

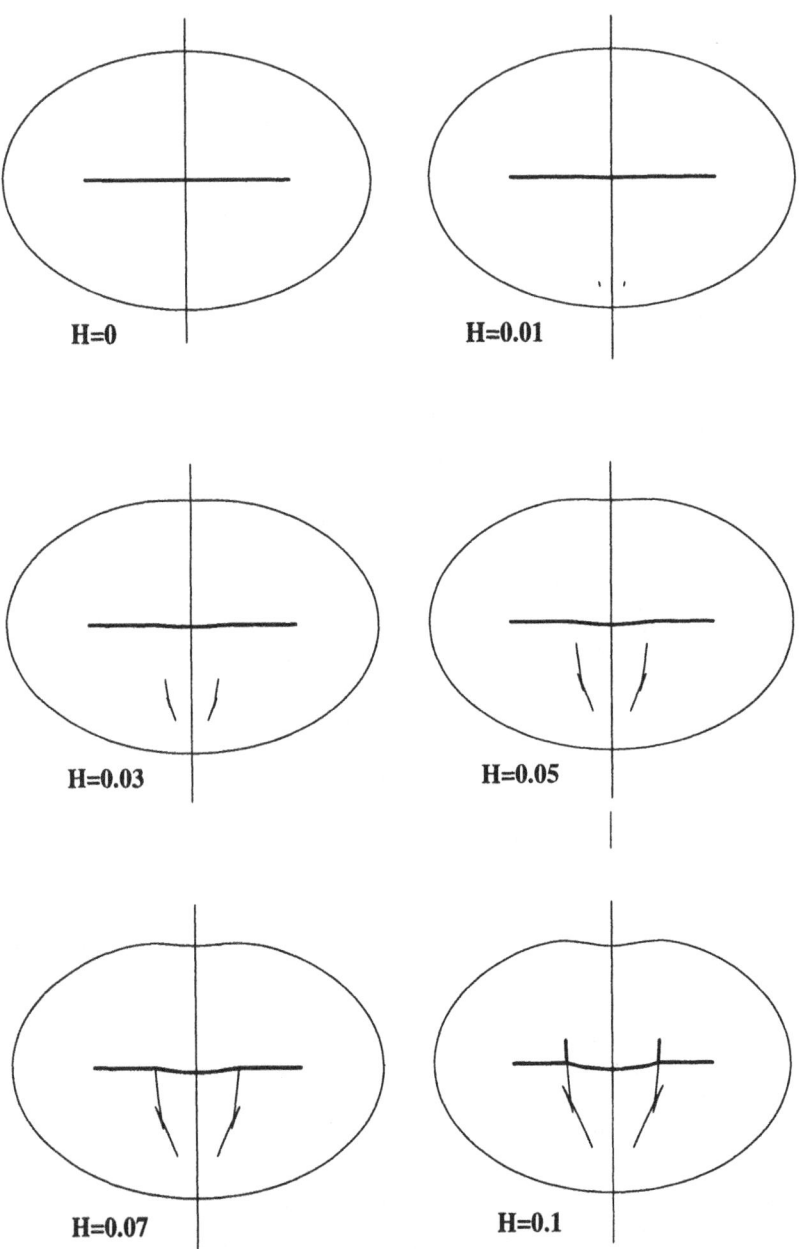

Fig. 1. An ellipse developing a symmetrical dent. The medial axis is the thick line and the rest of the symmetry set is the thin line. This includes a straight vertical axis of symmetry. Around $H = 0.01$ four new vertices are created on the dented ellipse, and by $H = 0.1$ two of the symmetry set branches so produced have penetrated the medial axis to create extra branches of that.

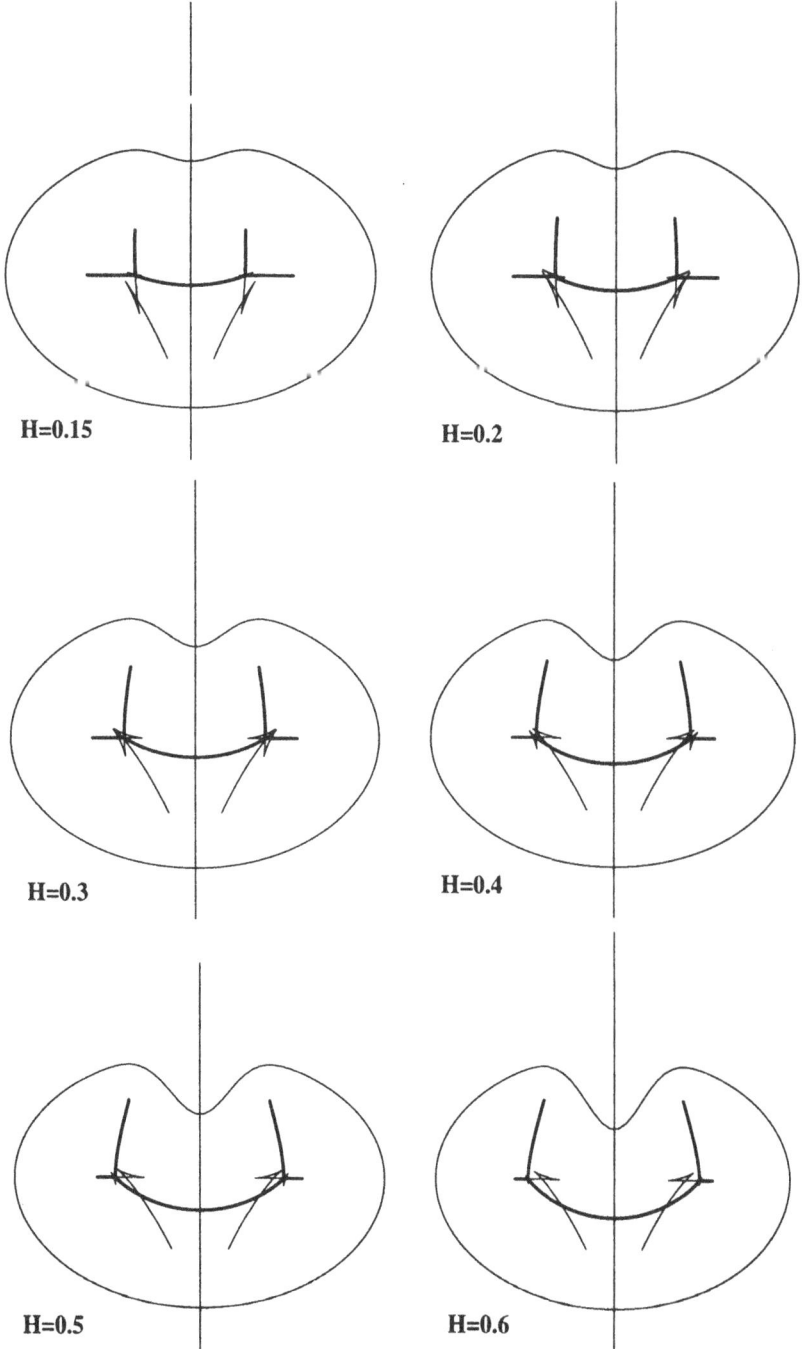

Fig. 2. The developing dent, continued. Between $H = 0.3$ and $H = 0.4$ a critical 'exchange of cusp branches' has taken place (see Figure 4 for a closeup). This enables a piece of the symmetry set to detach itself: by $H = 0.6$ this piece is ready to float clear of the medial axis.

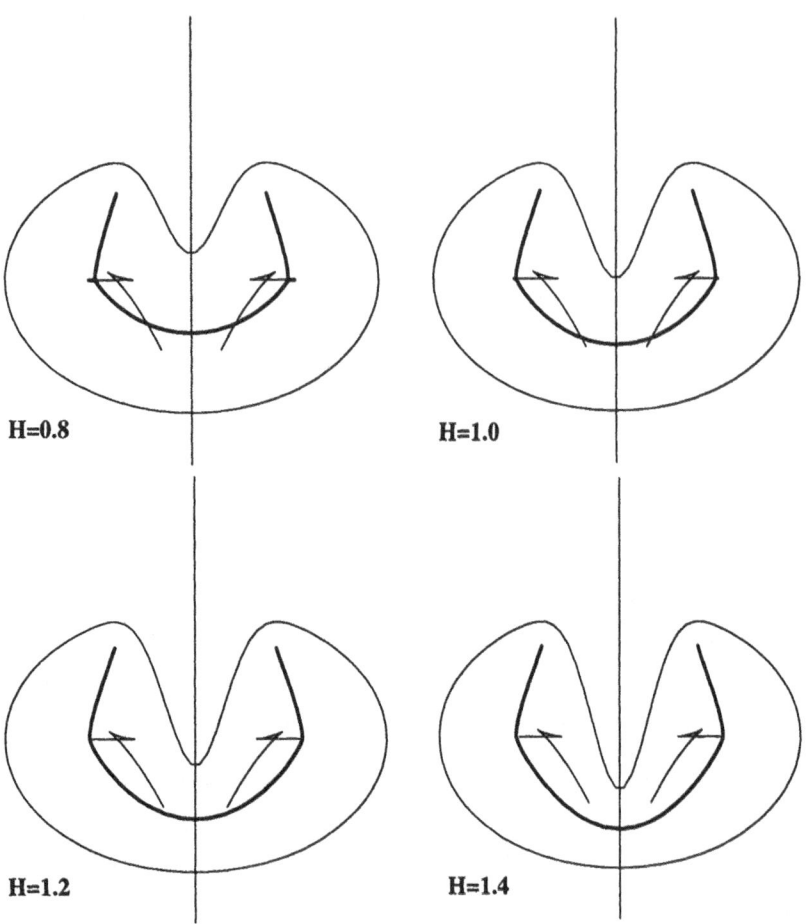

Fig. 3. The developing dent, concluded. Two (lower) branches of the symmetry set pull clear of the medial axis without incident, but the other two (pointing west and east) cause a short segment of the medial axis to disappear. Altogether the medial axis has acquired a U shape by a complex series of moves which essentially involves the symmetry set.

Somewhere around $H = 0.01$ it is clear that two small branches appear on the symmetry set while the medial axis merely bends slightly. By $H = 0.07$ the detail of these branches is evident: two endpoints and two cusps. Each such branch is the result of what is called an 'A_4 transition' [2,10]. Consider the right-hand half of the evolving curve; the left-hand half behaves in the same way. Slightly before $H = 0.01$ the curvature function acquires a horizontal inflexion at around $u = 0.4$ and then, as H increases, two new turning points of curvature appear—these correspond with vertices on the curve. The symmetry set has an *endpoint* for each of these vertices, just as the symmetry set of the original ellipse has an endpoint for each of the four vertices of this ellipse.

The next event of significance occurs when the new branches of the symmetry set *penetrate* the medial axis. This is called an '$A_1 A_3$ transition' and what is happening is that the circle centred at one of those endpoints on the symmetry set—the circle of curvature at a vertex of the curve—suddenly becomes tangent elsewhere. (More about $A_1 A_3$ in §4 below.) A closeup of the symmetry set and medial axis for $H = 0.2$ is in the left part of Figure 4, and this shows that as the medial axis is penetrated two extra cusps form on the symmetry set. Between $H = 0.2$ and $H = 0.3$ a remarkable event occurs, known as an 'A_2^2 transition' [2,10]. A single circle, centred at the place where two cusps come into coincidence, has become the osculating circle at two points of the curve, and this creates a change, enlarged in Figure 4. There are four cusps before (left) and four cusps after the transition (right) but the four smooth branches coming into two of the cusps have 'changed partners'. This new configuration enables a piece of the symmetry set to drift away as a separate entity (Figure 3) and the medial axis to regain the smoothness which it lost around $H = 0.07$.

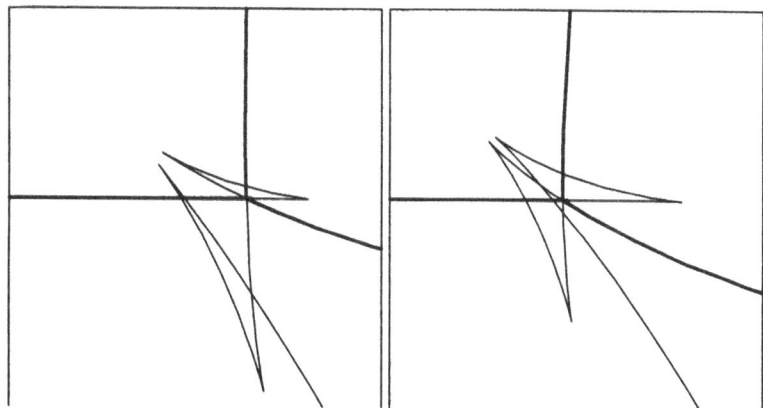

Fig. 4. Left: A closeup of $H = 0.2$ which shows (i) where the symmetry set penetrated the medial axis, at around $H = 0.07$, what is produced is a figure with two cusps (pointing north-west and east), (ii) the cusped branch of the symmetry set formed early on, before $H = 0.01$ (the cusps pointing north-north-west and south-south-east), has penetrated the medial axis without disturbing it at all. Right: a closeup of $H = 0.3$ which shows the remarkable 'exchange of cusps' transition, also called the 'nib'. This allows a branch of the symmetry set eventually to drift away, as is illustrated by Figure 3.

3   'Intrinsic' reconstruction via the radius function

3.1   Curve case

Consider two smooth curve segments A, B whose symmetry set is locally a smooth curve C. The radius of the bitangent circle will be denoted by r, which is therefore a function of the local parameter, say arclength s, on C. The function r can be regarded as a 'time' parameter: regarding C as being generated by the two segments A, B, and propagating at a constant rate down the normals to these two segments, the radius function measures (up to a constant multiplier) the time elapsed between starting from A or B and arriving at C. We can use r as a *parameter* on C, and write $C(r)$, provided r does not have a turning point with respect to arclength s on C, that is, provided the tangents to A and B at the corresponding points are not parallel. These maxima and minima of r are significant points on the symmetry set or medial axis, on an equal level with triple crossings or endpoints.

Let us assume that, as above, C is smooth, i.e. the bitangent circles are not osculating for any point of the local segment C of the symmetry set, and that dr/ds is not zero. The velocity with which the point on C moves, under the propagation from A and B is $v = ds/dr$. It is not difficult [16,9] to write down the reconstruction formula of A and B from C:

$$\mathbf{x}_i(r) = C(r) - \frac{r}{v}\mathbf{T} - (-1)^i\frac{r}{|v|}\sqrt{v^2 - 1}\mathbf{N}, \quad i = 1, 2, \tag{1}$$

where $\mathbf{T}$ is the unit tangent to C, $\mathbf{N}$ is the unit normal to C (oriented so that $\mathbf{T}, \mathbf{N}$ is anticlockwise). Note that $|v| > 1$; also $|v| = 1$ if and only if the two reconstructed points coincide, which occurs when C has an endpoint and the boundary $A \cup B$ has a vertex at the point of coincidence. In the case of curves, we can orient C and interpret $v > 0$ to mean that r is increasing in the chosen direction and $v < 0$ to mean that r is decreasing. In Figure 5, orienting C towards the right, $v > 0$ since it is clear that the radius will increase in this direction.

Using this 'intrinsic' reconstruction formula we can prove a relationship between the curvature κ of C and those of A and B, κ_1, κ_2 say. This has also been observed in [16] and [17]. One very attractive statement of this relationship is as follows. Let us orient the branches A and B by means of the bitangent circle; compare Figure 5. (This is the same convention as [9] and [17].) Writing $\overline{\kappa_i}$ for the curvatures of the parallels to A and B which pass through the chosen point of C, we have

$$\kappa = \frac{1}{2}(-\overline{\kappa_1} + \overline{\kappa_2})\sin\phi, \tag{2}$$

where ϕ is the angle as in Figure 5.

It is interesting to ask what happens to (2) at a vertex of the boundary curve, where the symmetry set has an endpoint and the two branches A and B come together at the vertex. The angle ϕ tends to 0 there, but the curvatures

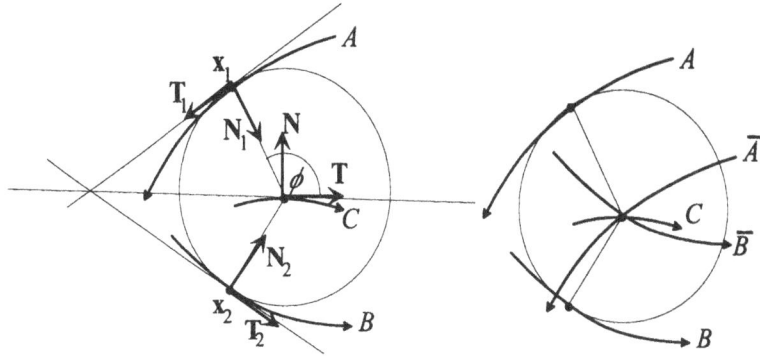

Fig. 5. Two smooth branches A and B giving rise to a segment of symmetry set or medial axis C. In this situation the radius is increasing as the point of C moves to the right, that is, the velocity v is positive. Right: the parallel curves $\overline{A}$ and $\overline{B}$ through a selected point of C

$\overline{\kappa_i}$ both tend to *infinity* since the corresponding point of C is actually the centre of curvature at the vertex and the parallel to A or B through this point is singular. So (2) is of the form $(\infty - \infty) \times 0$ and is so indeterminate. The limiting value of r at the endpoint is just $1/\kappa_1 = 1/\kappa_2$, and the limiting value of v is 1: see (1) where the two solutions come together when $v = 1$. It turns out that, when the symmetry set is a smooth curve with an endpoint, the limiting 'acceleration' $a = dv/dr$ is nonzero, and the limiting curvature κ of the symmetry set depends on the *third* derivative of κ_1 at the vertex of A. The formula, which does not appear to be very enlightening, is $\kappa = -4a^2 \kappa_1'''/15\kappa_1^5$, derivatives here being with respect to arclength on A.

Suppose now that $C(r_0)$ is the centre of a *maximal* circle, that is one whose radius equals the smallest distance from the centre to the boundary curve. It is clear that the curve segments A and B must be locally *outside* the circle centre $C(r_0)$, that is to say $1 - \kappa_1 r_0 > 0$ and $1 - \kappa_2 r_0 > 0$. Furthermore, if these hold, then this and nearby bitangent circles will be locally outside A and B: so far as the branches A and B are concerned, the circle is maximal. (Possibly some other part of a closed boundary curve will intrude inside this circle.) The condition comes to (see [9,16]):

$$r_0 \kappa v \sqrt{v^2 - 1} < v^2 - 1 + \frac{r_0 a}{v}, \tag{3}$$

where a and v are evaluated at $r = r_0$ and κ at $C(r_0)$. Note that $\sin \phi = \sqrt{v^2 - 1}/v$ and $\cos \phi = -1/v$ so the occurrences of v in the above inequality can be replaced by trigonometric functions of ϕ.

3.2   Surface case

Now consider two smooth surface patches A, B whose symmetry set is locally a smooth surface C. The radius of the bitangent sphere will be denoted by

r, so that this is a function of the local parameters on C. We are assuming, since C is smooth, that *these spheres are not spheres of curvature for any point of C*: otherwise, the symmetry set acquires a cuspidal edge. We should like to use r as one of the (curvilinear) coordinates on C. This is possible so long as the curves $r = $ constant are smooth curves on C, and it is not hard to check that the latter holds provided *the tangent planes to A, B at the corresponding points are not parallel*. (When they *are* parallel, the function r has a critical point, and the level set of r through that point of C will generally be an isolated point or a crossing.) Granted these assumptions, we can parametrize C by r and another parameter t say, where we take the level sets of t to be orthogonal trajectories of the level sets of r, that is $C_r \cdot C_t = 0$. We could assume that C_t is a unit vector for $r = r_0$, some specified value of r, and all t close to some specified t_0.

Now consider the family of spheres: $||\mathbf{x} - C(r, t)||^2 = r^2$. This will have envelope equal to the two sheets A and B. The envelope is defined by this equation and by differentiating with respect to r and t:

$$(\mathbf{x} - C) \cdot C_r = -r; \quad (\mathbf{x} - C) \cdot C_t = 0.$$

Let $\mathbf{x} - C = \lambda C_r + \mu C_t + \nu \mathbf{N}$, where $\mathbf{N}$ is the normal to the surface C. Using the three envelope equations shows that the surfaces A $(i = 1)$ and B $(i = 2)$ are given by

$$\mathbf{x}_i = C - \frac{r}{v}\mathbf{T} - (-1)^i \frac{r\sqrt{v^2 - 1}}{v}\mathbf{N}, \tag{4}$$

where v is the 'velocity' $||C_r||$ and $\mathbf{T}$ is the unit vector in the direction of C_r, that is $C_r = v\mathbf{T}$. Compare (1). Note that in the surface case it is probably more convenient to choose coordinates in this way, making $v > 0$ by definition. The vectors $\mathbf{T}$ and C_r are along the line joining the midpoint of the chord $\mathbf{x}_1\mathbf{x}_2$ to the point $C(r, t)$ of the symmetry set, and $\mathbf{T}$ is in the direction of the gradient of the radius function r, as a function on C. Thinking of r as a 'time' variable—the time from propagation at A or B to arrival at C—(4) is the intrinsic reconstruction formula for A and B from C.

In [18] there are formulae for the Gauss and mean curvatures of the surface C in terms of those for the surfaces A and B; naturally these are very complicated and do not appear to give insight into the relationship between C and the parallel surfaces to A and B passing through a given point of C, as in the curve case.

However, following Siersma (personal communication) we can draw quite a close analogy with the curve case, using (4). Figure 5 still gives the relevant notation, except that we need an extra vector $\mathbf{U}$ which points directly into the plane of the figure. That is, $\mathbf{T}_i, \mathbf{U}, \mathbf{N}_i$ are a right-handed triad for $i = 1, 2$, the first two forming a basis for the tangent plane to A or B and $\mathbf{N}_i$ being normal to A or B. Also $\mathbf{T}, \mathbf{U}$ and $\mathbf{N}$ form a right-handed triad.

As in the curve case, we shall consider the parallel $\overline{A}$ to A, or $\overline{B}$ to B passing through a particular point $C(r_0, t_0)$. This will be parametrized

$$\mathbf{y}_i = \mathbf{x}_i + r_0 \mathbf{N}_i = C - R\mathbf{N}_i,$$

where we use R to denote $r - r_0$ and we are particularly interested in the surfaces $\overline{A}, \overline{B}$ at the point $C(r_0, t_0)$. We can use R, t as local parameters on the surfaces $\overline{A}, \overline{B}$ and C, the 'base-point' on all three of them being then $C(0, t_0)$. We can clearly also assume that $C_t(0, t)$ is a unit vector for all t, in particular for $t = t_0$. Straightforward calculations show that

$$\mathbf{T}_i = (-1)^i \frac{\sqrt{v^2 - 1}}{v}\mathbf{T} - \frac{1}{v}\mathbf{N} = (-1)^i \sin \phi \mathbf{T} + \cos \phi \mathbf{N},$$

$$\mathbf{N}_i = \frac{1}{v}\mathbf{T} + (-1)^i \frac{\sqrt{v^2 - 1}}{v}\mathbf{N} = -\cos \phi \mathbf{T} + (-1)^i \sin \phi \mathbf{N},$$

and, at the parameter values $R = 0, t = t_0$,

$$\mathbf{y}_{iR} = (-1)^i \sqrt{v^2 - 1}\mathbf{T}_i, \quad \mathbf{y}_{it} = C_t.$$

If s is arclength along the curve $t = t_0$ on C, then $ds/dR = ||C_R|| = v$ so that

$$\frac{d\mathbf{y}_i}{ds} = (-1)^i \frac{\sqrt{v^2 - 1}}{v}\mathbf{T}_i = (-1)^i \sin \phi \mathbf{T}_i.$$

Every curve on C can be 'lifted' to A or B: we take the envelope points $\mathbf{x}_i$ where the spheres are centred at points of the curve on C. The above equations show the following interesting facts.

- For a curve on C tangent to the vector $\mathbf{T}$ at $C(0, t_0)$, the corresponding curves on $\overline{A}$ and $\overline{B}$ have tangents at $C(0, t_0)$ in the direction $\mathbf{T}_i$.
- Consider a curve on C tangent to the vector C_t at $C(0, t_0)$ (in Figure 5 this is parallel to $\mathbf{U}$, into the plane of the paper). Then the corresponding curves on $\overline{A}$ and $\overline{B}$ have tangents at $C(0, t_0)$ also in the direction C_t.

Note that the same is *not* true for the corresponding curves on A and B; that is why it is much better to consider the parallel surfaces through $C(0, t_0)$.

We can now differentiate $\mathbf{N} = (-\mathbf{N}_1 + \mathbf{N}_2)/(2 \sin \phi)$ in the two directions $\mathbf{T}$ and $\mathbf{U}$ and obtain a connexion between the second fundamental form matrix II of C with respect to this basis and the second fundamental form matrices II_1 of A with respect to $-\mathbf{T}_1, \mathbf{U}$ (note the minus sign!) and II_2 of B with respect to $\mathbf{T}_2, \mathbf{U}$, all at the point $C(0, t_0)$.

The result is the following.

Theorem (D.Siersma)

$$II = \frac{1}{2 \sin \phi}P(-II_1 + II_2)P, \text{ where } P = \begin{pmatrix} \sin \phi & 0 \\ 0 & 1 \end{pmatrix}.$$

Clearly this has a lot in common with (2) above.

It would be very interesting to have a simple analogue of the inequality (3) in the surface case of a medial axis. Is there a simple necessary and sufficient condition in terms of curvatures, velocities and accelerations, for a point to belong to the medial axis rather than merely to the symmetry set? It also remains to carry out a detailed examination of the function r on the symmetry set, at points where the above fails, that is when r has a critical point, and at boundary edges or singularities of the symmetry set.

4   Some medial axis geometry in the surface case

In §2 we saw, among other things, the effect of a branch of the symmetry set penetrating the medial axis in an 'A_1A_3 transition'—this occurs, for example, around $H = 0.07$ in Figure 1 and, in reverse, around $H = 1.2$ in Figure 3. The A_1A_3 situation for a curve occurs generically only in a family—if the circle of curvature at a vertex is tangent to the curve elsewhere then a small perturbation of the curve will remove all such exceptional tangencies. We can change the shape of the curve in a generic way so as to pass through this situation momentarily, but we do not expect to meet it on a single generic plane curve.

In three dimensions, however, the corresponding situation is generic for a *single* surface M. Suppose that P is a point of M where one of the principal curvatures, say κ_1, has a maximum or minimum along the corresponding line of curvature. Then P is called a *ridge point* of M—one of several definitions; this one is intrinsic to the surface and does not depend on choosing a preferred direction in space. See [7] for some other definitions. Ridges—loci of ridge points—are curves on M, and have been studied extensively [15,13,4,5]. Think of the appropriate sphere of curvature at points of a ridge: its radius is $1/\kappa_1$. As P moves along the ridge, the centre of the sphere traces a curve in space called by Porteous the *rib*: the symmetry set has a boundary edge along the rib, and the focal surface of M—the locus of all principal centres of curvature—has a cuspidal edge along the same rib. In addition, the sphere will grow or shrink in size as P moves along the ridge, and we can generically expect that for some special position P_0 the sphere will become tangent to M elsewhere, say at Q_0. The contact of the sphere at a ridge point is decribed as 'type A_3' and the situation where it has become tangent elsewhere is 'A_1A_3 contact', the A_1 referring to ordinary contact of the same sphere at the other point Q_0.

It is not a coincidence that the same name is used to describe the plane transition and the 3-space generic occurrence. Consider the symmetry set C of the surface M in the vicinity of an A_1A_3 point; C is a singular surface, that is it has self-intersections, cusp edges, boundaries and other singularities. It is a general result of singularity theory [2] that a suitable family of sections of C gives an accurate picture of the changes in the symmetry set of a *family of curves* passing through an A_1A_3 transition. In fact there are two qualitatively different families of sections but only one of these arises in the evolution of symmetry sets of curves[1]. The relevant family of symmetry sets in the plane and the corresponding local picture of the symmetry set in 3-space are shown in Figure 6. The 'swallowtail' surface consists of the centres of spheres tangent to M in two places, one near P_0 and one near Q_0. The other sheet, which appears as half a vertical plane in Figure 6, right, with boundary edge furthest from the viewer, consists of the centres of spheres which are tangent to M in two places, both of which are near P_0. *On* the boundary edge the two nearby tangencies have come into coincidence.

What is particularly fascinating about this configuration is that two curves of significance on the symmetry set come together. Let C_0 be the centre of the A_1A_3 sphere. The rib passes through C_0 but only one direction of the rib can be in the medial axis. The curve of points which are centres of *tritangent* spheres, that is, spheres tangent at three distinct points of M, comes to a complete end at C_0. This 'A_1^3 set' can be seen in Figure 6, right, as the intersection of three sheets of the symmetry set, two belonging to the 'swallowtail' surface and the third to the other component of the symmetry set.

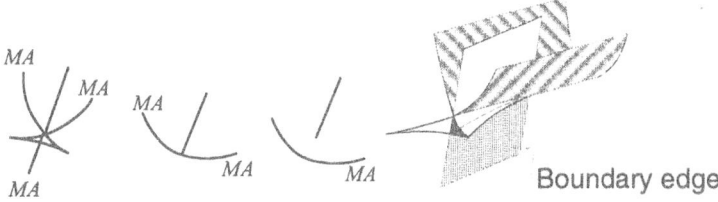

Fig. 6. Left: three snapshots from an evolving symmetry set of a family of plane curves, passing through an A_1A_3 transition. The branches labelled '*MA*' are part of the medial axis, as far as the triple intersection in the left-most picture. Right: the local structure of the symmetry set of of a *surface*, near an A_1A_3 point, with the medial axis shaded. Most boundary edges are simply clipping, but the one marked is the rib line and a genuine boundary. The pictures on the left are slices of the figure on the right by a family of parallel planes moving from lower left to upper right with normals to these planes pointing always to the upper right.

Rather than give general calculations [12], I shall present an example. Consider a parabolic surface A, of the form $z = (ay + b)x^2$, $b > 0$, and a plane B, with equation $z = cx + dy + e$. See Figure 7. We are to imagine these as part of a single surface M. The surface A has a ridge 'along the bottom', that is along the y-axis; the principal curvatures at $(0, y, 0)$ are

---

[1] It has recently been discovered by my student Paul Holtom that the other family of sections arises naturally when we consider symmetry sets defined in an *affinely invariant* way.

$\kappa_1 = 2(ay + b)$ and $\kappa_2 = 0$, and the principal directions are parallel to the x- and y-axes respectively. We can imagine a sphere of varying radius $1/\kappa_1$ rolling along the trough; this sphere has A_3 contact with A and its centre traces the corresponding rib. We can choose e so that for $y = 0$ the sphere, which is tangent to A at $P_0 = (0, 0, 0)$, is also tangent to the plane at say Q_0. (This requires in fact $2be = 1 + \sqrt{c^2 + d^2 + 1}$.) This sphere is the $A_1 A_3$ sphere, and for $b > 0$ it will be maximal, that is its centre will be a point of the medial axis.

The ridge consists of points $(0, y, 0)$ and the rib, that is the locus of corresponding centres of principal curvature, is the points $\gamma(y) = (0, y, 0) + (1/(2ay + 2b))(0, 0, 1) = (0, y, 1/(2ay + 2b))$. This is the boundary edge—the A_3 set—on the symmetry set, and for $y = 0$ the tangent is in the direction $(0, 1, -a/2b^2)$. But is the direction *into* the medial axis plus or minus this vector? Consider the function $f(y) = \|\gamma(y) - Q_0\|^2 - 1/\kappa_1^2$, which is 0 at $y = 0$. If $f'(0) > 0$ then f is instantaneously increasing, hence > 0 for small y, which means that the spheres continue to be maximal and we are moving into the medial axis rather than out of it. Calculating the derivative we find that the rib over the positive y-axis lies in the medial axis if and only if $2b^2 d + a(1 + \sqrt{1 + c^2 + d^2}) > 0$.

What about the triple-tangency curve—the A_1^3 set—on the symmetry set? By construction this passes through $\gamma(0)$. To find its tangent there, consider first the general situation of a sphere, centre C, tangent to M in three points. These three points lie in a plane, and it is not hard to see that the tangent to the A_1^3 set at C is perpendicular to this plane[2]. What happens when two of the contact points are very close, as in our example where they are close to the ridge? Call these points P_1, P_2 and forget the third tangency point for a moment. For a sphere of radius r we have $P_1 + r\mathbf{N}_1 = P_2 + r\mathbf{N}_2$; here $\mathbf{N}_i$ is simply the normal to M at P_i. Thus $P_1 - P_2, \mathbf{N}_1$ and $\mathbf{N}_2$ are coplanar. But when P_1 and P_2 are very close this means that the line joining them is approximately in a *principal direction* on M[3]. So, as P_1 and P_2 tend to coincidence at a ridge point, the line joining them has limiting direction which is principal. It is evident in our example, and is true generally, that the principal direction is actually the one associated with the ridge itself: that is perpendicular to the y-axis in our example. Thus the A_1^3 set has tangent at $\gamma(0)$ perpendicular to the plane joining the third contact point Q_0 with the x-axis. This direction works out as $(0, 1 + \sqrt{1 + c^2 + d^2}, d)$. Recall that the A_1^3 set *ends* at $\gamma(0)$.

---

[2] This becomes fairly evident if we replace M by its three tangent planes at the three contact points. The direction in which the centre of the tritangent sphere starts to move away from its starting position is unaffected by this replacement.

[3] If we move away from a point of a surface in a principal direction then the derivative of the unit normal is in the plane of the normal and the direction of motion—hence is in fact in the direction of motion.

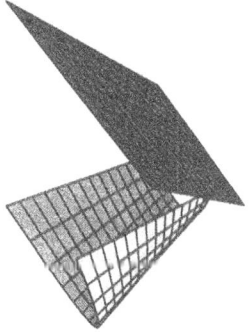

Fig. 7. Two surface pieces where an A_1A_3 situation can occur. One is a parabolic surface $z = (ay + b)x^2$ and the other a plane $z = cx + dy + e$. The y-axis slopes upwards to the right along the trough (ridge) of the parabolic surface.

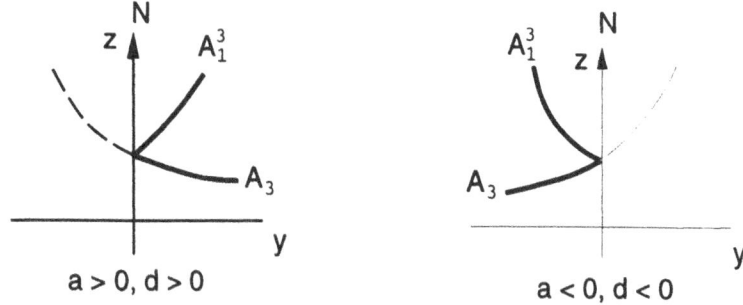

Fig. 8. For the example of Figure 7, taking $a > 0, d > 0$, these are the configurations of rib line (A_3), with the solid part being in the medial axis, triple-contact line (A_1^3), and normal **N** to the surface M. The y-axis is along the ridge of the parabolic surface. The gradient of r is also parallel to **N**, that is vertical for all y in the figure.

The gradient of the radius function r on the symmetry set C is in the direction **T** in Figure 5. At a rib point the two corresponding points $\mathbf{x}_1, \mathbf{x}_2$ on M as in (4) have coincided, and the normal to M is also in the direction **T**. Let us suppose for an example that a and d have the same sign; then the quantity $2b^2d + a(1 + \sqrt{1 + c^2 + d^2})$ occurring above also has this sign. The local picture of the A_3 and A_1^3 lines and the normal (parallel to $\nabla(r)$) are shown in Figure 8. Note that the rib line in the medial axis and the triple-tangent line are on the same side of the normal. It can be shown that this is always true.

There are other features of the symmetry set and medial axis of a surface M which have a rich geometry [11,12]. For instance, it is generic for a sphere to have contact with M at *four* points, and the structure of the medial axis near the centre of such a sphere is similar to the configuration of plane sectors joining a point C inside a tetrahedron to the six edges of the tetrahedron. These six sectors intersect along four rays joining C to the four vertices of the tetrahedron, and the flow of the radius function is restricted along these rays. In fact r can increase towards C only along two, three or four of these rays.

So this is just the beginning of a description of the symmetry set and medial axis in all its geometrical detail. As I said at the beginning of the article, this is very much work in progress.

References

1. T.F.Banchoff and P.J.Giblin 'Global theorems for symmetry sets of smooth curves and polygons in the plane', *Proc. Royal Soc. Edinburgh* 106A (1987), 221–231.
2. J.W.Bruce and P.J.Giblin 'Growth, motion and one-parameter families of symmetry sets', *Proc. Royal Soc. Edinburgh* 104A (1986), 179–204.
3. J.W.Bruce , P.J.Giblin and C.G.Gibson 'Symmetry sets', *Proc. Royal Soc. Edinburgh* 101A (1985), 163–186.
4. J.W.Bruce, P.J.Giblin and F.Tari, 'Ridges, crests and sub-parabolic lines of evolving surfaces', *Int. J. Computer Vision.* 18 (1996), 195–210.
5. J.W.Bruce, P.J.Giblin and F.Tari, 'Families of surfaces: focal sets, ridges and umbilics', *Math. Proc. Camb. Phil. Soc* 125 (1999), 243–268.
6. R.Cipolla and P.J.Giblin, *Visual Motion of curves and surfaces*, viii + 184pp. Cambridge University Press, 1999.
7. D.Eberly, *Ridges in Image and Data Analysis*, Kluwer Academic Publishers 1996.
8. P.J.Giblin and S.A.Brassett 'Local symmetry of plane curves', *Amer. Math. Monthly* 92 (1985), 689–707.
9. P. J. Giblin and B. B. Kimia, 'On the Intrinsic Reconstruction of Shape from its Symmetries', *Proceedings of the IEEE Computer Society Conference on Computer Vision and Pattern Recognition (CVPR99)* IEEE Computer Society Press (1999), 79–84.
10. P. J. Giblin and B. B. Kimia, 'On the Local Form and Transitions of Symmetry Sets, and Medial Axes, and Shocks in 2D', *Proceedings of the Fifth International Conference on Computer Vision (ICCV99)* IEEE Computer Society Press (1999), 385–391.
11. P.J.Giblin and B.Kimia, 'A formal classification of 3D medial axis points and their local geometry', to appear in the Proceedings of CVPR (IEEE Computer Society Conference on Computer Vision and Pattern Recognition), 2000.
12. P.J.Giblin and B.Kimia, 'Symmetry sets and medial axes of surfaces', in preparation.
13. P.L.Hallinan, G.G.Gordon, A.L.Yuille, P.Giblin and D.Mumford, *Two and three dimensional patterns of the face*, viii+262 pages, Natick, Massachusetts: A.K.Peters 1999.

14. R.J.Morris, 'The use of computer graphics for solving problems in singularity theory', *Visualization in Mathematics*, ed. H.-C.Hege and K.Polthier, Springer-Verlag (1997), 53–66.

15. I.R.Porteous, *Geometric Differentiation*, Cambridge University Press 1994.

16. D.Shaked and A.M.Bruckstein, 'The curve axis', *Computer Vision and Image Understanding* 63 (1996), 367–379.

17. D.Siersma, 'Properties of conflict sets in the plane', *Geometry and Topology of Caustics*, ed. S. Janeczko and V.M.Zakalyukin, Banach Center Publications Vol. 50, Warsaw 1999, pp. 267–276.

18. J.Sotomayor, D.Siersma and R.Garcia, 'Curvatures of conflict surfaces in Euclidean 3-space', *Geometry and Topology of Caustics*, ed. S. Janeczko and V.M.Zakalyukin, Banach Center Publications Vol. 50, Warsaw 1999, pp. 277–285.

Symmetry - A Research Direction in Curve and Surface Modelling; Some Results and Applications

H. E. Bez

Loughborough University, Department of Computer Science,
Loughborough, Leicestershire LE11 3TU, England.

Summary. Many parametric curve and surface methods, including polynomial and spline, require sets of 'parameter' functions to be specified in addition to control-, or interpolation- point sets. It is shown here that symmetry is a powerful tool for the analysis of this class of curve function and can, for example, be applied to provide complete answers to fundamental questions such as:

(i) if the control point set is held fixed, under what conditions do different sets of parameter functions determine the same curve?

and the related question:

(ii) what properties are required of the parameter functions to ensure invariance of curve shape with respect to a given set of geometric transformations of the control point set?

1   Introduction

Transformation groups and geometry have been strongly coupled mathematical disciplines since the fundamental contributions of Lagrange, Abel, Galois, Klein and Lie to the understanding of symmetry and its relationship to group theory. Symmetry and group theory are now applied in a vast number of disparate subject areas including elementary particle physics, classical mechanics, chemistry and biology. In general if Θ is a mathematical object of a particular category and μ is an invertible mapping such that $\mu * \Theta$ is an object of the same category, then μ is said to be a symmetry of Θ if $\mu * \Theta = \Theta$; ie if Θ is a fixed point (or an invariant) of μ. The set of all symmetries of an entity Θ determines a group.

It is usual to consider symmetries of a particular type - eg we can consider the group of affine symmetries of a triangle $\mathcal{T}$, or the subgroup of linear symmetries of $\mathcal{T}$. To give some simple examples, let $\mathcal{T}$ be defined in the two dimensional plane by the vertex set $(v_0, v_1, v_2) \in \mathbb{R}^2 \times \mathbb{R}^2 \times \mathbb{R}^2$. The graph, $G(\mathcal{T})$, of $\mathcal{T}$ is the point set determined by the straight line segments $[v_0, v_1], [v_1, v_2]$ and $[v_2, v_0]$ in $\mathbb{R}^2$. If g is a transformation of the plane, we denote by $g\mathcal{T}$ the triangle with vertices gv_0, gv_1 and gv_2 and define the symmetry group of $\mathcal{T}$ to be those g for which

$$G(g\mathcal{T}) = G(\mathcal{T})$$

ie the transformations that leave the graph of $\mathcal{T}$ invariant. For example if $\mathcal{T}$ is the isosceles triangle defined by $v_0 = (1,0), v_1 = (0,1)$ and $v_2 = (0,-1)$ then the group of linear symmetries of $\mathcal{T}$ is

$$\left\{ \begin{bmatrix} 1,0 \\ 0,1 \end{bmatrix}, \begin{bmatrix} -1,0 \\ 0,1 \end{bmatrix} \right\}.$$

A transformation g of $\mathbf{R}^2$ is said to be affine if it is of the form $gv = Av + a$ where $A \in GL(2, \mathbf{R})$ and $a \in \mathbf{R}^2$. Writing affine transformations as pairs (A, a), the group of affine symmetries of $\mathcal{T}$ is $\{(A_i, a_i) : 1 \le i \le 6\}$ where $\{A_1, \ldots, A_6\}$ and $\{a_1, \ldots, a_6\}$ are given respectively by

$$\left\{ \begin{bmatrix} 1,0 \\ 0,1 \end{bmatrix}, \begin{bmatrix} 1, 0 \\ 0, -1 \end{bmatrix}, \begin{bmatrix} -1/2, 1/2 \\ 3/2, 1/2 \end{bmatrix}, \right.$$

$$\left. \begin{bmatrix} -1/2, -1/2 \\ 3/2, -1/2 \end{bmatrix}, \begin{bmatrix} -1/2, 1/2 \\ -3/2, 1/2 \end{bmatrix}, \begin{bmatrix} -1/2, -1/2 \\ -1/2, 3/2 \end{bmatrix} \right\}$$

and

$$\{(0,0), (0,0), (1/2, -1/2), (1/2, -1/2), (1/2, 1/2), (1/2, -1/2)\}.$$

If $\mathcal{T}$ is the equilateral triangle defined by $v_0 = (1,0), v_1 = (1/2, \sqrt{3}/2)$ and $v_2 = (-1/2, -\sqrt{3}/2)$ then its group of linear symmetries is

$$\left\{ \begin{bmatrix} 1,0 \\ 0,1 \end{bmatrix}, \begin{bmatrix} -1/2, \sqrt{3}/2 \\ -\sqrt{3}/2, 1/2 \end{bmatrix}, \begin{bmatrix} -1/2, -\sqrt{3}/2 \\ \sqrt{3}/2, -1/2 \end{bmatrix}, \right.$$

$$\left. \begin{bmatrix} -1,0 \\ 0,1 \end{bmatrix}, \begin{bmatrix} 1/2, \sqrt{3}/2 \\ \sqrt{3}/2, -1/2 \end{bmatrix}, \begin{bmatrix} 1/2, -\sqrt{3}/2 \\ -\sqrt{3}/2, -1/2 \end{bmatrix} \right\}$$

- which is isomorphic to the permutation group on three symbols S_3.

Equivalently, if $\mathcal{T}$ is a triangle and G is a transformation group acting on the plane, we can define a transformation of the graph $G(\mathcal{T})$ of $\mathcal{T}$ by

$$g * G(\mathcal{T}) = G(g^{-1}\mathcal{T}) \quad \text{for} \quad g \in G.$$

A subgroup $G_{\mathcal{T}}$ of G is a symmetry group of $\mathcal{T}$ if

$$g * G(\mathcal{T}) = G(\mathcal{T}) \quad \text{for all} \quad g \in G_{\mathcal{T}}.$$

ie $G(\mathcal{T})$ is a fixed point of $G_{\mathcal{T}}$.

The above generalises to polygons in the plane and to polyhedra in three dimensional space [3].

If $v = (v_0, v_1, v_2)$, then $G(\mathcal{T})$ can be regarded as the 'path' $p[v]$ where

$$p[v](t) = (1 - t)v_0 + tv_1 \;;\; 0 \le t \le 1$$
$$= (2 - t)v_1 - (t - 1)v_2 \;;\; 1 < t \le 2$$
$$= (3 - t)v_2 + (t - 3)v_0 \;;\; 2 < t \le 3$$

it follows that $G(g\mathcal{T})$ is the path

$$
\begin{aligned}
p[gv](t) &= (1-t)gv_0 + tgv_1 \; ; \; 0 \leq t \leq 1 \\
&= (2-t)gv_1 - (t-1)gv_2 \; ; \; 1 < t \leq 2 \\
&= (3-t)gv_2 + (t-3)gv_0 \; ; \; 2 < t \leq 3
\end{aligned}
$$

and the symmetries of $\mathcal{T}$ can be described as those g for which the paths $p[v]$ and $p[gv]$ are equivalent. In this view, p can be regarded as a function $p : V \to \mathcal{P}$ where $V = \mathbf{R}^2 \times \mathbf{R}^2 \times \mathbf{R}^2$ and $\mathcal{P}$ is the set of paths in $\mathbf{R}^2$. A group of transformations of $\mathbf{R}^2$ acts on the path functions as $g * p[v] = p[g^{-1}v]$ and the symmetries of a given path $p[v]$ are those g for which $p[v]$ and $p[g^{-1}v]$ are equivalent (or $p[gv]$ equivalent to $p[v]$).

Alternative group actions on p are possible, for example we could define $(g * p)[v] = gp[g^{-1}v]$; under this action any affine transformation of $\mathbf{R}^2$ is a symmetry of any triangle - or any polygon. This action does not therefore capture the visual symmetries of regular polygons. However, this paper is concerned with the symmetries of general path functions and the implications for geometric modelling; in this application it is the extension of this alternative group action to arbitrary path functions that is appropriate. To illustrate, let $V = \mathbf{R}^2 \times \cdots \times \mathbf{R}^2$ let $v = (v_0, \ldots, v_n) \in V$, and let $p[v]$ be the Bezier path of degree n given by:

$$
p[v](t) = \sum_{k=0}^{n} {}^n C_k t^k (1-t)^{n-k} v_k \; ; \; 0 \leq t \leq 1.
$$

Defining the transformed path $g * p$ by

$$
(g * p)[v] = gp[g^{-1}v],
$$

it is easy to show that $g * p = p$ if and only if g is an affine transformation of the plane. The condition $g * p = p$ is equivalent to $p[gv] = gp[v]$, and the geometric interpretation of this is that a Bezier path on the transformed control vectors gv is identical to the transformed path on v, for all affine transformations g of the plane - ie Bezier paths are geometrically invariant under affine transformations of the control vectors, in the same way that all polygon paths are invariant with respect to this affine action.

Throughout the paper we work with paths in the two dimensional plane, everything generalises readily to paths in $\mathbf{R}^3$.

More generally, if G is a transformation group of $\mathbf{R}^2$, then all path functions, $p : V \to \mathcal{P}$, fixed by the group action $p \to g * p$ are geometrically invariant under the transformations of G. It is universally accepted that, for applications in CAD, CAGD and computer graphics, path functions should be invariant, in the sense described, at least with respect to Euclidean transformations. This ensures that the geometry of the path is independent of the underlying coordinate system, and that Euclidean transformations of the shape may be performed efficiently. Bezier paths are an example of the

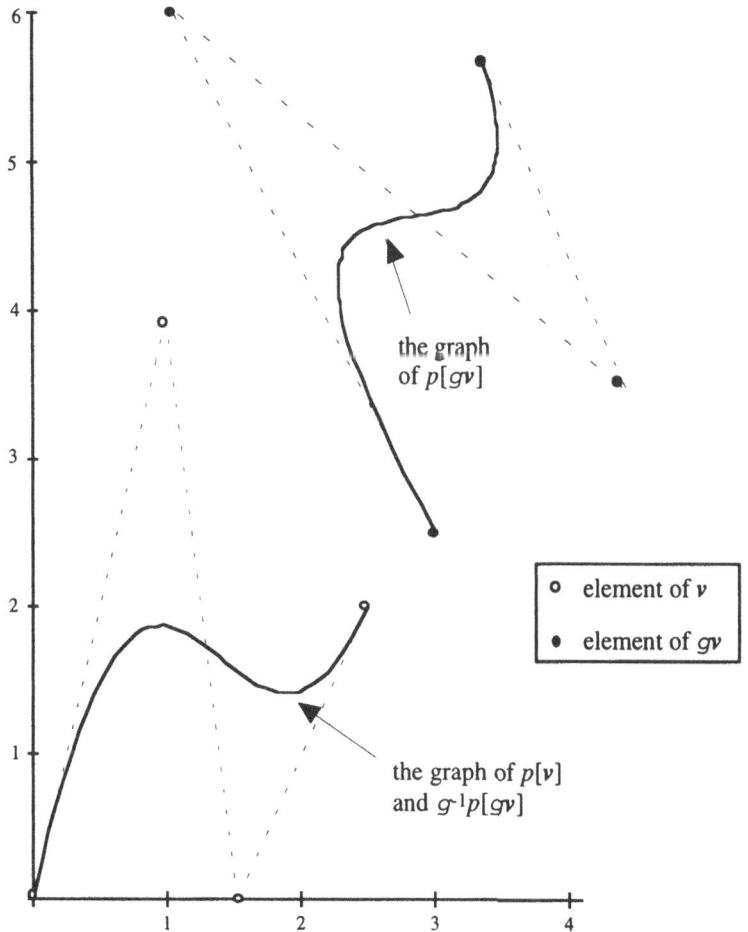

Fig. 1. Invariance of Bezier paths under Euclidean transformation

simple class $p : V \to \mathcal{P}$ that are invariant under Euclidean transformations - as illustrated in Fig. 1. Fig. 2 shows a path function of the same class but with variant behaviour with respect to Euclidean transformations. Not all path functions used in approximation theory, CAGD and other application domains are of this simple class; for example, parametric spline paths and Lagrange polynomial paths are of a more general type. For these it is necessary to specify a set $v = (v_0, \ldots, v_n)$ of vectors and a set $\tau = (\tau_0, \ldots, \tau_n)$ of parameters. These parameters are, in the simplest case, real numbers with the property $\tau_0 < \ldots < \tau_n$. The appropriate domain for path functions of this type is therefore $V \times T$, where T is the set of all (n+1)-tuples of real numbers, τ_i , such that $\tau_0 < \ldots < \tau_n$.

Path functions $p : V \times T \to \mathcal{P}$ on the product domain $V \times T$ are of particular interest in this paper. In general, for a fixed control point set v,

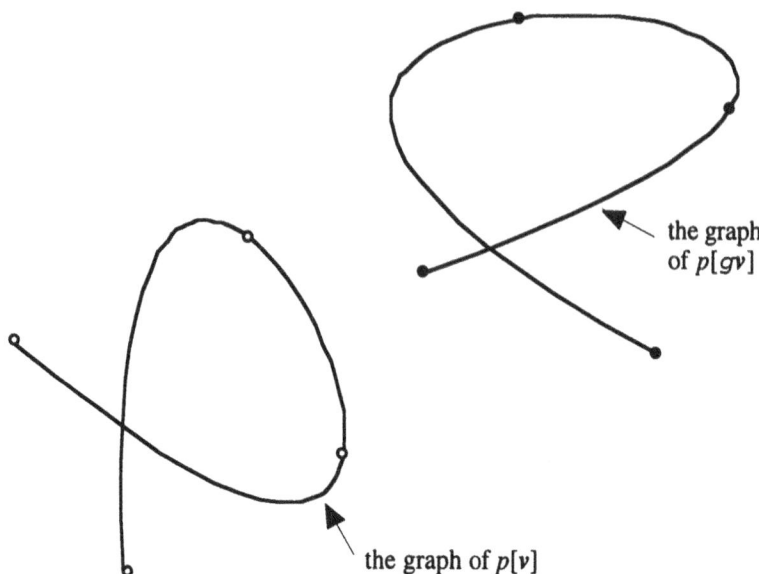

the graph
of $p[g\boldsymbol{v}]$

the graph of $p[\boldsymbol{v}]$

Fig. 2. A path function of the type $p : V \to \mathcal{P}$ for which $g * p \neq p$ for Euclidean transformations

the shape of the curve is different for different choices of $\tau \in T$ - see Fig. 3. For example, we prove later that the shape of a Lagrange path with $\tau_i = i$, ie

$$\sum_{k=0}^{n} \prod_{j=0, j\neq k}^{n} [\frac{t-j}{k-j}] \, \boldsymbol{v}_k \; ; \text{ for } 0 \leq t \leq n$$

is different to that with $\tau_i = i^2$, ie

$$\sum_{k=0}^{n} \prod_{j=0, j\neq k}^{n} [\frac{t-j^2}{k^2-j^2}] \, \boldsymbol{v}_k \; ; \text{ for } 0 \leq t \leq n^2.$$

However, distinct τ can also give rise to paths having the same shape - for example it is shown later that Lagrange paths with $\tau_i = i$, are identical to those with $\tau_i = 2i + 1$. Investigating the conditions under which path shapes remain invariant under a change of parameters is one of the objectives of the paper. This leads naturally to the consideration of symmetry with respect to the T domain. In applications, constant parameter values are not normally acceptable. Both experience and theory [1] suggest that parameters that depend on the control vector set $\boldsymbol{v}$ give better results. For example, chord-length parameter functions

$$\tau_0(\boldsymbol{v}) = 0,$$
$$\tau_i(\boldsymbol{v}) = \tau_{i-1}(\boldsymbol{v}) + \|\boldsymbol{v}_i - \boldsymbol{v}_{i-1}\|_2 \; ; \; i \geq 1$$

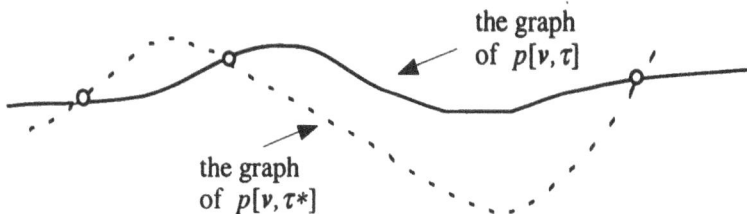

Fig. 3. In general, different τ produce different paths for functions of the class $p : V \times T \to \mathcal{P}$

where $\|\cdot\|_2$ denotes the Euclidean norm, are often used. It is therefore natural to define $T(V)$ to be (n+1)-tuples, written $(\tau_0, \ldots, \tau_n)$, of functions on V with the property $\tau_i(v) < \tau_{i+1}(v)$ for all $v \in V$ and to consider the class of path functions $p : V \times T(V) \to \mathcal{P}$ on the domain $V \times T(V)$.

The shape of a path of the simple type $p : V \to \mathcal{P}$, can be changed only by modifying $v \in V$, but shapes produced by the more general class $p : V \times T(V) \to \mathcal{P}$ can change when τ is altered. Further the behaviour under transformation of the general class depends on τ. It is therefore important to obtain answers to the following questions for the general class:

(i) under what conditions does a change of shape occur when τ is changed?

and

(ii) for which τ are the shapes produced by p invariant with respect to particular groups of transformations of V - minimally Euclidean transformations?

The purpose of the paper is to show that symmetry provides the key to obtaining comprehensive answers to these fundamental questions. Fig. 4 illustrates how an inappropriate choice of $\tau \in T(V)$ can lead to undesirable variance of shape under Euclidean transformation for path functions of the class $p : V \times T(V) \to \mathcal{P}$; an illustration of the general issue raised in (ii). Specifically, it is shown that a transformation group H may be associated with the domain T such that paths produced at $\tau \in T$ and $\tau^* \in T$ by a path function $p : V \times T \to \mathcal{P}$ have the same shape if and only if

$$\tau^* = \pi(\tau) \; ; \quad \text{for } \pi \in H.$$

Further, it is shown that a transformation group G may be associated with the domain V such that the paths produced by $p : V \times T(V) \to \mathcal{P}$ at $\tau \in T(V)$ are invariant under a subgroup G_τ of G if and only if τ transforms as:

$$\tau(gv) = \pi[g, v]\tau(v) \; ; \quad \text{for all } g \in G_\tau$$

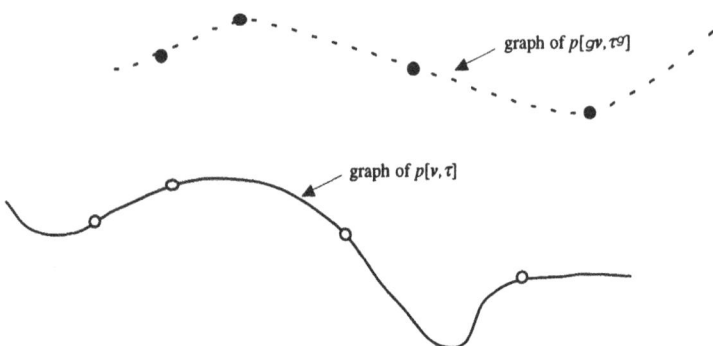

graph of $p[gv, \tau g]$

graph of $p[v, \tau]$

Fig. 4. For some choices of $\tau \in T(V)$, functions of the class $p : V \times T(V) \to \mathcal{P}$ are such that $g * p \neq p$ for Euclidean transformations

for some $\pi : G_\tau \times V \to H$, satisfying the conditions of a group cocycle.

Parametric Lagrange paths provide a case study to illustrate the ideas developed. A number of applications of the results are given. Much of the work generalises to other path functions, including parametric splines.

2   Paths, curves and curve functions

The introduction gives an overview of the aims and results of the paper without giving proper definitions of some of the concepts discussed. To derive the results it is necessary to be more precise. In particular it is important to distinguish clearly between the concepts of path and curve. A path in $\mathbf{R}^2$ is a C^1 function $p : I \to \mathbf{R}^2$ where I is an interval of $\mathbf{R}$ and $p' \neq 0$ on I. The condition $p' \neq 0$ ensures that the tangent space is well defined at each point $p(t)$ of p.

A curve is an equivalence class of paths in the following sense; we say that two paths p_1 and p_2 are related if there is a C^1 function $\phi : I_1 \to I_2$ such that $\phi' \neq 0$ on I_1 and $p_1 = p_2 \circ \phi$. The condition $\phi' \neq 0$ means that ϕ is strictly monotone; ϕ is therefore invertible and has C^1 inverse ϕ^{-1} with $(\phi^{-1})' \neq 0$. Hence the relation defined above is an equivalence relation on the set of paths of $\mathbf{R}^2$ and the equivalence classes are the curves of $\mathbf{R}^2$. If p is a path, the class of all paths equivalent to p can be written $<p>$; $<p>$ is therefore a curve having p as a representative path, and if p and $p*$ are equivalent paths then $<p>=<p*>$. In this paper we denote the set of all paths of $\mathbf{R}^2$ by $\mathcal{P}$ and the set of all curves in $\mathbf{R}^2$ by $\mathcal{C}$. The condition $\phi' \neq 0$ implies either $\phi' > 0$ or $\phi' < 0$, ie ϕ either strictly increasing or strictly decreasing on its interval of definition. We can, without loss of generality assume that ϕ is strictly increasing; if M denotes the set of all strictly increasing C^1 functions on $\mathbf{R}$, then M is a group under functional composition. The graph of a path $p : I \to \mathbf{R}^2$ is the point set $\{p(t) : t \in I\}$; it follows that if p_1 and p_2 are equivalent then their graphs are identical. The curves of $\mathbf{R}^2$ can therefore be loosely described as equivalence classes of paths in $\mathbf{R}^2$ having the same shape, and having parametrisations that define a tangent space at each point.

In the introduction, path functions with range $\mathcal{P}$ were introduced. As the primary interest is in shape, it is functions having range $\mathcal{C}$ that are significant. However, in computations with curves we work with representative paths and ensure that results are independent of representation. For example if $v = (v_0, v_1)$ then

$$c[v](t) = (1 - t)v_0 + tv_1 , \ 0 \leq t \leq 1,$$

defines $c[v]$ as the straight line segment between v_0 and v_1. Strictly, we should first define the path function $p : V \to \mathcal{P}$ by

$$p[v](t) = (1 - t)v_0 + tv_1 , \ 0 \leq t \leq 1$$

and then define $c : V \to \mathcal{P}$ by $c[v] = <p[v]>$. However, to avoid the notation getting out of hand we define c directly in terms of representative paths, and remain mindful that any path equivalent to $p[v]$ could be used for the definition of $c[v]$.

This approach is adopted throughout paper; therefore, because it is the properties of the underlying curves that are of interest, the monotone functions ϕ that relate representative paths feature in many of the definitions and derivations.

3   Group restrictions, actions and symmetries

The symmetries of triangles were discussed earlier with respect to restricted types of transformations of the plane - the linear symmetries and the affine symmetries were considered. The same approach is suited to curve and surface functions. Three distinct symmetries can be defined for path functions of the types $p : V \times T \to \mathcal{P}$ and $p : V \times T(V) \to \mathcal{P}$. For each symmetry, suitable restrictions on the transformations are required.

In this paper symmetries with respect to the range $\mathcal{P}$ are referred to as class symmetries, symmetries with respect to the domain V are referred to as geometric symmetries and symmetries with respect to the T domain are referred to as parametric symmetries. Answers to the fundamental questions posed for p in the introduction follow immediately from a complete knowledge of these symmetries and the relationships between them.

Consider class symmetry; if p is a path with domain $\mathbf{R}$ and $f : \mathbf{R} \to \mathbf{R}$ we can define the composite function $p \circ f$; however $p \circ f$ is only a path on $\mathbf{R}$ if $f \in M$. We therefore restrict f to M and define an action of M on p by

$$(f * p)[v] = p[v] \circ f^{-1}.$$

If p is a path on the finite interval I_p and $f \in M$, we define the interval $I_{f,p}$ by $I_{f,p} = f(I_p)$ and a path $f * p$ on $I_{f,p}$ by

$$(f * p)[v] = p[v] \circ f_p^{-1}.$$

where f_p denotes the restriction of f to I_p. A simple calculation verifies that $*$ is a well-defined action of M on $\mathcal{P}$. By definition, f is a symmetry of p if the class of p is invariant under $*$, ie $<f * p> = <p>$. From the definition of $<>$, it follows that M is the class symmetry group for all paths p.

Now consider geometric symmetry; in curve and surface modelling, translation, rotation and scale transformations are routinely applied to the geometric data sets $v \in V$. Rarely are any other transformations required. In identifying the geometric symmetries of p it is therefore appropriate to restrict attention to the actions of the affine group, and its subgroups, on V. Hence, let g be affine on $\mathbf{R}^2$ and define an action of g on $c : V \times T \to C$ by

$$(g * p)[v, \tau] = gp[g^{-1}v, \tau] \; ; \quad \text{where } c = <p> .$$

The geometric symmetries of c are the affine transformations g that fix the class of c ie

$$<g * p> = <p> .$$

The expression of this in terms of representative paths requires the existence of a class symmetry $\phi_{g,\tau} \in M$ such that

$$gp[v, \tau] = p[gv, \tau] \circ \phi_{g,\tau}.$$

We should note that this generalises the commutativity condition $gp[v] = p[gv]$, derived from the group action $(g * p)[v] = gp[g^{-1}v]$ on the simpler type $c : V \to C$; further, the same geometric interpretation applies - ie invariance of path shape with respect to transformation of v by g.

For the class $c : V \times T(V) \to C$ the domains are coupled by the dependence of τ on v and the appropriate g action is

$$(g * p)[v, \tau] = gp[g^{-1}v, \tau^g]$$

where $\tau^g \in T(V)$ is defined by $\tau^g(v) = \tau(gv)$. The geometric symmetries of c are such that $<g * p> = <p>$, ie all affine g for which there exists a $\phi_{g,\tau} \in M$ with

$$gp[v, \tau] = p[gv, \tau^g] \circ \phi_{g,\tau}.$$

Finally consider parametric symmetry. If $f : \mathbf{R} \to \mathbf{R}$, then f extends to a mapping f on the T domain as $f(\tau) = (f(\tau_0), ..., f(\tau_n))$. Unless f is strictly monotone increasing there will exist a $\tau \in T$ for which $f(\tau) \notin T$, but if $f \in M$ then $f(\tau) \in T$ for all $\tau \in T$; hence transformations of T are restricted to M and the parametric symmetry group of any $c : V \times T \to C$ will be a subgroup of M. Similarly for $T(V)$. Hence for $p \in M$ we define the action

$$(\pi * p)[v, \tau] = p[v, \pi^{-1}(\tau)],$$

and identify the parametric symmetries of c as all π for which

$$<\pi * p> = <p> \, .$$

In terms of representative paths, this is the requirement for a $\phi_\tau \in M$ to exist such that

$$p[v, \tau] = p[v, \pi(\tau)] \circ \phi_\tau$$

for all v and τ.

4   A case study - parametric Lagrange curves

In this section we present symmetry results and their applications for curves defined by Lagrange paths. We consider parametric symmetries first; for Lagrange curves with $\tau = (0, 1, 2, \ldots, n)$ and $\tau^* = (0, 2, 4, \ldots, 2n)$, ie

$$p_L[v, \tau](t) = \sum_{k=0}^{n} \prod_{j=0, j \neq k}^{n} [\frac{t - j}{k - j}] \, v_k \, ; \text{ for } 0 \leq t \leq n$$

and

$$p_L[v, \tau^*](t) = \sum_{k=0}^{n} \prod_{j=0, j \neq k}^{n} [\frac{t - 2j}{2k - 2j}] \, v_k \, ; \text{ for } 0 \leq t \leq 2n$$

the condition $p_L[v, \tau] = p_L[v, \tau^*] \circ \phi$ is satisfied, with $\phi : [0, n] \to [0, 2n]$, defined by $\phi(t) = 2t$, and we have $<p_L[v, \tau]> = <p_L[v, \tau^*]>$. However if $\tau^* = (0, 1, 4, \ldots, n^2)$ then no suitable ϕ exists to relate the paths $p_L[v, \tau]$ and $p_L[v, \tau^*]$. In this case they have distinct graphs. Necessary and sufficient conditions for $<p_L[v, \tau]> = <p_L[v, \tau^*]>$ for Lagrange paths with $\tau, \tau^* \in T$ are given in the following proposition. The result establishes the affine subgroup of M, denoted $A(1)$, as the parameter symmetry group for Lagrange paths.

Proposition 1. *If $\tau, \tau^* \in T$, and $c_L : V \times T \to C$ is the Lagrange curve function; then the paths $p_L[v, \tau]$ and $p_L[v, \tau^*]$ are equivalent for every v if and only if the parameter sets τ and τ^* are affine related; ie if there are real numbers $a(> 0)$ and b, such that $\tau_i^* = a\tau_i + b$ for all i.*

Proof (i) If $\tau_i^* = a\tau_i + b$ then set $\phi(t) = at + b$ to get $p_L[v, \tau] = p_L[v, \tau^*] \circ \phi$ as required.
Conversely: (ii) if $p_L[v, \tau] = p_L[v, \tau^*] \circ \phi$, with $\phi \in M$ then

$$\sum_{k=0}^{n} \prod_{j=0, j \neq k}^{n} [\frac{t - \tau_j}{\tau_k - \tau_j}] \, v_k = \sum_{k=0}^{n} \prod_{j=0, j \neq k}^{n} [\frac{\phi(t) - \tau_j^*}{\tau_k^* - \tau_j^*}] \, v_k$$

for all v and τ. If $D(\tau)$ denotes the Vandermonde matrix

$$D(\tau) = \begin{bmatrix} 1, \tau_0, \ldots, \tau_0^n \\ 1, \tau_1, \ldots, \tau_1^n \\ \vdots \\ 1, \tau_n, \ldots, \tau_n^n \end{bmatrix}$$

then

$$\sum_{k=0}^{n} \prod_{j=0, j \neq k}^{n} \left[\frac{t - \tau_j}{\tau_k - \tau_j}\right] v_k = [1, t, t^2, \ldots, t^n] D^{-1}(\tau) \begin{bmatrix} v_0 \\ v_1 \\ \vdots \\ v_n \end{bmatrix}$$

and we have

$$[1, t, \ldots, t^n] D^{-1}(\tau) \begin{bmatrix} v_0 \\ v_1 \\ \vdots \\ v_n \end{bmatrix} = [1, \phi(t), \ldots, \phi(t)^n] D^{-1}(\tau^*) \begin{bmatrix} v_0 \\ v_1 \\ \vdots \\ v_n \end{bmatrix}$$

for all $v_0, \ldots, v_n$. It follows that

$$[1, t, \ldots, t^n] D^{-1}(\tau) = [1, \phi(t), \ldots, \phi(t)^n] D^{-1}(\tau^*),$$

ie

$$[1, \phi(t), \phi(t)^2, \ldots, \phi(t)^n] = [1, t, t^2, \ldots, t^n] D^{-1}(\tau) D(\tau^*).$$

The matrix $D^{-1}(\tau) D(\tau^*)$ is a function of τ and τ^* only and, by equating components, it follows that $\phi(t)$ is a polynomial in t of degree $\leq n$, $\phi(t)^2$ is a polynomial in t of degree $\leq n$,..., $\phi(t)^n$ is a polynomial in t of degree $\leq n$. In particular $\phi(t)$ is a polynomial in t of degree $\leq n$ and $\phi(t)^n$ is a polynomial in t of degree $\leq n$; clearly if the degree of ϕ is > 1 then the degree of $\phi^n > n$; hence $\phi(t)$ is a polynomial in t of degree ≤ 1; ie ϕ is of the form $\phi(t) = A(\tau, \tau^*)t + B(\tau, \tau^*)$. The coefficients $A(\tau, \tau^*)$ and $B(\tau, \tau^*)$ can be determined by substituting $v_0 = (1, 0, 0)$ and

$$\sum_{k=0}^{n} \prod_{j=0, j \neq k}^{n} \left[\frac{t - \tau_j}{\tau_k - \tau_j}\right] v_k = \sum_{k=0}^{n} \prod_{j=0, j \neq k}^{n} \left[\frac{\phi(t) - \tau_j^*}{\tau_k^* - \tau_j^*}\right] v_k$$

this gives a functional equation

$$\prod_{j=1}^{n} \left[\frac{t - \tau_j}{\tau_k - \tau_j}\right] = \prod_{j=1}^{n} \left[\frac{\phi(t) - \tau_j^*}{\tau_k^* - \tau_j^*}\right] \quad \text{for all} \quad t$$

for ϕ. Substituting $\phi(t) = A(\tau, \tau^*)t + B(\tau, \tau^*)$ and comparing coefficients gives

$$\phi(t) = \left[\frac{\tau_n^* - \tau_0^*}{\tau_n - \tau_0}\right] t + \left[\frac{\tau_0^* \tau_n - \tau_n^* \tau_0}{\tau_n - \tau_0}\right]$$

and

$$\tau_i^* = \phi(\tau_i).$$

As $\tau_n > \tau_0$ and $\tau_n^* > \tau_0^*$ it follows that ϕ has degree 1 and $\tau^* = a\tau + b$, with $a > 0$, as required. The following proposition identifies the geometric symmetries of Lagrange paths with constant parameters as the full affine group of $\mathbf{R}^2$.

Proposition 2. If $\tau, \tau^* \in T$, and $c_L : V \times T \to C$ is the Lagrange curve function then $gp_L[v, \tau]$ and $p_L[gv, \tau]$ are equivalent for every v and τ if and only if g is an affine transformation of $\mathbf{R}^n$.

Proof This follows immediately from the observation that the representative paths of the Lagrange curve function are of the generic form:

$$\sum_{k=0}^{n} b_{n,k}(\tau, t)v_k \quad \text{with} \quad \sum_{k=0}^{n} b_{n,k}(\tau, t) \equiv 1$$

and the following standard result on vector spaces: if W is a vector space over $\mathbf{R}$ and $g : W \to W$, then $g(\sum_i a_i x_i) = \sum_i a_i g(x_i)$, for all finite vector sets $\{x_i\}$ and corresponding scalars $\{a_i\}$ with $\sum_i a_i \equiv 1$ if and only if g is affine.

For $c_L : V \times T \to C$ there is no interaction between the symmetry groups; we now consider the class $c_L : V \times T(V) \to C$, for which this is not the case. The generalisation of proposition 1 to paths on $V \times T(V)$ is as follows:

Proposition 3. If $\tau, \tau^* \in T(V)$, and $c_L : V \times T(V) \to C$ is the Lagrange curve function then; $p_L[v, \tau]$ and $p_L[v, \tau^*]$ are equivalent for every v if and only there exists a positive valued function $a : V \to \mathbf{R}$ and a function $b : V \to \mathbf{R}$, such that $\tau_i^*(v) = a(v)\tau_i(v) + b(v)$ for all i.

Consider now the geometric symmetries of these paths, ie affine g for which

$$gp_L[v, \tau] = p_L[gv, \tau^g] \circ \phi_{g,\tau} , \quad \text{where} \quad \tau^g(v) = \tau(gv).$$

It follows, from proposition 3, that $gp_L[v, \tau] = p_L[gv, \tau^g] \circ \phi_{g,\tau}$ if and only if τ transforms as

$$\tau^g(v) = a(gv)\tau(v) + b(gv).$$

We have shown that G is a geometric symmetry group for Lagrange paths on $V \times T(V)$ if and only if:

(i) G is a subgroup of the affine group $A(2)$ of $\mathbf{R}^2$

and

(ii) there exists a function

$$\pi : G \times V \to A(1)$$

such that

$$\tau(gv) = \pi[g, v]\tau(v).$$

It is easy to show that when such a π exists it must satisfy the conditions for a group cocycle (or multiplier), ie

$$\pi[e, v](t) = t \ , \text{ where } e \text{ is the identity transformation of } G$$

and

$$\pi[gh, v] = \pi[g, hv] \circ \pi[h, v] \quad \text{for all} \quad g, h \in G \text{ and } v \in V.$$

Cocycles occur in classical invariant theory and in the theory of group representations and their applications [4].

The geometric symmetries of these curves therefore depend critically on the parameters τ. The following simple examples show that; for some τ, invariance under the full affine group is imparted to c_L, whilst for others the geometric symmetries of c_L are significantly reduced.

Example 1 constant parameters; we have $\tau(gv) = \tau(v)$ for all affine transformations g of V, the relationship $\tau(gv) = \pi[g, v]\tau(v)$ is satisfied with π trivial, the paths are therefore geometrically invariant under the whole of $A(2)$,

Example 2 parameters $\tau_\alpha, (\alpha \geq 0)$, defined by

$$\tau_{\alpha,0}(v) = 0;$$
$$\tau_{\alpha,i}(v) = \tau_{\alpha,i-1}(v) + \|v_i - v_{i-1}\|_2^\alpha \ ;$$

(a) if g is Euclidean, then $\tau(gv) = \tau(v)$ and the relationship $\tau(gv) = \pi[g, v]\tau(v)$ is satisfied with π trivial, it follows that the paths with these parameters are geometrically invariant under the Euclidean subgroup of $A(2)$,

(b) if g is the uniform-scale transformation $gv = e^\lambda v$ of V, then $\tau^g(v) = e^{\alpha\lambda}\tau(v)$ and the relationship $\tau(gv) = \pi[g, v]\tau(v)$ is satisfied with $\pi[g, v](t) = e^{\alpha\lambda}t$; and the paths are therefore geometrically invariant under uniform-scale transformations,

$\alpha = 0$, gives equal-interval parameters, $\alpha = 1$ gives chord-length and $\alpha = 1/2$, gives centripetal parameters [2],

Example 3 parameters defined by $\tau_{f,i}(v) = f(v)\tau_i(v)$ where $\tau_i(v)$ is chord length and $f(v) > 0$; for g Euclidean or uniform-scale, we have

$$\tau_f(gv) = \frac{f(gv)}{f(v)} \ \tau_f(v)$$

and the paths are geometrically invariant under Euclidean and uniform-scale transformations of $\mathbf{R}^2$,

Example 4 parameters of the form

$$\tau_0(v) = 0,$$
$$\tau_i(v) = \tau_{i-1}(v) + \|v_i - v_{i-1}\|_1$$

where $\|\cdot\|_1$ denotes the L^1 norm; the invariance condition on τ is not satisfied for all rotations, but is satisfied for all translation and uniform-scale transformations. If the L^∞ norm is substituted then different shaped paths with the same geometric invariance properties are produced,

Example 5 parameters defined by

$$\tau_0(v) = 0,$$
$$\tau_i(v) = \tau_{i-1}(v) + d(v_i, v_{i-1}),$$

(a) when $d(v_i, v_{i-1}) = \|v_i - v_{i-1}\|_1/(1 + \|v_i - v_{i-1}\|_1)$; c_L is translation invariant but not invariant under all rotations or scale transformations,

(b) when $d(v_i, v_{i-1}) = \|v_i\|_1 \|v_{i-1}\|_1/(1 + \|v_i\|_1 \|v_{i-1}\|_1)$; c_L is only invariant under a finite subgroup of the rotation group of $\mathbf{R}^2$.

There are a number of other applications of the results; these are given as corollaries without proof.

Corollary 1. *Curves of the general type* $c : V \times T(V) \to C$ *with:*

(a) representative paths of the form

$$\sum_{k=0}^{n} b_{n,k}(\tau, t) v_k$$

and

(b) $A(1)$ *as parameter-invariance group,*

are invariant with parameters $\tau \in T(V)$, *under a subgroup* G *of the affine group of* $\mathbf{R}^2$ *if and only if*

(i)

$$\sum_{k=0}^{n} b_{n,k}(\tau, t) \equiv 1$$

and

(ii) there exists a function $\pi : G \times V \to A(1)$ such that τ satisfies

$$\tau_i(gv) = \pi[g,v]\tau_i(v), \quad \text{for all} \quad g, v, \text{ and } i.$$

Corollary 1 is applicable to many common curve functions - including splines and Lagrange curves; it provides a simple test for geometric invariance of curve functions and generalises a well-known result [5], ie that curves of the simpler type

$$\sum_{k=0}^{n} b_{n,k}(t)v_k$$

are affine invariant if and only if

$$\sum_{k=0}^{n} b_{n,k}(t) \equiv 1.$$

The results can be applied in other ways; for example if $\tau, \tau^* \in T(V)$, then combinations of τ and τ^* provide new parameters. The arithmetic mean $(\tau + \tau^*)/2$, or convex blends $(1-\lambda)\tau + \lambda\tau^*$ (for $0 < \lambda < 1$), or the geometric mean $(\tau\tau^*)^{1/2}$ (when $\tau, \tau^* > 0$) can be used [2].

Corollary 2. *For any curve function c of the type in corollary 1; if c is invariant under G at τ and τ^* then:*

(i) c is invariant under G at the geometric mean of τ and τ^*

(ii) c is invariant under convex combinations of τ and τ^* if and only if τ and τ^* have the same group multiplier.

Examples If τ is equal-interval and τ^* is chord-length, then τ and τ^* define curves invariant with respect to uniform-scale, but the combination $(\tau+\tau^*)/2$ does not; the combination $(\tau + \tau^*/\tau_n^*)/2$ does give scale-invariant curves, as does any combination of the form $(\tau + \tau^*/D)/2$ where D satisfies $D(e^\lambda v) = e^\lambda D(v)$. However the parametrisations $(\tau + \tau^*/\tau_n^*)/2$ and $(\tau + \tau^*/D)/2$ are not, in general, affinely related and do not therefore produce the same curves.

5   Summary and conclusions

It has been shown that symmetry groups may be associated with the V and T domains of curve functions of the class $c : V \times T \to C$. It has been demonstrated that a full understanding of these symmetries provides insight into the behaviour of the curve functions with respect to modification of the parametrisation and transformation of the geometric input data. For the more general class $c : V \times T(V) \to C$ it is clear that the components of the domain are closely coupled. It has been shown in the paper that this coupling manifests itself as a coupling between the symmetry groups of the V and T domains by means of group cocycles. This relationship provides key information concerning the behaviour of curve functions of this general class with respect to geometric transformations.A number of other ways in which the results can be used have been presented. Lagrange paths provided a case study, but the methods developed are quite general and may be applied to: arbitrary curve functions on the domains $V \times T$ and $V \times T(V)$, curve functions on any domain of the form $V \times W$, and surface functions. Proposition 1 provides the foundation on which the symmetry ideas can be effectively applied to parametric Lagrange paths. A similar result can be proved for some parametric spline curves. Currently, alternative ways of computing the parametric symmetry groups for other curve types are under investigation. It is hoped that general theorems can be developed for this purpose - to avoid detailed examination of each type. A deeper study of the role of cocycles and their applications to curve and surface functions is also in progress.

References

1. Epstein M. P. (1976) On the influence of parametrisation in parametric interpolation. SIAM J. Num. Analysis. **13**, 261–268
2. Lee E. T. Y. (1989) Choosing nodes in parametric curve interpolation. Computer Aided Design. **21**, 363–370
3. Neumann P. M., Stoy G. A. and Thompson E. C. (1994) Groups and geometry. Oxford Science Publications, Oxford University Press
4. Olver P. J. (1999) Classical invariant theory. London Mathematical Society, Student Texts **44**, Cambridge University Press
5. Bez H. E. (1989) An analysis of invariant curves. Computer Aided Geometric Design. **6**, 265–277

Functions and Methods to Analyze and Construct Developable Hull Surfaces

Salvatore Miranda, Claudio Pensa, and Fabrizio Sessa

Università degli Studi di Napoli "Federico II", Dipartimento di Ingegneria Navale, Via Claudio 21, 80125 Naples, Italy.

Summary. In this paper the authors present a way to study the analytical properties of a ship hull surface in order to modify it in a new one simpler to build for a shipyard but that also preserves some properties of the original surface.

The authors present a real case study of a ship with the functions used to modify the surface and with the hydrodynamic performances of the new ship.

Even if this is a "trial run" developed by the Dipartimento di Ingegneria Navale of Naples, the results are very interesting. However the main aim of this work is to optimize a numerical procedure for the above transformations.

For this reason the authors present the "tools" used to modify the surface in an algorithmic way so they can be easily implemented in a CAD software.

1 Introduction

The increasing competition in the shipbuilding industry requires the construction of ships with minimized production costs and times and obtaining, in the same time, good quality products. However many shipyards have to build hull surfaces of non trivial technological realization due, for example, to an high percentage of non developable surfaces – in the critical zones of the bow and the stern – that require very high construction times with an increase of the above costs. On the contrary a developable surface can be easily formed by bending or rolling a planar surface without stretching or tearing. In other words it can be unrolled isometrically into a plane.

For this reason developable surfaces are of big interest in Naval Architecture so they have been widely treated in literature and some advantages in using them have been presented, for example, by Norsko-Laurisen [1] in his work developed at the Burmeister Shipyard.

Among the others, Hoschek [2], Maekawa [3], Pottmann [4] have written high quality papers about developable surfaces giving also useful techniques to design or to approximate a surface with a developable one. Particularly, Randrup's PhD thesis [5] is a very interesting work because it shows also a new approach to plates manufacturing.

However, in the present work the authors treat the most common case in shipbuilding problems: they have a preliminary ship project that satisfies technical and hydrodynamical requests and must be changed into a new ship with an higher percentage of developable surface but with similar performance characteristics.

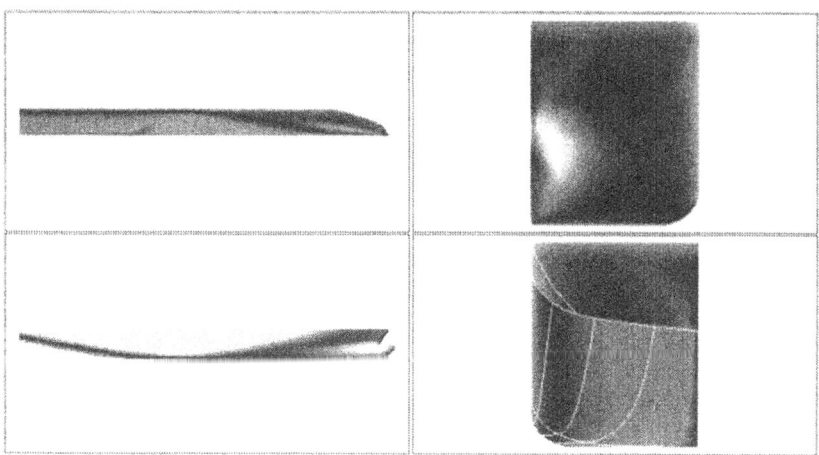

Fig. 1. The original ship surface

Therefore the paper shows the main points of this transformation:

- to find the zones of the ship surface that must be transformed;
- to transform these zones into a new ones that can be easily built by the shipyard;
- to modify globally the final surface in order to obtain the desired design requests, retaining an easy to build geometry.

2 Analysis of the original ship surface

The original ship surface S is described in Fig. 1 while the main dimensions of S are reported in Table 1. S is represented, in the CAD system used for this work, as a NURBS surface with 15×21 (transverse×longitudinal) control points.

Table 1. Main dimensions of the ship surface S

Length	185.8 m
Width (on half ship)	12.6 m
Height	15.0 m
Area (on half ship)	4150 m²

The ship is supposed to be in a frame of reference where the X axis is arranged in the longitudinal way, the Y axis in the transverse way while the

Table 2. Ship surface subdivision

Zone	Lenght (projection over X axis)	Area (percentage)
A	39.2 m	780 m² (19%)
B	139.3 m	3250 m² (80%)
C	7.3 m	80 m² (1%)

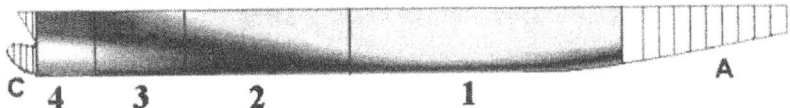

Fig. 2. Surface subdivision

Z axis is vertical. S can be divided into 3 zones: A, B and C, described in Table 2 and shown in Fig. 2.

The authors do not work on A because it can be considered completely developable (a cylinder) and has been obtained in a previous work as explained in [6].

Similarly the C zone cannot be optimized only for technological motivations because it has a very high influence on the hydrodynamical performances of the ship so it will be left unchanged.

The transformation will affect only the zone B that can be divided into 4 parts with similar values of the Gaussian curvature K.

For each part, the gaussian curvature of the surface has been calculated and represented by means of a colour code plot. These plots help to find the developable surface patches characterized by the $K = 0$ condition.

However this condition is too restrictive because a shipyard can assume "developable" also a surface where $0 < K < \varepsilon$. In fact, these ones can be modelled with a fast and low cost manufacture of the steel plates during their final arrangement over the ship structure.

So it is very important to choose a "technologically realistic" ε that also depends by the shipyard machinery and staff and can considerably modify the result. In [3] a similar problem has been illustrated and the authors have chosen a value of $\varepsilon = 1.0 \times 10^{-5}$ but, unfortunately, without giving further technical motivations.

Here the authors have chosen an higher tolerance of $\varepsilon = 0.5 \times 10^{-3}$ but, to assume a surface with $0 < K < \varepsilon$ "easy to build", they have also requested another condition: an "almost" constant principal curvature direction. This constraint reduce the K importance and assures that it is not necessary to modify many times the plates displacement during the bending process or to

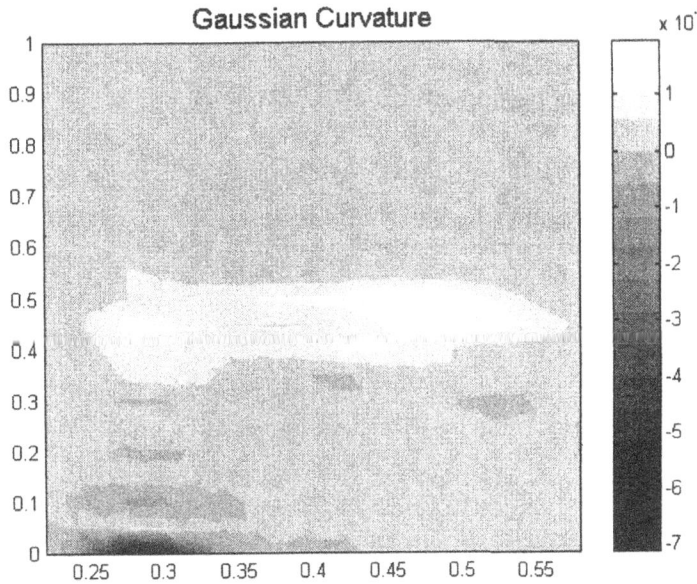

Fig. 3. Gaussian curvature of the B.1 zone

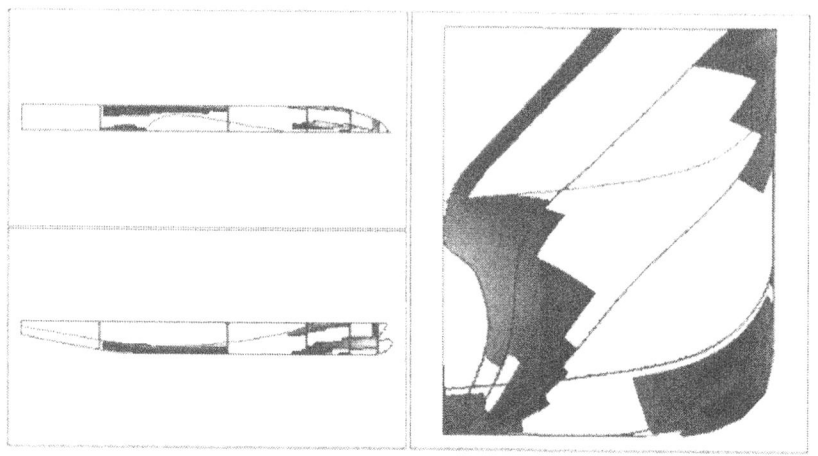

Fig. 4. Non–developable surface patches

make many different steel patches with an increase of the production costs. Besides, increasing the ε tollerance, the analysis gives results not affected by unwanted little bumps which cannot be avoided in a real ship.

In Fig. 3 is reported the K plot for the B.1 zone. The axes are the parameters that characterize the zone.

In Fig. 4 is shown the non–developable patches according to the constraint $K > \varepsilon = 0.5 \times 10^{-3}$.

In Table 3 is given the characterization of the non–developable patches of the B.2 (Area: $900\,\mathrm{m}^2$) zone obtained with the tolerances $K > \varepsilon = 1.0 \times 10^{-5}$ and $K > \varepsilon = 0.5 \times 10^{-3}$.

Table 3. Results obtained with a different tolerance value for the B.2 zone

Surface with $K > \varepsilon = 1.0 \times 10^{-5}$	35%
Surface with $K > \varepsilon = 0.5 \times 10^{-3}$	10%

The Table 4 gives the final characterization of the half ship surface.

Table 4. Ship surface characterization

Surface with $0 < K < \varepsilon = 0.5 \times 10^{-3}$ with a constant principal curvature direction	51%
Surface with $0 < K < \varepsilon = 0.5 \times 10^{-3}$ without a constant principal curvature direction	31%
Surface with $K > \varepsilon = 0.5 \times 10^{-3}$	18%

3   Transformation technique

The transformation of the ship surface has been realized in the following way:

- calculation of the ship form coefficients;
- detection of an opportune water line WL1;
- detection (by Naval Architecture criterions) of an altitude up to which extruding WL1, in order to have a cylindrical surface with a top boundary that is a new water line WL2;
- construction of a ruled surface between LW2 ad the top boundary of the hull;
- construction of a surface to fill the knee;
- construction of a cylindrical surface to connect the bulb with the hull;
- calculation of the new form coefficients;

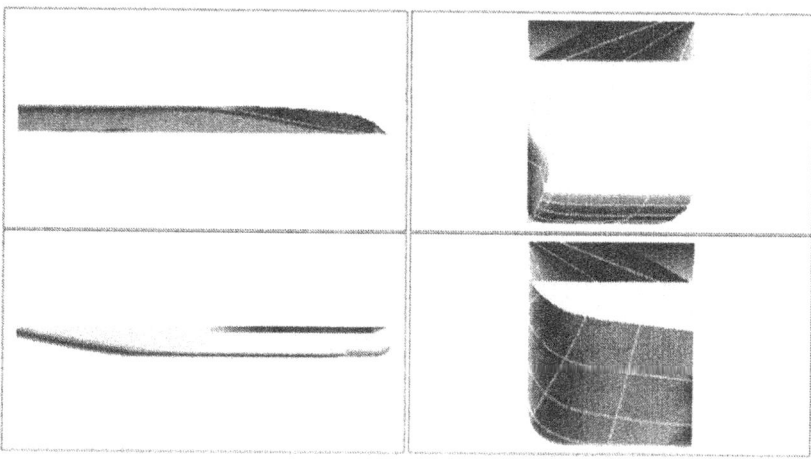

Fig. 5. The new ship

- modification of the hull surface with the Lackemby's transformation [7], in order to have form coefficients similar to the old ones.

This process has been implemented at the Dipartimento di Ingegneria Navale of Naples using the ThinkDesign CAD system according to the following procedure:

- developable and non–developable areas calculation according to the given tolerance ε;
- water lines WL1 and WL2 construction;
- surfaces generation;
- bulb attachment;
- Lackemby's transformation.

In order to follow these steps the authors have developed some applications (written in Visual C++) for the ThinkDesign CAD:

- SG: surface grid generation
- SE: surface curvature evaluation over the grid
- LA: Lackemby's transformation

4   The new ship

The new ship surface S is described in Fig. 5 while its main dimensions are similar to the old ones.

S can be divided into 7 (see Fig. 6) zones: A, B, C, D, E, F, G of which only the zones C and E require an analysis similar to the one previously done for the initial ship, because:

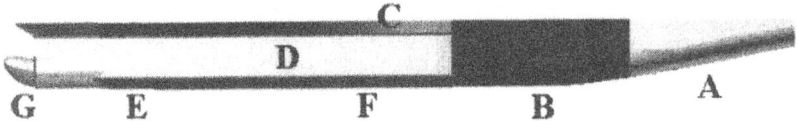

Fig. 6. Surface subdivision

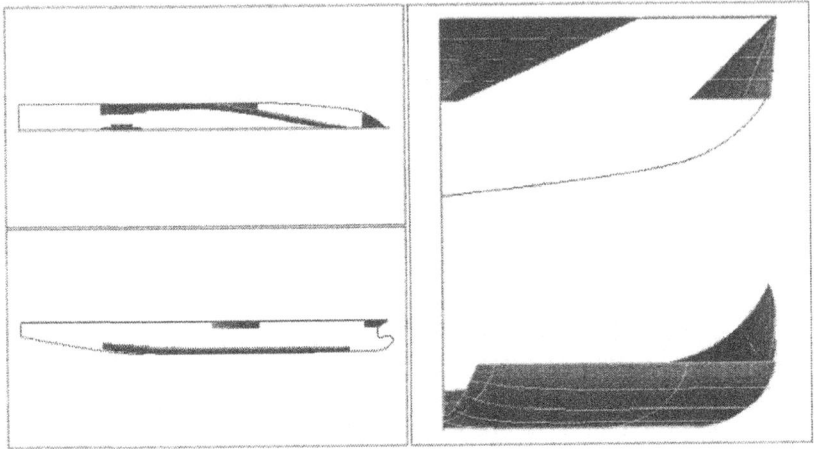

Fig. 7. Non–developable surface patches

- A and B are unchanged;
- B, D and F are completely developable (cylinders);
- G is not developable at all and coincides with the original bulb.

By the curvature plots and doing the same analysis of the principal curvatures, it is possible to find the non–developable zones of the new ship (see Fig. 7).

A final modification of the ship bow with a conical surface gives the hull characterization described in Table 5.

5 Results analysis

The Table 6 summarizes the differences between the two ships.

As the Table 6 shows, the performed process guarantees an important increment of the pure developable surface of the ship hull minimizing the surfaces with $K < \varepsilon$ but without a constant principal curvature direction. To evaluate the real differences it has to be said that the old ship already had an high percentage of developable surfaces because modified according to [6].

The Table 6 shows an increase of the 17% for the "simple" developable surface. This, on a 8300m^2 total surface, means about 1400m^2 of simpler surface.

Table 5. Ship surface characterization

Surface with $0 < K < \varepsilon = 0.5 \times 10^{-3}$ with a constant principal curvature direction	68%
Surface with $0 < K < \varepsilon = 0.5 \times 10^{-3}$ without a constant principal curvature direction	14%
Surface with $K > \varepsilon = 0.5 \times 10^{-3}$	18%

Table 6. Differences between the two ships

	Old Ship	New Ship
Surface with $0 < K < \varepsilon = 0.5 \times 10^{-3}$ with a constant principal curvature direction	51%	68%
Surface with $0 < K < \varepsilon = 0.5 \times 10^{-3}$ without a constant principal curvature direction	31%	14%
Surface with $K > \varepsilon = 0.5 \times 10^{-3}$	18%	18%

The tolerance ε used in the surface analysis can be fixed by the shipyard staff according to the machinery available on site.

A unique type of developable surface is used in the proposed transformation technique. Therefore the necessity of building different developable patches to be merged later is avoided, minimizing the costs.

Besides, the application of Lackemby's process allows to mantain the ship form coefficients similar to the original ones. Previous works on this topic give some algorithms to approximate the ship surface with a developable one but they do not assure the form coefficient preservation and the obtained ship can have a different behaviour.

Finally the transformation process should have also changed the longitudinal position of the bulb respect to the forward perpendicular (the extreme bow of the ship). Although this hydrodynamical optimization is not presented here, it is very easy to achieve the desired position of the bulb without a significant modification of the other surfaces because of the cylinder used to connect the bulb with the hull.

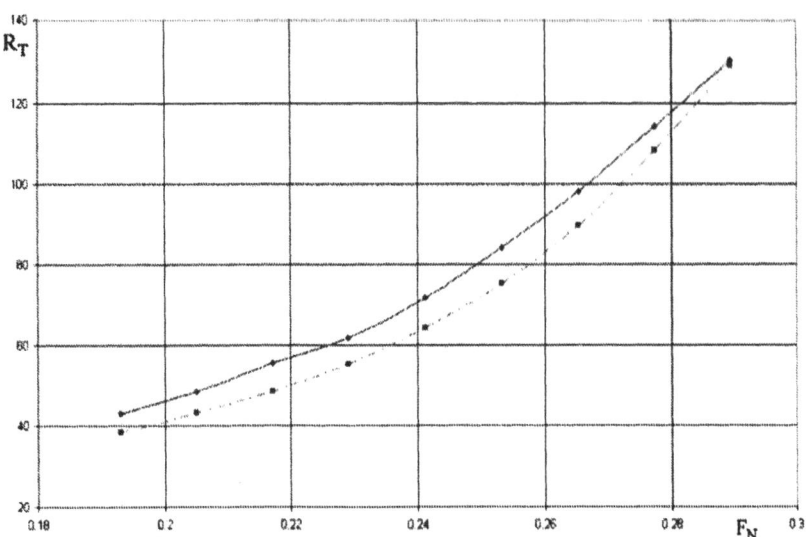

Fig. 8. Total hull resistance (Full - new hull; Dash dotted - old hull)

The Fig. 8 shows the experimental values for the total hull resistance of both the ships.

6   Conclusions and future research

The paper shows the study started at the Dipartimento di Ingegneria Navale of Naples on developable surfaces and their applications to ship hulls. The aim of this research is to build a numerical procedure that can be easily implemented and used by a shipyard. For this reason the transformation process uses simple surfaces and analytical tools well suited to the shipyard reality. The next steps of the research will be conducted in co–operation with Fincanieri, the greatest italian shipyard. The main problems will regard the study of a better definition for the tolerance ε of the gaussian curvature knowing that, according to Fincantieri's experience, an usual steel sheet (about 5 meters long), can be easily forced to assume non developable forms if the modification is up to 5 cm.

Besides, it will be studied the real influence on the production costs of the manufacture of non developable patches. Unfortunately only few data are available on this topic [1], but according to Fincantieri's experience the costs can be considerable. In fact, Fincantieri has built special bulk ships with only developable patches and the realization times where of about 243000 working hours against 367000 working hours needed for the same ship but with traditional forms. This is due to the fact that, for example, a "difficult–to–realize" patch 5 meters long may need up to 5 days of 2 men works to be realized against the few hours for a developable one.

References

1. O. Norsko-Laurisen (1985) Practical application of single curved hull definition background, application, software and experience. In: P. Banda, C. Kuo. (Eds.) Computer Applications in the Automation of Shipyard Operation and Design, Volume 11, Elsevier Science, North Holland, 485-491.

2. Hoschek J., Schneider M. (1997) Interpolation and Approximation with Developable Surfaces. In: Le Mhaut, C. Rabut, Schumaker L. (Eds.) Curves and Surfaces with Applications in CAGD, Vanderbilt University Press, Nashville, TN, 185-202.

3. Maekawa T., Chalfant J., (1998) Design for Manufacturing Using B-Spline Developable Surfaces, Journal of Ship Research, 3:207-215.

4. Pottmann H., Farin G. E., (1995) Developable rational Bezier Bspline Surfaces, Computer Aided Geometric Design, 12:513-531.

5. Randrup T., (1988) Reverse Engineering Methodology to Recognize and Enhance Manufacturability of Surfaces - with Examples from Shipbuilding, PhD. Thesis, TU Denmark.

6. Fiorentino A., Miranda S., Pensa C., (1997) Esperienza di rimorchio Carena C9703 (in Italian), Technical Report, Università degli Studi di Napoli "Federico II", Dipartimento di Ingegneria Navale, Naples.

7. Lackenby H., (1950) On the Systematic Geometrical Variation of Ship Forms, Transactions of The Institute of Naval Architects, 92:289-316

Bipolar and Multipolar Coordinates

Rida T. Farouki and Hwan Pyo Moon

Department of Mechanical and Aeronautical Engineering,
University of California, Davis, CA 95616, USA

Summary. Bipolar or multipolar coordinates offer useful insights and advantages over Cartesian coordinates in certain geometrical problems. In bipolar coordinates (r_1, r_2) the "simplest" curves are the conics, ovals of Cassini, Cartesian ovals, and their special cases, which are characterized by linear or hyperbolic relations in the (r_1, r_2) plane. As a natural extension of these classical examples, we consider the full range of curves characterized by conic (r_1, r_2) loci. A further useful generalization involves the extension of the curve equations to (redundant) multipolar coordinates $(r_1, \ldots, r_n)$, taking the n-th roots of unity as "canonical" poles. We survey two key applications of these methods, in geometrical optics and the Minkowski geometric algebra of complex sets, and explore the formulation of geometric design schemes using planar and spatial bipolar or multipolar coordinates.

1 Introduction

The design and analysis of free–form shapes is usually based upon parametric or implicit curve and surface equations, expressed in Cartesian coordinates. For certain applications, it may be advantageous to use polar coordinates in the plane [21,24,25] or spherical or cylindrical polar coordinates in space. In the description and analysis of curves and surfaces, the choice of coordinate system can play a key role in simplifying definitions, formulating algorithms, and revealing symmetries, characteristic dimensions, and other features.

In ordinary polar coordinates, a point is specified by its distance from a fixed "pole" or origin, and one or more angles measured with respect to fixed axes through this origin. Thus, unlike Cartesian coordinates, ordinary polar coordinates are dimensionally inhomogeneous. Alternately, we may dispense with angular variables and specify a point by its distance from *several* poles.

In two dimensions, for example, the distances r_1, r_2 from two distinct poles can serve to (almost) uniquely specify each point in the plane. Although they have received scant attention in computer geometry, such *bipolar coordinates* [17,18] provide very attractive descriptions and algorithms for certain types of geometrical problems. In a space of dimension $n \geq 3$, we require at least n distinct poles to define multipolar coordinates. However, we need not confine ourselves to just n poles in n–dimensional space — in the plane, for example, it may be advantageous in some problems (see §3) to use a *redundant* system of multipolar coordinates, based on $n \geq 3$ distinct poles.

Our goal in this paper is to survey the basic properties and applications of bipolar and multipolar coordinates. We begin by defining bipolar coordinates and identifying the constraints they satisfy in §2. The "simplest" equations in bipolar coordinates define some well-known classical curves — namely, the ellipse/hyperbola, ovals of Cassini, circles of Apollonius, and Cartesian ovals. These curves may be regarded as "bipolar conics" — i.e., the loci defined by a quadratic implicit equation $f(r_1, r_2) = 0$ — and we give examples of such curves that have not been systematically studied before. In §3 we show that all these curves admit intuitive multipolar generalizations, based on the n-th roots of unity as "natural" poles for an n-polar coordinate system. Two key applications of bipolar/multipolar forms, geometrical optics and Minkowski geometric algebra of complex sets, are then sketched in §4 and §5. Finally, §6 briefly discusses algorithms for designing free-form bipolar geometries, and §7 indicates some promising avenues for further research.

2 Bipolar coordinates

It is well known that if p_1, p_2 are distinct points in the plane, the locus traced by a moving point whose distances r_1, r_2 from those points satisfy

$$r_1 + r_2 = k \tag{1}$$

is, for any positive constant $k > |p_2 - p_1|$, an *ellipse* with foci p_1 and p_2.

The values (r_1, r_2) define *bipolar coordinates* in the plane, with respect to the *poles* p_1 and p_2. Without loss of generality we may choose[1] $p_1 = (+1, 0)$ and $p_2 = (-1, 0)$: the bipolar coordinates are then non-negative values that must satisfy (see Figure 1) the constraints

$$r_1 + r_2 \geq 2 \quad \text{and} \quad |r_1 - r_2| \leq 2. \tag{2}$$

Note that these constraints are satisfied with *equality* by points on the x-axis (the former for $-1 \leq x \leq +1$, and the latter for $x \leq -1$ or $x \geq +1$).

For the chosen poles we obtain, in terms of "ordinary" polar coordinates (r, θ) about the origin,

$$r_1 = \sqrt{r^2 - 2r \cos \theta + 1} \quad \text{and} \quad r_2 = \sqrt{r^2 + 2r \cos \theta + 1}. \tag{3}$$

Solving these relations for (r, θ) in terms of (r_1, r_2), we have

$$r = \sqrt{\frac{r_1^2 + r_2^2 - 2}{2}} \quad \text{and} \quad \theta = \cos^{-1} \frac{r_2^2 - r_1^2}{\sqrt{8(r_1^2 + r_2^2 - 2)}}. \tag{4}$$

Note that a bipolar coordinate pair (r_1, r_2) actually specifies *two* points, that are images of each other under reflection in the x-axis. A curve described in

---

[1] Note, however, that distinct finite poles preclude limiting cases where one pole moves to infinity (the parabola) or the two poles coincide (the limaçon of Pascal).

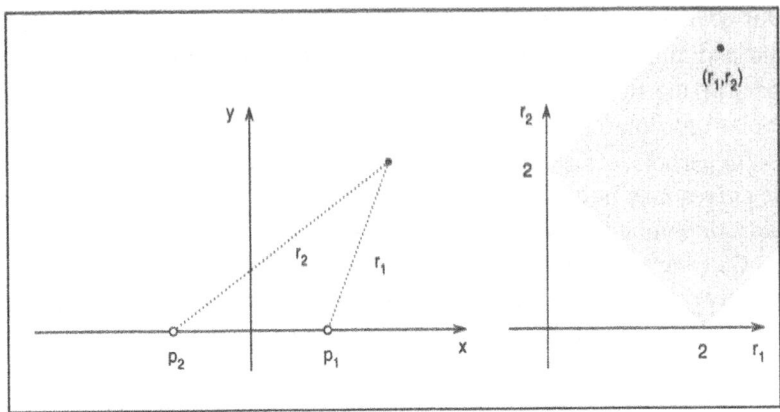

Fig. 1. Bipolar coordinates (r_1, r_2) with respect to the two poles $\mathbf{p}_1 = (+1, 0)$ and $\mathbf{p}_2 = (-1, 0)$ — the region of valid (r_1, r_2) values is shown shaded on the right.

bipolar coordinates is therefore symmetric about the x–axis, unless separate equations are specified for the lower and upper half–planes.

From (4) we can easily obtain the transformation from bipolar coordinates (r_1, r_2) to Cartesian coordinates $(x, y) = (r \cos \theta, r \sin \theta)$ as

$$(x, y) = \left(\frac{r_2^2 - r_1^2}{4}, \pm \frac{\sqrt{8(r_1^2 + r_2^2 - 2) - (r_1^2 - r_2^2)^2}}{4} \right). \tag{5}$$

Thus, for a parametric curve $(r_1(t), r_2(t))$ specified in bipolar coordinates, we can easily determine the tangent, curvature, etc., by substituting $r_1(t)$, $r_2(t)$ in the above expression, and differentiating with respect to the parameter t. Conversely, for a curve defined by an implicit (polynomial) equation

$$f(r_1, r_2) = 0$$

in bipolar coordinates, we can invoke the methods of algebraic geometry on substituting $r_1 = \sqrt{x^2 + y^2 - 2x + 1}$ and $r_2 = \sqrt{x^2 + y^2 + 2x + 1}$ into f, and repeatedly squaring the resulting expression to eliminate all radicals.

2.1   Ellipse and hyperbola

By re–arranging terms and squaring twice, equation (1) can be written as

$$(r_1^2 - r_2^2)^2 - 2k^2(r_1^2 + r_2^2) + k^4 = 0. \tag{6}$$

Actually, this equation describes not only (1), but all the loci defined by

$$r_1 \pm r_2 = \pm k. \tag{7}$$

For any $k < |\mathbf{p}_2 - \mathbf{p}_1|$, the loci $r_1 - r_2 = \pm k$ (characterized by the fact that the *difference*, rather than the sum, of the bipolar coordinates is constant) define

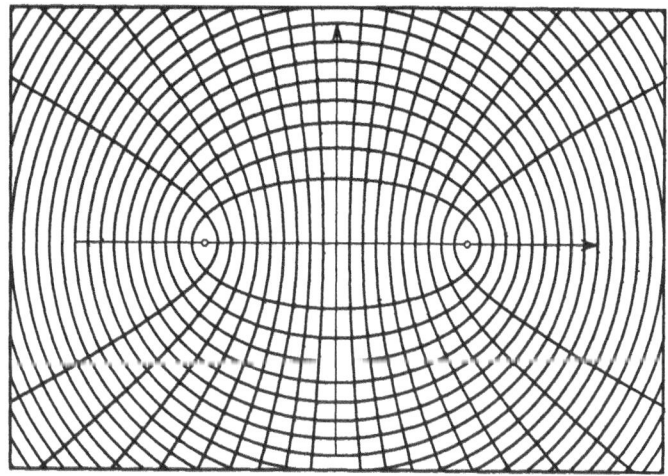

Fig. 2. The confocal family of conics (ellipses $r_1+r_2 = k$ with $k > 2$, and hyperbolas $r_1 - r_2 = \pm k$ with $k < 2$) defined by the bipolar equations (7).

the two branches of a hyperbola with foci $\mathbf{p}_1, \mathbf{p}_2$. The equation $r_1 + r_2 = -k$, on the other hand, defines a vacuous real locus for any $k > 0$.

Substituting from (3) into (6) and simplifying gives the polar equation

$$(16\cos^2\theta - 4k^2)r^2 + k^4 - 4k^2 = 0. \tag{8}$$

If $k > 2$, this gives real r values for each θ, and the curve is an ellipse. For $k < 2$, however, real r values occur only for $-\cos^{-1}\frac{1}{2}k < \theta < +\cos^{-1}\frac{1}{2}k$ and $\pi - \cos^{-1}\frac{1}{2}k < \theta < \pi + \cos^{-1}\frac{1}{2}k$, and the curve is a hyperbola. In terms of Cartesian coordinates $(x,y) = (r\cos\theta, r\sin\theta)$, equation (8) becomes

$$\frac{x^2}{k^2} + \frac{y^2}{k^2 - 4} = \frac{1}{4}. \tag{9}$$

The family of ellipses/hyperbolas defined by (8) and (9) is shown in Figure 2.

The ellipse can be traced using a simple mechanical device. We press two pins into a sheet of paper at the foci $\mathbf{p}_1, \mathbf{p}_2$ and wrap a loop of string, of total length $\ell > 2|\mathbf{p}_2 - \mathbf{p}_1|$, around them. The point of a pencil that pulls the loop taut as it moves will then trace out the ellipse (1), with $k = \ell - |\mathbf{p}_2 - \mathbf{p}_1|$. In §3 we generalize both this "mechanical" definition of an ellipse and the analytical definition (1) to cases with $n \geq 3$ foci: we shall find that these two generalizations yield quite different curves.

2.2   Ovals of Cassini, circles of Apollonius

Suppose we now require the *product* of the bipolar coordinates (r_1, r_2), rather than their sum or difference, to remain constant:

$$r_1 r_2 = k. \tag{10}$$

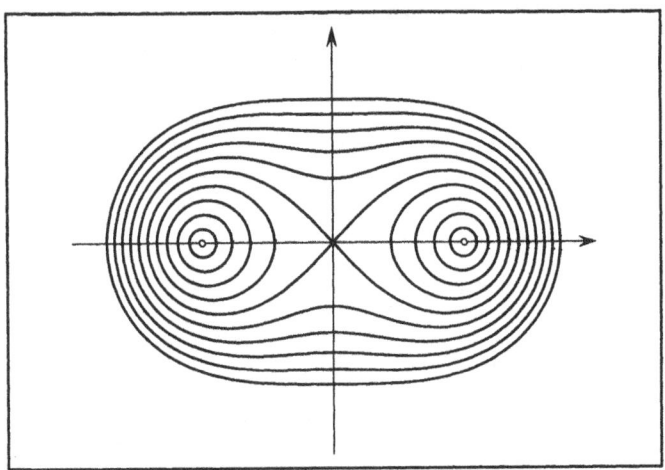

Fig. 3. The ovals of Cassini (10) for values $k = 0.2, 0.4, \ldots, 2.0$. The curve comprises one or two loops according to whether $k > 1$ or $k < 1$, while the singular case $k = 1$ (the "figure–of–eight" curve) is a lemniscate of Bernoulli.

The curve thus defined is known as the *ovals of Cassini*. Substituting from (3) and squaring, we obtain the implicit equation

$$r^4 - 2r^2 \cos 2\theta + 1 = k^2 \tag{11}$$

in ordinary polar coordinates (r, θ). When $k > 1$, this equation admits real solutions r for each angle θ, and the curve consists of a single loop containing both the poles $\mathbf{p}_1, \mathbf{p}_2$. When $k < 1$, however, real r values are obtained only for $-\frac{1}{2} \sin^{-1} k \leq \theta \leq +\frac{1}{2} \sin^{-1} k$ and $\pi - \frac{1}{2} \sin^{-1} k \leq \theta \leq \pi + \frac{1}{2} \sin^{-1} k$, and the curve comprises two disjoint loops containing the poles $\mathbf{p}_1, \mathbf{p}_2$ individually. In the exceptional case $k = 1$, the two loops pinch together at the origin to yield a "figure–of–eight" curve. This case, in which equation (11) simplifies to give $r = \sqrt{2 \cos 2\theta}$, is called the *lemniscate of Bernoulli*.

Figure 3 illustrates the three morphologically distinct forms for the ovals of Cassini. In Cartesian coordinates, the polar equation (11) becomes

$$(x^2 + y^2)^2 - 2x^2 + 2y^2 - k^2 + 1 = 0.$$

The ovals of Cassini[2] were called the *spiric sections of Perseus* by the ancient Greeks, who recognized them to be planar sections of a torus. The arc length of the lemniscate can be described [16] in terms of the integral

$$F(x) = \int_0^x \frac{dt}{\sqrt{1 - t^4}},$$

which played a key role in the historical development of the theory of elliptic functions by Gauss, Abel, and Jacobi [30].

---

[2] Cassini, the discoverer of a gap in Saturn's rings, rejected Kepler's description of planetary orbits as ellipses — he proposed orbits described by the curves (11).

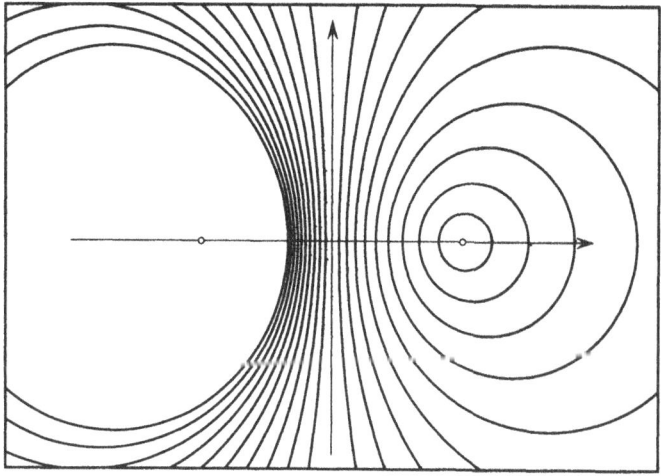

Fig. 4. Circles of Apollonius (12) for values $k = 0.1, 0.2, \ldots, 0.9$ (in the right half of the plane) and for $k = 1.1, 1.2, \ldots, 2.0$ (in the left half of the plane).

Finally, having considered loci defined by a constant sum, difference, and product of bipolar coordinates, we now set their *ratio* equal to a constant:

$$\frac{r_1}{r_2} = k. \tag{12}$$

We require $k > 0$ here, since r_1 and r_2 are positive by definition. Substituting from (3), clearing radicals, and setting $c = (k^2 + 1)/(k^2 - 1)$, this gives

$$r^2 + 2cr \cos\theta + 1 = 0$$

or, in Cartesian coordinates,

$$(x + c)^2 + y^2 = a^2$$

where $a = 2k/|k^2 - 1|$. For any $k > 0$, this defines a *circle of Apollonius*. For $k < 1$ the circle contains the pole $\mathbf{p}_1 = (+1, 0)$, while for $k > 1$ it contains $\mathbf{p}_2 = (-1, 0)$. In the degenerate case $k = 1$, the circle is of infinite radius: it is the perpendicular bisector of $\mathbf{p}_1$ and $\mathbf{p}_2$, i.e., the y–axis (see Figure 4).

2.3 Cartesian ovals

To further generalize the bipolar ellipse equation (1), we must introduce some new parameters. Suppose, for example, we choose (positive) proportionality constants m and n for r_1 and r_2, and write

$$mr_1 + nr_2 = 1 \tag{13}$$

(we set $k = 1$ on the right–hand side, since any value $k \neq 1$ can be absorbed into m and n). This equation defines a *Cartesian oval*.

m and n	Cartesian oval equations
$m < n < \frac{1}{2}$ or $m < \frac{1}{2} < n$	$mr_1 + nr_2 = +1$, $mr_1 - nr_2 = -1$
$n < m < \frac{1}{2}$ or $n < \frac{1}{2} < m$	$mr_1 + nr_2 = +1$, $mr_1 - nr_2 = +1$
$m > \frac{1}{2}$ and $n > \frac{1}{2}$	$mr_1 - nr_2 = +1$, $mr_1 - nr_2 = -1$

Table 1. The appropriate members from equations (15) defining the two loops of a Cartesian oval for various m and n values (with $m \neq n$ and $m, n \neq \frac{1}{2}$).

By squaring twice and re–arranging, we can re–write (13) as an equation involving only r_1^2 and r_2^2:

$$(m^2 r_1^2 - n^2 r_2^2)^2 - 2(m^2 r_1^2 + n^2 r_2^2) + 1 = 0.$$

Substituting from (3) then gives the equation

$$(\alpha r^2 - 2\beta r \cos\theta + \alpha)^2 - 2(\beta r^2 - 2\alpha r \cos\theta + \beta) + 1 = 0 \qquad (14)$$

in ordinary polar coordinates (r, θ), with $\alpha = m^2 - n^2$, $\beta = m^2 + n^2$. As with (8), the above equation describes not only (13), but all the loci defined by

$$mr_1 \pm nr_2 = \pm 1. \qquad (15)$$

Now in the (r_1, r_2) plane, equations (15) define four lines passing through the points $(\pm 1/m, 0)$ and $(0, \pm 1/n)$. When m and n are positive, $mr_1 + nr_2 = -1$ obviously defines a vacuous locus. Of the remaining three lines, one can easily see (Figure 5) that only two possess segments within the valid domain (2) for bipolar coordinates. In general, the Cartesian oval comprises two nested loops — depending upon m and n, the equations from (15) that individually define these loops are identified in Table 1.

In Cartesian coordinates, equation (14) becomes

$$(\alpha x^2 + \alpha y^2 - 2\beta x + \alpha)^2 - 2(\beta x^2 + \beta y^2 - 2\alpha x + \beta) + 1 = 0.$$

Ordinarily, this quartic curve has a double point at each of the circular points at infinity but no other singular points, and is therefore of genus 1. However, if $m = \frac{1}{2}$ or $n = \frac{1}{2}$, one of the poles $(-1, 0)$ or $(+1, 0)$ is also a double point. This special (rational) form is called a *limaçon of Pascal*.

Now equations (11) and (14) are both quartic in r. However, the former is more convenient for curve plotting, since it is simply a *biquadratic* equation. We can obtain a simpler form than (14) by choosing poles at $\mathbf{p}_1 = (2, 0)$ and $\mathbf{p}_2 = (0, 0)$, so that $r_1 = \sqrt{r^2 - 4r\cos\theta + 4}$ and $r_2 = r$, instead of the values (3) associated with the "standard" poles $\mathbf{p}_1 = (+1, 0)$ and $\mathbf{p}_2 = (-1, 0)$. This corresponds to a shift by $\Delta x = 1$ along the x–axis. The polar equation

$$(n^2 - m^2)r^2 + 2(2m^2 \cos\theta \pm n)r - 4m^2 + 1 = 0 \qquad (16)$$

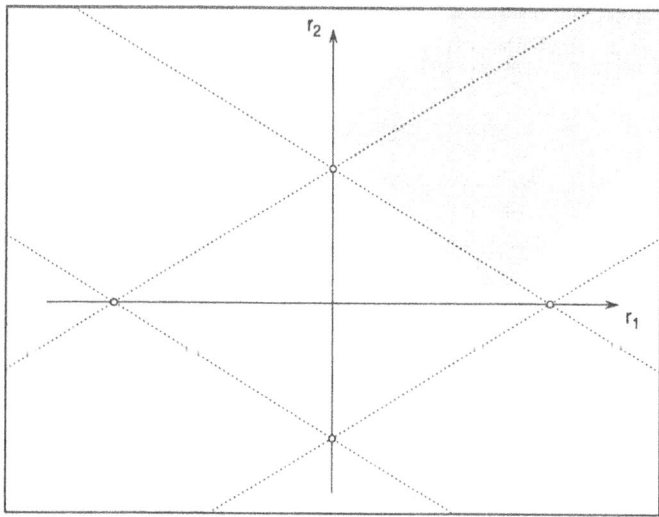

Fig. 5. The four lines in the (r_1, r_2) plane defined by (15), of which only two cross the valid domain (shaded region) for bipolar coordinates when $m \neq n$. These two line segments define the two nested loops of a Cartesian oval.

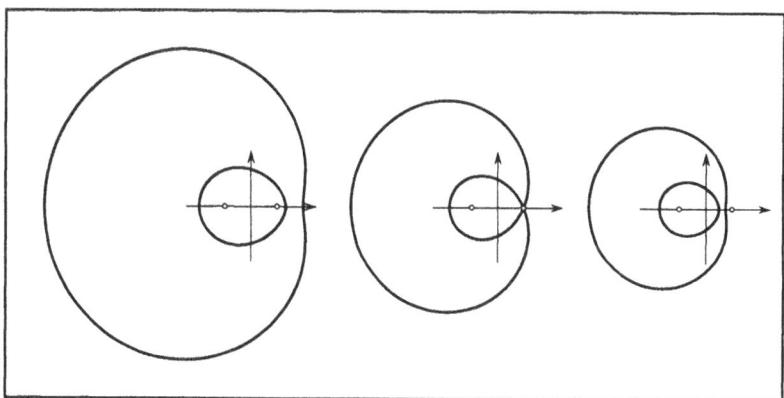

Fig. 6. Examples of the Cartesian oval with $m = 0.2$ and $n = 0.4, 0.5, 0.6$ — the case $n = 0.5$, with a double point at one pole, is a limaçon of Pascal.

is then just a *quadratic* in r — its discriminant

$$\Delta = 4m^2(4n^2 \pm 4n \cos \theta + 1 - 4m^2 \sin^2 \theta)$$

is positive for all θ when $n > m$, and the two real solutions to (16) define the "inner" and "outer" loops of the Cartesian oval.

A remarkable feature of Cartesian ovals is that the bipolar representation is not unique: one can find *three* different choices [14] for the poles p_1, p_2 and corresponding proportionality constants m, n that yield precisely the same Cartesian oval. In addition to $p_1 = (+1, 0)$ and $p_2 = (-1, 0)$, we may define

$$p_3 = \left(\frac{1 - 2m^2 - 2n^2}{2m^2 - 2n^2}, 0 \right).$$

Then, instead of p_1, p_2, we may use bipolar coordinates with respect to p_2, p_3 or p_3, p_1 as poles to describe the Cartesian oval (with appropriate choices for the proportionality constants m and n). Note that the limaçon of Pascal ($m = \frac{1}{2}$ or $n = \frac{1}{2}$) corresponds to the case where the "new" pole p_3 coincides with either p_1 or p_2. As $m \to n$, on the other hand, p_3 moves off to infinity, and we recover the ellipse/hyperbola with foci p_1, p_2.

Although it is not widely known — some catalogs [27] of special curves do not even mention it — the Cartesian oval is, from the bipolar perspective, the most "natural" generalization of the conics. It is of fundamental importance in the description of reflection/refraction in geometrical optics (see §4), and the algebra of point sets in the complex plane (see §5).

2.4   Generalized ovals of Cassini

To obtain the Cartesian oval (15) as a generalization of the ellipse/hyperbola (7), we multiply the bipolar coordinates r_1 and r_2 by factors m and n in their sum or difference. An analogous generalization for the ovals of Cassini (10) is obtained by setting the product of the bipolar coordinates r_1 and r_2, raised to (different) exponents m and n, equal to a constant:

$$r_1^m r_2^n = k. \tag{17}$$

Figure 7 shows these curves for $m = 2$, $n = 3$, and a sequence of k values.

2.5   The complete family of "bipolar conics"

In general, any locus in the valid domain (2) for bipolar coordinates defines a curve that is symmetric about the x axis. The ellipse, hyperbola, circle of Apollonius, limaçon of Pascal, and Cartesian oval are the "simplest" bipolar curves, since they are defined by *straight lines* in the (r_1, r_2) plane. The ovals of Cassini, on the other hand, correspond to a hyperbola in the (r_1, r_2) plane — they comprise one or two loops, according to whether the hyperbola has one or two segments within the valid domain.

We may also consider all bipolar curves that are defined by (segments of) conic loci within the valid domain of the (r_1, r_2) plane — we call such curves *bipolar conics*. Since the "bipolar lines" and "bipolar hyperbolas" are famous classical curves, it is natural to enquire into the nature of "bipolar ellipses" (of which bipolar circles are special cases). Figure 8 shows examples of these curves, which do not appear to have been systematically investigated before. For $k_1, k_2 > 0$ the locus

$$k_1 (r_1 - a_1)^2 + k_2 (r_2 - a_2)^2 = 1 \tag{18}$$

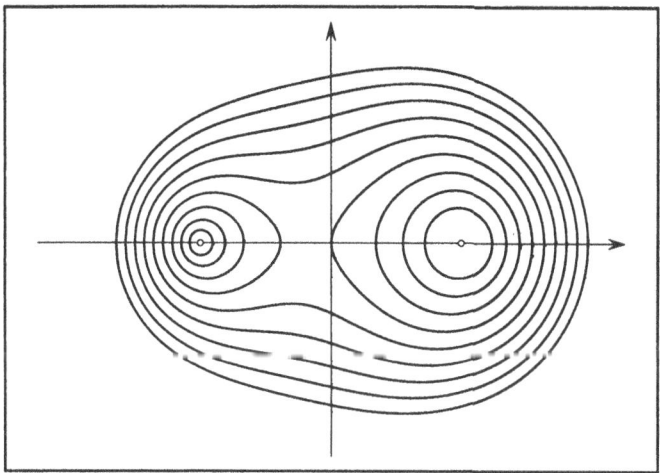

Fig. 7. Generalized ovals of Cassini, as defined by the bipolar equation (17) with exponents $m = 2$, $n = 3$ and values $0.4, 0.6, \ldots, 2.0$ for the constant k.

describes an ellipse in the (r_1, r_2) plane, centered on the point (a_1, a_2). With $r_1 = \sqrt{x^2 + y^2 - 2x + 1}$ and $r_2 = \sqrt{x^2 + y^2 + 2x + 1}$, we need to square *twice* to eliminate radicals — the resulting equation

$$\left[\,[\,k_1(r_1^2 + a_1^2) + k_2(r_2^2 + a_2^2) - 1\,]^2 - 4(k_1^2 a_1^2 r_1^2 + k_2^2 a_2^2 r_2^2)\right]^2$$
$$= 64 k_1^2 k_2^2 a_1^2 a_2^2 r_1^2 r_2^2$$

defines an algebraic curve of degree 8, the case $k_1 = k_2$ being a bipolar circle.

Figure 9 shows further examples of these curves. One family corresponds to a sequence of concentric circles ($k_1 = k_2 = k$, say) of increasing radius $R = 1/k$, with center $(a_1, a_2) = (2, 2)$, in the (r_1, r_2) plane. The other family corresponds to sequence of concentric ellipses, with k_1 fixed and k_2 increasing.

3 Multipolar coordinates

Let $\mathbf{p}_1, \mathbf{p}_2, \ldots, \mathbf{p}_n$ be distinct points in the plane. Any point in the plane may be specified by its distances $r_1, r_2, \ldots, r_n$ from these fixed points. Of course, such a specification is redundant — only *two* of $r_1, r_2, \ldots, r_n$ are independent. We call $(r_1, r_2, \ldots, r_n)$ the *multipolar coordinates* of a point, with respect to the *poles* $\mathbf{p}_1, \mathbf{p}_2, \ldots, \mathbf{p}_n$. With multipolar coordinates, it is convenient to use the complex–number representation $\mathbf{p} = x + iy$ of points in the plane.

We may, in principle, select any n distinct points as poles. However, we are mainly concerned with the most natural generalization of the "canonical" bipolar poles $\mathbf{p}_1 = +1$, $\mathbf{p}_2 = -1$ — namely, the n-th roots of unity

$$\mathbf{p}_k = \exp(i2\pi(k-1)/n) \qquad \text{for } k = 1, \ldots, n. \qquad (19)$$

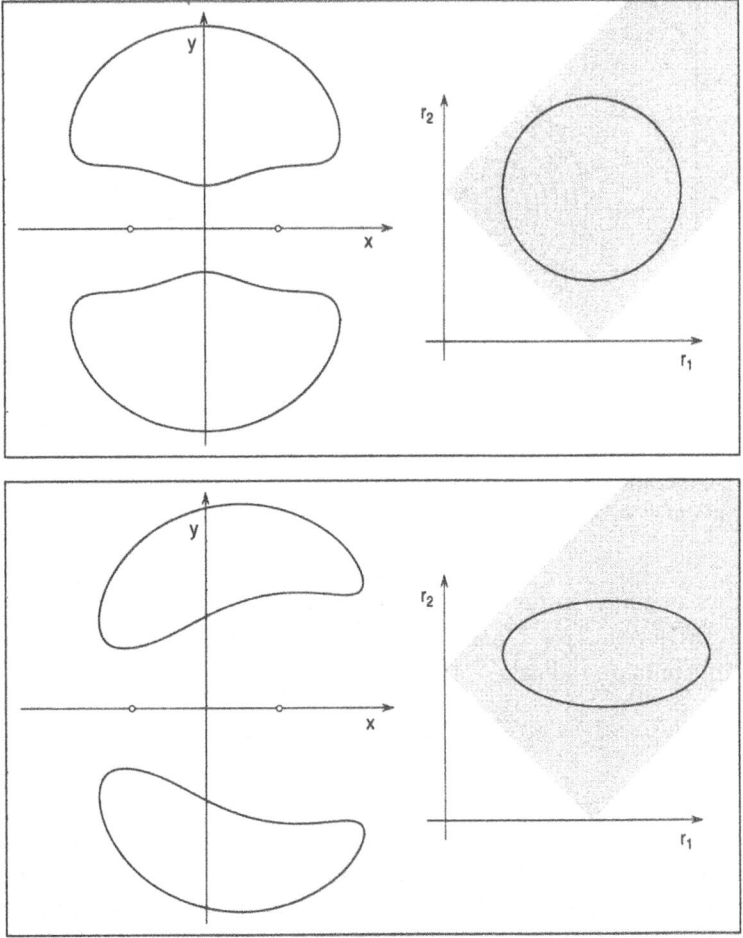

Fig. 8. Two higher–order curves, defined by a circle (upper), and an ellipse (lower), within the valid domain of the (r_1, r_2) plane for bipolar coordinates.

The multipolar coordinates $r_1, r_2, \ldots, r_n$ are then given by

$$r_k = \sqrt{r^2 - 2r\cos(\theta - 2\pi(k-1)/n) + 1} \qquad \text{for } k = 1, \ldots, n. \qquad (20)$$

Apart from their redundancy, a set of values $(r_1, \ldots, r_n)$ must satisfy certain constraints (as with bipolar coordinates — see §2) to define valid multipolar coordinates. For example, one can easily see that we require

$$r_1 + r_2 + \cdots + r_n \geq n.$$

Furthermore, the magnitudes $|r_j - r_k|$ of each of the differences of multipolar coordinates must satisfy certain fixed bounds.

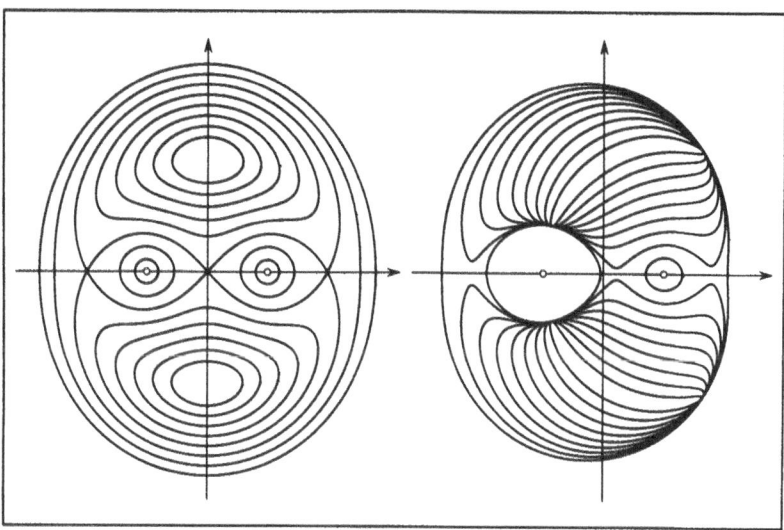

Fig. 9. Curve families defined by concentric circles (left) and ellipses (right) in the (r_1, r_2) plane — the bipolar equation (18) encompasses all these curves.

3.1   Multipolar ellipse

As a natural extension of (1) we define a multipolar (or *multifocal*) ellipse to be the locus traced by a point that moves so as to maintain a constant value for the sum of its distances $r_1, \ldots, r_n$ from n fixed poles $\mathbf{p}_1, \ldots, \mathbf{p}_n$:

$$r_1 + r_2 + \cdots + r_n = k. \tag{21}$$

To define a non–vacuous real locus, the constant k must exceed the value

$$k_{\min} = \min_{\mathbf{p}} \sum_{i=1}^{n} |\mathbf{p} - \mathbf{p}_i|,$$

which is realized at a unique point $\mathbf{p}$, provided $\mathbf{p}_1, \ldots, \mathbf{p}_n$ are not collinear. For poles at the n–th roots of unity, we clearly have $k_{\min} = n$.

In the case $n = 3$ with poles at the cube roots of unity, for example, we can formulate the tripolar ellipse $r_1 + r_2 + r_3 = k$ as an algebraic curve in Cartesian coordinates by substituting $r_1^2 = x^2 + y^2 - 2x + 1$, $r_2^2 = x^2 + y^2 + x - \sqrt{3}y + 1$, and $r_3^2 = x^2 + y^2 + x + \sqrt{3}y + 1$ into

$$\left[r_1^4 + r_2^4 + r_3^4 - 2(r_1^2 r_2^2 + r_2^2 r_3^2 + r_3^2 r_1^2) - 2k^2(r_1^2 + r_2^2 + r_3^2) + k^4 \right]^2$$
$$= 64 k^2 r_1^2 r_2^2 r_3^2$$

This defines an algebraic curve of degree 8, and we require $k \geq 3$ for it to have a non–vacuous real locus. Examples of these curves are shown in Figure 10 — note the exceptional case with $k = 2\sqrt{3}$, for which the trifocal ellipse passes through (and is singular at) each of the three poles.

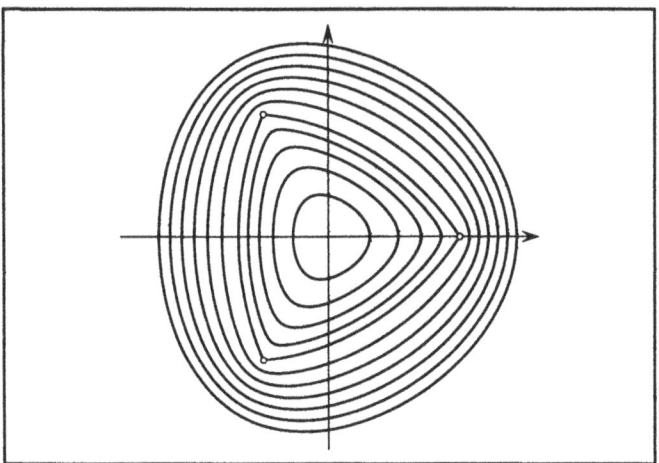

Fig. 10. Tripolar ellipses $r_1 + r_2 + r_3 = k$ with poles $\mathbf{p}_1, \mathbf{p}_2, \mathbf{p}_3$ at the cube roots of unity, for equal increments in k. In the exceptional case $k = 2\sqrt{3}$, the curve pass through (and has a singular point at) the three poles $\mathbf{p}_1, \mathbf{p}_2, \mathbf{p}_3$.

The multipolar ellipse is an attractive choice for design applications that require a closed curve defined by a single analytic expression (e.g., the design of cams). The approximation of closed, convex curves by multifocal ellipses has been considered by Weiszfeld [31] and by Erdös and Vincze [6].

3.2   Generalized mechanical ellipse

We recall from §2.1 the mechanical approach to drawing an ellipse, based on wrapping a loop of string of length $\ell > 2|\mathbf{p}_2 - \mathbf{p}_1|$ around pins pressed into a sheet of paper at the two foci $\mathbf{p}_1 = (+1, 0)$ and $\mathbf{p}_2 = (-1, 0)$. Suppose we now insert n pins into the paper at each of the n–th roots of unity (19). The pins define the vertices of a regular n–gon with perimeter

$$L_n = n\sqrt{2(1 - \cos 2\pi/n)}.$$

If we wrap a loop of string of length $\ell > L_n$ around these pins, and pull it taut with a pencil, the pencil will trace a "generalized mechanical ellipse" as it moves. This is *not* the same as the generalized ellipse defined by equation (21): as we shall presently see, it is actually a *composite* of several (ordinary) ellipse segments that are smoothly pieced together.

Consider first the case $n = 3$. The cube roots of unity $\mathbf{p}_1, \mathbf{p}_2, \mathbf{p}_3$ define the vertices of an equilateral triangle with side $\sqrt{3}$ and perimeter $L_3 = 3\sqrt{3}$. By extending the sides of this triangle indefinitely, we divide its exterior into six regions, adjacent to each vertex and each side of the triangle (Figure 11). The generalized mechanical ellipse comprises segments from six different ellipses, one in each of these regions, that meet with tangent continuity.

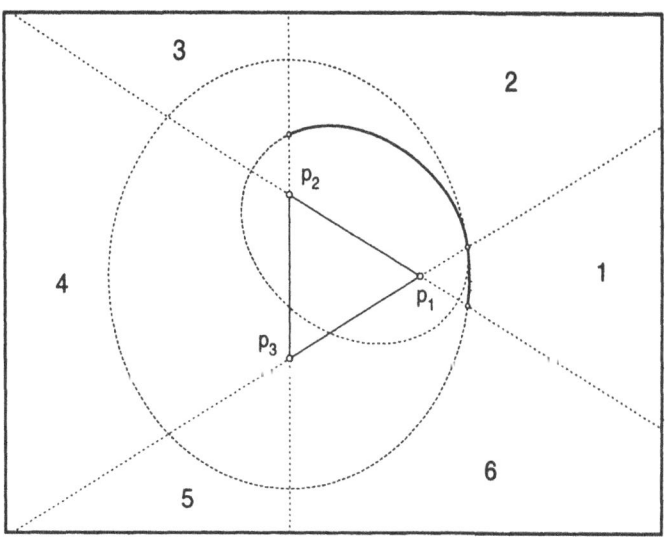

Fig. 11. Construction of $n = 3$ generalized mechanical ellipse. In region 1, adjacent to vertex $\mathbf{p}_1$, we take the segment of the ellipse defined by $r_2 + r_3 = k$, where r_2, r_3 are distances from the foci $\mathbf{p}_2, \mathbf{p}_3$ and $k = \ell - \sqrt{3}$ is the free length of a loop of string of length ℓ constrained by side $\mathbf{p}_2$–$\mathbf{p}_3$. In region 2, adjacent to side $\mathbf{p}_1$–$\mathbf{p}_2$, we take the segment of the ellipse $r_1 + r_2 = k'$, where r_1, r_2 are distances from the foci $\mathbf{p}_1, \mathbf{p}_2$ and $k' = \ell - 2\sqrt{3}$ is the free length of the loop constrained by sides $\mathbf{p}_2$–$\mathbf{p}_3$ and $\mathbf{p}_3$–$\mathbf{p}_1$. Similarly for the regions 3,4 and 5,6.

Referring to Figure 11, we see that in region 1 (adjacent to vertex $\mathbf{p}_1$) the loop of string is constrained only by side $\mathbf{p}_2$–$\mathbf{p}_3$ of the triangle. Thus, in this region, a pencil that pulls the loop taut traces an ellipse with foci $\mathbf{p}_2, \mathbf{p}_3$, the sum of distances from these foci to each point being equal to the free length of string, $k = \ell - \sqrt{3}$. In polar coordinates, this ellipse is described by

$$(12 \sin^2 \theta - 4k^2) r^2 - 4k^2 r \cos \theta - 4k^2 + k^4 = 0.$$

In region 2 (adjacent to the side $\mathbf{p}_1$–$\mathbf{p}_2$), on the other hand, the loop of string is constrained by *two* sides of the triangle, namely, $\mathbf{p}_2$–$\mathbf{p}_3$ and $\mathbf{p}_3$–$\mathbf{p}_1$. In this region, the locus traced by the pencil is an ellipse corresponding to foci $\mathbf{p}_1, \mathbf{p}_2$ and free length $k' = \ell - 2\sqrt{3}$. The polar equation of this ellipse is

$$[6(\cos \theta - \sqrt{3} \sin \theta) \cos \theta - 4k'^2 + 3] r^2$$
$$+ 2k'^2 (\cos \theta + \sqrt{3} \sin \theta) r - 4k'^2 + k'^4 = 0.$$

One can easily determine angular intervals for these equations, delineating the appropriate ellipse segments in regions 1 and 2, in terms of the parameter k (since they are rather cumbersome, we refrain from presenting them here). Finally, the ellipse segments for the regions 3,4 and 5,6 may be obtained by rotating the segments in regions 1,2 through angles $2\pi/3$ and $4\pi/3$. Examples of the resulting "ellipse splines" are shown in Figure 12.

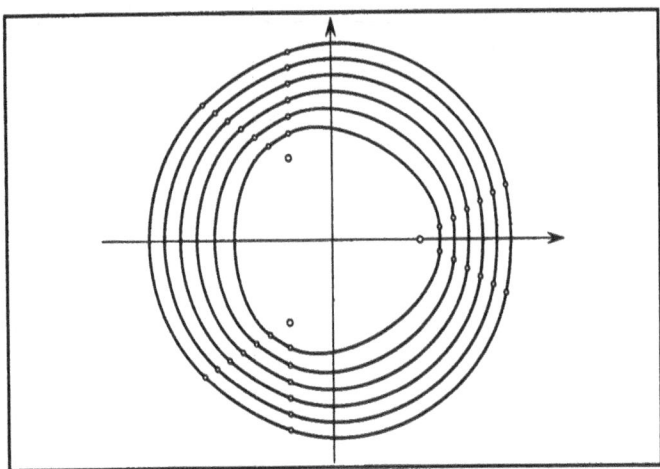

Fig. 12. Examples of $n = 3$ generalized mechanical ellipses for successive ℓ values. The six constituent (ordinary) ellipse segments are delineated by dots — note that these individual ellipse segments meet with tangent continuity.

The above construction can be easily generalized to $n > 3$. With $n = 4$, for example, extending the sides of the square with vertices at ± 1 and $\pm i$ divides its exterior into eight regions, and the generalized mechanical ellipse comprises segments of different (ordinary) ellipses in each of them. If $n \geq 5$, however, extending the sides of the regular n-gon with vertices at the n-th roots of unity gives a more intricate division of its exterior. This can be seen in Figure 13 for the case $n = 5$. In region 1, adjacent to vertex p_1, we take the segment of an ellipse with foci p_2, p_5 and free length $k = \ell - (3/5)L_5$, since the loop is constrained by *three* sides of the pentagon. In region 2, adjacent to side p_1–p_2, the loop is constrained by *four* sides, the appropriate ellipse segment having foci p_1, p_2 and $k = \ell - (4/5)L_5$. Finally, in region 3, which is not adjacent to the pentagon, the loop is constrained by just *two* sides, and the ellipse segment has foci p_3, p_5 and $k = \ell - (2/5)L_5$.

3.3   Multipolar ovals of Cassini and Cartesian ovals

We can also generalize the ovals of Cassini (10), by considering the locus of a moving point that maintains a constant value for the *product* of its distances $r_1, \ldots, r_n$ from n fixed poles $p_1, \ldots, p_n$:

$$r_1 r_2 \cdots r_n = k. \tag{22}$$

For the poles (19) at the n-th roots of unity, we can transform equation (22) into ordinary polar coordinates (r, θ) by noting [10] that

$$r_1 r_2 \cdots r_n = |z - p_1| |z - p_2| \cdots |z - p_n| = |z^n - 1|,$$

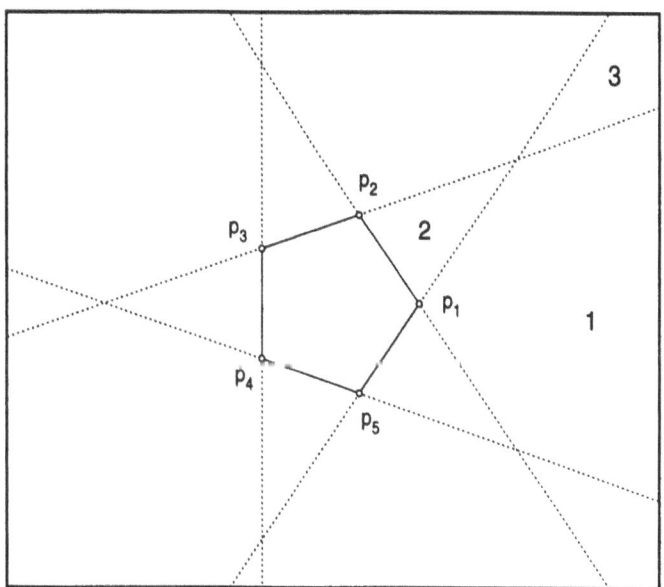

Fig. 13. By extending the sides of a regular pentagon, we divide its exterior into 15 (rather than 10) regions. For the generalized mechanical ellipse with $n = 5$ we always take, in region 1, the segment of an ellipse with foci p_2, p_5. Depending on the value of ℓ, however, we take the segment of *either* an ellipse with foci at p_1, p_2 in region 2 *or* of an ellipse with foci at p_3, p_5 in region 3.

where $z = re^{i\theta}$. Substituting the final expression, with $z^n = r^n e^{in\theta}$, into (22) and squaring then gives

$$r^{2n} - 2r^n \cos n\theta + 1 = k^2 . \tag{23}$$

We see that (11) corresponds to the special case $n = 2$ of this equation. When $k = 1$, equation (23) reduces to $r = \sqrt[n]{2\cos n\theta}$, which defines the *generalized lemniscate of Bernoulli*. Figure 14 shows examples of the generalized ovals of Cassini, for various k values, in the cases $n = 3$ and $n = 4$.

Curves defined by an equation of the form (22) are important in analyzing the convergence of polynomial approximations, constructed by interpolating discrete values sampled from a given function — see Davis [4], who calls such curves "lemniscates" with radius $\sqrt[n]{k}$ and foci $p_1, \ldots, p_n$.

Likewise, the circles of Apollonius (12) generalize to the loci defined by

$$\frac{\rho_1 \rho_2 \cdots \rho_m}{r_1 r_2 \cdots r_n} = k, \tag{24}$$

where $\rho_1, \rho_2, \ldots, \rho_m$ and $r_1, r_2, \ldots, r_n$ represent distances from two given sets of fixed poles, $q_1, q_2, \ldots, q_m$ and $p_1, p_2, \ldots, p_n$. Complex–variable methods can help elucidate the geometry of such curves (see, for example, §3.3). The use of complex–valued functions of a real parameter to describe plane curves

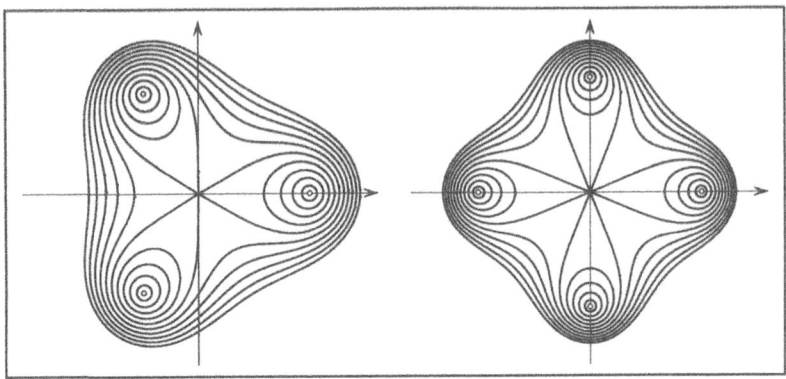

Fig. 14. The generalized ovals of Cassini (22), with the n–th roots of unity (19) as poles, for $n = 3$ and 4 with values $0.2, 0.4, \ldots, 2.0$ of the constant k.

was first systematically explored by Siebeck [28]. In his 1873 treatise *Sur une Classe Remarquable de Courbes et de Surfaces Algébriques et sur la Théorie des Imaginaires* [3], Darboux considered loci traced by a point $\mathbf{z}$ satisfying

$$\frac{|(\mathbf{z} - \mathbf{a}_1) \cdots (\mathbf{z} - \mathbf{a}_m)|}{|(\mathbf{z} - \mathbf{b}_1) \cdots (\mathbf{z} - \mathbf{b}_n)|} = k$$

for fixed points $\mathbf{a}_1, \ldots, \mathbf{a}_m$ and $\mathbf{b}_1, \ldots, \mathbf{b}_n$ in the complex plane, and a real constant $k > 0$. These curves encompass the ovals of Cassini and circles of Apollonius, and their generalization (24). A comprehensive treatment of the complex–variable approach to plane curves may be found in Zwikker [32] — unfortunately, out of print. Complex variable representations are especially pertinent in a key application of bipolar and multipolar forms, the Minkowski geometric algebra of complex sets (see §5 below). Further useful sources on the "geometry of complex numbers" are [5,22,26].

Finally, we may also generalize the Cartesian oval (13) to the case of $n \geq 3$ poles, as the locus traced by a moving point that maintains a weighted sum of its distances $r_1, \ldots, r_n$ from n fixed poles $\mathbf{p}_1, \ldots, \mathbf{p}_n$ equal to a constant:

$$m_1 r_1 + m_2 r_2 + \cdots + m_n r_n = 1.$$

As with equation (14), we must allow each "+" to be replaced with "±" on the left hand side (and 1 with ±1 on the right), if we wish to describe this generalized Cartesian oval as a single irreducible algebraic curve.

4   Geometrical optics

Bipolar coordinates play an important role in characterizing the geometry of spherical waves that are reflected or refracted by a smooth interface between two homogeneous media [8,9]. This characterization employs a theoretical construct, first proposed by Jakob Bernoulli [1], called[3] the *anticaustic* for the system of reflected/refracted wavefronts.

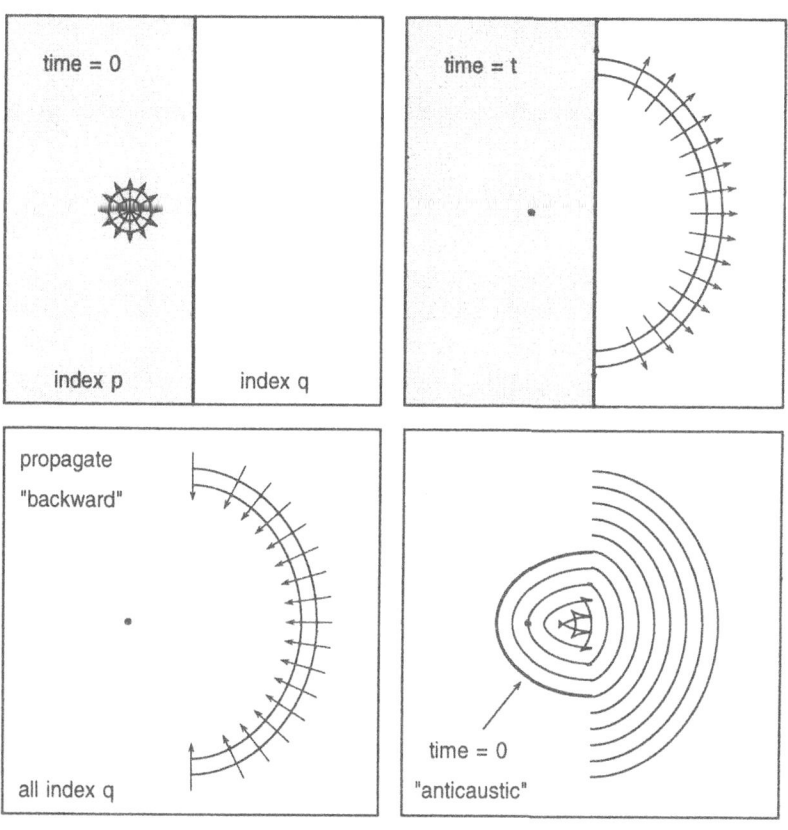

Fig. 15. Definition of the anticaustic (an ellipse) for refraction of spherical waves by a planar interface between media with refractive indices p and q.

Huygens' principle [29] governs the propagation of wavefronts in a single homogeneous medium: given an "initial" wavefront W_0 at time 0, it describes the wavefront W at any subsequent time t as an *offset* or "parallel" to W_0, at distance $d = ct$ from it (c being the wavespeed). In the presence of a smooth refracting or reflecting surface between different media, the wavefronts before and after the reflection or refraction are not members of a single family of offsets. Nevertheless, we may still invoke Huygens' principle to characterize the reflected/refracted wavefronts as follows:

---

[3] The anticaustic goes by many alternate names in the geometrical optics literature, e.g., the *secondary caustic* [2,23], *orthotomic* [15], and *archetypal wavefront* [29].

surface	mode	anticaustic	wavefront
plane	reflect	point	degree 2
plane	refract	ellipse/hyperbola	degree 8
sphere	reflect	limaçon of Pascal	degree 10
sphere	refract	Cartesian oval	degree 14

Table 2. Nature of the anticaustics for reflection/refraction of spherical waves by planar or spherical surfaces. The degree of the offsets to these anticaustics, which describe the reflected or refracted wavefronts, is also indicated above.

Suppose a spherical wave emanates from a point source at time $t = 0$ and, after reflection or refraction at a smooth surface $\mathcal{A}$ between two homogeneous media, subsequently assumes shape W at time t. By propagating W *backward in time in a single homogeneous medium*, we obtain a (hypothetical) "initial" wavefront W_0 at $t = 0$. The significance of W_0 is that its uniform propagation via Huygens' principle (without reflection/refraction by the surface $\mathcal{A}$) yields the true reflected/refracted wavefront W at the prescribed time t.

This hypothetical "initial" wavefront W_0 is the anticaustic for reflection or refraction of a spherical wave by the surface $\mathcal{A}$. Its name arises from the fact that it is an involute of the *caustic* (i.e., the envelope of the reflected/refracted rays, which are normals to the reflected/refracted wavefronts). The caustic — from the Greek for "burning" — was thus named by Ehrenfried Walther von Tschirnhaus. Figure 15 illustrates the concept of the anticaustic.

For axisymmetric configurations of the light source and the surface $\mathcal{A}$, it suffices to restrict the problem to a plane of symmetry. As shown in Table 2, curves described by linear equations in the bipolar coordinates (r_1, r_2) feature prominently among "simple" anticaustics (for planar/spherical surfaces $\mathcal{A}$).

5   Minkowski geometric algebra

Minkowski geometric algebra [10,12,13] is concerned with the complex sets

$$\mathcal{A} \oplus \mathcal{B} = \{\, \mathbf{a} + \mathbf{b} \mid \mathbf{a} \in \mathcal{A} \text{ and } \mathbf{b} \in \mathcal{B} \,\},$$
$$\mathcal{A} \otimes \mathcal{B} = \{\, \mathbf{a} \times \mathbf{b} \mid \mathbf{a} \in \mathcal{A} \text{ and } \mathbf{b} \in \mathcal{B} \,\}, \tag{25}$$

populated by the sums and products of all pairs of complex numbers selected from complex–set operands[4] $\mathcal{A}$ and $\mathcal{B}$. The Minkowski sum and product (25) are natural generalizations of (real) *interval arithmetic* operations [19,20] to sets of *complex* numbers. The geometry of connected two–dimensional sets is, of course, incomparably richer than that of one–dimensional intervals, and

---

[4] Here we denote real variables by italic characters, complex variables by bold characters, and sets of complex numbers by upper–case calligraphic characters.

the Minkowski geometric algebra is thus of as much interest for its geometrical content as its formal algebraic methods. It offers an intuitive language for the description, generation, and analysis of planar shapes, and for basic wavefront reflection/refraction problems in geometric optics; it is likewise of interest in stability analysis of indeterminate dynamical systems [11].

As a generalization of Minkowski sums and products, one may consider [12,13] "implicitly–defined sets" of the form

$$A \circledf B = \{\, f(a,b) \mid a \in A,\ b \in B \,\} \tag{26}$$

i.e., sets of complex values generated by a bivariate function with arguments a and b selected from given complex sets, A and B. Whereas the Minkowski sums and products can be regarded as unions of translated or scaled/rotated instances of one set, expression (26) is a much more powerful concept — it can be interpreted as the union of a family of *conformal mappings* of one set. Equation (26) clearly subsumes (25) as special cases, and under suitable conditions on f the Minkowski sum/product algorithms can be extended [13] to perform boundary evaluation for sets of the form (26).

Curves defined in bipolar and multipolar coordinates play a central role in the Minkowski geometric algebra of "simple" sets (i.e., lines and circles). The Minkowski product of a line and a circle, for example, is the region bounded by an ellipse or hyperbola, while the Minkowski product of two circles is the region bounded by the two loops of a Cartesian oval [12]. In the Minkowski geometric algebra, problems of an algebraic nature are of as much interest as those phrased in geometrical terms. Since the Minkowski product operation is commutative and associative, we may speak of the n–th Minkowski power

$$\bigotimes{}^{n}A = \underbrace{A \otimes A \otimes \cdots \otimes A}_{n \text{ times}} = \{\, z_1 z_2 \cdots z_n \mid z_i \in A \text{ for } i = 1,\ldots,n \,\}$$

of a given set A. Conversely, we may seek a complex set that corresponds to an n–th Minkowski root $\bigotimes{}^{1/n}A$ of the set A, defined by the property that

$$\{\, z_1 z_2 \cdots z_n \mid z_i \in \bigotimes{}^{1/n}A \text{ for } i = 1,\ldots,n \,\} = A,$$

i.e., the n–th Minkowski power of $\bigotimes{}^{1/n}A$ is identical to the given set A.

When A is a circular disk that does not contain the origin, its Minkowski square root $\bigotimes{}^{1/2}A$ is a single loop of the ovals of Cassini. Furthermore, the Minkowski n–th root $\bigotimes{}^{1/n}A$ is a single loop of the n–polar ovals of Cassini. However, if the disk A contains the origin, the situation is more complicated: a *composite* curve is required to define a valid Minkowski root [10].

The Minkowski product $A \otimes B$ can be regarded as the union of scaled and rotated copies of set B, defined by multiplying it by each point $z = re^{i\phi}$ of set A (or vice–versa). In Figure 16 we illustrate how the region between the two loops of a Cartesian oval is thus generated, in two different ways, as a one–parameter family of circles. Similarly, Figure 17 shows how a single loop

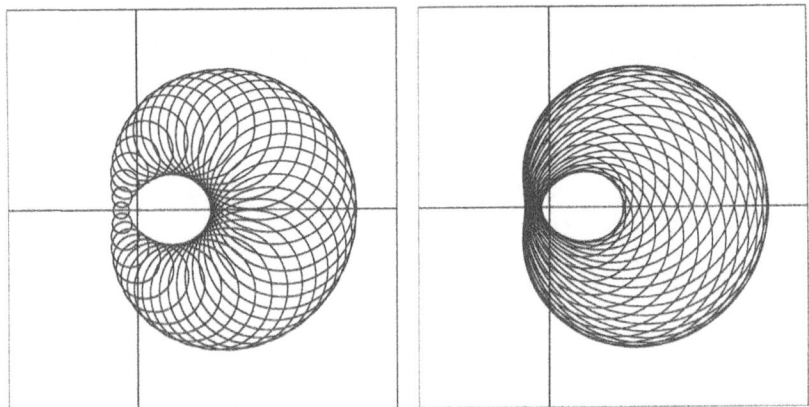

Fig. 16. The area bounded by the two loops of a Cartesian oval, generated as the Minkowski product of two circles A and B that do not pass through the origin. The illustrations show the one–parameter families of circles that are defined by the scaling/rotation of circle A by each point of circle B, and vice–versa.

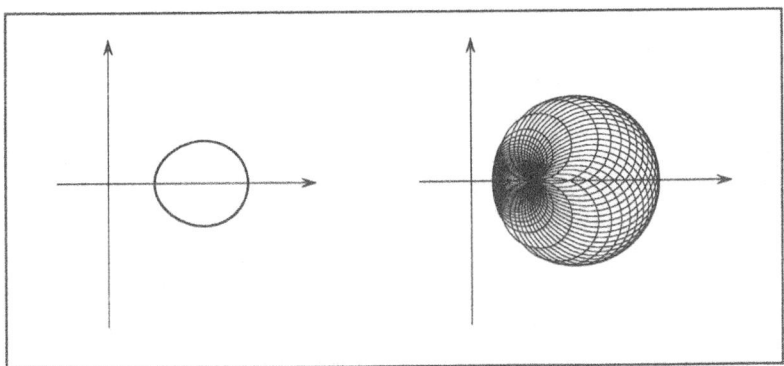

Fig. 17. A single loop of the ovals of Cassini as the Minkowski square root of a circular disk. The one–parameter family of scalings and rotations of this loop defined by each point on it covers the interior of the circular disk.

of the ovals of Cassini is the Minkowski square root of a circular disk (i.e., the Minkowski product of this loop with itself yields the disk). Note that this is valid *only* for disks that do not contain the origin — the determination of the Minkowski square root for a disk with the origin as an interior point is a much more complicated matter; see [10] for further details.

The Minkowski geometric algebra offers elegant characterizations of the optical constructions described in §4 — for example, the anticaustic for the refraction of spherical waves by a surface A is simply the boundary $\partial(A \otimes C)$ of the Minkowski product of a circle C and (a medial section of) A [12].

6   Curve design/construction methods

In §2 we considered curves defined by simple *implicit* equations $f(r_1, r_2) = 0$ in bipolar coordinates, such as (6), (10), (15), (17), and (18). For geometric design, the use of a *parametric* bipolar representation

$$(r_1(t), r_2(t)) \quad \text{for } t \in [0, 1] \tag{27}$$

is advantageous. Many standard curve design/construction methods [7] can then be easily translated into the bipolar context. For example, there exists a unique cubic interpolant to the first–order Hermite data

$$(r_1(0), r_2(0)),\ (r_1'(0), r_2'(0)) \quad \text{and} \quad (r_1(1), r_2(1)),\ (r_1'(1), r_2'(1)),$$

and this can be used to formulate a C^2 bipolar spline interpolating an ordered sequence of points, $(\rho_{1,k}, \rho_{2,k})$ for $k = 1, \ldots, N$. Note that, in these schemes, the functions (27) defining the constructed curve must be checked against the basic constraints (2) on bipolar coordinates (alternately, one may modify the algorithms to automatically satisfy these constraints).

A more design–oriented approach is based on the bipolar Bézier form

$$(r_1(t), r_2(t)) = \sum_{k=0}^{n} (\rho_{1,k}, \rho_{2,k}) \binom{n}{k} (1-t)^{n-k} t^k \tag{28}$$

defined by control points $(\rho_{1,k}, \rho_{2,k})$ for $k = 1, \ldots, n$. This defines a proper locus in the (x, y) plane when all the control points, and their convex hull, lie within the valid domain (2) of the (r_1, r_2) plane. Figure 18 shows an example of a bipolar Bézier cubic — note that, although the locus defined by (28) in the (r_1, r_2) plane is an "ordinary" Bézier curve, its image in the (x, y) plane is a more interesting curve. We see from the coordinate transformation (5), for example, that the latter is not ordinarily a rational curve.

Analogs of familiar Bézier curve algorithms (subdivision, degree elevation, etc.) and properties (convex hull, variation diminishing, rational forms, etc.) can be associated with (28). As noted in §2.3, an arbitrary line segment in the (r_1, r_2) plane defines a Cartesian oval segment in the (x, y) plane, and hence the "ordinary" control polygon of (28) is transformed into a control structure comprising $n + 1$ points connected by n Cartesian oval segments (see Figure 18). This leads, by the de Casteljau algorithm, to the remarkable observation that a bipolar Bézier curve can be approximated to any specified accuracy by a convergent sequence of piecewise–Cartesian–ovals. Moreover, the extension to B–spline forms should not pose much difficulty.

The use of such bipolar design schemes can be advantageous whenever one desires explicit control over the variation of the distance of a point along a locus from two prescribed fixed points (the poles).

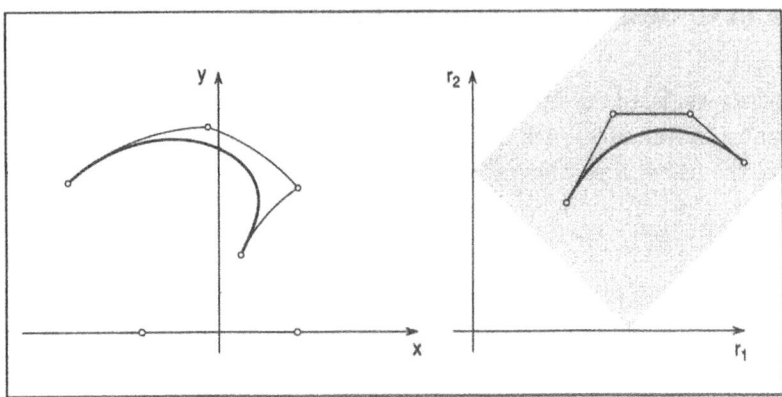

Fig. 18. A bipolar Bézier cubic, defined by an "ordinary" cubic in the valid domain (2) of the (r_1, r_2) plane, yields the locus shown on the right in the (x, y) plane, with a "control polygon" composed of Cartesian oval segments.

7 Closure

The methods described herein can be extended to describe three–dimensional surfaces as well as planar curves. For a surface that is rotationally symmetric about an axis, two poles on the axis suffice. For a more general surface, three poles are necessary (the surface will be symmetric with respect to reflection in the plane defined by the three poles, unless we specify separate equations "above" and "below" this plane). However, the complex–variable approach cannot be used in this context, and there is no "natural" choice — analogous to the n-th roots of unity (19) — for the poles of a spatial multipolar system.

References

1. Bernoulli, J. (1692), Lineæ cycloidales, evolutæ, ant–evolutæ, causticæ, anti-causticæ, peri–causticæ, *Acta Eruditorum*, May 1692.
2. Cayley, A. (1857), A memoir upon caustics, and Supplementary memoir upon caustics, *Phil. Trans. R. Soc. London* **147**, 273–312 and **157**, 7–16.
3. Darboux, G. (1873), *Sur une Classe Remarquable de Courbes et de Surfaces Algébriques et sur la Théorie des Imaginaires*, Gauthier-Villars, Paris.
4. Davis, P. J. (1963), *Interpolation and Approximation*, Blaisdell Publishing Co., New York.
5. Deaux, R. (1956), *Introduction to the Geometry of Complex Numbers* (translated from the French by H. Eves), F. Ungar, New York.
6. Erdös, P. and Vincze, I. (1982), On the approximation of convex, closed plane curves by multifocal ellipses, *J. Applied Probability* **19a**, 89–96.
7. Farin, G. (1997), *Curves and Surfaces for Computer Aided Geometric Design* (4th Edition), Academic Press, Boston.

8. Farouki, R. T. and Chastang, J–C. A. (1992), Curves and surfaces in geometrical optics, in *Mathematical Methods in Computer Aided Geometric Design II*, (T. Lyche & L. L. Schumaker, eds.), Academic Press, 239–260.

9. Farouki, R. T. and Chastang, J–C. A. (1992), Exact equations of "simple" wavefronts, *Optik* **91**, 109–121.

10. Farouki, R. T., Gu, W., and Moon, H. P. (2000), Minkowski roots of complex sets, *Geometric Modeling and Processing 2000*, IEEE Computer Society Press, 287–300.

11. Farouki, R. T. and Moon, H. P. (2000), Minkowski geometric algebra and stability of characteristic polynomials I. Coefficients in complex disks, preprint.

12. Farouki, R. T., Moon, H. P., and Ravani, B. (2000), Minkowski geometric algebra of complex sets, *Geometriae Dedicata*, to appear.

13. Farouki, R. T., Moon, H. P., and Ravani, B. (2000), Algorithms for Minkowski products and implicitly–defined complex sets, *Adv. Comp. Math.*, to appear.

14. Gomes Teixeira, F. (1971), *Traité des Courbes Spéciales Remarquables Planes et Gauches*, Tome I, Chelsea (reprint), New York.

15. Herman, R. A. (1900), *A Treatise on Geometrical Optics*, Cambridge University Press.

16. Lawden, D. F. (1989), *Elliptic Functions and Applications*, Springer, New York.

17. Lawrence, J. D. (1972), *A Catalog of Special Plane Curves*, Dover, New York.

18. Lockwood, E. H. (1967), *A Book of Curves*, Cambridge University Press.

19. Moore, R. E. (1966), *Interval Analysis*, Prentice–Hall, Englewood Cliffs, NJ.

20. Moore, R. E. (1979), *Methods and Applications of Interval Analysis*, SIAM, Philadelphia.

21. Neamtu, M., Pottmann, H., and Schumaker, L. L. (1998), Designing NURBS cam profiles using trigonometric splines, *ASME J. Mech. Design* **120**, 175–180.

22. Needham, T. (1997), *Visual Complex Analysis*, Oxford University Press.

23. Salmon, G. (1960), *A Treatise on the Higher Plane Curves*, Chelsea, New York (reprint).

24. Sanchez–Reyes, J. (1990), Single–valued curves in polar coordinates, *Comput. Aided Design* **22**, 19–26.

25. Sanchez–Reyes, J. (1992), Single–valued spline curves in polar coordinates, *Comput. Aided Design* **24**, 307–315.

26. Schwerdtfeger, H. (1979), *Geometry of Complex Numbers*, Dover, New York.

27. Von Seggern, D. H. (1993), *CRC Standard Curves and Surfaces*, CRC Press, Boca Raton.

28. Herr Siebeck (1857), Über die graphische Darstellung imaginärer Funktionen, *J. Reine Angew. Math.* **55**, 221–253.

29. Stavroudis, O. N. (1972), *The Optics of Rays, Wavefronts, and Caustics*, Academic Press, New York.

30. Stillwell, J. (1989), *Mathematics and Its History*, Springer, New York.

31. Weiszfeld, E. (1937), Sur le point pour lequel la somme des distances de n points donnés minimum, *Tohoku Math. J.* **43**, 355–386.

32. Zwikker, C. (1963), *The Advanced Geometry of Plane Curves and Their Applications*, Dover, New York.

Polar Curves and Surfaces

Kenji Ueda

RICOH Company Ltd., 1-1-17, Koishikawa, Bunkyo-ku, Tokyo 112-0002, Japan

Summary. Polar curves with respect to proper conics and polar surfaces with respect to proper quadrics are investigated. Polar curves (surfaces) are defined as the envelope of the polar lines (planes) of the points on a given curve (surface) with respect to a quadratic curve (surface). Polar curves and surfaces are collineations of dual curves and surfaces. Polar curves and surfaces with respect to real unit spheres are negative dual curves and surfaces and are obtained as the inverse of the pedal curves and surfaces with respect to the centers of the sphere.

1 Introduction

Projective geometry is a useful tool in computer graphics [7,9] and computer aided design [3]. Rational parametric curves [5] are based on projective geometry.

The principle of duality in projective geometry [4,15] gives the concept of dual curves and surfaces [8,10]. The applications of dual curves and surfaces, e.g., developable surfaces and offset-rational curves and surfaces, are surveyed in [13].

Polarity with respect to proper quadrics is another dual relationship between points and lines or planes. Polarity also derives a curve or surface from a given curve or surface. The polar curves and surfaces are investigated in this paper. First, polarity and conjugate points are introduced. Polar curves with respect to conics and polar surfaces with respect to quadrics are presented and investigated in sections 2 and 3, respectively. Finally, special cases of polar curves and surfaces are shown.

2 Polarity and Conjugate Points

Conics are represented as the quadratic form

$$ax^2 + by^2 + d + 2lx + 2my + 2gxy = 0, \tag{1}$$

and the form is rewritten, using a symmetric matrix Q, into

$$\mathbf{x}^T Q \mathbf{x} = 0, \qquad \mathbf{x} = \begin{bmatrix} 1 \\ x \\ y \end{bmatrix}, \qquad Q = \begin{bmatrix} d & l & m \\ l & a & g \\ m & g & b \end{bmatrix}. \tag{2}$$

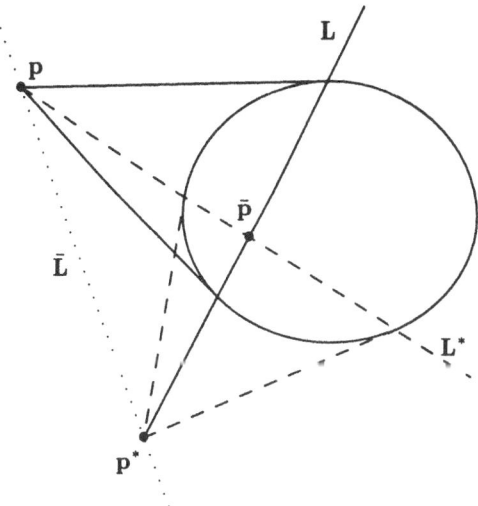

Fig. 1. Polarity with respect to a conic.

The conics are proper (non-degenerated), if the determinant $\det(Q)$ is not equal to 0,

In the homogeneous coordinate system, a point $\mathbf{p} = (X/W, Y/W)^T$ is expressed as

$$\mathbf{p} = \begin{bmatrix} W \\ X \\ Y \end{bmatrix}. \tag{3}$$

In projective geometry, the line $\mathbf{P}\mathbf{x} = W + Xx + Yy = 0$ is represented by

$$\mathbf{P} = [W \quad X \quad Y], \tag{4}$$

and is called the *dual line* $\mathbf{P}$ of the point $\mathbf{p}$.

The *polar line* $\mathbf{L}$ of a point $\mathbf{p}$ with respect to a proper conic Q is defined as

$$\mathbf{L} = \mathbf{p}^T Q. \tag{5}$$

Geometrically, the line $\mathbf{L}$ is the straight line joining the points of contact of the tangent that can be drawn from the point $\mathbf{p}$ to the conic Q. The point $\mathbf{p}$ is called the *pole* of the polar line $\mathbf{L}$, and is obtained as [3,15]:

$$\mathbf{p} = Q^{-1} \mathbf{L}^T. \tag{6}$$

Two points such that each lies on the polar of the other are *conjugate points* with respect to a conic. Any point $\mathbf{p}^*$ on the polar line $\mathbf{L}$, that satisfies the equation $\mathbf{L}\mathbf{p}^* = 0$, and the point $\mathbf{p}$ are conjugate points, namely,

$$\mathbf{p}^T Q \mathbf{p}^* = 0. \tag{7}$$

The intersecting point $\bar{\mathbf{p}}$ between the polar lines $\mathbf{L}$ of the point $\mathbf{p}$ and $\mathbf{L}^*$ of the point $\mathbf{p}^*$ is also the pole of the line $\bar{\mathbf{L}}$ passing through $\mathbf{p}$ and $\mathbf{p}^*$, i.e., $\bar{\mathbf{p}} = Q^{-1}\bar{\mathbf{L}}$. Hence, the triangle $\triangle(\mathbf{p}, \mathbf{p}^*, \bar{\mathbf{p}})$ is called a *self-polar triangle*, as illustrated in Figure 1.

The dual line $\mathbf{P}$ of the point $\mathbf{p}$ is the polar line with respect to the imaginary circle $x^2 + y^2 = -1$. namely,

$$\mathbf{P} = \mathbf{p}^T I = [W \quad X \quad Y], \tag{8}$$

where I is an identity matrix of order three.

On a projective plane, the intersection of two lines $\mathbf{L}$ and $\mathbf{M}$ is given by the *cross product* $\mathbf{L} \wedge \mathbf{M}$. Therefore, the situation in Figure 1 is expressed as

$$\bar{\mathbf{p}} = \mathbf{L} \wedge \mathbf{L}^*, \qquad \mathbf{p}^* = \mathbf{L} \wedge \bar{\mathbf{L}}, \qquad \mathbf{p} = \bar{\mathbf{L}} \wedge \mathbf{L}^*. \tag{9}$$

According to the principle of duality, the line passing through two points $\mathbf{p}$ and $\mathbf{q}$ is also given by the cross product. As we distinguish the cross product ($\wedge$) between two lines from the cross product ($\sqcap$) between two points in this paper, the situation in Figure 1 is also expressed as

$$\bar{\mathbf{L}} = \mathbf{p} \sqcap \mathbf{p}^*, \qquad \mathbf{L}^* = \mathbf{p} \sqcap \bar{\mathbf{p}}, \qquad \mathbf{L} = \bar{\mathbf{p}} \sqcap \mathbf{p}^*. \tag{10}$$

The definitions of the cross products $\wedge$ and $\sqcap$ are given by [7].

$$\mathbf{L} \wedge \mathbf{M} = \det \begin{bmatrix} \mathbf{e}_0 & \mathbf{e}_1 & \mathbf{e}_2 \\ L_0 & L_1 & L_2 \\ M_0 & M_1 & M_2 \end{bmatrix}, \qquad \mathbf{e}_j = \begin{bmatrix} \delta_{0,j} \\ \delta_{1,j} \\ \delta_{2,j} \end{bmatrix}, \tag{11}$$

$$\mathbf{p} \sqcap \mathbf{q} = \det \begin{bmatrix} \mathbf{E}_0 & p_0 & q_0 \\ \mathbf{E}_1 & p_1 & q_1 \\ \mathbf{E}_2 & p_2 & q_2 \end{bmatrix}, \qquad \mathbf{E}_j = [\delta_{0,j} \quad \delta_{1,j} \quad \delta_{2,j}], \tag{12}$$

where $\delta_{i,j}$ is the Kronecker delta. The following properties for the cross product are easily obtained.

$$\mathbf{p} \sqcap \mathbf{p} = 0, \qquad \mathbf{p} \sqcap \mathbf{q} = -\mathbf{q} \sqcap \mathbf{p}, \qquad (\mathbf{p} \sqcap \mathbf{q})\,\mathbf{p} = (\mathbf{p} \sqcap \mathbf{q})\,\mathbf{q} = 0. \tag{13}$$

Relationships between poles and polar lines are expressed as

$$\frac{(\mathbf{p}^T Q) \wedge (\mathbf{q}^T Q)}{\det(Q)} = Q^{-1}(\mathbf{p} \sqcap \mathbf{q})^T, (Q^{-1}\mathbf{L}^T) \sqcap (Q^{-1}\mathbf{M}^T)$$

$$= \frac{(\mathbf{L} \wedge \mathbf{M})^T Q}{\det(Q)}. \tag{14}$$

Using a geometric equivalence operator $\hat{=}$, they become the following geometrical relationships.

$$(\mathbf{p}^T Q) \wedge (\mathbf{q}^T Q) \hat{=} Q^{-1}(\mathbf{p} \sqcap \mathbf{q})^T, (Q^{-1}\mathbf{L}^T) \sqcap (Q^{-1}\mathbf{M}^T) \hat{=} (\mathbf{L} \wedge \mathbf{M})^T Q. \tag{15}$$

For example, the first equation means that the intersection of the polar lines of points $\mathbf{p}$ and $\mathbf{q}$ is the pole of the line passing through the points $\mathbf{p}$ and $\mathbf{q}$ geometrically.

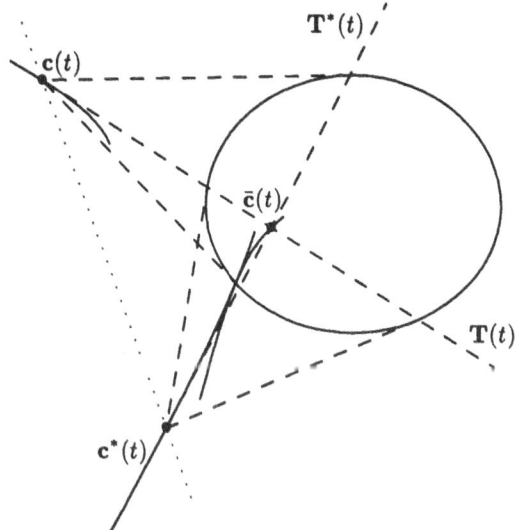

Fig. 2. A polar curve.

3   Polar Curves with Respect to Conics

A rational curve $\mathbf{c}(t)$ and its derivative in homogeneous space are given by

$$\mathbf{c}(t) = \begin{bmatrix} W(t) \\ X(t) \\ Y(t) \end{bmatrix}, \qquad \mathbf{c}'(t) = \begin{bmatrix} W'(t) \\ X'(t) \\ Y'(t) \end{bmatrix}. \tag{16}$$

As the tangent line to the rational curve $\mathbf{c}(t)$ is expressed as

$$1 + \frac{Y'(t)W(t) - W'(t)Y(t)}{X'(t)Y(t) - Y'(t)X(t)} x + \frac{W'(t)X(t) - X'(t)W(t)}{X'(t)Y(t) - Y'(t)X(t)} y = 0, \tag{17}$$

the tangent line $\mathbf{T}(t)$ in homogeneous space is given by

$$\mathbf{T}(t) = \begin{bmatrix} X'(t)Y(t) - Y'(t)X(t) \\ Y'(t)W(t) - W'(t)Y(t) \\ W'(t)X(t) - X'(t)W(t) \end{bmatrix}^T = \mathbf{c}'(t) \sqcap \mathbf{c}(t). \tag{18}$$

The *polar curve* $\mathbf{c}^*(t)$ of the curve $\mathbf{c}(t)$ is defined as the envelope of the polar line of $\mathbf{c}(t)$. The polar curve $\mathbf{c}^*(t)$ is also obtained as the locus of the pole of the tangent line $\mathbf{T}(t)$ to $\mathbf{c}(t)$.

$$\mathbf{c}^*(t) = Q^{-1}\,\mathbf{T}(t)^T. \tag{19}$$

The polar curve is a rational curve of degree $2(n-1)$, because the degree of the tangent line $\mathbf{T}(t)$ is $2(n-1)$ [17].

As the curve $\mathbf{c}(t)$ and its polar curve $\mathbf{c}^*(t)$ satisfy the statement

$$\mathbf{c}(t)^T Q\, \mathbf{c}^*(t) = 0, \tag{20}$$

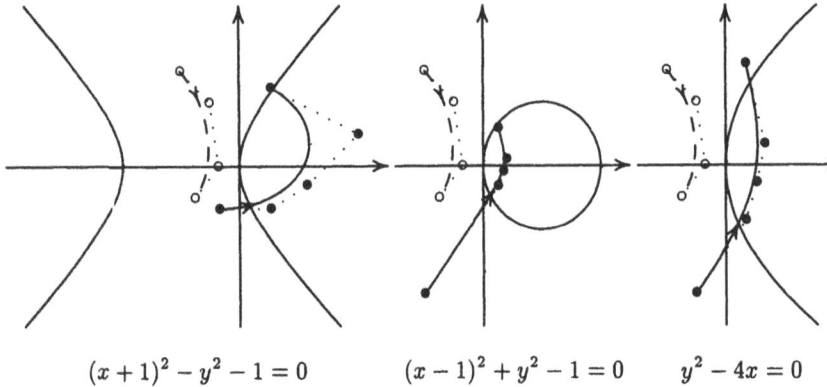

$$(x+1)^2 - y^2 - 1 = 0 \qquad (x-1)^2 + y^2 - 1 = 0 \qquad y^2 - 4x = 0$$

Fig. 3. Polar curves of a Bézier curve with respect to conics.

they are *conjugate curves* with respect to the conic Q.

The transformation constructing the polar curve from a given curve is a self-inverse operation.

$$\mathbf{c}^{**}(t) = \mathbf{c}(t). \tag{21}$$

As the curve $\mathbf{c}(t)$, and its polar curve $\mathbf{c}^*(t)$ and the intersection between their tangent lines $\mathbf{T}(t) \wedge \mathbf{T}^*(t)$ form a self-polar triangle, the locus of the intersection defines a new curve $\bar{\mathbf{c}}(t)$, as illustrated in Figure 2. The new curve is also the pole of the straight line $\mathbf{c}(t) \sqcap \mathbf{c}^*(t)$.

$$\bar{\mathbf{c}}(t) = \mathbf{T}(t) \wedge \mathbf{T}^*(t) \mathrel{\hat{=}} Q^{-1} \left(\mathbf{c}(t) \sqcap \mathbf{c}^*(t) \right)^T. \tag{22}$$

This curve is a rational curve of degree $3(n-1)$.

If the conic Q is the imaginary unit circle centered at the origin, i.e., $Q = I$, then the polar curve is equivalent to the *dual curve* of a given curve [8,10,12]. Therefore, polar curves are *collineations* [4,5] of dual curves as:

$$\mathbf{c}^*(t) = Q^{-1} \mathbf{T}(t)^T = Q^{-1} \left(I \, \mathbf{T}(t)^T \right). \tag{23}$$

A collineation maps collinear points to collinear points. It is also known that the dual line of a double point on a curve is a bitangent to the dual curve of the curve and an inflection point on a curve transformed to a cusp of the dual curve.

Figure 3 illustrates three polar curves of the same cubic integral Bézier curve with respect to a hyperbola, a circle and a parabola. The polar curves, which are quartic rational Bézier curves, are collineations of each other.

3.1   Polar Bézier Curves

Suppose a curve $\mathbf{c}(t)$ is a rational Bézier curve of degree n.

$$\mathbf{c}(t) = \sum_{i=0}^{n} \mathbf{b}_i B_i^n(t), \qquad B_i^n(t) = \binom{n}{i}(1-t)^{n-i} t^i. \tag{24}$$

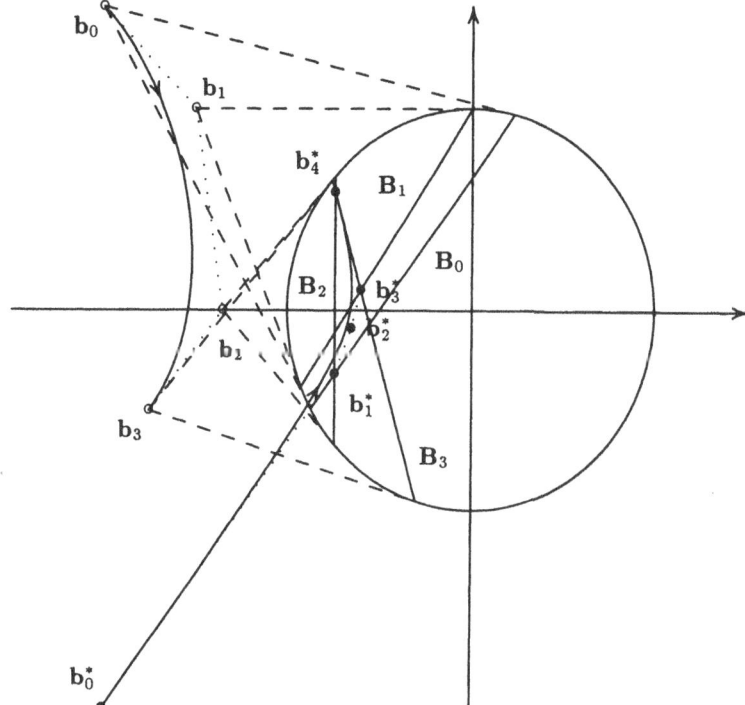

Fig. 4. Relationship between poles and polars of Bézier points.

The curve $\mathbf{c}(t)$ is rewritten into

$$\mathbf{c}(t) = \mathrm{B}_0^1(t)\mathbf{c}_0(t) + \mathrm{B}_1^1(t)\mathbf{c}_1(t), \qquad \mathbf{c}_k(t) = \sum_{i=0}^{n-1} \mathbf{b}_{i+k}\mathrm{B}_i^{n-1}(t). \tag{25}$$

These two points $\mathbf{c}_0(t)$ and $\mathbf{c}_1(t)$ lie on the tangent line to the curve $\mathbf{c}(t)$. The polar lines $\mathbf{B}_0(t)$ and $\mathbf{B}_1(t)$ of the points are expressed as

$$\mathbf{B}_k(t) = \sum_{i=0}^{n-1} \mathbf{B}_{i+k}\mathrm{B}_i^{n-1}(t) = \sum_{i=0}^{n-1} (\mathbf{b}_{i+k}^T Q)\mathrm{B}_i^{n-1}(t) = \mathbf{c}_k(t)^T Q. \tag{26}$$

The polar curve $\mathbf{c}^*(t)$ is obtained as the intersection of the polar lines $\mathbf{B}_0(t)$ of the point $\mathbf{c}_0(t)$ and $\mathbf{B}_1(t)$ of $\mathbf{c}_1(t)$.

$$\mathbf{c}^*(t) = \sum_{k=0}^{2n-2} \mathrm{B}_k^{2n-2}(t)\mathbf{b}_k^* \mathrel{\hat{=}} \mathbf{B}_0(t)^T \wedge \mathbf{B}_1(t)^T, \tag{27}$$

where

$$\mathbf{b}_k^* = \frac{1}{\binom{2n-2}{k}} \sum_{\substack{i+j=k \\ 0 \le i,j \le n-1}} \binom{n-1}{i}\binom{n-1}{j} \mathbf{B}_i \wedge \mathbf{B}_{j+1}. \tag{28}$$

Four outer Bézier points are defined as the intersections of polar lines of two Bézier points of the curve $c(t)$.

$$\mathbf{b}_0^* = \mathbf{B}_0 \wedge \mathbf{B}_1 = (\mathbf{b}_0^T Q) \wedge (\mathbf{b}_1^T Q) = Q^{-1}(\mathbf{b}_0 \sqcap \mathbf{b}_1)^T,$$

$$\mathbf{b}_1^* = \frac{\mathbf{B}_0 \wedge \mathbf{B}_2}{2} = \frac{(\mathbf{b}_0^T Q) \wedge (\mathbf{b}_2^T Q)}{2} = Q^{-1}\frac{(\mathbf{b}_0 \sqcap \mathbf{b}_2)^T}{2},$$

$$\mathbf{b}_{2n-3}^* = \frac{\mathbf{B}_{n-2} \wedge \mathbf{B}_n}{2} = \frac{(\mathbf{b}_{n-2}^T Q) \wedge (\mathbf{b}_n^T Q)}{2} = Q^{-1}\frac{(\mathbf{b}_{n-2} \sqcap \mathbf{b}_n)^T}{2},$$

$$\mathbf{b}_{2n-2}^* = \mathbf{B}_{n-1} \wedge \mathbf{B}_n = (\mathbf{b}_{n-1}^T Q) \wedge (\mathbf{b}_n^T Q) = Q^{-1}(\mathbf{b}_{n-1} \sqcap \mathbf{b}_n)^T. \tag{29}$$

Figure 4 illustrates a Bézier cubic and its polar Bézier curve with the polar lines of the Bézier points of the cubic curve.

3.2   Polar Conics with Bitangency

It is well known that conic sections are represented as rational quadratic Bézier curves. Consider the following quadratic Bézier curve $\hat{c}(t)$ with its Bézier points p_i $(i = 0, 1, 2)$ and the weight $\hat{w}$ of the central Bézier point p_1.

$$\hat{c}(t) \doteq p_0 B_0^2(t) + \hat{w}\, p_1 B_1^2(t) + p_2 B_2^2(t), \quad p_i = \begin{bmatrix} 1 & x_i & y_i \end{bmatrix}^T. \tag{30}$$

This curve can be implicitized into a quadratic form by calculating the resultant [3,16] of the polynomials in (30). The elements of the matrix Q of the implicit form of the conic section are obtained as

$$
\begin{aligned}
a &= (y_2 - y_0)^2 - 4\hat{w}^2(y_2 - y_1)(y_1 - y_0), \\
b &= (x_2 - x_0)^2 - 4\hat{w}^2(x_2 - x_1)(x_1 - x_0), \\
d &= (x_0 y_2 - y_0 x_2)^2 - 4\hat{w}^2(x_2 y_1 - x_1 y_2)(x_1 y_0 - x_0 y_1), \\
g &= -(x_2 - x_0)(y_2 - y_0) + 2\hat{w}^2((x_1 - x_0)(y_2 - y_1) + (x_2 - x_1)(y_1 - y_0)), \\
l &= (y_0 x_2 - x_0 y_2)(y_2 - y_0) \\
  &\quad -2\hat{w}^2((x_1 y_0 - x_0 y_1)(y_2 - y_1) + (x_2 y_1 - x_1 y_2)(y_1 - y_0)), \\
m &= (x_0 y_2 - y_0 x_2)(x_2 - x_0) \\
  &\quad -2\hat{w}^2((x_0 y_1 - x_1 y_0)(x_2 - x_1) + (x_1 y_2 - x_2 y_1)(x_1 - x_0)).
\end{aligned}
\tag{31}
$$

Here we can compute polar curves of a Bézier curve $c(t)$ with the same Bézier points and a different central weight w.

$$c(t) = p_0 B_0^2(t) + w\, p_1 B_1^2(t) + p_2 B_2^2(t). \tag{32}$$

The polar curve $c^*(t)$ of the curve $c(t)$ with respect to the conic $\hat{c}(t)$ becomes

$$c^*(t) \doteq p_0 B_0^2(t) + w^*\, p_1 B_1^2(t) + p_2 B_2^2(t), \tag{33}$$

where the weight w^* of the central Bézier point satisfies the equation:

$$w^* w = \hat{w}^2. \tag{34}$$

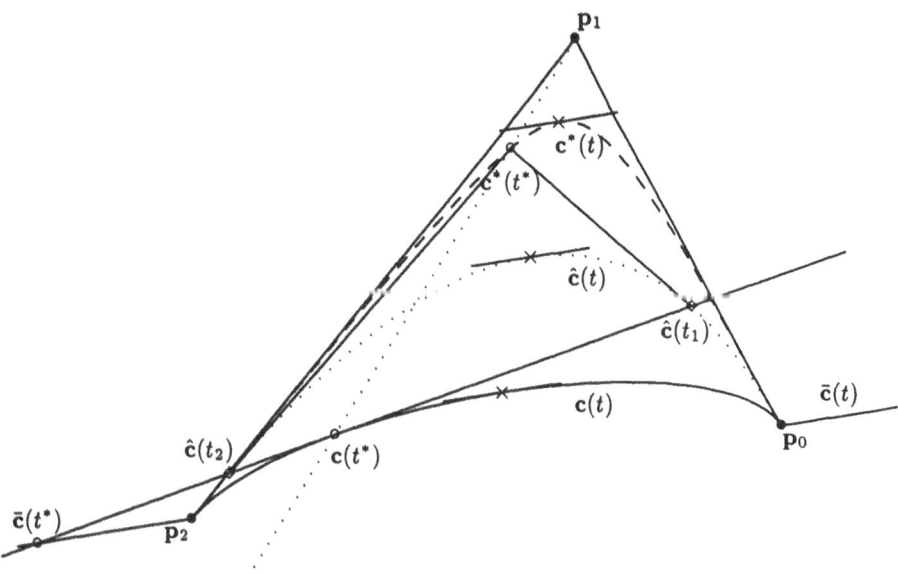

Fig. 5. Polar Bézier quadratics with the same Bézier points.

The curve $\bar{\mathbf{c}}(t)$ defined by the intersection of the tangent lines of the curves $\mathbf{c}(t)$ and $\mathbf{c}^*(t)$ becomes the straight line passing through the end points $\mathbf{p}_0$ and $\mathbf{p}_2$ of the curves.

$$\bar{\mathbf{c}}(t) = \mathbf{T}(t) \wedge \mathbf{T}^*(t) \hat{=} -\mathbf{p}_0 B_0^2(t) + \mathbf{p}_2 B_2^2(t). \tag{35}$$

Figure 5 illustrates an example of such curves defined with the same Bézier points. There is the following relationship between the parameter values t_1 and t_2 for the conic $\hat{\mathbf{c}}(t)$ and the parameter t^* for the curves $\mathbf{c}(t)$, $\mathbf{c}^*(t)$ and $\bar{\mathbf{c}}(t)$.

$$0 < t_1 < t^* = \frac{t_1 t_2 - \sqrt{(1-t_1)t_1(1-t_2)t_2}}{t_1 + t_2 - 1} < t_2 < 1. \tag{36}$$

In the case where $t_1 + t_2 = 1$, the parameter value t^* is equal to $1/2$ and the tangent lines to the curves are parallel to the straight line passing through the end points.

4   Polar Surfaces with Respect to Quadrics

Quadrics are represented as the following quadratic form.

$$ax^2 + by^2 + cz^2 + d + 2eyz + 2fzx + 2gxy + 2lx + 2my + 2nz = 0. \tag{37}$$

The form is rewritten as

$$\mathbf{x}^T Q \mathbf{x} = 0, \qquad \mathbf{x} = \begin{bmatrix} 1 \\ x \\ y \\ z \end{bmatrix}, \qquad Q = \begin{bmatrix} d & l & m & n \\ l & a & g & f \\ m & g & b & e \\ n & f & e & c \end{bmatrix}. \tag{38}$$

As the *polar plane* of a point $\mathbf{p} = [W \ \ X \ \ Y \ \ Z]^T$ with respect to the quadric Q satisfies

$$\mathbf{p}^T Q \mathbf{x} = 0, \tag{39}$$

the polar plane $\mathbf{P}$ is expressed as

$$\mathbf{P} = \mathbf{p}^T Q. \tag{40}$$

A surface $\mathbf{s}(u, v)$ and the partial derivatives in projective space are expressed as

$$\mathbf{s}(u, v) = \begin{bmatrix} W(u, v) \\ X(u, v) \\ Y(u, v) \\ Z(u, v) \end{bmatrix}, \mathbf{s}_u(u, v) = \begin{bmatrix} W_u(u, v) \\ X_u(u, v) \\ Y_u(u, v) \\ Z_u(u, v) \end{bmatrix}, \mathbf{s}_v(u, v) = \begin{bmatrix} W_v(u, v) \\ X_v(u, v) \\ Y_v(u, v) \\ Z_v(u, v) \end{bmatrix}. \tag{41}$$

The tangent plane $\mathbf{T}(u, v)$ to the surface $\mathbf{s}(u, v)$ is given by

$$\begin{aligned} &\mathbf{T}(u, v) \\ &= \begin{bmatrix} (Y_v Z_u - Z_v Y_u)X + (Z_v X_u - X_v Z_u)Y + (X_v Y_u - Y_v X_u)Z \\ (Z_v Y_u - Y_v Z_u)W + (W_v Z_u - Z_v W_u)Y + (Y_v W_u - W_v Y_u)Z \\ (X_v Z_u - Z_v X_u)W + (Z_v W_u - W_v Z_u)X + (W_v X_u - X_v W_u)Z \\ (Y_v X_u - X_v Y_u)W + (W_v Y_u - Y_v W_u)X + (X_v W_u - W_v X_u)Y \end{bmatrix}^T \\ &= \mathbf{s}_v(u, v) \sqcap \mathbf{s}_u(u, v) \sqcap \mathbf{s}(u, v), \end{aligned}$$

where the cross products among three components are defined as [7]

$$\mathbf{L} \wedge \mathbf{M} \wedge \mathbf{N} = \det \begin{bmatrix} \mathbf{e}_0 & \mathbf{e}_1 & \mathbf{e}_2 & \mathbf{e}_3 \\ L_0 & L_1 & L_2 & L_3 \\ M_0 & M_1 & M_2 & M_3 \\ N_0 & N_1 & N_2 & N_3 \end{bmatrix}, \qquad \mathbf{e}_j = \begin{bmatrix} \delta_{0,j} \\ \delta_{1,j} \\ \delta_{2,j} \\ \delta_{3,j} \end{bmatrix}, \tag{42}$$

$$\mathbf{p} \sqcap \mathbf{q} \sqcap \mathbf{r} = \det \begin{bmatrix} \mathbf{E}_0 & p_0 & q_0 & r_0 \\ \mathbf{E}_1 & p_1 & q_1 & r_1 \\ \mathbf{E}_2 & p_2 & q_2 & r_2 \\ \mathbf{E}_3 & p_3 & q_3 & r_3 \end{bmatrix}, \quad \mathbf{E}_j = [\delta_{0,j} \ \ \delta_{1,j} \ \ \delta_{2,j} \ \ \delta_{3,j}]. \tag{43}$$

The cross products have the following properties.

$$(\mathbf{p} \sqcap \mathbf{q} \sqcap \mathbf{r})\mathbf{p} = (\mathbf{p} \sqcap \mathbf{q} \sqcap \mathbf{r})\mathbf{q} = (\mathbf{p} \sqcap \mathbf{q} \sqcap \mathbf{r})\mathbf{r} = 0, \tag{44}$$

$$\mathbf{p} \sqcap \mathbf{p} \sqcap \mathbf{q} = \mathbf{p} \sqcap \mathbf{q} \sqcap \mathbf{p} = \mathbf{q} \sqcap \mathbf{p} \sqcap \mathbf{p} = 0, \tag{45}$$

$$p \sqcap q \sqcap r = q \sqcap r \sqcap p = r \sqcap p \sqcap q = -p \sqcap r \sqcap q = -q \sqcap p \sqcap r = -r \sqcap q \sqcap p. \quad (46)$$

Poles and polars have the following relationships.

$$(p^T Q) \wedge (q^T Q) \wedge (r^T Q) \hateq Q^{-1}(p \sqcap q \sqcap r)^T, \quad (47)$$

$$(Q^{-1} L^T) \sqcap (Q^{-1} M^T) \sqcap (Q^{-1} N^T) \hateq (L \wedge M \wedge N)^T Q. \quad (48)$$

The *polar surface* $s^*(u, v)$ of a surface $s(u, v)$ with respect to a quadric is defined as the envelope of the polar plane of the surface $s(u, v)$, that is, the pole of the tangent plane $T(u, v)$ to the surface. Hence, the polar surfaces are defined as

$$s^*(u, v) = Q^{-1} T(u, v)^T. \quad (49)$$

The surface $s(u, v)$ and its polar surface $s^*(u, v)$ are the conjugate points with respect to the quadric Q.

$$s(u, v)^T Q s^*(u, v) = 0. \quad (50)$$

4.1   Polar Bézier Triangles

A Bézier triangle of degree n represents a rational surface $s(u, v, w)$.

$$s(u, v, w) = \sum_{\substack{i+j+k=n \\ 0 \le i,j,k}} b_{i,j,k} \, b_{i,j,k}^n(u, v, w), \quad b_{i,j,k}^n(u, v, w) = \frac{n!}{i!j!k!} u^i v^j w^k. \quad (51)$$

The Bézier triangle is also expressed as

$$s(u, v, w) = u \, s_{100}(u, v, w) + v \, s_{010}(u, v, w) + w \, s_{001}(u, v, w), \quad (52)$$

where

$$s_{pqr}(u, v, w) = \sum_{\substack{i+j+k=n-1 \\ 0 \le i,j,k}} b_{i+p,j+q,k+r} \, b_{i,j,k}^{n-1}(u, v, w). \quad (53)$$

The three points $s_{100}(u, v, w)$, $s_{010}(u, v, w)$ and $s_{001}(u, v, w)$ are on the tangent plane to the surface $s(u, v, w)$. The polar planes $P_{pqr}(u, v, w)$ of the points $s_{pqr}(u, v, w)$ ($0 \le p, q, r$, $p + q + r = 1$) with respect to a quadric Q are given by

$$P_{pqr} = s_{pqr}(u, v, w)^T Q = \sum_{\substack{i+j+k=n-1 \\ 0 \le i,j,k}} (b_{i+p,j+q,k+r}^T Q) \, b_{i,j,k}^{n-1}(u, v, w). \quad (54)$$

Hence, the polar planes are expressed as the convex combinations of the polar planes $B_{i,j,k} = b_{i,j,k}^T Q$ of the Bézier points $b_{i,j,k}$ of the surface $s(u, v, w)$.

The polar surface, that is the envelope of the polar planes of the surface $s(u, v, w)$, is expressed as the locus of the intersection of the polar planes $\mathbf{P}_{pqr}$ $(0 \le p, q, r, \; p + q + r = 1)$.

$$\mathbf{s}^*(u, v, w) = \sum_{\substack{i+j+k=3n-3 \\ 0 \le i,j,k}} \mathbf{b}^*_{i,j,k} \mathbf{b}^{3n-3}_{i,j,k}(u, v, w)$$

$$= \mathbf{P}_{100}(u, v, w)^T \wedge \mathbf{P}_{010}(u, v, w)^T \wedge \mathbf{P}_{001}(u, v, w)^T. \qquad (55)$$

The Bézier points $\mathbf{b}^*_{i,j,k}$ are expressed as

$$\sum \frac{\binom{n-1}{i_i,j_i,k_i}\binom{n-1}{i_j,j_j,k_j}\binom{n-1}{i_k,j_k,k_k}}{\binom{3n-3}{i,j,k}} \mathbf{B}_{i_i+1,j_i,k_i} \wedge \mathbf{B}_{i_j,j_j+1,k_j} \wedge \mathbf{B}_{i_k,j_k,k_k+1}, \quad (56)$$

where the summation is taken for indices such that

$$\begin{array}{lll} i_i + i_j + i_k = i, & j_i + j_j + j_k = j, & k_i + k_j + k_k = k, \\ i_i + j_i + k_i = n - 1, & i_j + j_j + k_j = n - 1, & i_k + j_k + k_k = n - 1, \\ 0 \le i_i, j_i, k_i, & 0 \le i_j, j_j, k_j, & 0 \le i_k, j_k, k_k. \end{array} \quad (57)$$

The polar surface $\mathbf{s}^*(u, v, w)$ is a rational Bézier triangle of degree $3(n-1)$. The Bézier point at each corner of the Bézier triangle is the pole of the tangent plane at each corner.

$$\begin{aligned} \mathbf{b}^*_{3n-3,0,0} &= \mathbf{B}_{n,0,0} \wedge \mathbf{B}_{n-1,1,0} \wedge \mathbf{B}_{n-1,0,1}, \\ \mathbf{b}^*_{0,3n-3,0} &= \mathbf{B}_{1,n-1,0} \wedge \mathbf{B}_{0,n,0} \wedge \mathbf{B}_{0,n-1,1}, \qquad\qquad (58) \\ \mathbf{b}^*_{0,0,3n-3} &= \mathbf{B}_{1,0,n-1} \wedge \mathbf{B}_{0,1,n-1} \wedge \mathbf{B}_{0,0,n}. \end{aligned}$$

5   Polar Curves and Surfaces with Respect to Real Spheres

In this section, polar curves and surfaces with respect to the real unit circle S^1 $(x^2 + y^2 = 1)$ and sphere S^2 $(x^2 + y^2 + z^2 = 1)$ are investigated. The matrices Q^1 for the circle and Q^2 for the sphere are expressed by

$$S^1 : Q^1 = \begin{bmatrix} -1 & 0 & 0 \\ 0 & 1 & 0 \\ 0 & 0 & 1 \end{bmatrix}, \qquad S^2 : Q^2 = \begin{bmatrix} -1 & 0 & 0 & 0 \\ 0 & 1 & 0 & 0 \\ 0 & 0 & 1 & 0 \\ 0 & 0 & 0 & 1 \end{bmatrix}. \qquad (59)$$

As the polar curves and surfaces with respect to the imaginary unit circle $x^2 + y^2 = -1$ and unit sphere $x^2 + y^2 + z^2 = -1$ are dual curves and surfaces [8], the polar curves and surfaces with respect to the real circle and sphere are negative dual curves and surfaces.

Polarity with respect to a circle or a sphere is closely related to the *inversion* operation, as illustrated in Figure 6. The foot $\mathbf{q}$ of the perpendicular from a pole $\mathbf{p}$ to the polar line of $\mathbf{p}$ is the inverse of the pole $\mathbf{p}$ in the circle.

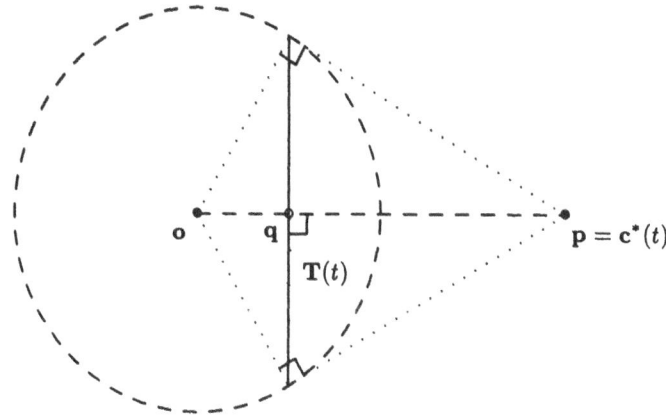

Fig. 6. Pedal point construction and inversion.

On the other hand, the locus of the foot of the perpendicular from the point
o to the tangent line $\mathbf{T}(t)$ of a curve $\mathbf{c}(t)$ is called the *pedal curve* $\pi\,\mathbf{c}(t)$ of
the curve $\mathbf{c}(t)$ with respect to the point o. Therefore, the polar curve $\mathbf{c}^*(t)$
with respect to a circle is recognized as the inverse of the pedal curve with
respect to the center of the circle [15]. This relationship is expressed by the
following diagram.

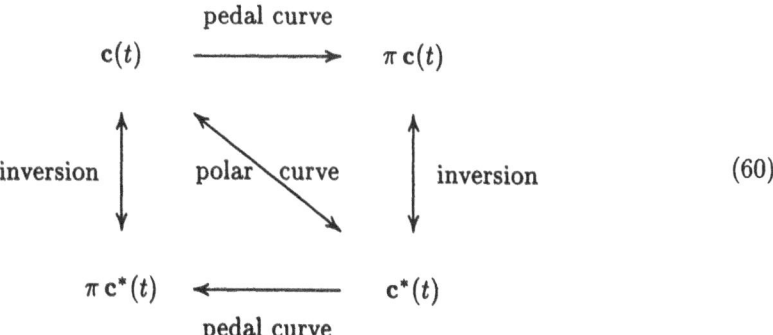

$$(60)$$

The polar surface $\mathbf{s}^*(u, v)$ with respect to a sphere is also the inverse of the
pedal surface [1] of $\mathbf{s}(t)$ with respect to the center of the sphere. Conversely,
pedal curves or surfaces can be obtained as the inverses of polar curves or
surfaces with respect to real circles or spheres.

5.1   Offset-Rational Curves and Surfaces

The *hodograph* $\mathbf{h}(t)$ of a curve $\mathbf{c}(t)$ and the *normal* vector $\mathbf{n}(u,v)$ to a surface $\mathbf{s}(u,v)$ are defined by

$$\mathbf{h}(t) = \frac{d}{dt}\mathbf{c}(t), \qquad \mathbf{n}(u,v) = \frac{\partial}{\partial u}\mathbf{s}(u,v) \times \frac{\partial}{\partial v}\mathbf{s}(u,v). \tag{61}$$

The hodograph $\mathbf{h}(t)$ and the normal vector $\mathbf{n}(u,v)$ are expressed in the projective space as

$$\mathbf{h} = \begin{bmatrix} W(t)^2 \\ X'(t)W(t) - W'(t)X(t) \\ Y'(t)W(t) - W'(t)Y(t) \end{bmatrix}, \tag{62}$$

$$\mathbf{n} = \begin{bmatrix} W^3 \\ (Y_u Z_v - Z_u Y_v)W + (Z_u W_v - W_u Z_v)Y + (W_u Y_v - Y_u W_v)Z \\ (W_u Z_v - Z_u W_v)X + (Z_u X_v - X_u Z_v)W + (X_u W_v - W_u X_v)Z \\ (Y_u W_v - W_u Y_v)X + (W_u X_v - X_u W_v)Y + (X_u Y_v - Y_u X_v)W \end{bmatrix}. \tag{63}$$

As the matrices Q^1 and Q^2 are self-inverse, the polar curve and surfaces with respect to the real spheres become $\mathbf{c}^*(t) = Q^1\mathbf{T}(t)$ and $\mathbf{s}^*(u,v) = Q^2\mathbf{T}(u,v)$. Comparing Eq. (18) of $\mathbf{T}(t)$ with Eq. (62) and Eq. (42) of $\mathbf{T}(u,v)$ with Eq. (63), we can find the following properties.

$$\frac{\mathbf{h}(t)}{|\mathbf{h}(t)|} = \frac{\mathbf{c}^*(t)}{|\mathbf{c}^*(t)|}, \qquad \frac{\mathbf{n}(u,v)}{|\mathbf{n}(u,v)|} = \frac{\mathbf{s}^*(u,v)}{|\mathbf{s}^*(u,v)|}. \tag{64}$$

This equations mean that the polar curve of a Pythagorean-hodograph (PH) curve [6] is a Pythagorean curve [5] with respect to the circle S^1 and the polar surface of a Pythagorean-normal (PN) surface is a Pythagorean surface with respect to the sphere S^2, and vice versa. As described in [10–12], the polar curve $\mathbf{c}^*(t)$ of a Pythagorean curve $\mathbf{c}(t)$ of the following form with respect to S^1

$$\mathbf{c}(t) = \begin{bmatrix} p(t) \\ q(t)f(t) \\ q(t)g(t) \end{bmatrix}, \qquad f(t)^2 + g(t)^2 = e(t)^2, \tag{65}$$

is a PH curve, and the polar surface $\mathbf{s}^*(u,v)$ of a Pythagorean surface $\mathbf{s}(u,v)$ of the following form with respect to S^2

$$\mathbf{s}(u,v) = \begin{bmatrix} p(u,v) \\ q(u,v)f(u,v) \\ q(u,v)g(u,v) \\ q(u,v)h(u,v) \end{bmatrix}, \qquad f(u,v)^2 + g(u,v)^2 + h(u,v)^2 = e(u,v)^2, \tag{66}$$

is a PN surface.

As the *reciprocal* transformation:

$$\begin{bmatrix} W(t) \\ X(t) \\ Y(t) \end{bmatrix} \mapsto \begin{bmatrix} X(t)Y(t) \\ W(t)Y(t) \\ W(t)X(t) \end{bmatrix} \tag{67}$$

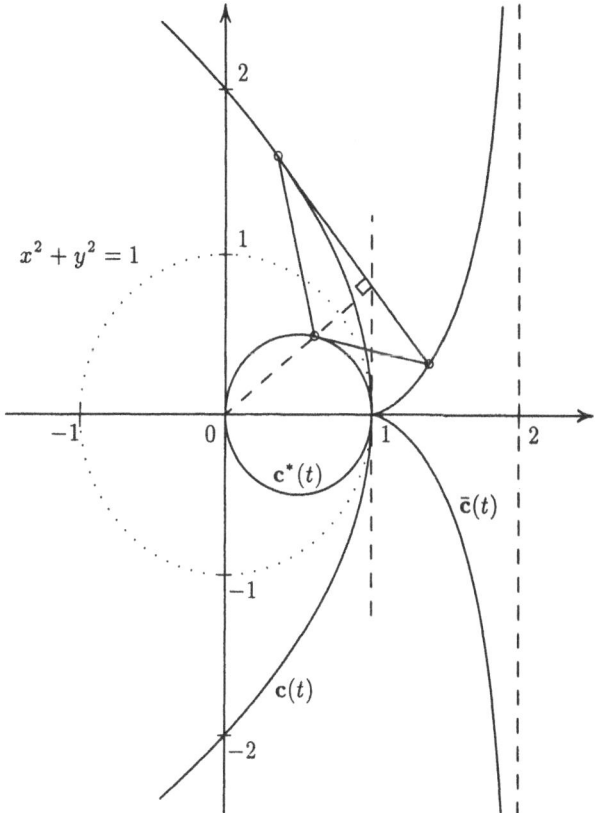

Fig. 7. A self-polar triangle with respect to $x^2 + y^2 + 1$.

preserves Pythagorean curves, the reciprocal curve $\underline{c}(t)$ of the curve $c(t)$

$$\underline{c}(t) = \begin{bmatrix} q(t)q(t)f(t)g(t) \\ p(t)q(t)g(t) \\ p(t)q(t)f(t) \end{bmatrix} \doteq \begin{bmatrix} q(t)f(t)g(t) \\ p(t)g(t) \\ p(t)f(t) \end{bmatrix}, \quad f(t)^2 + g(t)^2 = e(t)^2, \text{(68)}$$

is also a Pythagorean curve.

5.2   Examples

Polar Curves with Respect to S^1

Figure 7 illustrates a parabola $c(t)$, a circle $c^*(t)$ and a cissoid of Diocles $\bar{c}(t)$ $((x-2)y^2 + (x-1)^3 = 0)$.

$$c(t) = \begin{bmatrix} 1 \\ 1-t^2 \\ 2t \end{bmatrix}, \quad c^*(t) = \begin{bmatrix} 1+t^2 \\ 1 \\ t \end{bmatrix}, \quad \bar{c}(t) = \begin{bmatrix} 1+t^2 \\ 1+2t^2 \\ t^3 \end{bmatrix}. \quad \text{(69)}$$

The three curves for a parameter value t form a self-polar triangle with respect to S^1. The circle $c^*(t)$ is the polar curve of the parabola $c(t)$, and vice versa.

The pedal curve of the parabola $c(t)$ is the straight line $[1, 1, t]^T$ [18], and its inverse curve is the circle $c^*(t)$. The parabola $c(t)$ is a Pythagorean curve and the circle $c^*(t)$ is a rational PH curve.

The reciprocal curve $\underline{c}(t)$ of the parabola $c(t)$ is also Pythagorean.

$$\underline{c}(t) = \begin{bmatrix} 2t(1-t^2) \\ 2t \\ 1-t^2 \end{bmatrix}, \qquad 4y^2(1-x) + x = 0. \tag{70}$$

Polar Surfaces with Respect to S^2

A cubic cyclide is represented as [14]

$$s(u, v) = \begin{bmatrix} u^2 + v^2 + 1 \\ pu^2 + qv^2 \\ u(p + (p-q)v^2) \\ v(q + (q-p)u^2) \end{bmatrix}, \tag{71}$$

and the surface has the following implicit form.

$$qy^2 + pz^2 = x(y^2 + z^2 + (x-p)(x-q)). \tag{72}$$

The polar surface $s^*(u, v)$ of the cubic cyclide $s(u, v)$ with respect to S^2 is obtained as

$$s^*(u, v) = -(q - u^2(p-q))(p - v^2(q-p)) \begin{bmatrix} pu^2 + qv^2 \\ u^2 + v^2 - 1 \\ 2u \\ 2v \end{bmatrix}. \tag{73}$$

Figure 8 illustrates an example of the polar surface $s^*(u, v)$.

As a cubic cyclide is a PN surface, the polar surface, which is a quadratic rational Pythagorean surface, is a quartic surface of the implicit form

$$x^2 + y^2 + z^2 = \left(x + \frac{p}{2}y^2 + \frac{q}{2}z^2\right)^2. \tag{74}$$

In 3D space, the polar surfce $s(u, v)^*$ is expressed as $(u^2 + v^2 + 1)/(pu^2 + qv^2) S^2(u, v)$, where $S^2(u, v)$ is a parametric form of the unit sphere S^2:

$$S^2(u, v) = \frac{1}{u^2 + v^2 + 1} \begin{pmatrix} u^2 + v^2 - 1 \\ 2u \\ 2v \end{pmatrix}. \tag{75}$$

Hence, the pedal surface of the cubic cyclide with respect to the origin is given by $(pu^2 + qv^2)/(u^2 + v^2 + 1) S^2(u, v)$, which is the inverse surface of the polar surface $s^*(u, v)$.

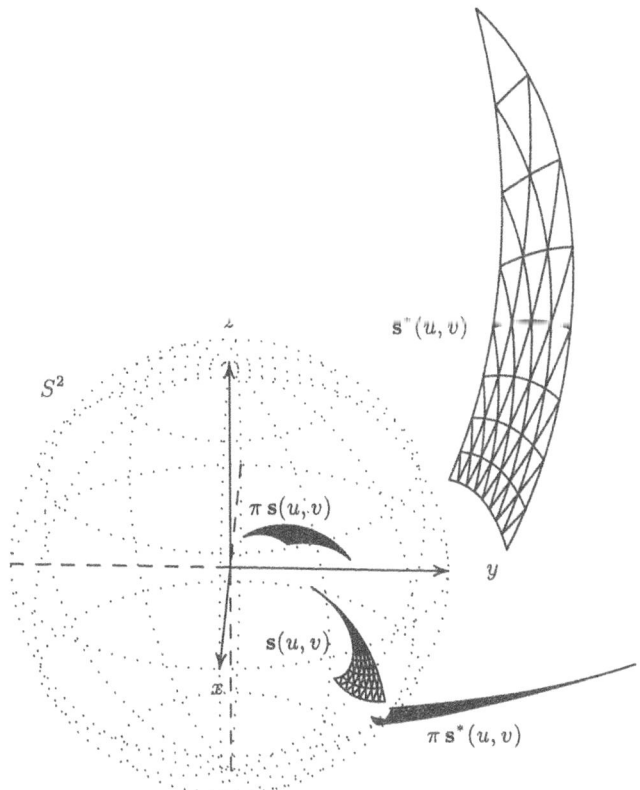

Fig. 8. A cubic cyclide ($p = 1$, $q = 1/4$) patch, its polar surface with respect to S^2 and their pedal surfaces with respect to the origin.

6   Conclusion

The polar curves and surfaces with respect to proper quadratic curves and surfaces have been investigated. Polar curves and surfaces are constructed from given curves and surfaces by using the polarity with respect to quadrics. Dual curves and surfaces are special cases of polar curves and surfaces.

While it is considered that a curve or surface and its dual are in different spaces, polar curves or surfaces can be constructed geometrically in the same space by using the polarity with respect to quadrics. The concepts of polar curves and surfaces are useful to investigate the geometric properties of rational curves and surfaces.

References

1. Berger M. (1987) Geometry. Springer-Verlag

2. Blinn J. (1997) The algebraic properties of second-order surfaces. In Bloomenthal J. (Ed.) Introduction to Implicit Surfaces. Morgan Kaufmann, 52–97
3. Boehm W., Prautzsch H. (1990) Geometric Concepts for Geometric Design. A K Peters
4. Cederberg J.N. (1989) A Course in Modern Geometries. Springer-Verlag
5. Farin G. (1999) NURBS – From Projective Geometry to Practical Use, 2nd edn. A K Peters
6. Farouki R.T., Pottmann H. (1996) Polynomial and rational Pythagorean-hodograph curves reconciled, In Mullineux G. (Ed.) The Mathematics of Surfaces VI, Oxford University Press, 355–378
7. Herman I. (1991) The Use of Projective Geometry in Computer Graphics. LNCS 564, Springer-Verlag
8. Hoschek J. (1983) Dual Bézier curves and surfaces, In Barnhill R.E., Boehm W. (Eds.) Surfaces in Computer Aided Geometric Design. North-Holland, 147–156
9. Penna M., Patterson R. (1986) Projective Geometry and its Applications to Computer Graphics. Prentice-Hall
10. Pottmann H. (1994) Applications of the dual Bézier representation of rational curve and surfaces. In Laurent P.J., Méhauté A.L., Schumaker L.L. (Eds.) Curves and Surfaces in Geometric Design. A K Peters, 377–384
11. Pottmann H. (1995) Rational curves and surfaces with rational offsets. Computer Aided Geometric Design 12:175–192
12. Pottmann H. (1995) Curve design with rational Pythagorean-hodograph curves. Advances in Computational Mathematics 3:147–170
13. Pottmann H. (1995) Studying NURBS curves and surfaces with classical geometry. In Dæhlen M., Lyche T., Schumaker L.L. (Eds.) Mathematical Methods for Curves and Surfaces. Vanderbilt University Press, 413–438
14. Pratt M.J. (1998) On a class of Pythagorean-normal surfaces with planar lines of curvature. In Cripps R. (Ed.) The Mathematics of Surfaces VIII. Information Geometers, 281–295
15. Samuel P. (1988) Projective Geometry, Springer-Verlag
16. Sederberg T.W., Goldman R.N. (1986) Algebraic geometry for computer-aided geometric design. IEEE Computer Graphics and Applications 4(6):52–59
17. Sederberg T.W., Wang X. (1987) Rational hodographs. Computer Aided Geometric Design 4:333–335
18. Ueda K. (1997) A sequence of Bézier curves generated by successive pedal-point constructions. In Le Méhauté A., Rabut C., Schumaker L.L. (Eds.) Curves and Surfaces with Applications in CAGD. Vanderbilt University Press, 427–434

Boundary Representation Models: Validity and Rectification

Nicholas M. Patrikalakis, Takis Sakkalis, and Guoling Shen

Massachusetts Institute of Technology, Cambridge MA 02139-4307, USA

Summary. Model validity, especially of manifold boundary representation (B-rep) models, has long been recognized as an important problem. This paper reviews issues on model validity of existing models. In particular, we present a set of sufficient conditions for representational validity of a typical B-rep data structure; we propose a rectify-by-reconstruction approach to the B-rep model rectification problem, and present results on the inherent complexity of the corresponding boundary reconstruction problem. Further, we develop the concept of an interval solid model, associated with a solid, for achieving numerical robustness. Finally, we present a set of sufficient conditions so that an interval solid model is "approximately equal" to its associated solid.

1 Introduction

Model validity, especially of manifold boundary representation (B-rep), has long been recognized as an important problem in solid modeling theory and CAD/CAM/CAE practice. B-rep defines solids by providing explicit information of solid boundaries. A manifold B-rep model is valid if it describes a 2-manifold without boundary [1], which bounds the intended solid. As is well-known, the validity of B-rep models is not self-guaranteed [2–4].

B-rep contains topological and geometric specification of the boundary of a solid [8]. Topological information, in general, is represented by a graph describing incidence and adjacency relations between topological boundary entities. Geometric specification involves equations for points, curves and surfaces. The distinction between topological and geometric specification is useful for several reasons [8]: a) In design of part families of the same topological but different geometric specifications, this distinction leads to significant data storage reduction. b) In design, many operators are geometric in nature (scaling, dilation, rotation, translation, digitizing, sculpting, tweaking), but involve only a single topological specification thereby reducing system complexity. c) The distinction allows use of the Euler operators [8,7], which reduce CAD system implementation complexity because of the bookkeeping they provide concerning at least partial coverage of the well-formedness conditions.

Defects in B-rep models are representational entities that do not conform to modeling constraints due to topological errors and/or inconsistencies between topological and geometric specification. In manifold B-rep models, defects appear as gaps, inappropriate intersections, dangling entities, internal walls and inconsistent orientations. Causes of defects exist throughout the entire life cycle of a model, including pathological behavior of geometric algorithms (e.g. geometric approximations), poor implementation of modeling operations (eg. during parametric model changes), precision limitation of the computer, data exchange and poor user practices. Consequences of defects could be severe. CAD data with defects could make modeling systems fail, generate useless CAE analysis results, produce defective products, and need tremendous rework at the receiving end of data exchange.

This paper studies validity verification and defect rectification (correction) of existing models. Section 2 gives definitions of solids and their boundaries, presents a complete set of conditions on representational nodes of a typical B-rep data structure and theorems on model validity. Section 3 proposes a rectify-by-reconstruction approach to the B-rep model rectification problem, and presents a theorem on the inherent complexity nature of the boundary reconstruction problem. Section 4 develops the concept of interval solid models for achieving robustness in a floating point computational environment, defines the notion of approximate equality between an interval solid and its intended solid, and provides sufficient conditions for approximate equality. For conciseness of the paper, proofs are omitted, but can be found in Shen [5]. A more detailed description on the material of Section 2 can also be found in Sakkalis et al [6].

2   Representational Validity of Manifold B-Rep Models

2.1   Prior work

Manifold B-rep describes real solids whose mathematical abstractions are r-sets with 2-manifold boundary [2,1,3]. An r-set is a bounded, regular and semianalytic subset in $\mathbf{R}^3$. B-rep uses a collection of 2-dimensional entities, called *faces*, properly connected and consistently oriented, to define solid boundaries. A B-rep model is valid if its faces form a 2-manifold without boundary. Specifically, a B-rep model is valid if the following conditions are satisfied [1]: (1) Faces may intersect only at common edges or vertices. (2) Each edge is shared by exactly two faces. (3) Faces around each vertex can be arranged in a cyclical sequence such that each consecutive pair share an edge incident to the vertex. Similar conditions were also presented in terms of a simplicial complex methodology by Hoffmann [4]. The above conditions for validity require topological integrity, meaning that it is possible to assign each topological entity a well-behaved geometry such that the overall geometry is a 2-manifold without boundary [1], and geometric integrity, meaning that geometric representation conforms to all topological relations presented in the topological structure [7]. Orientations of faces must obey *Moebius' Rule* [8,1]; that is, each edge is traversed exactly once in each direction.

Topological integrity can be assured by using Euler operators as primitive modeling operators. Euler operators were introduced by Baumgart [9] in his work on computer vision, and were extensively studied by many others [10,11,7,12,1]; see also Braid *et al.* [13] for historical notes on Euler operators. The theoretical basis is the well-known Euler-Poincaré formula, which holds for polyhedra in $\mathbf{R}^3$. In Braid et al [7], a topological structure is called admissible if certain conditions are satisfied, mainly the Euler-Poincaré formula. Admissibility does not imply validity because the Euler-Poincaré formula is necessary but not sufficient, meaning that a polyhedral model satisfying this formula may not be a 2-manifold [11,7,4]. Mäntylä [1] further proved that any topological polyhedron constructed using Euler operators is topologically valid. For models created using a modeling kernel whose operations are not based on Euler operators, topological structures could be invalid even if the Euler-Poincaré formula is satisfied.

Geometric integrity can be verified by two tests; see Braid *et al.* [7] for polyhedral models. Local test verifies if underlying curves of edges and coordinates of vertices are consistent with surfaces. Global test checks if there are any surface intersections in the interiors of faces.

2.2   Solids and their Boundaries

The definition of r-set captures mathematical characteristics of most real solids of interest, such as mechanical parts. However, this definition allows non-manifold features, such as the contacting edge between two cubes, a feature which is non-manufacturable (see Fig. 1(b)). It is for this reason that many prefer a solid to be a topological manifold [3]. In view of this, we may define:

Definition 21 *Let $S \subset \mathbf{R}^3$. S shall be called a solid if it is a compact and connected 3-manifold with boundary, so that:*

1. *Its boundary ∂S is a 2-manifold without boundary, and*
2. *$\partial S \subset \bigcup_{j=1}^{k} R_j$, where $\{R_j\}$ are almost smooth (AS) surfaces[1].*

Figure 1 illustrates several solids and non-solids. A solid is a triangulable manifold with boundary, and thus it is *orientable* [14]. The latter implies that we can distinguish its interior as well as its exterior.

The boundary of a solid S, $\partial S = S - int(S)$, may not be connected if the solid has cavities.

Definition 22 *Let S be a solid. A shell C of S is a connected component of ∂S [1]. For such a C, we define the inner (outer) part of C, $C^I(C^O)$, as the interior of the unique bounded (unbounded) connected component of $\mathbf{R}^3 - C$, respectively.*

---

[1] A differentiable surface Φ is smooth at a point (u_0, v_0, w_0) if there exists a tangent plane at Φ at (u_0, v_0, w_0). We also call Φ an almost smooth (AS) surface if Φ is smooth for all points in its domain, except at a set of points of measure zero.

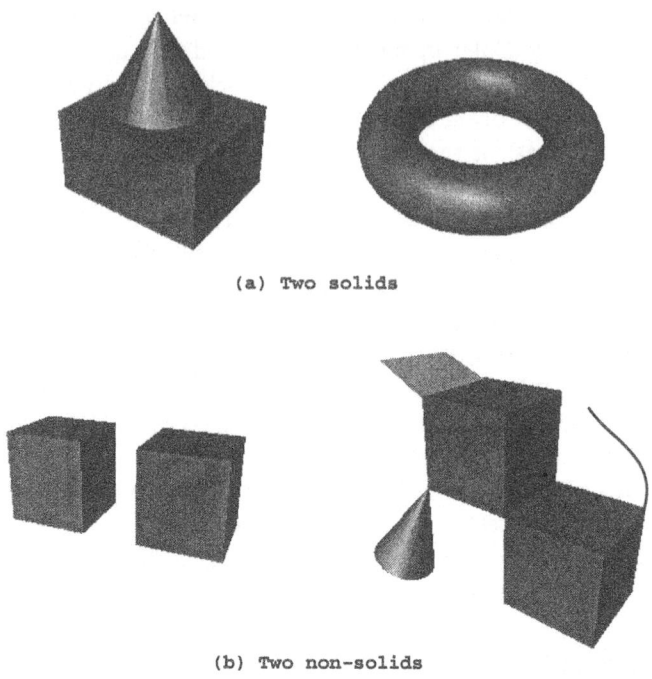

(a) Two solids

(b) Two non-solids

Fig. 1. Examples of solids and non-solids

Since C is a 2-manifold without boundary sitting in $\mathbf{R}^3$, it is orientable, and thus $\mathbf{R}^3 - C$ has precisely two connected components, one bounded and one unbounded.

Remark 21 *Let S be a solid. Then, there exists a unique shell C_e, called external shell, so that every other shell of S is contained in its inner part C_e^I.*

All shells of S, except C_e shall be called *internal*. Note from the above remark that

$$S = \overline{C_e^I} - \bigcup_{i=1}^{k} C_i^I , \tag{1}$$

where C_i are the internal shells of S, and $\overline{C_e^I}$ denotes the closure of C_e^I.

We next define the so-called *faces* of a solid, in order to achieve a boundary representation of solids. Face can be defined in an analogous way to a solid; see Mäntylä [1]. The following definition gives a clear description of face topology.

Definition 23 *A face f of a solid S is a non-void subset of ∂S having the following properties:*

1. f is a subset of one, and only one, AS surface.
2. The interior $Int(f)$ in ∂S is a connected 2-manifold, and
3. f is homeomorphic to a (topological) sphere Σ minus a finite number of mutually disjoint (nondegenerate) open disks $\Delta_i \subset \Sigma$, $i = 0, 1, \cdots, k, k \geq 0$, so that for $i \neq j$, $\partial\Delta_i \cap \partial\Delta_j$ is either empty or a single point.

Figure 2 illustrates several faces and non-faces. A face has no handles. The following remark comes directly from definition 23 and is useful for face representation.

Remark 22 Let f be a face of a solid S, and $k \geq 0$ be as in definition 23(3) Then,

- f is homeomorphic to F, where F is the closed unit disk $D = D_0$ minus k mutually disjoint (nondegenerate) open disks D_i, so that $\overline{D_i} \subset D$, for all $1 \leq i \leq k$, and $\partial D_i \cap \partial D_j$ is either empty or a single point, for $i \neq j, 0 \leq i, j \leq k$, and
- The interior of any simple closed curve $c \in \cup_{i=0}^{k}\partial D_i$ in F, is a subset of $\cup_{i=1}^{k} D_i$.

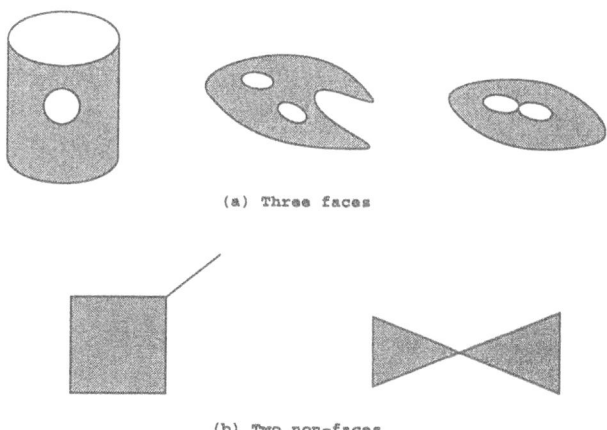

(a) Three faces

(b) Two non-faces

Fig. 2. Faces and non-faces

The reason that disks may meet at a single point as in Figure 2a (3rd case) is that there are 2-manifold boundary surfaces with such faces, which we do not wish to exclude from modeling. For example, consider a cube from which we subtract two small hemispheres with their equatorial planes on a single face of the cube and their equator curves tangent at a single point.

We define the boundary of f as $Bd(f) = f - Int(f)$. Once a homeomorphism between f and F has been established, it is easy to see what the boundary of f is. Indeed, let $g : f \to F$ be a homeomorphism of f and F. Then

$$Bd(f) = g^{-1}(\partial D) \cup \left(\bigcup_{i=1}^{k} \{g^{-1}(\partial D_i)\} \right) . \tag{2}$$

Definition 24 *Let f, F, g be as above. A* loop *of f is one of the simple closed curves $g^{-1}(\partial D)$ and $g^{-1}(\partial D_i)$.*

We may now define the notions of external/internal loops of a face f. Let f, F be as above. We define the external loop of f, L_e to be $L_e = g^{-1}(\partial D)$. We shall call all other loops L_i of f *internal.* Even though the definition of external/internal loops of a face is not unique, we may use that notion to 'reconstruct' a face from its boundary loops. To achieve the latter, we will have to define the notion of a region bounded by a loop in f. First define $R_e = g^{-1}(D)$, and $R_i = g^{-1}(D_i)$. Then, we have

$$f = R_e - \bigcup_{i=1}^{k} Int(R_i) . \tag{3}$$

Notice the similarity of the above formula to formula (1).

We may now expand the above discussion of the notion of the *region bounded by loops.* Let $\Sigma : \Phi : [0,1] \times [0,1] \to \mathbf{R}^3$ be an *AS* surface and let $c_j, j = 1, \cdots, m$ be simple closed curves that belong to Σ, with the property that $c_r \cap c_i$ is either empty or a single point. We also assume that in a neighborhood N_j of each c_j, Φ is 1–1. Let also Σ as well as the c_j's come with a specific orientation. Then, it is possible to define the *region bounded by the closed curves c_j, R_c^Σ.* To do that, we consider the curves $l_j = \Phi^{-1}(c_j)$. Then, since Φ is $1 - 1$ over N_j, each l_j is a simple closed curve.

The given orientation(s) on Σ and c_j induce–via Φ^{-1}–an orientation on $[0,1] \times [0,1]$ and each l_j. We can then agree that each of the above oriented closed curves define a region in $[0,1] \times [0,1]$. For each l_j we define the region bounded by l_j, $R(l_j)$ as the subset of $[0,1] \times [0,1]$ that lies to the left of l_j as one is walking on the positive side of $[0,1] \times [0,1]$, along l_j with respect to its orientation. Finally, we define

$$R_c^\Sigma = \begin{cases} R^* = \cap_j \Phi(R(l_j)), & \text{if } Bd(R^*) = \cup_j c_j \text{ and } R^* \text{ is connected} \\ \emptyset & \text{otherwise} \end{cases} \tag{4}$$

A loop can be decomposed into edges, and each edge is bounded by vertices.

Definition 25 *An edge e of a face f is a subset of $Bd(f)$ such that*

1. *It is also a subset of a loop, and*
2. *It is homeomorphic to either a simple closed curve or to an open simple curve.*

A vertex v of an edge e is a component of $Bd(e)$, if $Bd(e) \neq \emptyset$, or an arbitrary point in $Bd(e)$ if $Bd(e) = \emptyset$.

Definition 26 *A boundary representation of a solid S is a collection of faces f_i, $1 \leq i \leq n$, such that*

1. *$\cup_{i=1}^{n} f_i = \partial S$, and*
2. *For any $i \neq j$, $f_i \cap f_j = Bd(f_i) \cap Bd(f_j)$.*

2.3 Representational Validity Conditions

A typical data structure for B-rep models has a graph structure as shown in Fig. 3. The data structure is an adaptation of the STEP file structure [15], and is able to hold non-manifold features, as we assume a given representation could be topologically incorrect.

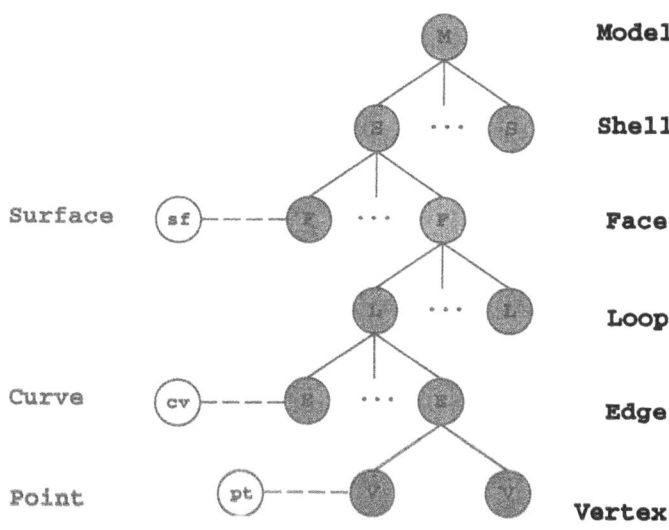

Fig. 3. Graph representation of the data structure

Representational validity verification, for a certain data structure like this one, is a process of verifying that each node in an instance of the data structure is a valid representation of the corresponding topological entity. The following sufficient conditions [6] are compiled from prior work such as [8,7,1,3,4,15], and the discussion in Section 2.2.

C 21 Vertex node validity *A vertex node is a valid representation of a vertex if it is assigned with the coordinates specifying the position of the vertex in $\mathbf{R}^3$.*

C 22 Edge node validity *An edge node is a valid representation of an edge if*

1. *Its two vertex nodes represent points on the underlying curve.*
2. *There exists a path on the curve, connecting these two points in the given orientation.*
3. *The curve does not self-intersect in the interior of the path.*

C 23 Loop node validity *A loop node represents a valid loop if*

1. *The directed graph constructed by taking vertices as nodes and oriented edges as directed arcs, is a simple directed cycle.*
2. *Any two adjacent edges intersect only at their common vertices. Any two non-adjacent edges do not intersect.*

C 24 Face node validity *A face node represents a valid face if*

1. *All its loop nodes represent loops on the surface.*
2. *The surface does not self-intersect in the interior of the region bounded by the loops, see (3).*
3. *The region bounded by the loops satisfies Remark 22.*

C 25 Shell node validity *A shell node represents a valid shell C if*

1. *Each face of C has exactly one adjacent face, which also belongs to C, through each of its edges.*
2. *Each vertex $v \in C$ has a finite number of incident faces $f_i \in C$ and (nondegenerate) edges $e_i \in C$, $1 \leq i \leq m$. These incident faces and edges can be arranged in the form of $f_1 - e_1 - f_2 - \cdots - f_m - e_m - f_{m+1} = f_1$, and e_i is a common edge of f_i and f_{i+1}, for $i = 1, \cdots, m$.*
3. *Any two adjacent faces only intersect at their common vertices and/or edges. Any two non-adjacent faces do not intersect.*
4. *Faces are consistently oriented.*

C 26 Model node validity *A model node represents a valid B-rep model if*

1. *There exists exactly one external shell.*
2. *Each of the inner shells is properly oriented such that the region it bounds is in the region bounded by the external shell.*
3. *None of the inner shells is in the region bounded by another inner shell.*
4. *Shells do not intersect.*

Using algebraic topology methods [14,16], the following theorems can be proven [5,6]:

Theorem 21 *Let C be a set that comes from a shell node satisfying conditions* **C21** *through* **C25**. *Then C is a compact oriented triangulable 2-manifold without boundary.*

Corollary 22 *Let S be a set that comes from a model node satisfying* **C26**. *Then S is a solid.*

3   Analysis of Manifold B-Rep Model Rectification

3.1   Prior work

Research on model rectification (repairing) has been done mainly on triangulated models, specifically, STL models for rapid prototyping, which represent solids using oriented triangles [17]. Defects in STL models are gaps due to missing triangles, inconsistently oriented triangles, inappropriate intersections in the interiors of triangles. Most algorithms [18,10] identify erroneous triangle edges, string such edges to form hole boundaries, and then fill holes with triangles. As pointed out in [18], topological ambiguities are resolved by intuitive heuristics. These algorithms [18,19] use local topology (incidence and adjacency) to rectify defects and are successful in the majority of candidate models, but may create undesirable global topological and geometric changes. In addition, they introduce ill-shaped triangles to fill gaps, which are inappropriate for many applications.

Barequet and Sharir [20] developed a global gap-closing algorithm for polyhedral models, using a partial curve matching technique. In their method, gap boundaries are discretized. Each match between any two parts of gap boundaries is given a score based on the closeness of their discrete points. They have shown that finding a consistent set of partial curve matches with maximum score, a subproblem of their repairing process, is NP-hard. Using this algorithm, Barequet and Kumar [21] developed a model repairing system, and the system was improved by Barequet et al [22] in terms of efficiency, and extended to models with regular arrangement of entire NURBS surface patches. However, this extension does not handle trimmed patches with intersection curve boundaries and general B-rep models involving non-regular arrangement of surface patches.

A different type of global algorithm is based on spatial subdivision. Murali and Funkhouser [23] developed an algorithm which handles defects such as intersecting and overlapping polygons and mis-oriented polygons. The algorithm first subdivides $\mathbf{R}^3$ into convex cells using planes on which polygons sit, and then uses heuristic rules to identify convex cells which are supposed to belong to the intended polyhedral solid. A major advantage of this algorithm is that it always outputs a solid. One limitation is that it may mishandle missing polygons and add cells which do not belong to the model.

Very little research on curved B-rep model verification and rectification has been published. Some commercial software is now marketed including model verification and rectification tools. Such tools, in general, are capable of identifying a wide variety of local defects. Rectification could be topological (e.g. removing an edge of under-tolerance size) or geometric (e.g. altering trimmed surfaces to bring gaps within tolerance), and may involve user assistance, see [46] for a partial list of such products.

3.2   Discussion

Current B-rep model rectification methods identify and rectify local defects using certain criteria. Global consistency and optimality are left unresolved. The ultimate goal of model rectification is to find the model intended by the designer but misrepresented. However, without user assistance, any solution resulting from the erroneous model is only an educated guess. It comes down to the question "What do we trust about a model that contains defects?" This is the most fundamental question we must answer before we can move any further. For example, if a B-rep model has correct topological information, while there is no appropriate geometry embedded in the underlying surfaces, what is the intended boundary? Should the topological information be modified to accommodate the geometry, or should the geometry (surfaces) be perturbed to have a boundary consistent with the topological information?

To answer such questions, we need to classify the pieces of information in a boundary representation: those initialized or selected by the designer and those induced by the system. In a solid modeling system, the only entities the designer can directly manipulate are the underlying surfaces. All others, topological information and individual topological and geometric entities, are computed by the system. For this reason, we opt to believe that *surfaces* are the basic information to be used in a rectification method. Another reason for such a hypothesis is that surfaces are specifically designed to fulfill certain performance requirements and functionalities, and therefore, should not be subjected to any modification without the designer's permission. In addition to the above, another piece of information which can be used in the rectification process is the *genus* of the intended solid boundary. Genus captures some aspects of the designer's intent and can be computed using the well-known Euler's formula [11].

In this context, model rectification becomes a model *reconstruction* problem. An algorithm for model rectification searches for the intended boundary in the union of all the surfaces, and rebuilds all necessary topological and geometric entities. However, there may exist many potential solid boundaries, or none at all, resulting from the surfaces. Without additional information, it would be difficult to make a choice. Since an erroneous model in a neutral format (e.g. STEP [15]) is computed with reasonable precision in its native system, the information in the model could help clarify such ambiguities, although it should be used with caution. Roughly speaking, a desirable solution is a model which describes a boundary somewhere "near" the object described by the original model, both topologically and geometrically.

3.3   Boundary reconstruction problem

Problem statement A typical data structure for B-rep models consists of topological structure and geometric representation. For a simplified version, see Fig. 3. The topological structure, shaded in Fig. 3, is a graph which describes adjacency and incidence relations[2] (represented by arcs) between topological entities (represented by nodes). In typical implementations more detailed topological relations are explicitly stored than those implied by Fig. 3 (e.g. adjacency relations between a face and all its neighboring faces). The geometric representation includes points, curve and surface equations, which are associated with appropriate topological entities. A model, thus, is an *instance* of the data structure, and is *valid* if it describes a solid boundary.

Let m_o be a model with topological structure $G(m_o)^3$. $G(m_o)$ is valid if it is possible to assign each topological entity a geometry whose interior is a manifold, such that the overall geometry bounds a solid. That is, if $G(m_o)$ is valid, there exists a nonempty set

$$\mathcal{M} = \{m|m \text{ is a valid model and has topological structure } G(m_o)\} \quad (5)$$

For simplicity, in the following analysis, we assume that *models have only one shell*. It will be clear at the end of this section that the same result applies to models with multiple shells. With this assumption, for any $m_1, m_2 \in \mathcal{M}$, M_1 is homeomorphic to M_2, because they have the same genus. Therefore, in case that the geometric representation of m_o is inconsistent with $G(m_o)$, if a reconstructed model m_n has topological structure $G(m_o)$, m_n is topologically equivalent to the model incorporating the design intent. If $G(m_n)$ is different from $G(m_o)$, the topological equivalence between m_n and m_o can be imposed by requiring that the genus of ∂M_n is equal to that of ∂M, where model $m \in \mathcal{M}$. We simply denote this by $g(m_n) = g(m_o)$, because both genera can be computed by applying the Euler's formula to $G(m_n)$ and $G(m_o)$, respectively.

Geometrically, two objects are close to each other if each one is in a neighborhood of the other. Some form of distance function could be used as a measure for this purpose, either the maximum distance or a well-defined average distance. An alternative, arguably more suitable for boundary rectification, is the boundary area change before and after rectification, because both the rectified and the original models use the same set of underlying surfaces. A correspondence can be established between a rectified face and an old face if they both have the same underlying surface, and the area difference between them measures the geometric change. No matter what measure is used, it should approach zero as the erroneous model becomes the exact model.

Let ϕ be a function which evaluates the geometric difference between ∂M_o and ∂M_n, and ε be a user-specified tolerance for the geometric change. Let

---

[2] Two topological entities of different dimensionalities have an incidence relation if one is a proper subset of the other. Two topological entities of same dimensionality have an adjacency relation if their intersection is a lower dimensional entity that has an incidence relation with each of them.

[3] We denote models and face nodes by lowercase letters, and the point-sets they represent by uppercase letters. For example, model m_o represents solid M_o. The topological structure of a node is denoted by $G(node)$.

$m_o, G(m_o)$ be as above. An ideal boundary reconstruction algorithm should follow the following procedure:

1. Find a new model m_n, such that m_n has topological structure $G(m_o)$ and $\phi(\partial M_o, \partial M_n) \leq \varepsilon$.
2. If there exist a number of such new models, select the one with the minimal ϕ value.
3. Otherwise, find a new model m_n, such that $G(m_n)$ is different from $G(m_o)$ but $g(m_n) = g(m_o)$, and $\phi(\partial M_o, \partial M_n) \leq \varepsilon$.
4. If there exist a number of such new models, select the one with the minimal topological structure change, e.g. the difference of the total numbers of arcs and nodes in $G(m_n)$ and $G(m_o)$ is minimal.
5. Otherwise, find a new model m_n with $\phi(\partial M_o, \partial M_n) \leq \varepsilon$. If there exist more than one such new boundaries, select the one with the minimal topological change (i.e. minimal genus change); otherwise, no new model is reconstructed.

In the next section, we study the following subproblem which is essential to this reconstruction process:

Boundary reconstruction (BR) problem: Given a B-rep model m_o, whose geometric representation is inconsistent with its topological structure, reconstruct a new model m_n using only the information in m_o, such that: (1) $g(m_n) = g(m_o)$ and (2) $\phi(\partial M_o, \partial M_n)$ is minimal.

Problem Complexity Before moving onto the boundary reconstruction (BR) problem, we first study a lower-dimensional problem which not only gives us insight on the nature of such reconstruction problems, but also plays a crucial role in understanding the complexity of the BR problem.

A face node in a B-rep model represented in a certain format such as STEP [15] is likely to have inconsistent geometric features such as edges whose underlying curves are not on the underlying surface of the face. Topological errors such as open loops make a clear definition of the face geometry even more elusive. Similar to boundary reconstruction, face reconstruction builds a new face node f_n from an erroneous face node f_o, using only the information in the given model, such that f_n is not only valid but also close to the object described by f_o in both topology and geometry. We formulate the face reconstruction problem as an analog to the BR problem.

A valid face is homeomorphic to a closed disk minus k mutually disjoint open disks, and has no handles. If f_o is valid, $G(f_o)$ is a planar graph consists of simple cycles. Any two of these cycles may share at most one common nodes. Two graphs are homeomorphic if both can be obtained from the same graph by a sequence of subdivisions of arcs [24]. However, two homeomorphic graphs may have different geometric embeddings, and thus define two faces which are not homeomorphic. See Fig. 4. In order to capture the design intent topologically, the component containing the outer loop in $G(f_n)$ also needs to be homeomorphic to that in $G(f_o)$, in addition to that $G(f_n)$ is homeomorphic to $G(f_o)$, so that the above situation is prevented. In this case we say that $G(f_n)$ is homeomorphic to $G(f_o)$ in the *strong sense*. In the following problem statement, ϕ_f is a function evaluating the geometric change before and after rectification.

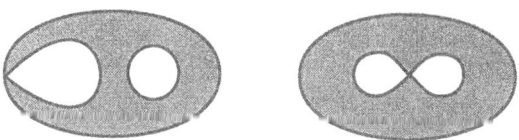

Fig. 4. Two homeomorphic graphs with different geometric embeddings

Face reconstruction (FR) problem: Given a face node f_o, whose geometric representation is inconsistent with its topological structure, in a B-rep model m_o, reconstruct a valid face node f_n, using only the information in m_o, such that $G(f_n)$ is homeomorphic to $G(f_o)$ in the strong sense and $\phi_f(F_n, F_o)$ is minimal.

It can be proven that a restricted FR problem is NP-hard, and therefore, the FR problem is also NP-hard. We then have the following theorem:

Theorem 31 *The BR problem is NP-hard.*

The proof of this theorem can be done by converting the restricted FR problem to the BR problem; see [5]. For models with multiple shells, the same result holds, because the boundary reconstruction problem of models with one shell is a special case of that of models with multiple shells.

4   Topological and Geometric Properties of Interval Solid Models

4.1   Motivation

Even with the existence of polynomial algorithms, building ideal boundary models still remains beyond the reach of current computing technology, because most geometries in solid modeling cannot be represented exactly due to the precision limitation of the computer [4,25]. In addition, pathological behavior of geometric algorithms (particularly, geometric approximations) introduce significant computational errors. Together with round-off errors in numerical computations, such errors make topological decisions inconsistent, especially when geometric degeneracies occur. In boundary model representations, these errors leave gaps along edges and thus create inconsistency between topological and geometric information.

One solution widely adopted by CAD/CAM practice is the use of tolerances. However, a tolerance often only represents accuracy of geometric computations. Moreover, semantics of computational tolerancing and its influence on validity and topological properties of models have not been fully studied.

Research efforts have been mainly directed to develop new arithmetic systems in which numbers are representable in the floating point environment and operations are closed. Essentially, these are approximation methods. Instead of exact geometries, approximations with certain guaranteed properties on numerical reliability and stability, are constructed. Typical examples are integer and rational arithmetic (viewed here as approximations of real arithmetic), and interval and lazy arithmetic, see [26–35] for more detail.

As an efficient and powerful extension of traditional floating point representation, interval arithmetic [36,37] and interval geometric representation [38–41] not only increase numerical stability but also assist in achieving model validity by defining gap-free boundaries. Hu et al [32,33] introduced a method for robust solid modeling using interval arithmetic, and developed a data structure and Boolean operations for manifold and non-manifold interval boundary models. The required curve and surface intersection algorithms using interval arithmetic were presented in [42,43]. However, topological issues involved in interval solid modeling were not studied in these papers [32,33,42,43]. As an interval boundary model defines a family of infinitely many boundaries, maintaining topological invariance of such boundaries is a complex problem. This section addresses such topological issues. Although the results presented here are primarily useful in boundary model reconstruction, they are also relevant in Boolean operations and boundary evaluation of interval models.

4.2   Covering Manifolds with Boxes

Throughout this section, a *box b* refers to a non-degenerate rectangular (closed) parallelepiped in $\mathbf{R}^3$, whose edges are parallel to the axes. A box defines a region in $\mathbf{R}^3$, and can be represented by an interval vector (i.e. a vector whose components are interval numbers.) The size of a box is the maximum length of its edges. Operations on boxes using rounded interval arithmetic [44] preserve enclosure, meaning that the computed boxes always contain exact boxes under the same operations. In this section, we present some results concerning the topology of a finite union of boxes. In particular, we shall be interested in the topology of a manifold when covered with boxes. The motivation behind this is our way of constructing interval solid models. We assume familiarity with basic topology [16].

Definition 41 *Let $\mathcal{B}$ be a finite collection of boxes, and let $A \subseteq \mathbf{R}^3$. We define $\mathbf{B} = \cup\{b|b \in \mathcal{B}\}$ and $A^{\mathcal{B}} = A \cup \mathbf{B}$.*

We begin with a general result. Let $\mathcal{B}$ be as above so that the following is satisfied:

A. Let $b_i, b_j \in \mathcal{B}$. Whenever $b_i \cap b_j \neq \emptyset$, then $b_{ij} = b_i \cap b_j$ is a box.

Then we have,

Proposition 41 *Let the collection $\mathcal{B}$ satisfy condition **A**. Then, **B** is a compact 3-manifold with boundary. Moreover, if **B** is connected, then **B** is a solid.*

Now let $V \subset \mathbb{R}^3$ be a compact connected orientable 2 manifold without boundary, and $\mathcal{B}$ as above. Suppose that, in addition to condition **A**, $\mathcal{B}$ satisfies the following:

B1. $V \subset \mathbf{B}$, that is, $\mathcal{B}$ covers V, and
B2. $b \cap V \neq \emptyset$, for every $b \in \mathcal{B}$.

We then have:

Corollary 41 *For $\mathcal{B}$ and $V^{\mathcal{B}}$ as above, $V^{\mathcal{B}}$ is a solid.*

Since V is connected and orientable, its complement has precisely two connected components, V_I, V_O; we may assume that V_I is interior to V, while V_O lies in its exterior. An argument similar to the one used in the proof of Proposition 41 shows that

Remark 41 $V_I^{\mathcal{B}}$ *is a solid.*

Evidently, the above construction can be applied to solids as well. Indeed, if M is a solid, and $\mathcal{B}$ satisfies conditions **A, B1** and **B2** when V is replaced by ∂M, then the above result shows that $M^{\mathcal{B}} = M \cup \mathbf{B}$ is a solid. In that case we shall call $M^{\mathcal{B}}$ the *interval solid* generated by M and $\mathcal{B}$. Figure 5 shows a 2D example of a solid M and its associated interval solid $M^{\mathcal{B}}$.

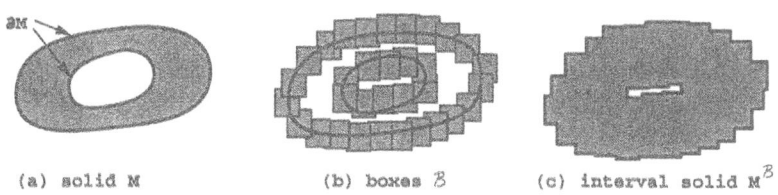

(a) solid M   (b) boxes $\mathcal{B}$ (c) interval solid $M^{\mathcal{B}}$

Fig. 5. An example of an interval solid

Consider now $S = V \cup V_I$. Then, obviously, S is a compact connected 3-manifold whose boundary is precisely V, i.e. S is a solid. For the following generic conditions (see Fig. 6):

C1. $\{Int(b_i), \ b_i \in \mathcal{B}\}$ is a cover of V.

C2. Each member b of $\mathcal{B}$ intersects V generically; that is, $b \cap V$ is a nondegenerate closed disk that separates b into two nondegenerate closed balls, B_b^+ and B_b^-.

C3. Whenever $b_i \cap b_j \neq \emptyset$, then $b_{ij} = b_i \cap b_j$ is a box that satisfies **C2**, for $b_i, b_j \in \mathcal{B}$.

we have:

Lemma 42 *S is homeomorphic to $V_I^{\mathcal{B}}$.*

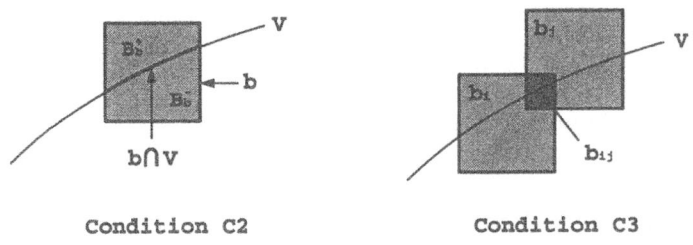

Fig. 6. 2D version of condition **C2** and **C3**

Now we are ready to state the main result of this section. Let M be a compact connected 3-manifold with boundary, and let $\mathcal{B}$ be a finite collection of boxes in $\mathbf{R}^3$ that satisfies conditions **C1-C3**, when V is replaced by ∂M. Then we have:

Theorem 43 *Let M, $\mathcal{B}$ be as above. Recall that $M^{\mathcal{B}} = M \cup \mathbf{B}$. Then, $M^{\mathcal{B}}$ is a 3-manifold with boundary that is homeomorphic to M. In fact, $M^{\mathcal{B}}$ is a solid [5].*

Corollary 44 *If M is a solid and $\mathcal{B}$ satisfies conditions **C1** through **C3**, then $M^{\mathcal{B}}$ and $M - \mathbf{B}$ are interval solids that are homeomorphic to M.*

Observe that in Corollary 44 condition **C1** is very natural in order for $M^{\mathcal{B}}$ to be an interval solid. The fact that $\mathcal{B}$ has to satisfy **C2** says that for every connected component C_i of ∂M, every member of $\mathcal{B}_i$ has to be in an open neighborhood N_i of C_i, and these N_i's are mutually disjoint. The latter puts a constraint on how big the boxes can be. In fact, we may confine each member of $\mathcal{B}_i$ to be in a *tubular neighborhood* T_i of C_i. Condition **C3** puts some constraints on the structure of $\mathcal{B}$. It can be shown that such a collection of boxes can be constructed using interval surface intersection (ISI) algorithms (e.g. [43]) with sufficiently tight resolution.

We close this section with the following definition which is motivated by the above results:

Definition 42 *Let M be a solid, and $\mathcal{B}$ be a finite collection of boxes. We say that the interval solid $M^{\mathcal{B}}$ is approximately equal to M if $M^{\mathcal{B}}$, as well as $M - \mathbf{B}$, are homeomorphic to M and the size of each $b \in \mathcal{B}$ is small.*

Figure 5 gives an example of an $M^{\mathcal{B}}$ which is approximately equal to M.

5   Example

The model shown in Fig. 7 is one part of a shaver handle, and created using a commercial CAD system. The size of the model is roughly $0.04\,m \times 0.06\,m \times 0.14\,m$. The underlying surfaces of the model consist of 16 integral and rational B-spline surfaces, 3 cylindrical surfaces and 5 planes. The global uncertainty measure is given by the designer as $10^{-6}\,m$. The model has $V = 40$ vertices, $E = 62$ edges, $F = 24$ faces, no inner loop, and one shell. The topological structure satisfies the sufficient conditions presented in Section 2.3. Therefore, from the Euler-Poincaré formula $V - E + F = 2(1 - G)$, we can deduce that the genus $G = 0$ and that the model is homeomorphic to a sphere.

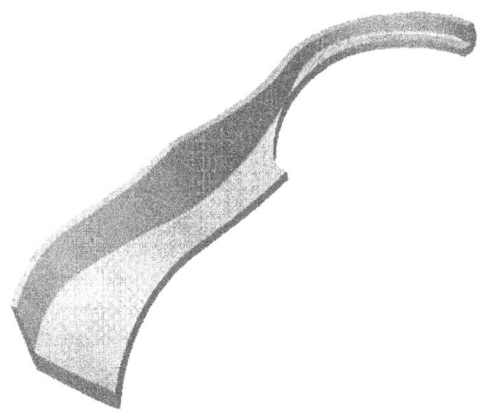

Fig. 7. Part of a shaver handle

We convert this model into an interval solid model. Because edges in the original model are reasonably computed, initial conversion of each face is performed by growing the width of underlying curves of its edges by a given resolution. If such growth gives a valid face, the conversion of the face is finished, unless adjacency relations with neighboring faces are violated. For the latter cases, the reconstruction process uses surface intersections.

The experiment starts with resolution $10^{-6}\,m$, given in the original STEP file. Eight faces have edges not on their underlying surfaces. Figure 8 shows one face with 2 edges partially on the underlying surface and one edge not overlapping the surface at all. Further computation reveals that the original face becomes valid at resolution $2 \times 10^{-4} \, m$. Figure 9 shows the valid face boundary. Overall, the model becomes valid at resolution $5 \times 10^{-4} \, m$.

Fig. 8. An invalid face with resolution $10^{-6} \, m$

Fig. 9. The same face as in Fig. 8 becomes valid at resolution $2 \times 10^{-4} \, m$.

We now test whether it is possible to reconstruct an interval model at the given resolution $10^{-6} \, m$, i.e. when all the surfaces in the model become interval surfaces with width $10^{-6} \, m$. For the face in Fig. 8, because its underlying surface, now an interval surface, does not intersect one of the surfaces on which its adjacent faces are embedded, no valid face boundary can be constructed from surface intersections. If we further grow the width of all the surfaces to $10^{-5} \, m$, the face can then be reconstructed. The remaining seven invalid faces can be rectified at various resolutions, and an interval model can be constructed at resolution $5 \times 10^{-5} \, m$.

6   Concluding Remarks

Further research should focus on developing actual rectification algorithms, perhaps including some form of approximation, a standard practice in NP-hard problems, see some preliminary results in Shen et al [45]. It is expected that such algorithms would be time-consuming, especially when computing surface intersections with high accuracy. Therefore, more efficient surface intersection and intersection curve data reduction methods are desirable.

In model representation methods and data exchange standards, tangential and higher order contact of two adjacent faces or edges may also be symbolically recorded and transmitted, because such information not only helps surface intersection computation, but also retains design intent. Additional topological information, such as the genus of each shell of a model, may also be included to assist in model verification and rectification. Finally, longer term research to develop modellers based on exact arithmetic methods is recommended.

Acknowledgements

NSF (DMI-9215411), ONR (N00014-96-1-0857), Kawasaki chair at MIT.

References

1. Mäntylä M. (1984) A note on the modeling space of Euler operators. Computer Vision, Graphics and Image Processing. **26**, 45–60
2. Requicha, A. A. G. (1980) Representations of Solid Objects - Theory, Methods and Systems. ACM Computing Surveys. **12**, 437–464
3. Mäntylä, M. (1988) An Introduction to Solid Modeling. Computer Science Press, Rockville, Maryland
4. Hoffmann, C. M. (1989) Geometric and Solid Modeling: An Introduction. Morgan Kaufmann Publishers Inc., San Mateo, California
5. Shen, G. (2000) Analysis of Boundary Representation Model Rectification. PhD Thesis, Massachusetts Institute of Technology, Cambridge, Massachusetts
6. Sakkalis, S., Shen, G., Patrikalakis, N. M. (2000) Representational Validity of Boundary Representation Models. Computer Aided Design. In press.
7. Braid, I. C., Hillyard, R. C., Stroud, I. A. (1980) Stepwise Construction of Polyhedra in Geometric Modeling. In: Brodlie K. W. (Ed.) Mathematical Methods in Computer Graphics and Design. Academic Press, London, 123–141
8. Eastman, C., Weiler, K. (1979) Geometric Modeling using Euler Operators. In: Proceedings of the First Annual Conference on Computer Graphics in CAD/CAM Systems. 248–259
9. Baumgart, B. (1975) A Polyhedron Representation for Computer Vision. In: Proceedings of National Computer Conference. AFIPS Press, Montvale, New Jersey, 589–596
10. Eastman, C., Lividini, J., Stoker, D. (1975) A database for designing large physical systems. In: Proceedings of National Computer Conference. AFIPS Press, Montvale, New Jersey, 603–611
11. Braid, I. C. (1979) Notes on a Geometric Modeller. CAD Group Document Number 101, Computer Laboratory, University of Cambridge, England
12. Mäntylä, M., Sulonen, R. (1982) GWB – A Solid Modeller with Euler Operators. IEEE Computer Graphics and Applications. **2**, 17–32

13. Braid, I. C. (1993) Boundary Modeling. In: Piegl, L. (Ed.) Fundamental Development of Computer-Aided Geometric Modeling. Academic Press, San Diego, CA, 165–184

14. Moise, E. E. (1977) Geometric Topology in Dimensions 2 and 3. Springer-Verlag, New York

15. U.S. Product Data Association (1994) ANS US PRO/IPO-200-042-1994: Part 42 – Integrated Geometric Resources: Geometric and Topological Representation

16. Munkres, J. R. (1975) Topology: a First Course. Prentice-Hall, Englewood Cliffs, New Jersey

17. 3D Systems Inc. (1988) Stereolithography Interface Specification

18. Bohn, J. H., Wozny, M. J. (1993) A Topology-Based Approach for Shell-Closure. In: Wilson, P. R., Wozny, M. J., Pratt, M. J. (Ed.) Geometric Modeling for Product Realization. Elsevier Science Publishers BV, 297–318

19. Mäkelä, I., Dolenc, A. (1993) Some Efficient Procedures for Correcting Triangulated Models. In: Proceedings of Solid Freeform Fabrication Symposium. University of Texas at Austin, 126–134

20. Barequet, G., Sharir, M. (1995) Filling gaps in the boundary of a polyhedron. Computer Aided Geometric Design. **12**, 207–229

21. Barequet, G., Kumar, S. (1997) Repairing CAD Models. In: Yagel, R., Hagen, H. (Ed.) Proceedings of IEEE Visualization Conference. Phoenix, Arizona, 363–370

22. Barequet, G., Duncan, C. A., Kumar, S. (1998) RSVP: A Geometric Toolkit for Controlled Repair of Solid Models. IEEE Transactions on Visualization and Computer Graphics. **4**, 162–177

23. Murali, T. M., Funkhouser, T. A. (1997) Consistent Solid and Boundary Representations from Arbitrary Polygonal Data. In: Van Dam, A. (Ed.) Proceedings of 1997 Symposium on Interactive 3D Graphics. ACM Press, New York, 155–162

24. Harary, F. (1969) Graph Theory. Addison-Wesley, Reading, Massachusetts

25. Hoffmann, C. M. (1989) The Problems of Accuracy and Robustness in Geometric Computation. Computer. **22**, 31–41

26. Milenkovic, V. (1993) Robust Polygon Modelling. Computer Aided Design. **25**, 546–566

27. Fortune, S. (1997) Polyhedral Modeling with Multiprecision Integer Arithmetic. Computer Aided Design. **29**, 123–133

28. Keyser, J., Krishnan, S., Manocha, D. (1999) Efficient and accurate B-rep generation of low degree sculptured solids using exact arithmetic: I–representations. Computer Aided Geometric Design. **16**, 841–859

29. Keyser, J., Krishnan, S., Manocha, D. (1999) Efficient and accurate B-rep generation of low degree sculptured solids using exact arithmetic: II–computation. Computer Aided Geometric Design. **16**, 861–882

30. Salesin D., Stolfi J., Guibas L. (1989) Epsilon Geometry: Building robust algorithms from imprecise calculations. ACM Annual Symposium on Computational Geometry at Saarbruecken, Germany. ACM Press, 208–217

31. Salesin, D. (1991) Epsilon geometry: Building robust algorithms from imprecise computations. PhD Thesis, Stanford University, California

32. Hu, C.-Y., Patrikalakis, N. M., Ye, X. (1996) Robust Interval Solid Modeling: Part I, Representations. Computer Aided Design. **28**, 807–817

33. Hu, C.-Y., Patrikalakis, N. M., Ye, X. (1996) Robust Interval Solid Modeling: Part II, Boundary evaluation. Computer Aided Design. **28**, 819–830

34. Benouamer, M., Michelucci, D., Peroche, B. (1994) Error-Free Boundary Evaluation based on a Lazy Rational Arithmetic: A detailed Implementation. Computer Aided Design. **26**, 403–415

35. Agrawal, A. (1995) A General Approach to the Design of Robust Algorithms for Geometric Modeling. PhD Thesis, University of Southern California, Los Angeles, California

36. Moore, R. (1966) Interval Analysis. Prentice-Hall, Englewood Cliffs, New Jersey

37. Alefeld, G., Herzberger, J. (1983) Introduction to Interval Computations. Academic Press, New York

38. Sederberg, T. W., Farouki, R. T. (1992) Approximation by Interval Bézier Curves. IEEE Computer Graphics and Applications. **12**, 87–95

39. Tuohy, S. T., Patrikalakis, N. M. (1993) Representation of geophysical maps with uncertainty. In: Thalmann, N., Thalmann, D. (Ed.) Communicating with Virtual Worlds, Proceedings of CG International '93 at Lausanne, Switzerland. Springer, Tokyo, 179–192

40. Tuohy, S. T., Patrikalakis, N. M. (1996) Nonlinear Data Representation for Ocean Exploration and Visualization. Journal of Visualization and Computer Animation. **7**, 125–139

41. Shen, G., Patrikalakis, N. M. (1998) Numerical and Geometric Properties of Interval B-Splines. International Journal of Shape Modeling. **1/2**, 35–62

42. Hu, C.-Y., Maekawa, T., Sherbrooke, E. C., Patrikalakis, N. M. (1996) Robust Interval Algorithm for Curve Intersections. Computer Aided Design. **28**, 495–506

43. Hu, C.-Y., Maekawa, T., Patrikalakis, N. M., Ye, X. (1997) Robust Interval Algorithm for Surface Intersections. Computer Aided Design. **29**, 617–627

44. Abrams, S., Cho, W. et al (1998) Efficient and Reliable Methods for Rounded Interval Arithmetic. Computer Aided Design. **30**, 657–665

45. Shen, G., Sakkalis, T., Patrikalakis, N. M. (2000) Manifold Boundary Representation Model Rectification. Proceedings of the 3rd International Conference on Integrated Design and Manufacturing in Mechanical Engineering, C. Mascle, C. Fortin, J. Pegna, editors. pp. 199 and CDROM. Presses internationales Polytechnique, Montreal, Canada, May 2000.

46. WWW-sites: www.cadiq.com, www.spatial.com, www.fegs.co.uk, www.iti-oh.com, www.theorem.co.uk, www.coretech-int.com.

Interval and Affine Arithmetic for Surface Location of Power- and Bernstein-Form Polynomials

Irina Voiculescu[1], Jakob Berchtold[2], Adrian Bowyer[3], Ralph R. Martin[4], and Qijiang Zhang[4]

[1] Oxford University Computing Laboratory, Oxford
[2] Spatial Technology Ltd, Cambridge
[3] Department of Mechanical Engineering, University of Bath, Bath
[4] Department of Computer Science, Cardiff University, Cardiff

Summary. This paper describes a problem of interest in CSG modelling, namely the location of implicit polynomial surfaces in space. It is common for surfaces defined by implicits to be located using interval arithmetic. However, the method only gives conservative bounds for the values of the function inside a region of interest. This paper gives two possible ways of producing tighter bounds. One involves using a Bernstein-form representation of the implicit polynomials used as input to the method. The other fine-tunes the method itself by employing careful use of affine arithmetic—a more sophisticated version of interval arithmetic. As both methods contribute significant improvements, we speculate about combining the two into a fast and accurate method for surface location.

1 Introduction

1.1 Surface location

CSG systems often use implicit polynomials to represent curved surfaces in space. Although the implicit representation is useful in many respects, it offers no information about the location of the surface it represents. Hence a common problem that needs to be solved is whether the surface cuts a given region of interest. This is important for any surface manipulation and rendering algorithms which may need to be implemented in a CSG modeller.

One of the most common surface location methods uses interval arithmetic. This paper considers how the straightforward approach, based on computing interval polynomials defined in the power basis, can be improved by two different techniques. The first is the use of the Bernstein basis instead of the power basis, and the second is the use of a careful affine arithmetic method for computing the intervals. These improvements also have the potential to be used together.

1.2   Power– and Bernstein–form polynomials

A univariate power–form polynomial of degree $n \in \mathbb{N}$ in the variable x is defined by:

$$p(x) = \sum_{k=0}^{n} a_k x^k \, ,$$

where $a_k \in \mathbb{R}$. The equation $p(x) = 0$ is the *implicit equation* corresponding to the polynomial $p(x)$.

Another possible representation of implicit polynomials is in terms of the Bernstein basis. For a given $n \in \mathbb{N}$ the univariate Bernstein basis functions of degree n on the interval $[\underline{x}, \overline{x}]$ (see also Lorentz [14]) are defined by:

$$B_k^n(x) = \binom{n}{k} \frac{(x - \underline{x})^k (\overline{x} - x)^{n-k}}{(\overline{x} - \underline{x})^n} \, , \qquad \forall x \in [\underline{x}, \overline{x}] \, , \qquad k = 0, 1, \ldots, n \, ,$$

and the Bernstein representation equivalent to $p(x)$ is:

$$b_p(x) = \sum_{k=0}^{n} m_k^n B_k^n(x) \, ,$$

where m_k^n are the Bernstein coefficients corresponding to the degree-n bases. The two univariate representations $p(x)$ and $b_p(x)$ are equivalent on the interval $[\underline{x}, \overline{x}]$ and conversion between them is fairly straightforward (see, for example, Farouki [9]).

The generalisation of Bernstein bases to multivariate polynomials is not immediate. The power form of a polynomial in d variables is written in terms of $x_1, \ldots, x_d$ like this:

$$p(x_1, \ldots, x_d) = \sum_{0 \leq k_1 + \cdots + k_d \leq n} a_{(k_1, \ldots, k_d)} x_1^{k_1} \cdots x_d^{k_d} \, ,$$

where the coefficients $a_{(k_1, \ldots, k_d)} \in \mathbb{R}$. Again, the equation $p(x_1, \ldots, x_d) = 0$ is the *implicit equation* corresponding to the *implicit polynomial* $p(x_1, \ldots, x_d)$. By convention, the degree of each term is $k_1 + \cdots + k_d$, and the degree of the polynomial is the maximum of all degrees of its terms.

The multivariate Bernstein form is defined recursively as a polynomial whose main variable is x_d and whose coefficients are multivariate Bernstein–form polynomials in $x_1, \ldots, x_{d-1}$.

Conversion between the power– and the Bernstein representation is possible regardless of the number of variables (Geisow [13], Garloff [12,24], Patrikalakis [17], Sherbrooke [20]). The examples presented in this paper make use of a conversion method from power into Bernstein form devised by some of the authors (Berchtold et al. [2]).

Geometrically, $p(x_1, \ldots, x_d)$ represents a surface in d–dimensional space. The results in this paper have been designed mainly for geometric applications in three–dimensional space, and so have been applied and implemented in a three–dimensional geometric modeller [3]. Generalisations can be made easily to any number of dimensions; however, most examples here will be given for bivariate polynomials as they are easier to depict.

1.3   Interval arithmetic in geometry

Any implicit trivariate polynomial expression $f(x, y, z)$ completely defines one or more (proper or degenerate) surfaces in space. A problem that must be solved if implicits are to be used for surface definition is that of locating the surface in space. It is very easy to membership–test a point against an implicit function—all that is required is to evaluate the function at the point. By convention in our modeller [3], a negative result indicates that the point is inside the solid, and a positive that it is outside. This simple way of distinguishing between the inside and the outside of an object is the principal reason for the use of implicits in CSG modelling schemes.

Interval arithmetic provides an efficient way to extend the idea of point membership testing to the testing of whole regions of space. Specifically, the point is replaced by a box defined by three intervals. Given the axially–aligned box $[\underline{x}, \overline{x}] \times [\underline{y}, \overline{y}] \times [\underline{z}, \overline{z}]$, the range of values that the function f takes inside it decides the type of the box. The range may be entirely positive (representing an 'air' box), entirely negative (representing a 'solid' box), or may straddle zero and so the box may be cut by the surface.

Thus if the three variables of the surface expression $f(x, y, z)$ are replaced by the three interval coordinates $[\underline{x}, \overline{x}]$, $[\underline{y}, \overline{y}]$ and $[\underline{z}, \overline{z}]$ respectively, this substitution produces an interval expression. By applying interval arithmetic rules, the evaluation of the interval expression $f([\underline{x}, \overline{x}], [\underline{y}, \overline{y}], [\underline{z}, \overline{z}])$ results in an interval which gives conservative bounds of the values of f in the box.

The best-known work on interval arithmetic is the textbook by Moore [16]. (He was mainly concerned with the use of intervals in numerical analysis, to bound errors in floating-point arithmetic.) For more geometrical applications of interval arithmetic the reader is referred to Snyder's book [21], to Bowyer et al. [4], and to Bowyer and Woodwark [5].

Philip Milne's thesis [15] (which is available as a technical report) also has a useful chapter of interval definitions and discussion, and Stolfi and Figueiredo [22] give a great deal of useful information.

Other applications of interval arithmetic to geometry have used intervals to represent geometric uncertainty in curves and surfaces [18,19,23]. This is done by using intervals for the *coefficients* of the polynomials in a suitable basis. In the current work, we consider these coefficients to be well-known, and only use intervals to represent ranges (boxes) of interest in space, within which we wish to localise a surface or curve.

2 Limitations of the interval arithmetic technique

2.1 The conservativeness problem

As described above (Section 1.3), once the interval coordinates of a box have been substituted for the variables of an implicit expression, and the interval arithmetic calculations have been performed, it may be the case that the resulting interval straddles zero.

At first sight this may seem to correspond to a situation where the box contains some surface. However, in reality, the box may be entirely solid or entirely air. Interval arithmetic is unable to classify all cases properly. This can be easily observed and generalised to any number of dimensions, and is known as the *conservativeness* problem.

Consider the following three functions f, g and h. Despite the fact that they take only positive values over the interval $[0, 1]$, the interval function evaluation outputs intervals containing zero. Each of these intervals is a superset of the function's image[1].

$$
\begin{array}{llll}
f, g, h & : & [0, 1] \to \mathbb{R}, & f(x) = g(x) = h(x) \\
f(x) & = & 4x^2 - 12x + 9 & \text{power form} \\
g(x) & = & (4x - 12)x + 9 & \text{Horner form} \\
h(x) & = & 9(x - 1)^2 - 6x(x - 1) + x^2 & \text{Bernstein form}
\end{array}
$$

$$
\begin{aligned}
f([0, 1]) &= 4([0, 1])^2 - 12[0, 1] + 9 \\
&= [-3, 13] & \text{resulting interval} \\
&\supset [1, 9] = Im_f & \text{actual image}
\end{aligned}
$$

$$
\begin{aligned}
g([0, 1]) &= (4[0, 1] - 12)[0, 1] + 9 \\
&= [-3, 9] & \text{resulting interval} \\
&\supset [1, 9] = Im_g & \text{actual image}
\end{aligned}
$$

$$
\begin{aligned}
h([0, 1]) &= 9([0, 1] - 1)^2 - 6[0, 1]([0, 1] - 1) + [0, 1]^2 \\
&= [0, 16] & \text{resulting interval} \\
&\supset [1, 9] = Im_h & \text{actual image}
\end{aligned}
$$

The functions f, g and h take the same values everywhere and have equivalent implicit expressions, so they must have the same image in the range—namely the interval $[1, 9]$. However, the interval arithmetic method gives predictions for the image which are wider intervals *including* it. This phenomenon is known as *interval swell*.

The example above illustrates the typical situation where the interval resulting from the location method straddles zero but does not indicate the presence of surface in the box: the box itself lies entirely in the positive

---

[1] $f : \mathbb{R}^3 \to \mathbb{R}$, $Im_f = \{y \in \mathbb{R} | \exists x \in [0, 1] \text{ s.t. } f(x) = y\}$

halfspace. This is why, when an interval straddling zero results from the calculation, the box involved cannot be classified as actually containing surface and is merely labelled "unknown".

The boxes that interval arithmetic does label as solid or air are always properly identified. Unfortunately this is only a cautious box classification as the technique cannot fully determine the type of *all* the boxes in a given region of interest.

It might be thought that the more variable occurrences there are in the implicit expression, the larger the interval swell, but this is not a general rule. There are other aspects (such as the presence of even exponents and the order in which the arithmetic operations are carried out) which contradict this assumption, as will be shown in Section 3.

2.2   Geometrical counterexample

In the rest of this paper the examples will be two–dimensional functions in rectangular regions of the plane, and we will use an analysis technique that we described in our previous paper [2]. A unit area of interest is subdivided recursively until a minimum box size of $2^{-6} \times 2^{-6}$ is reached. Individual sub-boxes are membership tested using interval arithmetic; boxes straddling zero are subdivided.

As an illustration of the conservativeness problem, let us consider a two-dimensional example (see also Berchtold et al. [2]):

$$s(x,y) = 0.945\,x\,y - 9.43214\,x^2\,y^3 + 7.4554\,x^3\,y^2 + y^4 - x^3 \; .$$

The function's graph is shown on the left–hand side of Figure 1. The graph on the right–hand side illustrates how many of the boxes could not be classified as either solid or air. (The definitely classified ones are shaded dark.) The light–shaded boxes are those minimum–sized ones still labelled as *unknown*.

As seen in Section 2.1, the technique provides looser or tighter bounds on the values of the polynomial in the box where it is being studied depending on the form in which the polynomial is stored. This paper suggests possible ways of improving the box classification.

On the one hand one can improve the results by providing equivalent input in a different format. Specifically, we have found the stability of the *Bernstein-form polynomials* particularly suitable for use with the interval arithmetic technique (see Farouki and Rajan [8,9]). This is the subject of Section 3.

On the other hand, the method itself can be improved on. As is, the technique does not take into account the fact that the different occurrences of x in a polynomial refer to the same quantity. The related *affine arithmetic* technique (see Comba [7]) takes into consideration the correlations and dependencies between the sources of error. In this way it is able to produce much tighter and more accurate intervals than are produced by interval arithmetic, if care is taken. This will be dealt with in Section 4.

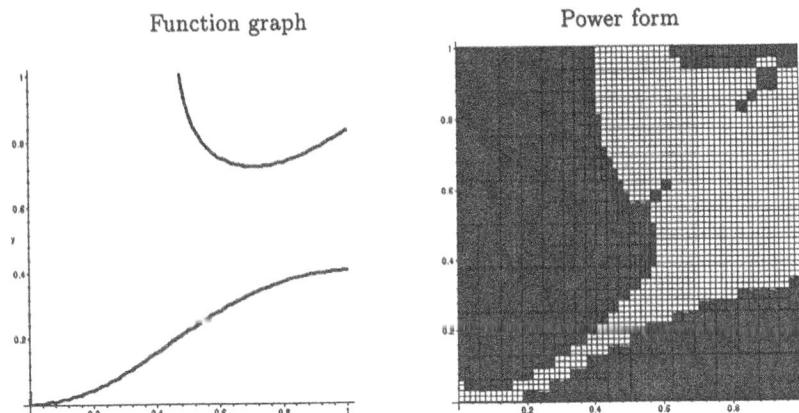

Fig. 1. Geometrical illustration of the conservativeness problem when using interval arithmetic; boxes giving entirely negative or entirely positive bound intervals are dark; light–shaded ones give intervals straddling zero.

3   Using the Bernstein–form of polynomials

3.1   Improvements of the box classification

In their well-known papers [8] and [9], Farouki and Rajan propose the following definition: *the numerical condition of a problem is a measure of the sensitivity of its solution to perturbations in its input parameters.* They argue and give evidence that the Bernstein form for a polynomial has a better numerical condition than the power form.

This property of Bernstein–form polynomials is especially interesting for geometric modelling. It is sometimes said that the conservativeness effect is a consequence of repeated occurrences of interval variables in expressions. In reality things are counter–intuitive. The Bernstein form of a polynomial has many occurrences of each of the variables—many more than the power form does. However, the former is better behaved than the latter from the point of view of interval arithmetic testing.

Let us consider once again the polynomial studied in Section 2.2, first in its power form

$$s(x, y) = 0.945\, x\, y - 9.43214\, x^2\, y^3 + 7.4554\, x^3\, y^2 + y^4 - x^3\ ,$$

and then in its Bernstein form in the unit box $[0, 1] \times [0, 1]$:

$$
\begin{aligned}
b_s(x, y) = &-x^3\, (1 - y)^4 \\
&+4(0.23625\, x\, (1 - x)^2 + 0.4725\, x^2\, (1 - x) - .76375\, x^3)\, y\, (1 - y)^3 \\
&+6(0.4725\, x\, (1 - x)^2 + 0.945\, x^2\, (1 - x) + 0.715066667\, x^3)\, y^2\, (1 - y)^2 \\
&+4\, (0.70875\, x\, (1 - x)^2 - 0.940535\, x^2\, (1 - x) + 1.078415\, x^3)\, y^3\, (1 - y) \\
&+((1 - x)^3 + 3.945\, x\, (1 - x)^2 - 4.54214\, x^2\, (1 - x) - 1.03174\, x^3)\, y^4\ .
\end{aligned}
$$

Obviously $b_s(x,y)$ is more complicated and has more variable occurrences than $s(x,y)$. However, it is evident from Figure 2 that it is worth using the polynomial in its Bernstein form for membership testing of boxes. The graph on the left (already encountered in Figure 1) shows the conservative results of interval arithmetic being applied to the power form. The graph on the right shows how much better behaved the Bernstein form is: using the same minimum box size, the profile of the implicit function is now clear, and many more of the boxes are classified definitely as *solid* or *air*.

Power form $s(x,y)$ Bernstein form $b_s(x,y)$

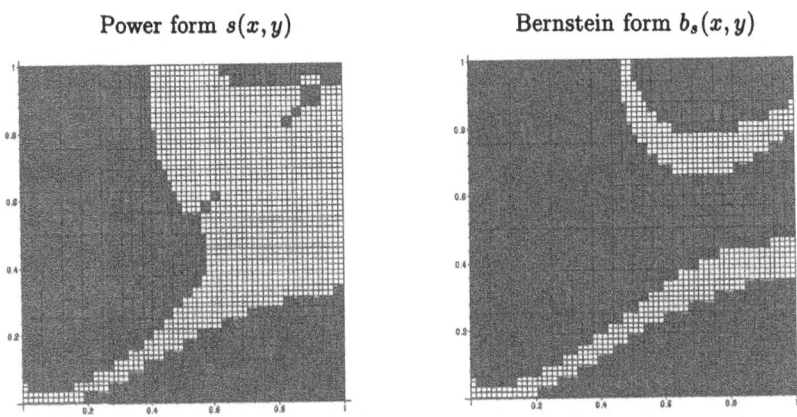

Fig. 2. Interval arithmetic tested on equivalent implicit polynomial forms

A numerical measure of the effectiveness of using different bases for localising a polynomial in a box can be obtained by considering the fractional area of the two–dimensional box which is definitely classified in each case. The relative areas definitely classified are shown below when applying interval arithmetic for the power and Bernstein bases:

	fractional area definitely classified
IA on power:	62.16 %
IA on Bernstein:	85.55 %

Many more examples of the differences in interval arithmetic performance between the power and the Bernstein form can be found in our previous paper [2] and in Berchtold [1]. The latter also considers the standard form (planar basis)[2].

---

[2] The standard form of a circle, for example, is $(x - x_c)^2 + (y - y_c)^2 - r^2 = 0$.

4   Using affine arithmetic

4.1   Definition

Affine arithmetic was proposed by Stolfi, Comba and others [7] in the early 1990s with the aim of tackling the conservativeness problem caused by standard interval arithmetic. Like interval arithmetic, affine arithmetic can be used to manipulate imprecise values and to evaluate functions over intervals. Also like interval arithmetic it provides guaranteed bounds for computed results, but affine arithmetic also takes into consideration the correlations or dependencies between the sources of error. In this way it is able to produce much tighter and more accurate intervals than interval arithmetic, especially in long chains of computations.

In affine arithmetic an uncertain quantity x is represented by an affine form $\hat{x}$ that is a first-degree polynomial of a set of noise symbols ε_i:

$$\hat{x} = x_0 + x_1\varepsilon_1 + \ldots + x_m\varepsilon_m = x_0 + \sum_{i=1}^{m} x_i\varepsilon_i .$$

Here the values of noise symbols ε_i are unknown but assumed to be in the range $[-1, 1]$. The corresponding coefficient x_i is a real number that determines the magnitude and sign of ε_i. Each ε_i stands for an independent source of error or uncertainty which contributes to the total uncertainty in the quantity x. One may make the number m as large as necessary in order to represent all the sources of error. These may well be input data uncertainty, formula truncation errors, arithmetic rounding errors, and so on.

This piece of reasoning is not restricted to the univariate case. On the contrary, given a polynomial expression in any number of variables, the dependencies between them can be easily expressed by using the same noise symbol ε_i wherever necessary. If the same noise symbol ε_i appears in two or more affine forms (e.g. in both $\hat{x}$ and $\hat{y}$) it indicates that some dependencies and correlations exist between the underlying quantities x and y.

Let $f(x, y)$ be a function representing a planar curve. Computing with affine forms is a matter of replacing each operation in $f(x, y)$ with an adequate operation on affine forms. This operation must take into account the relationships between the noise symbols in x and y. As with interval arithmetic, we do not detail them here. The reader will find them fully defined in Comba et al. [7].

One important thing to notice about the way affine arithmetic works is that it takes into account the fact that the same variable may appear more than once. Thus, when using affine arithmetic, similar terms are cancelled when they appear in an expression (e.g. $2\hat{x} + \hat{y} - \hat{x} = \hat{x} + \hat{y}$). This is not the case with interval arithmetic, as shown in Section 2.

4.2   Conversions between affine forms and intervals

Conversions between affine forms and intervals are defined in Comba and Stolfi [7], Figueiredo [10] and Figueiredo and Stolfi [11]:

Given an ordinary interval $[\underline{x}, \overline{x}]$ representing a quantity x, its affine form can be written as

$$\hat{x} = x_0 + x_1\varepsilon_x, \quad \text{where} \quad x_0 = \frac{\underline{x} + \overline{x}}{2}, \quad x_1 = \frac{\overline{x} - \underline{x}}{2}. \tag{1}$$

Conversely, given an affine form $\hat{x} = x_0 + x_1\varepsilon_1 + \ldots + x_m\varepsilon_m$, the range of possible values of its corresponding interval is

$$[\underline{x}, \overline{x}] = [x_0 - \xi, x_0 + \xi], \quad \text{where} \quad \xi = \sum_{i=1}^{m} |x_i|.$$

4.3   Practical applications

In our application of this to algebraic surface location, the polynomial representing the implicit surface needs to be evaluated on the intervals over which its variables range. In particular, in order to locate a planar curve we wish to evaluate a polynomial $f(x, y)$ over the ranges in x and y. These are $[\underline{x}, \overline{x}]$ and $[\underline{y}, \overline{y}]$ or their affine equivalents $\hat{x}$ and $\hat{y}$ respectively.

When represented in affine arithmetic each variable will have a representation containing just one noise symbol according to Equation 1. Let us suppose the general term of $f(x, y)$ is of the form $cx^a y^b$. The affine arithmetic evaluation of this term could be done directly using general affine arithmetic multiplication and addition rules. However practical experience with polynomials, other than those of lowest degree, shows that simply using the rules of affine arithmetic directly gives relatively little advantage over ordinary interval arithmetic when localising polynomials, due to rapid introduction of many new error symbols. Much better results can be obtained by taking more care (see Zhang and Martin [25]), as outlined in the rest of this section.

To evaluate a bivariate polynomial over intervals in x and y, we first convert the intervals to affine form using Equation 1, then compute the powers, next multiply the variables in each term, and finally add the terms.

Let us reason about the noise elements in the general product $x^a y^b$. By raising x to the power of a we get

$$\hat{x}^a = (x_0 + x_1\varepsilon_x)^a = x_0^a + \sum_{i=1}^{a} \binom{a}{i} x_0^{a-i} x_1^i \varepsilon_x^i. \tag{2}$$

Note that in the above polynomial (2), when i is odd ε_x^i will vary in the range $[-1, 1]$, and when i is even ε_x^i will vary in the range $[0, 1]$. In other words, the noise ε_x^i varies over the same range in alternate terms.

We now replace all ε_x^i, which must all have the same sign for $i = 1, 3, 5, \ldots$, by a new noise symbol ε_{xod}, and add together their signed coefficients into x_{od}. Similarly, for all $i = 2, 4, 6, \ldots$ we get $x_{ev}\varepsilon_{xev}$.

The result of the power operation $\hat{x}^a$ can be simplified from a degree-a polynomial of $a + 1$ terms to a degree-one polynomial of three terms and just two noise symbols: $x_0^a + x_{od}\varepsilon_{xod} + x_{ev}\varepsilon_{xev}$. Similarly let $y_0^b + y_{od}\varepsilon_{yod} + y_{ev}\varepsilon_{yev}$ be the polynomial result after $\hat{y}$ is raised to the power of b.

Note that while this does not lead to the narrowest possible intervals for the polynomials x^a and y^b over the range, it does provide quite tight intervals in many cases. Furthermore, the noise symbols produced are the same as those occurring in the other terms of $f(x, y)$.

A single term evaluates to:

$$
\begin{aligned}
\hat{x}^a\hat{y}^b &= (x_0^a + x_{od}\varepsilon_{xod} + x_{ev}\varepsilon_{xev})(y_0^b + y_{od}\varepsilon_{yod} + y_{ev}\varepsilon_{yev}) \\
&= x_0^a y_0^b + \\
&\quad (x_0^a y_{od})\varepsilon_{yod} + (x_0^a y_{ev})\varepsilon_{yev} + \\
&\quad (x_{od}y_0^b)\varepsilon_{xod} + (x_{ev}y_0^b)\varepsilon_{xev} + \\
&\quad (x_{od}y_{od})\varepsilon_{xod}\varepsilon_{yod} + (x_{ev}y_{od})\varepsilon_{xev}\varepsilon_{yod} + \\
&\quad (x_{od}y_{ev})\varepsilon_{xod}\varepsilon_{yev} + (x_{ev}y_{ev})\varepsilon_{xev}\varepsilon_{yev} .
\end{aligned}
\tag{3}
$$

After computing powers and multiplications, all the terms of $f(x, y)$ will be in the same form as shown above (3). Addition and subtraction of these terms is now straightforward, yielding an expression of the same form.

Let this final result be:

$$
\begin{aligned}
r = {}& r_0 + r_1\varepsilon_{yod} + r_2\varepsilon_{yev} + r_3\varepsilon_{xod} + r_4\varepsilon_{xev} + \\
& r_5\varepsilon_{xod}\varepsilon_{yod} + r_6\varepsilon_{xev}\varepsilon_{yod} + r_7\varepsilon_{xod}\varepsilon_{yev} + r_8\varepsilon_{xev}\varepsilon_{yev} .
\end{aligned}
$$

The term $\varepsilon_{xod}\varepsilon_{yev}$ and similar quantities can be treated as independent noise quantities. The products $\varepsilon_{xod}\varepsilon_{yev}$, $\varepsilon_{xev}\varepsilon_{yod}$ and $\varepsilon_{xod}\varepsilon_{yod}$ will still vary between $[-1, 1]$ as does a single ε_{xod} or ε_{yod}; the product $\varepsilon_{xev}\varepsilon_{yev}$ will vary between $[0, 1]$ as does a single ε_{xev} or ε_{yev}; so

$$
\begin{aligned}
r = {}& [r_0 - \zeta_{lo} , \ r_0 + \zeta_{hi}] , \qquad \text{where} \\
\zeta_{lo} = {}& |r_1| + |min(0, r_2)| + |r_3| + |min(0, r_4)| + \\
& |r_5| + |r_6| + |r_7| + |min(0, r_8)| , \\
\zeta_{hi} = {}& |r_1| + |max(0, r_2)| + |r_3| + |max(0, r_4)| + \\
& |r_5| + |r_6| + |r_7| + |max(0, r_8)| .
\end{aligned}
$$

When using affine arithmetic, boxes are membership-tested against an implicit function, just as when locating curves with normal interval arithmetic. If the interval $r = [r_0 - \zeta_{lo} , \ r_0 + \zeta_{hi}]$ resulting from the affine arithmetic is entirely negative then the box is inside the solid; if the interval is entirely positive—the box is air; and if the interval straddles zero the box is unknown.

4.4 Examples

Because the affine arithmetic form can be converted back into an interval, it can easily be used as an alternative to producing box classifications for power- or Bernstein-form polynomials using direct interval arithmetic rules.

To compare the relative merits of interval arithmetic and carefully evaluated affine arithmetic for curve drawing, we now present a practical example. We shall consider the function $s(x, y)$ studied in Section 3. The left hand side

Interval arithmetic on Bernstein form   Affine arithmetic on power form

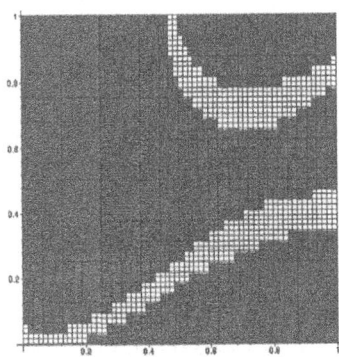
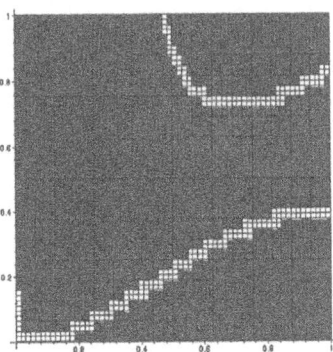

Fig. 3. Interval- and affine arithmetic classifications for $b_s(x, y)$ and $s(x, y)$

of Figure 3 represents the interval arithmetic classification of the Bernstein form, already encountered in Figure 2. The right hand side illustrates the result of applying affine arithmetic to the power-form polynomial $s(x, y)$.

As apparent from the figure, affine arithmetic definitely classifies comparatively a larger area than either case of interval arithmetic. The table below gives the respective surface area percentages in all the classifications considered in this paper:

fractional area definitely classified

IA on power form:	62.16 %
IA on Bernstein form:	85.55 %
AA on power form:	93.71 %

The complexity of each algorithm depends on the type of arithmetic used (i.e. classical interval arithmetic or affine arithmetic), as well as on the form of the input. The table below summarises the running times and the number of subdivisions in each case. (Note that the times are interesting to compare, but not relevant in absolute terms, as the implementation made use of an inefficient interval package in Maple [6].)

	time	subdivisions
IA on power form:	20.94 (s)	1854
IA on Bernstein form:	51.52 (s)	947
AA on power form:	10.07 (s)	433

For the example given here the affine arithmetic method produces results more quickly (and accurately) than either interval arithmetic method. The former involves slightly more calculations per box, but classifies big boxes in a very efficient manner. When interval arithmetic is applied there are less calculations per box than for affine arithmetic. Still, the Bernstein polynomial form is so much more complicated that the program runs much slower.

Regarding the number of subdivisions, interval arithmetic needs to subdivide boxes a lot finer for the power form than for the Bernstein form and ends up with a less accurate result. Affine arithmetic needs comparatively less subdivisions to reach a very accurate result.

5 Conclusions

Overall, we conclude that the conservativeness problem which occurs in surface location can be reduced in at least two ways[3]: either the input is given in Bernstein form instead of power form and interval arithmetic is used, or the calculations are carried out on the power form, but a careful strategy based on affine arithmetic is used instead of interval arithmetic.

When the Bernstein form is used the improvement is significant: boxes can be located much more accurately in a given region of interest. The shape of the surface is outlined in enough detail for location purposes.

When affine arithmetic is used as outlined above, our experimental results here demonstrate that curves are even more closely located. This occurs because the intervals produced during polynomial evaluation are tighter.

Affine arithmetic calculations are more complicated than interval arithmetic ones. In some cases, we have found it to be perhaps twice as slow as simple interval arithmetic, although this is strongly dependent on the implementation. However, speed advantages are present in some cases when interval arithmetic performs particularly badly. These advantages arise in the subdivision method because less boxes need to be considered, even though the amount of computation for any single box is greater.

To sum up, we may remark that we fully expect the benefits shown in curve drawing to also be applicable to other uses of solutions to implicit equations, such as surface intersection, surface location, etc. Although the examples shown here have used polynomials, similar approaches could also be used if non-polynomial functions are needed for modelling. Different suitable basis functions and affine evaluation methods will need to be found for such cases.

In principle, we could also, instead of computing the power form using the affine arithmetic approach, compute the Bernstein form using the affine arithmetic (rather than the interval arithmetic) approach. We hope that this will give even better results. This will be the subject of future research.

---

[3] There are others too, like storing some polynomials in standard form or in the planar basis; see [2].

References

1. Berchtold J. (2000) The Bernstein Form in Set-Theoretic Geometric Modelling. PhD Thesis, University of Bath.
2. Berchtold J., Voiculescu I., Bowyer A. (1998) Interval Arithmetic Applied to Multivariate Bernstein-Form Polynomials. Technical Report Number 31/98, School of Mechanical Engineering, University of Bath.
3. Bowyer, A. (1995) sVLIs : Set-Theoretic Kernel Modeller, Information Geometers Ltd.
4. Bowyer A., Berchtold J., Eisenthal D., Voiculescu I., and Wise K. (2000) Interval Methods in Geometric Modelling. In: Martin R., Wang W. (Eds.) Geometric Modeling and Processing 2000, IEEE Computer Society Press, 321–327.
5. Bowyer A., Woodwark J. (1993), Introduction to Computing with Geometry, Information Geometers Ltd.
6. Char B. W. [et al.] (1991) Maple V Language Reference Manual, Springer.
7. Comba J. L. D., Stolfi J. (1993) Affine Arithmetic and its Applications to Computer Graphics. Proceedings of the VII Sibgrapi (Brazilian Symposium on Computer Graphics and Image Processing), Recife, Brazil.
8. Farouki R. T., Rajan V. T. (1987) On the Numerical Condition of Polynomials in Bernstein Form. Computer Aided Geometric Design 4:191–216.
9. Farouki R. T., Rajan V. T. (1988) Algorithms for Polynomials in Bernstein Form. Computer Aided Geometric Design 5:1–26.
10. de Figueiredo L. H. (1996) Surface Intersection Using Affine Arithmetic. Proceedings of Graphics Interface'96, 168–175.
11. de Figueiredo L. H., Stolfi J. (1996) Adaptive Enumeration of Implicit Surfaces with Affine Arithmetic. Computer Graphics Forum 15(5):287–296.
12. Garloff J. (1985) Convergent Bounds for the Range of Multivariate Polynomials. In: Interval Mathematics 1985, Lecture Notes in Computer Science 212:37–56.
13. Geisow A. (1983) Surface Interrogations, PhD Thesis, University of East Anglia.
14. Lorentz G. G. (1986) Bernstein Polynomials. Chelsea Publishing Company, New York.
15. Milne P. S. (1990) On the Algorithms and Implementation of a Geometric Algebra System. University of Bath Computer Science Technical Report 90–40.
16. Moore R.E. (1979) Methods and Applications of Interval Analysis, SIAM.
17. Patrikalakis N. M., Kriezis G. A. (1989) Representation of Piecewise Continuous Algebraic Surfaces in Terms of B-Splines. The Visual Computer, Journal of the Computer Graphics Society. 5(6):360–374.
18. Sederberg T. W., Farouki R. T. (1992) Approximation By Interval Bezier Curves. CGA, 12(5):87–95.
19. Shen G., Patrikalakis N. M. (1998) Numerical and Geometric Properties of Interval B-Splines. International Journal of Shape Modeling. 4(1,2):35–62.
20. Sherbrooke E. C., Patrikalakis N. M. (1993) Computation of the Solutions of Nonlinear Polynomial Systems. Computer Aided Geometric Design. 10(5):379–405.
21. Snyder J. M. (1992) Generative Modeling for Computer Graphics and CAD. Academic Press.

22. Stolfi J., de Figueiredo L. H. (1997) Self-Validated Numerical Methods and Applications. Course notes for the 21st Brazillian Mathematics Colloquium, IMPA, July.
23. Tuohy S. T., Maekawa T., Shen G., Patrikalakis N. M. (1997) Approximation of Measured Data with Interval B-Splines. Computer Aided Design. 29(11):791–799.
24. Zettler M., Garloff J. (1998) Robustness Analysis of Polynomials with Polynomial Parameter Dependency using Bernstein Expansion. IEEE Transactions on Automatic Control, 43 (3):425–431.
25. Zhang Q., Martin R. R. (2000) Polynomial Evaluation using Affine Arithmetic for Curve Drawing. In: Proceedings Eurographics UK Conference, 2000, 49–56.

A Class of Bernstein Polynomials that Satisfy Descartes' Rule of Signs Exactly

Joab R. Winkler[1] and David L. Ragozin[2]

[1] The University of Sheffield, Department of Computer Science
   211 Portobello Street, Sheffield S1 4DP, United Kingdom
[2] The University of Washington, Department of Mathematics
   Box 354350, Seattle, WA 98195, USA

Summary. Let $Ta = b$ where $a = \{a_i\}_{i=0}^n$ and $b = \{b_i\}_{i=0}^n$ are the coefficients of a polynomial in the power and Bernstein bases respectively, and T is the transformation matrix between the bases. If USV^T is the singular value decomposition of T, it is shown that the Bernstein polynomial $p_r(x)$ whose coefficients are given by column r of the left singular matrix U, that is, $b = \{u_{ir}\}_{i=0}^n$, satisfies Descartes' rule of signs exactly because the number of sign changes of the coefficients b_i is exactly equal to the number of roots of $p_r(x)$ in the interval $[0, 1]$. This result provides a new interpretation of polynomial basis conversion because U also determines the numerical condition of the basis transformation equation $Ta = b$. This connection is established by showing that T is a totally non–negative matrix and TT^T is an oscillation matrix. Examples that illustrate the theoretical results are presented.

1   Introduction

Many problems in geometric modelling that are associated with computations on curves and surfaces require the determination of the real roots of a univariate polynomial on a bounded interval. Typical examples include the processing of the curve of intersection of parametric surfaces [5], ray–tracing parametric patches [10], and the computation of the points of intersection of parametric curves [11]. An interesting property of univariate Bernstein polynomials that unites two distinct topics in the theory of polynomials is considered in this paper. Specifically, it is shown that the singular value decomposition of the transformation matrix between the power and Bernstein bases is important in determining both the numerical condition of the transform between the bases and the number of roots of a polynomial that lie in the interval $I = \{x : 0 \leq x \leq 1\}$.

Polynomial basis conversion is an important requirement for high integrity data exchange between computer–aided design systems that use different representations of curves and surfaces. Although the linear algebraic equation that relates the coefficients of a polynomial $p(x)$ in the power and Bernstein polynomial bases may be ill–conditioned, computationally reliable answers can be obtained by regularisation [13–15]. This topic has been considered separately from the number of roots of $p(x)$ that lie in the interval I, but it is shown in this paper that these two topics are closely related.

Let the power and Bernstein basis coefficients of $p(x)$ be $a = \{a_i\}_{i=0}^n$ and $b = \{b_i\}_{i=0}^n$ respectively,

$$p(x) = \sum_{i=0}^n a_i x^i = \sum_{i=0}^n b_i \binom{n}{i} (1-x)^{n-i} x^i. \tag{1}$$

These coefficients are related by [6]

$$Ta = b, \tag{2}$$

where T, the polynomial basis transformation matrix, is a non–singular matrix of order $(n+1) \times (n+1)$ whose elements $\{t_{ij}\}_{i,j=0}^n$ are

$$t_{ij} = \begin{cases} \binom{i}{j} / \binom{n}{j}, & \text{if } i \geq j, \\ 0, & \text{if } i < j. \end{cases} \tag{3}$$

It is shown in this paper that the expansion of b with respect to the left singular vectors of T determines the numerical condition of (2) and yields bounds on the number of roots of $p(x)$ that lie in the interval I. This provides an interpretation of polynomial basis conversion that complements the established interpretation of ill–conditioned linear algebraic equations in terms of the discrete Picard condition.

The numerical condition of (2) is considered in section 2 using the singular value decomposition (SVD) of T, and it is shown in section 3 that T is a totally non–negative matrix and that TT^T is an oscillation matrix. This imposes restrictions on the signs of the elements of the eigenvectors of TT^T and $T^T T$, or equivalently, the left and right singular vectors of T. These singular vectors are obtained from the SVD of T, which enables the theoretical results in sections 2 and 3 to be combined. Examples that illustrate the results are presented in section 4, and section 5 contains a discussion of the results.

2   The numerical condition of $Ta = b$

The 2–norm condition number, $\kappa_2(T)$, of T is given by s_0/s_n, where $s_i, i = 0, \ldots, n$, are the singular values of T, arranged in non–increasing order. It is important to note that this measure of the condition of (2) is only a function of T and is independent of b. However in practice the numerical condition of this equation changes as the right hand side vector b changes, and it is therefore appropriate to define a condition number for a given vector b, such that as b changes, this condition number also changes. This will reveal those vectors b (and hence polynomials (1)) for which (2) is well–conditioned, and those vectors for which this equation is ill–conditioned. This revised condition number for a given b is called the *effective condition number* and denoted by $S(T, b)$. The distinction between the condition number $\kappa_2(T)$ and the effective condition number $S(T, b)$ follows clearly from their definitions,

$$\kappa_2(T) = \max_{\delta b, b \in \mathbb{R}^{n+1}} \frac{\Delta a}{\Delta b} \quad \text{and} \quad S(T, b) = \max_{\delta b \in \mathbb{R}^{n+1}} \frac{\Delta a}{\Delta b}. \tag{4}$$

Δa and Δb are the relative errors in a and b respectively,

$$\Delta a = \frac{\|\delta a\|_2}{\|a\|_2} \quad \text{and} \quad \Delta b = \frac{\|\delta b\|_2}{\|b\|_2},$$

where, from (2), $T\delta a = \delta b$. It follows from (4) that the condition number $\kappa_2(T)$ is the maximum of the effective condition number $S(T, b)$ taken over all vectors b,

$$\kappa_2(T) = \max_{b \in \mathbb{R}^{n+1}} S(T, b).$$

Using the singular value decomposition of T, it is easily shown that the effective condition number $S(T, b)$ is given by [12]

$$S(T, b) = \frac{1}{s_n} \frac{\|c\|_2}{\|S^{-1}c\|_2}, \qquad c = U^T b, \tag{5}$$

where USV^T is the SVD of T. U and V are orthogonal matrices and S is the diagonal matrix of the singular values $s_i, i = 0, \ldots, n$. A measure that is similar to the effective condition number is used in [3] to show that the projection of b onto the range space of T can strongly affect the sensitivity of a.

The effective condition number is approximately equal to $\kappa_2(T)$ when

$$\frac{\|c\|_2}{\|S^{-1}c\|_2} = \sqrt{\frac{\sum_{i=0}^n c_i^2}{\sum_{i=0}^n \frac{c_i^2}{s_i^2}}} \approx s_0,$$

and this equation is satisfied when

$$\frac{|c_i|}{s_i} \gg \frac{|c_{i+1}|}{s_{i+1}}, \qquad i = 0, \ldots, n-1. \tag{6}$$

Since the singular values are arranged in non–increasing order, it follows from (6) that the effective condition number is approximately equal to $\kappa_2(T)$ when the coefficients $|c_i|$ decrease to zero more rapidly than do the singular values s_i. The condition (6) is the discrete Picard condition [8], and its satisfaction is crucial for the regularisation of ill–conditioned linear algebraic equations by truncated singular value decomposition and Tikhonov regularisation in standard form [9]. It is noted that if $|c_i| = s_i, i = 0, \ldots, n$, and the singular values decay sufficiently rapidly towards zero, then (2) is ill–conditioned [16] but truncated singular value decomposition and Tikhonov regularisation in standard form cannot be used to regularise the equation because the discrete Picard condition is not satisfied, and as a consequence, the regularisation error is large [9].

It follows from (5) that

$$1 \leq S(T, b) \leq \kappa_2(T), \tag{7}$$

and there exists an elegant geometric interpretation of these bounds. In particular, it is easily verified that the lower and upper bounds in (7) are achieved when

$$c = c_l = \begin{bmatrix} 0 \dots 0 \, 0 \, c_n \end{bmatrix} \qquad \text{and} \qquad c = c_u = \begin{bmatrix} c_0 \, 0 \, 0 \dots 0 \end{bmatrix},$$

respectively. It follows from (5) that if $c = c_l$, then b is equal (up to a scalar multiplier) to the last column of U, and if $c = c_u$, then b is equal (up to a scalar multiplier) to the first column of U. Since $TT^T = US^2U^T$, it follows that (2) is well–conditioned if the dominant components of b lie along the columns of U that are associated with the small eigenvalues of TT^T, that is,

$$\frac{|c_i|}{s_i} \ll \frac{|c_{i+1}|}{s_{i+1}}, \qquad i = 0, \dots, n-1.$$

Similarly, (2) is ill–conditioned if the dominant components of b lie along the columns of U that are associated with the large eigenvalues of TT^T, such that (6) is satisfied.

It is easily shown that the effective condition number $S(T, b)$ obeys an uncertainty principle. In particular, the effective condition number of the equation $T^{-1}b = a$ is

$$S(T^{-1}, a) = s_0 \frac{\|d\|_2}{\|Sd\|_2},$$

where $d = V^T a$, and it follows from (5) and the singular value decomposition of T and T^{-1} that

$$\begin{aligned} S(T, b)S(T^{-1}, a) &= \frac{1}{s_n} \frac{\|c\|_2}{\|S^{-1}c\|_2} s_0 \frac{\|d\|_2}{\|Sd\|_2} \\ &= \frac{1}{s_n} \frac{\|b\|_2}{\|VS^{-1}U^T b\|_2} s_0 \frac{\|a\|_2}{\|USV^T a\|_2} \\ &= \frac{s_0}{s_n} \\ &= \kappa_2(T). \end{aligned} \qquad (8)$$

It follows that as the effective condition number of the transformation from one polynomial basis to the other increases, the effective condition number of the inverse transformation decreases, such that the product of the effective condition numbers is equal to the condition number of the transformation matrix between the bases. The result (8) emphasizes that a large condition number implies that a linear algebraic equation *may* be ill–conditioned, and it is evident that this result is valid for all linear algebraic equations that have a square non–singular coefficient matrix.

Example 2.1 Consider the Wilkinson polynomial in the power and Bernstein bases,

$$p(x) = \prod_{i=1}^{20} \left(x - \frac{i}{20} \right) = \sum_{i=0}^{20} a_i x^i = \sum_{i=0}^{20} b_i \binom{20}{i} (1-x)^{20-i} x^i. \qquad (9)$$

The Bernstein form of this polynomial was generated and the coefficients stored in the vector b. The singular value decomposition of the transformation matrix T for polynomials of degree 20 was calculated and the product $c = U^T b$ was evaluated. The effective condition number (5) of (2) for the polynomial (9) is equal to 1.186, which shows that this equation is very well–conditioned. This condition estimate must be compared with the condition number $\kappa_2(T) = 3.120 \times 10^9$, which is very unrealistic for the given right hand side vector b of (2). The uncertainty principle (8) implies that the equation $T^{-1}b = a$ is very ill–conditioned. This example is an extreme case but it clearly shows that the condition number of a matrix may significantly overestimate the true numerical condition of a linear algebraic equation.□

3   Totally non–negative and oscillation matrices

Totally non–negative (TNN) matrices are important in several areas of applied mathematics, including approximation theory, statistics and computer–aided geometric design. Restrictions on the properties of TNN matrices lead to oscillation matrices, which are important in finite difference approximations, the study of the small vibrations of mechanical systems and spline smoothing [1,4]. These two terms are now defined [7], pages 98 and 103 :

Definition 1 : A rectangular matrix is totally non–negative (totally positive) if all its minors of all orders are non–negative (positive).

Definition 2 : A square matrix A is oscillatory if it is TNN and there exists an integer $q > 0$ such that A^q is totally positive.

The following properties of TNN and oscillation matrices are required [1], proposition 1.1, and [7], page 105 :

1. The product of two or more TNN matrices is also a TNN matrix.
2. A square TNN matrix A is oscillatory if and only if it is non–singular and the elements on the main diagonal, first superdiagonal and first subdiagonal are greater than zero,

$$a_{i,j} > 0 \qquad \text{for} \qquad |i - j| \leq 1. \qquad (10)$$

3. An oscillatory matrix A of order n always has n positive distinct eigenvalues.

It is shown in the next sections that the basis transformation matrix T is TNN and that TT^T and $T^T T$ are oscillation matrices.

3.1   Totally non–negative matrices

The link between the numerical condition of (2) and the number of roots of the polynomial $p(x)$ in the interval I requires that the oscillatory nature of TT^T be established. This result will be obtained by showing that T, and therefore TT^T, are TNN and that the elements of TT^T satisfy the conditions (10).

It is readily verified from (3) that T is symmetric about the reverse diagonal, $t_{ij} = t_{n-j,n-i}$, and thus

$$KT = (KT)^T, \qquad K = K^{-1} = K^T, \tag{11}$$

where K is the reverse unit matrix. Since T is a lower triangular matrix such that $t_{00} = t_{nn} = 1$ for all n, T is not oscillatory because it has a double eigenvalue $\lambda = 1$. However examination of T for low polynomial orders shows that it is TNN, and the extension of this result to all polynomial orders will be proved by induction, for which it will be necessary to distinguish between $T^{(n)}$ and $T^{(n-1)}$, the transformation matrices for polynomials of degrees n and $n-1$ respectively.

It is shown in [17] that these two transformation matrices are related by

$$T^{(n)} = G \begin{bmatrix} T^{(n-1)} & 0 \\ 0 & 1 \end{bmatrix} H, \qquad n \geq 1, \tag{12}$$

where G is a square diagonal matrix of order $n+1$ with elements $g_{ii} = g_i, i = 0, \ldots, n$,

$$g_i = \begin{cases} 1, & \text{if } i = 0, \\ (n-i)/n, & \text{for } i = 1, \ldots, n-1, \\ 1, & \text{if } i = n, \end{cases}$$

for $n > 1$, and $G = I$ for $n = 1$. The elements $\{h_{ij}\}_{i,j=0}^n$ of H are given by

$$h_{ij} = \begin{cases} 1, \text{ if } i \geq j, \\ 0, \text{ otherwise.} \end{cases}$$

It is noted that if $T^{(n-1)}$ is TNN, then

$$\begin{bmatrix} T^{(n-1)} & 0 \\ 0 & 1 \end{bmatrix},$$

is also TNN. Since G and H are TNN and $T^{(0)} = 1$, it follows from (12) by induction that T is TNN for all polynomial orders n.

Example 3.1 Consider the recurrence relationship (12) for $n = 4$,

$$T^{(4)} = \begin{bmatrix} 1 & 0 & 0 & 0 & 0 \\ 0 & \frac{3}{4} & 0 & 0 & 0 \\ 0 & 0 & \frac{1}{2} & 0 & 0 \\ 0 & 0 & 0 & \frac{1}{4} & 0 \\ 0 & 0 & 0 & 0 & 1 \end{bmatrix} \begin{bmatrix} 1 & 0 & 0 & 0 & 0 \\ 1 & \frac{1}{2} & 0 & 0 & 0 \\ 1 & \frac{2}{3} & \frac{1}{3} & 0 & 0 \\ 1 & 1 & 1 & 1 & 0 \\ 0 & 0 & 0 & 0 & 1 \end{bmatrix} \begin{bmatrix} 1 & 0 & 0 & 0 & 0 \\ 1 & 1 & 0 & 0 & 0 \\ 1 & 1 & 1 & 0 & 0 \\ 1 & 1 & 1 & 1 & 0 \\ 1 & 1 & 1 & 1 & 1 \end{bmatrix} = \begin{bmatrix} 1 & 0 & 0 & 0 & 0 \\ 1 & \frac{1}{4} & 0 & 0 & 0 \\ 1 & \frac{1}{2} & \frac{1}{6} & 0 & 0 \\ 1 & \frac{3}{4} & \frac{1}{2} & \frac{1}{4} & 0 \\ 1 & 1 & 1 & 1 & 1 \end{bmatrix},$$

whose elements satisfy (3).$\Box$

3.2 Oscillatory matrices

The results of section 3.1 are extended to show that TT^T and T^TT are oscillation matrices, and thus the eigenvectors of these matrices (or the singular vectors of T) satisfy restrictions on the sign changes of their elements. In particular, since TT^T is non–singular and TNN, and its elements satisfy (10), it follows that it is oscillatory. The restrictions on the signs of the elements of the singular vectors of T are established by theorem 13, page 105 in [7].

Theorem 1. *Let u_i, $i = 0, \ldots, n$, be the left singular vectors, that is, columns of U, of the transformation matrix T for polynomials of degree n, where the singular values s_i are arranged in non–increasing order. Then*

1. *There are exactly r sign changes in the elements of the singular vector u_r.*
2. *For arbitrary real numbers $d_g, d_{g+1}, \ldots, d_h, 0 \leq g \leq h \leq n$, the number of sign variations in the coordinates of the vector*

$$r = \sum_{k=g}^{h} d_k u_k,$$

is between g and h.

Theorem 1 shows that the elements of the columns of the left singular matrix U of T have sign patterns that are well–defined, and it was shown in section 2 that U is also important in determining the numerical condition of (2). The application of this theorem and Descartes' rule of signs to the Bernstein polynomials $p_r(x)$ whose coefficients are the elements $\{u_{ir}\}_{i=0}^{n}$ of u_r,

$$p_r(x) = \sum_{i=0}^{n} u_{ir} \binom{n}{i} (1 - x)^{n-i} x^i, \qquad r = 0, \ldots, n, \tag{13}$$

suggests that there is a close connection between the number of roots of the polynomials $p_r(x)$ in the interval I, and the numerical condition of their transformation from the Bernstein basis to the power basis. The uncertainty principle (8) can then be used to deduce the numerical condition of their transformation from the power basis to the Bernstein basis, that is, the numerical condition of $T^{-1}b = a$.

4   Examples

This section considers examples that illustrate the theoretical results in sections 2 and 3. In particular, *computational* evidence that suggests that the columns of U satisfy stronger results than those in theorem 1, is presented.

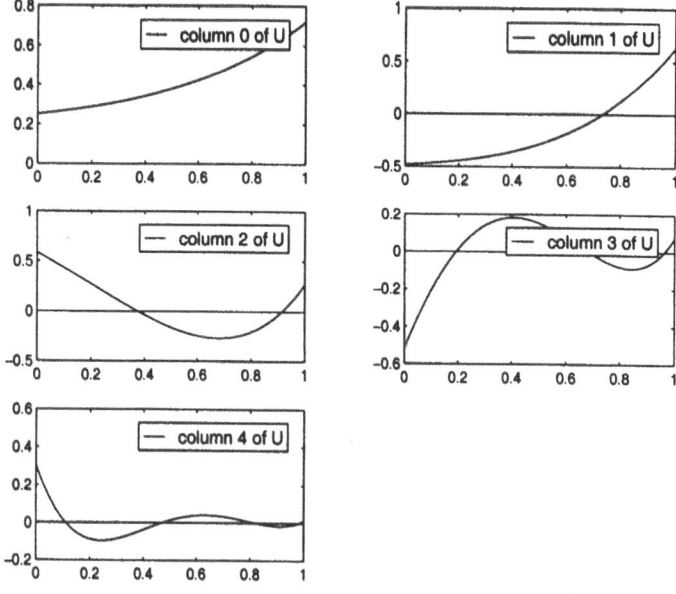

Fig. 1. The polynomials $f_r(x), r = 0, \ldots, 4$.

Example 4.1 Consider the sequences of Bernstein polynomials (13) formed from the five columns of U for polynomials of degree four,

$$f_r(x) = \sum_{i=0}^{4} u_{ir} \binom{4}{i} (1-x)^{4-i} x^i, \qquad r = 0, \ldots, 4, \tag{14}$$

and the eight columns of U for polynomials of degree seven,

$$g_r(x) = \sum_{i=0}^{7} u_{ir} \binom{7}{i} (1-x)^{7-i} x^i, \qquad r = 0, \ldots, 7. \tag{15}$$

These polynomials are shown in figures 1 and 2 respectively and it is seen that as the column index j increases, the polynomials $f_j(x)$ and $g_j(x)$ become more oscillatory in the interval I. Furthermore, *the number of roots of $f_j(x)$ and $g_j(x)$ in the interval I is exactly equal to the column index j*, and since the coefficients of these polynomials are derived from the left singular vectors of T, which satisfy theorem 1, *the polynomials satisfy an exact form of Descartes' rule of signs* [2].

Although the results for polynomials of degree four and seven are shown, identical results were observed for polynomials of other degrees.□

It was noted in section 2 that the discrete Picard condition is interpreted geometrically in terms of the orthonormal axes defined by the columns of U, and example 4.1 suggests that the numerical condition of (2) can also be interpreted in terms of the number of roots in the interval I of each of the $n+1$ polynomials (13). In particular, the polynomial $p_r(x)$, defined

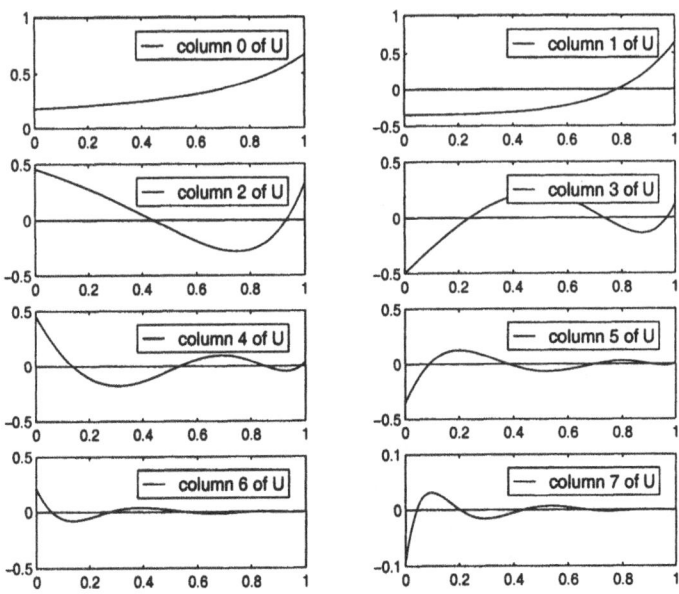

Fig. 2. The polynomials $g_r(x), r = 0, \ldots, 7$.

in (13), has exactly r roots in I and the effective condition number of its transformation from the Bernstein basis to the power basis is s_r/s_n.

Example 4.1 shows that the Bernstein polynomials that are formed from the columns of U have well–defined geometric properties. It is interesting to note that the power basis polynomials

$$h_r(x) = \sum_{i=0}^{n} u_{n-i,r} x^i, \qquad r = 0, \ldots, n, \tag{16}$$

display the same properties as the polynomials (13). Specifically, the number of roots of $h_r(x)$ in the interval I is exactly equal to the column index r, and thus theorem 1 shows that Descartes' rule of signs is satisfied exactly in this interval. It is noted that this rule is satisfied in I and not the positive real line, which is the more usual region in which it is defined for power basis polynomials.

The next example considers two interlacing properties of the polynomials (13).

Example 4.2 Table 1 shows that the roots, that lie in I, of the polynomials (13) for $n = 6$ interlace each other. This pattern was also observed for other polynomial degrees, and the results in the table are therefore typical.

The interlacing property of the zeros of the roots of the polynomials (13) also manifests itself in the polynomials $p_c^{(n)}(x)$ and $p_c^{(n+1)}(x)$, of degrees n and $n + 1$ respectively, that are formed from a given column c, where $2 \leq$

Table 1 : The roots of the polynomials $p_r(x), r = 1, \ldots, 6$.

Roots of polynomial $p_r(x), r = 1, \ldots, 6$					
$p_6(x)$	$p_5(x)$	$p_4(x)$	$p_3(x)$	$p_2(x)$	$p_1(x)$
0.99313					
	0.98948				
0.93581		0.98257			
	0.90298		0.96760		
0.78354		0.84296		0.92758	
	0.68371		0.72017		0.77051
0.53746		0.51578		0.42156	
	0.36411		0.22632		
0.26032		0.13112			
	0.08198				
0.05454					

$c \leq n$. More precisely, if $U^{(n)}$, with elements $u_{ij}^{(n)}, i, j = 0, \ldots, n$, and $U^{(n+1)}$, with elements $u_{ij}^{(n+1)}, i, j = 0, \ldots, n+1$, are the left singular matrices of the transformation matrix T for polynomials of degrees n and $n+1$ respectively, then the roots, that lie in I, of the polynomials

$$p_c^{(n)}(x) = \sum_{i=0}^{n} u_{ic}^{(n)} \binom{n}{i} (1-x)^{n-i} x^i,$$

and

$$p_c^{(n+1)}(x) = \sum_{i=0}^{n+1} u_{ic}^{(n+1)} \binom{n+1}{i} (1-x)^{n+1-i} x^i,$$

interlace.

Table 2 shows the roots of the polynomials $p_5^{(6)}(x)$ and $p_5^{(7)}(x)$ that are in I, and it is readily seen that they interlace.□

In general, the coefficients of a polynomial will not be equal to the entries of one column of U, but will be formed by a linear combination of the entries in some or all of the columns of U. For example, if the coefficients b_i are formed by a linear combination of columns r and $r + 1$, then it would be expected that the effective condition number of (2) for this polynomial is in the range $\left[\frac{s_{r+1}}{s_n}, \frac{s_r}{s_n}\right]$. This observation is quantified by noting that

$$\frac{\|c\|_2^2}{\|S^{-1}c\|_2^2} = \frac{\sum_{i=0}^{n} c_i^2}{\sum_{i=0}^{n} \left(\frac{c_i}{s_i}\right)^2},$$

and thus if $c_i = 0$ for $i < n_1$ and $i > n_2$, then

$$s_{n_2} \leq \frac{\|c\|_2}{\|S^{-1}c\|_2} \leq s_{n_1},$$

Table 2 : The roots of the polynomials $p_5^{(7)}(x)$ and $p_5^{(6)}(x)$.

Roots of $p_5^{(7)}(x)$	Roots of $p_5^{(6)}(x)$
0.99008	
	0.98948
0.90830	
	0.90298
0.69803	
	0.68371
0.38151	
	0.36411
0.08831	
	0.08198

which enables lower and upper bounds on the effective condition number to be established. It is interesting to determine if this result can be extended to the roots of $p(x)$, that is, if the coefficients of $p(x)$ are formed by a linear combination of columns $r, r+1, \ldots, s$, of U, then it may be expected that the number of roots of $p(x)$ in the interval I lies in the range $[r, s]$. This is considered in the following example.

Example 4.3 Consider the polynomial

$$p(x) = \sum_{r=0}^{n} \alpha_r p_r(x) = \sum_{r=0}^{n} \alpha_r \sum_{i=0}^{n} u_{ir} \binom{n}{i} (1-x)^{n-i} x^i,$$

where the polynomials $p_r(x)$ are defined in (13) and α_r is the weight attached to $p_r(x)$. Computational experiments showed that if $\alpha_r = 0$ for $r < n_1$ and $r > n_2$, then the number of roots of $p(x)$ in the interval I is in the range $[n_1, n_2]$, and the exact number of roots is dependent on the weights α_r. This result can be considered an extension of part 2 of theorem 1 from the number of sign changes of the coefficients to the number of roots of the polynomial. □

5   Discussion

It was shown in section 3 that T is TNN and TT^T is oscillatory. It follows immediately that $T^T T$ is also oscillatory and thus it satisfies theorem 13, page 105 in [7].

It was noted that the power basis polynomials that are defined in (16) display the same zero crossing properties as the polynomials (13). This feature of the power basis polynomials follows from (11). In particular, since

$$KUSV^T = (KUSV^T)^T,$$

it follows that $USV^T = (KV)S(KU)^T$ and thus $(KV) S (KU)^T$ is another singular value decomposition of T. Since the singular values of the oscillatory matrix TT^T are distinct, the SVD is unique up to sign changes in the columns of U and V. It follows that $Ku_r = \pm v_r$, or, in terms of coordinates for some choice of $\pm$ depending only on r,

$$v_{ir} = \pm u_{n-i,r}, \qquad r = 0, \ldots, n. \qquad (17)$$

On the other hand, the orthogonality of V leads to

$$Tv_r = USV^T v_r = USe_r = u_r s_r.$$

Comparison of this equation with (2) shows that the polynomial with power basis coefficients $\{v_{ir}\}_{i=0}^n$ has Bernstein basis coefficients $\{u_{ir}s_r\}_{i=0}^n$, and thus

$$\sum_{i=0}^n v_{ir} x^i = \sum_{i=0}^n u_{ir} s_r \binom{n}{i} (1-x)^{n-i} x^i = s_r p_r(x), \qquad r = 0, \ldots, n.$$

It follows from this equation and (17) that

$$s_r p_r(x) = \pm \sum_{i=0}^n u_{n-i,r} x^i,$$

and thus the polynomials $h_r(x)$ in (16) are, up to a scale factor of $\pm s_r$, the same as the polynomials (13), and hence they have the same roots.

A highly oscillatory polynomial $q(x)$ can be represented concisely by a set of basis functions $\{\phi_i(x)\}_{i=0}^n$ that possess many of the properties of $q(x)$. Example 4.1 shows that as the column index j increases, the polynomial $p_j(x)$ becomes more oscillatory because it has a larger number of roots in the interval I, and it therefore follows that $q(x)$ can be accurately represented by the polynomials $p_j(x)$ that are associated with large values of the index j. This result is consistent with the Wilkinson polynomial but it is important to note that it may not be true for all oscillatory polynomials. For example, the roots of the Wilkinson polynomial are uniformly distributed in I, but the result may not be true if the roots are highly clustered in a subinterval of I. More work must be performed in order to investigate this further, but it is likely that both the *number* and *distribution* of roots in I are required to accurately bound the effective condition number of the linear algebraic equation that defines the transformation between the power and Bernstein bases. For example, it was shown in example 2.1 that the effective condition number of (2) for the Wilkinson polynomial, whose roots form an arithmetic progression, is 1.186. This condition estimate must be compared with the effective condition number of (2) for the polynomial

$$p(x) = \prod_{i=1}^{10} \left(x - 2^{-i} \right), \qquad (18)$$

whose roots form a geometric progression. In particular, $\kappa_2(T) = 5.174 \times 10^4$ for polynomials of degree 10 and $S(T, b) = 2.30 \times 10^3$ for the polynomial (18), and thus the transformation of this polynomial from its Bernstein form to its power form is ill–conditioned.

It was shown in section 2 that the discrete Picard condition has an elegant geometric interpretation in terms of the eigenvectors of TT^T. However an interpretation of polynomial basis conversion in terms of the characteristics of the polynomial, for example, the location of its roots, may be more informative than an interpretation in terms of the discrete Picard condition, and this paper begins to address this issue. Although there exist unresolved issues, it has been shown that there is computational evidence to suggest that an interpretation of the numerical condition of the transformation of a polynomial between the power and Bernstein bases in terms of its roots may be possible.

References

1. Carnicer, J. M., Pēna, J. M. (1997) Bidiagonalization of oscillatory matrices, Linear and Multilinear Algebra **42**, 365–376.
2. Carnicer, J. M., Pēna, J. M. (1998) Characterizations of the optimal Descartes' rule of signs, Mach. Nachr. **189**, 33–48.
3. Chan, T. F., Foulser, D. E. (1988) Effectively well–conditioned linear systems, SIAM. J. Sci. Stat. Comput. **9**, 963–969.
4. Demmler, A., Reinsch, C. (1975) Oscillation matrices with spline smoothing, Numer. Math. **24**, 375–382.
5. Farouki, R. T. (1986) The characterization of parametric surface sections, Computer Vision, Graphics and Image Processing **33**, 209–236.
6. Farouki, R. T., Rajan, V. T. (1987) On the numerical condition of polynomials in Bernstein form, Computer Aided Geometric Design **4**, 191–216.
7. Gantmacher, F. R. (1959) The Theory of Matrices, volume 2, Chelsea Publishing Company, New York.
8. Hansen, P. C. (1990) The discrete Picard condition for discrete ill–posed problems, BIT **30**, 658–672.
9. Hansen, P. C. (1990) Truncated singular value decomposition solutions to discrete ill–posed problems with ill–determined numerical rank, SIAM J. Sci. Stat. Comput. **11**, 503–518.
10. Kajiya, J. T. (1982) Ray tracing parametric patches, ACM Computer Graphics **16**, 245–254.
11. Sederberg, T. W., Parry, S. R. (1986) Comparison of three curve intersection algorithms, Computer–aided design **18**, 58–63.
12. Winkler, J. R. (1994) The sensitivity of linear algebraic equations, Fifth SIAM Conference on Applied Linear Algebra, Snowbird, Utah, USA, 279–283.
13. Winkler, J. R. (1997) An ill–conditioned problem in computer aided geometric design, Neural, Parallel and Scientific Computations **5**, 179–200.
14. Winkler, J. R. (1997) Polynomial basis conversion made stable by truncated singular value decomposition, Applied Mathematical Modelling **21**, 557–568.

15. Winkler, J. R. (1997) Tikhonov regularisation in standard form for polynomial basis conversion, Applied Mathematical Modelling **21**, (1997) 651–662.

16. Winkler, J. R. (1997) The condition number of a matrix and the stability of linear algebraic equations, Memoranda in Computer and Cognitive Science CS–97–05, Department of Computer Science, The University of Sheffield.

17. Winkler, J. R. (1997) A recurrence relationship for the transformation matrix between the power and Bernstein polynomial bases, Memoranda in Computer and Cognitive Science CS–97–06, Department of Computer Science, The University of Sheffield.

On Approximation in Spaces of Geometric Objects

Helmut Pottmann and Martin Peternell

Institut für Geometrie, Technische Universität Wien
Wiedner Hauptstrase 8–10, A–1040 Wien, Austria

Summary. We present a concept for approximation in spaces of geometric objects. It is worked out in detail for approximation problems in the spaces of planes, lines and spheres. Applications include geometric computing with developable surfaces, ruled surfaces and canal surfaces.

Keywords: approximation, duality, line geometry, sphere geometry, developable surface, ruled surface, canal surface

1 Introduction

In advanced classical geometry, *point models* for spaces of simple geometric objects, such as planes, lines and spheres, play an important role. For example, the set of hyperplanes of a projective space is the point set of a projective space (the so-called dual projective space). To give a further example, we note that an appropriate point model for the set of lines in projective 3-space is the so-called Klein quadric Ω in projective 5-space P^5: Lines in P^3 are represented as points of Ω, pencils of lines in P^3 are seen as lines in Ω, ruled surfaces are visualized as curves in Ω, and so on [9].

These classical concepts have recently found a variety of applications in geometric computing. It is, for example, convenient to view a developable surface as envelope of a one-parameter family of planes and thus treat it as a curve in dual projective space [1,2,12,13,23,27]. There are also advantages of viewing a canal surface, which is defined as envelope of a one-parameter family of spheres, as a curve in $\mathbb{R}^4$ (as a point model for the set of spheres) [15–19,24]. Analogously, ruled surfaces can be treated as curves in the Klein quadric [9,20,26,28].

For computational applications the treatment of approximation problems is fundamental. Therefore, we will first discuss how to formulate approximation problems in the sets of spheres, planes and lines and apply these to canal surfaces, developable surfaces and ruled surfaces, respectively. In the final section, we will then summarize by formulating a more general strategy. Thereby, we are able to view the previously discussed cases (spheres, planes, lines) as a special case for approximation in a set of affinely equivalent geometric objects. Connections to kinematic mappings and motion design, and pointers to future research conclude our paper.

2   Approximation in the space of spheres

Recently, it could be shown that certain problems in geometric computing such as computations with canal surfaces, offsets, and medial axis, can be efficiently solved by using concepts from sphere geometry [4,5,15–19,24]. Thus, we will start our investigation with approximation in the space of spheres.

There are different types of classical sphere geometries. *Möbius geometry* incorporates planes as special case of spheres and will not be pursued here. *Laguerre geometry* works with oriented spheres and includes points as special cases of spheres. It will be (partially) used henceforth. *Lie geometry* subsumes both Laguerre and Möbius geometry as special cases.

An oriented sphere in Euclidean 3-space E^3 shall be given by its center (m_1, m_2, m_3) and its signed radius r. Vanishing radius $r = 0$ characterizes points (as degenerate spheres). $r > 0$ belongs to positively oriented spheres (unit normals pointing outside), $r < 0$ characterizes negatively oriented spheres (unit normals pointing inside). Laguerre geometry then also uses oriented planes as another type of fundamental objects and oriented contact as fundamental relation. This will not be needed in the sequel. We will just need the so-called *cyclographic model* for spheres. This is a point model for the set of oriented spheres obtained by the cyclographic mapping ζ, which maps a sphere S with center (m_1, m_2, m_3) and radius r to a point in $\mathbb{R}^4$ via

$$\zeta : S \mapsto (m_1, m_2, m_3, r) \in \mathbb{R}^4. \tag{1}$$

The cyclographic mapping is a well studied classical subject. For its relation to geometric computing, the reader may consult [15,19,24].

Using ζ we can transform a set of spheres into a set of points. In order to solve approximation problems for spheres, we need to come up with an appropriate *distance measure between spheres* and interpret it in the point model. It is easy to see that the natural distance measure of Laguerre geometry (tangential distance) is not useful for our purpose. Hence, we now present an alternative. It has an additional advantage over the tangential distance: whereas the latter leads to a pseudo-Euclidean metric, our distance measure will result in a Euclidean metric in the image space $\mathbb{R}^4$ of the set of spheres.

Given two oriented spheres A, B with

$$\zeta(A) = (a_1, a_2, a_3, a_4), \quad \zeta(B) = (b_1, b_2, b_3, b_4),$$

there exists a unique central similarity σ which maps A onto B. Let the centers of A and B be denoted by $\mathbf{a} = (a_1, a_2, a_3)$ and $\mathbf{b} = (b_1, b_2, b_3)$, the center of σ is

$$\mathbf{c} = \frac{1}{a_4 - b_4}(a_4\mathbf{b} - b_4\mathbf{a}).$$

It is at infinity for congruent spheres ($a_4 = b_4$). Let D be the difference vector of the point $x \in A$ and the image point $\sigma(x) \in B$ (see Fig. 1). Note

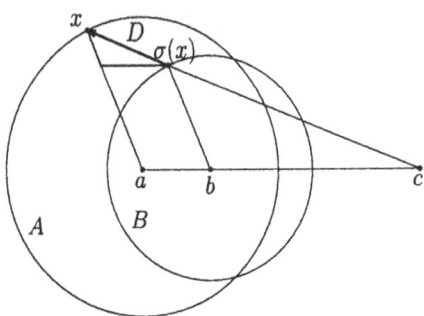

Fig. 1. Distance between two spheres

that all the lines $x\sigma(x)$ pass through **c**. For concentric spheres, these lines are orthogonal to both spheres, and we then measure orthogonal distances. For other cases, this is not true. However, we can still expect that the integral mean of $\|D\|^2$, viewed as a function on the unit sphere S^2, is a useful measure for the distance $d(A, B)$ of the two spheres,

$$d(A, B)^2 = \frac{1}{4\pi} \int_{S^2} \|D\|^2 d\omega, \tag{2}$$

with $d\omega$ as surface element of S^2.

Surprisingly, this distance measure leads to the canonical Euclidean metric in $\mathbb{R}^4$, since we have the following result.

Theorem 1. *The distance of two oriented spheres A (center $\mathbf{a}$ and radius a_4) and B (center $\mathbf{b}$ and radius b_4), defined via (2), is given by the Euclidean distance of their image points $\zeta(A)$ and $\zeta(B)$ in $\mathbb{R}^4$,*

$$d(A, B)^2 = \sum_{i=1}^{4} (a_i - b_i)^2. \tag{3}$$

Proof. Let the unit sphere S^2 be parametrized by

$$\mathbf{s}(u, v) = (s_1, s_2, s_3)(u, v) = (\cos u \cos v, \sin u \cos v, \sin v).$$

The difference vector $D = x - \sigma(x)$ between corresponding points on A and B and its squared length are given by

$$D(u, v) = \mathbf{a} - \mathbf{b} + (a_4 - b_4)\mathbf{s},$$
$$\|D\|^2 = \|\mathbf{a} - \mathbf{b}\|^2 + (a_4 - b_4)^2 + 2(a_4 - b_4)(\mathbf{a} - \mathbf{b}) \cdot \mathbf{s}(u, v).$$

With the surface element $d\omega = \cos v\, du\, dv$ we obtain

$$\int_{S^2} \|D\|^2 d\omega = 4\pi(\|\mathbf{a} - \mathbf{b}\|^2 + (a_4 - b_4)^2) = 4\pi \sum_{i=1}^{4} (a_i - b_i)^2.$$

3 Canal surfaces

A *canal surface* Φ is defined to be the envelope of a one-parameter family of (oriented) spheres

$$S(t) : (x_1 - m_1(t))^2 + (x_2 - m_2(t))^2 + (x_3 - m_3(t))^2 = r(t)^2, \qquad (4)$$

where t is a real parameter varying in an interval $[a, b] \subset \mathbb{R}$. The functions $m_i(t)$ and $r(t)$ are considered to be sufficiently often differentiable in $[a, b]$. The center curve (or spine curve) of the spheres shall be denoted by $M(t)$. The envelope Φ is tangent to the spheres $S(t)$ in points of the *characteristic curves* $c(t)$. These curves are obtained by intersecting $S(t)$ with the plane $\dot{S}(t)$,

$$\dot{S}(t) : \sum_{i=1}^{3} (x_i - m_i(t))\dot{m}_i(t) + r(t)\dot{r}(t) = 0. \qquad (5)$$

Thus, the family $c(t)$ consists of circles which form one family of principal curvature lines on Φ. The normal vector of $\dot{S}$ is $\dot{M}(t) = (\dot{m}_1, \dot{m}_2, \dot{m}_3)$. The circles $c(t)$ are real (consist of real points) if and only if the reality condition

$$\|\dot{M}(t)\|^2 - \dot{r}(t)^2 \geq 0 \qquad (6)$$

holds. Then, Φ is a real surface.

If equality holds in condition(6) for an isolated parameter value t_0 then the plane $\dot{S}(t_0)$ is tangent to the sphere $S(t_0)$ in an umbilic point of Φ. The sphere $S(t_0)$ has second order contact with Φ there. The canal surface Φ can

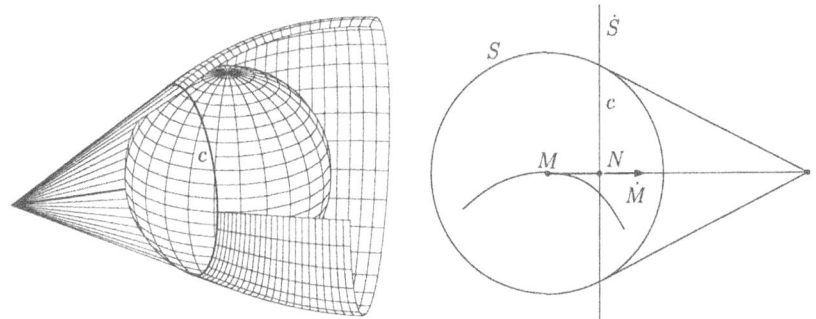

Fig. 2. Geometric properties of a canal surface

be represented by an equation which is obtained by eliminating parameter t from equations (4) and (5). A parametrization of Φ is constructed as follows. The circles $c(t)$ possess centers $N(t)$ and radii $r_1(t)$ with

$$N(t) = M(t) - \frac{r(t)\dot{r}(t)}{\|\dot{M}(t)\|^2}\dot{M}(t), \quad r_1(t) = \frac{r(t)}{\|\dot{M}(t)\|}\sqrt{\|\dot{M}(t)\|^2 - \dot{r}(t)^2}.$$

The centers $N(t)$ are the intersection points of the tangent lines of $M(t)$ and the planes $\dot{S}(t)$; see Fig. 2. The radii $r(t)$ are obviously real if (6) holds. The curve $M(t)$ possesses an orthonormal frame (T, E, F) with

$$T(t) = \frac{1}{\|\dot{M}(t)\|}\dot{M}(t), \; E(t) = \frac{1}{\|\dot{T}(t)\|}\dot{T}(t) \; \text{ and } \; F(t) = T(t) \times E(t),$$

where $X \times Y$ denotes the cross product of two vectors in $\mathbb{R}^3$. T denotes the unit tangent vector of M and E, F form an orthonormal basis of the normal plane. A parametrization of Φ is then obtained by

$$X(t, u) = N(t) + r_1(t)\cos(u)E(t) + r_1(t)\sin(u)F(t). \tag{7}$$

Instead of principal normal vector E and binormal vector F, any other orthonormal basis in the normal plane of the spine curve $M(t)$ may be used in this parameterization.

In particular, let $C(t)$ be an arbitrary piecewise polynomial curve of degree n in $\mathbb{R}^4$. It possesses a B-spline representation,

$$C(t) = \sum_{i=0}^{n} N_i^n(t)C_i, \tag{8}$$

with $N_i^n(t)$ as normalized B-spline functions of degree n over an appropriate knot vector and control points C_i. An analogous representation we have for piecewise rational curves. There, the C_i are homogeneous coordinate vectors of points in the projective extension of $\mathbb{R}^4$, see [22]. It is proved in [18] that the envelope Φ of the one parameter family of spheres $\zeta^{-1}(C(t))$ is a rational surface, thus representable as a rational tensor product spline surface. This result shows that we may use B-spline curves in $\mathbb{R}^4$ in order to compute with canal surfaces, which possess an exact NURBS representation.

Nevertheless, the computation of a rational parametrization is not straightforward: if $m_i(t)$ and $r(t)$ are rational functions, parametrization (7) is in general not rational.

3.1   Approximation of canal surfaces

Approximation algorithms will be based on the interpretation of canal surfaces as envelopes of one-parameter families of spheres $S(t)$. The mapping (1) allows to change the point of view in the sense that we will not consider the envelope Φ in the following but its image curve $\zeta(S(t))$ in the cyclographic model $\mathbb{R}^4$.

Applying the mapping ζ, interpolation or approximation problems concerning spheres and canal surfaces can be transformed into interpolation or approximation problems concerning points and curves in $\mathbb{R}^4$.

Let us consider the following problem. Given k spheres $\Sigma_1, \ldots, \Sigma_k$ and corresponding parameter values $\tau_1, \ldots, \tau_k$, approximate these spheres by a canal surface $\Phi = S(t)$ such that $S(\tau_j)$ is close to Σ_j for $j = 1, \ldots k$. Clearly, close is meant according to the distance (3) between corresponding spheres. We want to determine a canal surface $\Phi = S(t)$ which minimizes

$$F := \sum_{i=1}^{k} d(S(\tau_i), \Sigma_i)^2. \tag{9}$$

If $\zeta(S(t))$ is chosen to be a B-spline curve (8), functional F is quadratic in the coefficients of the unknown control points C_i. Thus, the minimization leads to a linear system.

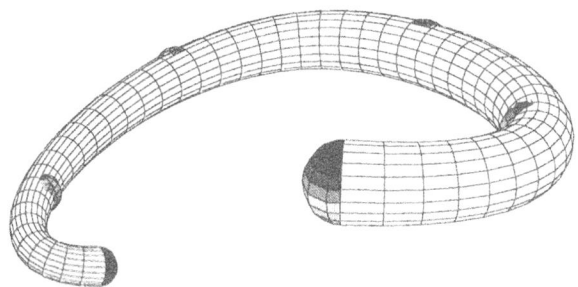

Fig. 3. Canal surface $S(t)$ approximating six spheres; $\zeta(S)$ is a cubic B-spline curve consisting of two segments

Since monotonicity of the radius function is an essential shape property of canal surfaces we will study the following problem. Consider spheres and parameter values as in the previously formulated problem, but additionally a monotonic sequence of radii of the spheres $\Sigma(\tau_i)$. We want to determine an approximating canal surface with monotonic radius function.

A sufficient condition for possessing monotonic radii is that the fourth coordinates c_{i4}, $i = 1, \ldots, k$ of the control points C_i are a monotonic sequence. Thus,

$$c_{04} \leq c_{14} \leq \ldots \leq c_{k-1,4} \leq c_{k4}. \tag{10}$$

Having computed an optimal approximation X^* by minimizing (9), we will look for an approximation X satisfying constraints (10) and being as close as possible to X^*. This is a constraint optimization problem and is solved by quadratic programming [7].

Since parameter values τ_i have to be chosen in advance and have a great influence on the resulting canal surface one can start with an initial guess and improve it by a parameter correction (see [11]).

4   Approximation in the space of planes

Our motivation for studying approximation in the set of planes comes from the computational geometry of developable surfaces. There, it turned out that viewing these surfaces as envelopes of planes yields computational advantages [1,2,12,13,23,27].

In order to solve approximation problems in the set of planes, it is necessary to introduce an appropriate *distance* between two planes. Euclidean geometry does not directly provide such a distance function. All invariants are expressed in terms of the angle between planes and are inappropriate for our purposes. In view of applications, we are interested in the distances of points of the two planes which are near some region of interest, and this distance can become arbitrarily large with the angle getting arbitrarily close to zero at the same time.

We use the following well-known facts from projective geometry. If we extend real Euclidean 3-space E^3 by ideal points (points at infinity), i.e., intersections of parallel lines, we obtain a model of real projective 3-space P^3. All ideal points form a plane in P^3, the so-called ideal plane. The set of planes in P^3 is a projective space itself, the dual projective space. It is isomorphic to P^3.

Analytically, one uses homogeneous Cartesian coordinates (x_0, x_1, x_2, x_3) for points. For points not at infinity, i.e., $x_0 \neq 0$, the corresponding inhomogeneous Cartesian coordinates will be denoted by

$$x = \frac{x_1}{x_0}, \quad y = \frac{x_2}{x_0}, \quad z = \frac{x_3}{x_0}.$$

A *plane* with equation $u_0 x_0 + u_1 x_1 + u_2 x_2 + u_3 x_3 = 0$, or, equivalently, $u_0 + u_1 x + u_2 y + u_3 z = 0$ can be represented by its *homogeneous plane coordinates* $U = (u_0, u_1, u_2, u_3)$.

We will later introduce a Euclidean metric in the dual space and thus first have to obtain the structure of an affine space. Given a projective space P, we obtain an affine space if we remove a hyperplane from P. Thus, we have to remove the points of a plane from the dual space. Viewed from the original space P^3, this means we have to remove a bundle of planes from P^3. Since we actually want to remove the ideal plane, this bundle must have a vertex at infinity. Hence, if we remove all planes passing through a fixed ideal point (for example, planes through the ideal point of the z-axis = planes parallel to the z-axis), we get a set of planes which has the structure of an affine space. This is easily seen in the analytic model. Planes, which are not parallel to the z-axis, can be written in the form

$$z = u_0 + u_1 x + u_2 y, \tag{11}$$

i.e., they have homogeneous plane coordinates $U = (u_0, u_1, u_2, -1)$. We see that (u_0, u_1, u_2) are affine coordinates in the resulting affine space A^* (of planes, which are not parallel to the z-axis).

We will now introduce a Euclidean metric in A^*. Thereby we make sure that the deviation between two planes shall be measured within some region of interest. This region shall be captured by its projection Γ onto the xy-plane.

For a positive measure μ in $\mathbb{R}^2$ we define the distance d_μ between planes $A = (a_0, a_1, a_2, -1)$ and $B = (b_0, b_1, b_2, -1)$ as

$$d_\mu(A, B) = \|(a_0 - b_0) + (a_1 - b_1)x + (a_2 - b_2)y\|_{L^2(\mu)}, \tag{12}$$

i.e., the $L^2(\mu)$-distance of the linear functions whose graphs are A and B. This, of course, makes sense only if the linear function which represents the difference between the two planes is in $L^2(\mu)$. We will always assume that the measure μ is such that all linear and quadratic functions possess finite integral.

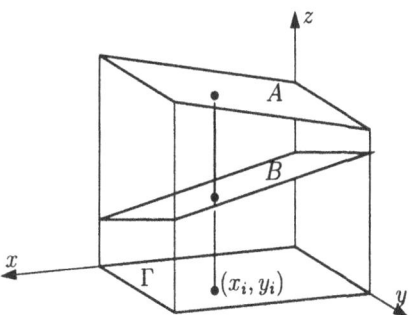

Fig. 4. To the definition of the deviation of two planes

A useful choice for μ is the Lebesgue measure $dxdy$ times the characteristic function χ_Γ of the *region of interest* Γ (Fig. 4). If $\mu = dxdy\chi_\Gamma$, we have

$$d_\mu(A, B)^2 = \int_\Gamma ((a_0 - b_0) + (a_1 - b_1)x + (a_2 - b_2)y)^2 dxdy. \tag{13}$$

We write $d_\Gamma(A, B)$ instead of $d_\mu(A, B)$. With $c_i := a_i - b_i$, equation (13) can be written as

$$d_\Gamma(A, B)^2 = (c_0, c_1, c_2) \cdot \begin{pmatrix} \int 1 & \int x & \int y \\ \int x & \int x^2 & \int xy \\ \int y & \int xy & \int y^2 \end{pmatrix} \cdot \begin{pmatrix} c_0 \\ c_1 \\ c_2 \end{pmatrix}. \tag{14}$$

This is a quadratic form, whose matrix depends on the domain of integration Γ for the integrals (where we omitted the differentials $dxdy$ for brevity).

Another possibility is that μ equals the sum of several point masses at points (x_j, y_j); see [12]. In this case we have

$$d_\mu(A, B)^2 = \sum_j ((a_0 - b_0) + (a_1 - b_1)x_j + (a_2 - b_2)y_j)^2. \tag{15}$$

Theorem 2. *The distance d_μ defines a Euclidean metric in the set of planes of type (11), if and only if μ is not concentrated in a straight line.*

Proof. See [27].

In this way, approximation problems in the set of planes are transformed into approximation problems in the set of points in Euclidean 3-space, whose metric is based on d_μ. In the next section, we will illustrate this at hand of developable surfaces.

5   Approximation algorithms for developable surfaces

5.1   Developable NURBS surfaces as envelopes of planes

Developable surfaces can be isometrically mapped (*developed*) into the plane, at least locally. When sufficient differentiability is assumed, they are characterized by vanishing Gaussian curvature. A non-flat developable surface is the envelope of its one parameter family of tangent planes. Such a developable surface locally is either a conical surface, a cylindrical surface, or the tangent surface of a twisted curve. Globally, of course, it can be a rather complicated composition of these three surface types. Thus, developable surfaces are ruled surfaces, but with the special property that they possess the same tangent plane at all points of the same generator (=*ruling*).

Because in all points of a generator line the tangent plane is the same, we can identify a developable surface with the one-parameter family of its tangent planes $U(t)$, or in other words, with a certain curve in dual projective space. If this curve is a NURBS curve

$$U(t) = \sum_{i=0}^{n} U_i N_i^k(t), \tag{16}$$

the original surface is a *developable NURBS surface*. Methods for computing a parameterization in standard NURBS tensor product form have been developed [23].

The symbol U_i denotes a homogeneous coordinate quadruple of the i-th *control plane* U_i. Of course the coordinate quadruple contains more information than just the plane as a point set, but for simplicity we just speak of the coordinates of the plane.

For the approximation algorithms discussed in this paper, we will restrict the class of developable surfaces we are working with: We only consider surfaces whose family of tangent planes is of the form

$$U(t) = (u_0(t), u_1(t), u_2(t), -1) \iff z = u_0(t) + u_1(t)x + u_2(t)y. \tag{17}$$

For NURBS surfaces this is equivalent to the choice of control planes $U_i = (u_{0,i}, u_{1,i}, u_{2,i}, u_{3,i})$ such that always $u_{3,i} = -1$. This means that for all possible planes U we no longer allow to choose an arbitrary coordinate quadruple describing U, but we restrict ourselves to the unique one whose last coordinate equals -1. This is not possible if the last coordinate is zero, so we have to exclude all surfaces with tangent planes parallel to the z-axis. In most cases this requirement is easily fulfilled by choosing an appropriate coordinate system.

Dual projective space with the bundle of planes $(u_0, u_1, u_2, 0)$ removed is an affine space and (u_0, u_1, u_2), describing the plane $(u_0, u_1, u_2, -1)$, are affine coordinates in it. The surfaces (17) become ordinary piecewise polynomial B-spline curves in this dual model.

Recently, algorithms for the computation with the dual representation, the conversion to the standard tensor product representation and the solution of interpolation and some approximation algorithms have been developed [12,13,23]. The general approximation scheme briefly outlined below is discussed in detail in [27]. We include it here since it is a typical example for our concept of geometric approximation.

5.2   Approximation of tangent planes

Consider the following approximation problem. Given m planes $V_1, \ldots, V_m$ and corresponding parameter values v_i, approximate these planes by a developable surface $U(t)$, such that $U(v_i)$ is close to the given plane V_i within an associated area of interest, where i ranges from 1 to m.

The meaning of 'close' is the following: There is a Cartesian coordinate system fixed in space such that all planes are graphs of linear functions of the xy-plane. Its third unit vector may be found as solution of a regression problem to the given plane normals. For all i there is a region of interest Γ_i, or, more generally, a measure μ_i, in the xy-plane. We want to minimize

$$F_1 := \sum_{i=1}^{m} d_{\mu_i}(V_i, U(v_i))^2, \tag{18}$$

for an unknown developable surface $U(t)$. If $U(t)$ is a NURBS surface of type (17), F_1 is a quadratic function in the unknown coordinates of the control planes U_i. These can then be found by solving a linear system of equations.

A good choice for μ_i would be $w_i \chi_{\Gamma_i} dx dy$. An example of this can be seen in Fig. 5. The positive weights w_i can be used to assign more or less importance to the single parameter values v_i. It would also be possible to choose different coordinate systems for different planes V_i, but this is not necessary, because it is equivalent to multiplying the weights w_i with appropriate factors. With $w_i = \sin^2 \gamma_i$, where γ_i is the Euclidean angle, which is enclosed between V_i and the z-axis, we can correct the influence of measuring distances in the z-direction of a fixed coordinate system for all i.

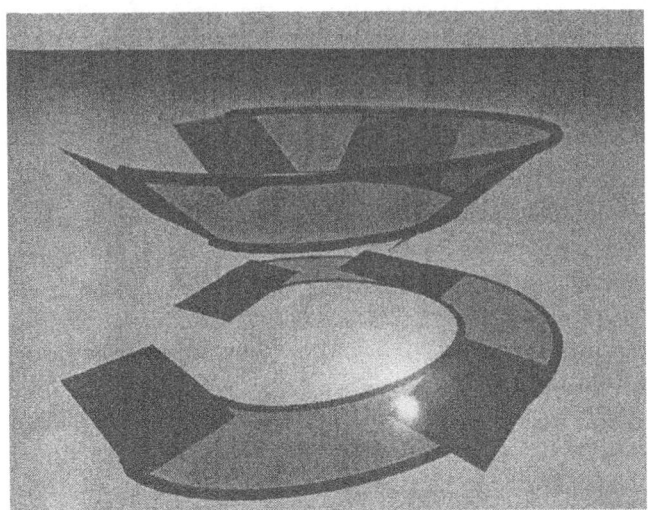

Fig. 5. Approximation of a set of planes by a developable surface

One may fix some boundary control planes in order to ensure a smooth join of subsequent surface segments. Note that the computation of the surface $U(t)$ is equivalent to a polynomial B-spline curve approximation problem using different Euclidean metrics at different points to be approximated. Working with the same μ or Γ for all planes, we get an ordinary curve approximation problem in Euclidean 3-space [6,11,22].

Since the parameters v_i have to be fixed in advance and another choice could have given better results, one will start with an initial guess and then improve it by *parameter correction*. With the Euclidean norms defined above, we can directly apply the known computational schemes [11].

For more details and extensions, such as approximation of points and generators and control of the curve of regression, we refer to Pottmann and Wallner [27].

6   Approximation in line space and applications

Recent research on surface reconstruction, kinematics of parallel manipulators and NC machining [3,25,31] led us to various approximation problems in line space. Thus we started to develop a concept for approximation in line space which will be outlined and illustrated in the sequel.

To understand the nature of the problem, let us briefly review a few facts from line geometry [9,10,26,28].

For two points with homogeneous coordinates $(x_0, \ldots, x_3)$ and $(y_0, \ldots, y_3)$, one defines the six homogeneous *Plücker coordinates* of the spanning line L as

$$(l_1, \ldots, l_6) := (l_{01}, l_{02}, l_{03}, l_{23}, l_{31}, l_{12}) \text{ with } l_{ij} := x_i y_j - x_j y_i. \qquad (19)$$

These coordinates do not depend on the choice of the two points on L and are related by the *Plücker identity*

$$l_1 l_4 + l_2 l_5 + l_3 l_6 = 0. \tag{20}$$

There is a bijective map between ordered, homogeneous 6–tuples $(l_1, \ldots, l_6) \neq (0, \ldots, 0)$ of real numbers and lines in real projective 3–space P^3. Therefore, one may view the six Plücker coordinates of a line L as homogeneous coordinates of a point $\gamma(L)$ in real projective 5–space P^5. The thereby defined *Klein mapping* γ provides a bijection between the set of lines $\mathcal{L}$ in P^3 and the set of points in a quadric $\Omega \subset P^5$ with equation (20), usually referred to as *Klein quadric*. We see that projective line space, which is clearly a four-dimensional manifold, has the structure of a quadric in P^5.

If we work in $\mathbb{R}^3$ and thereby rule out lines at infinity, we remove a 2–dimensional plane (γ–image of the lines at infinity) from the Klein quadric. Thus, approximation in the set $\mathcal{L}$ of lines in $\mathbb{R}^3$ means approximation in the Klein quadric with a plane Π being removed. Unfortunately the resulting set $\Omega' := \Omega \setminus \Pi$ does not have the structure of an affine space. However, it is well–known that removal of a cut with a tangential hyperplane Γ at some point C of a quadric gives the structure of an affine space. The mapping to an affine space is then realized by *stereographic projection* with center C. Stereographic projection of a quadric Φ from one of its points $C \in \Phi$ is the restriction to the quadric of a central projection with center $C \in \Phi$ and any image hyperplane (not through C). The affine structure in the image hyperplane H is obtained by removing from the projective hyperplane the points of the tangent hyperplane Γ of Φ at C. The reader may visualize this with help of the familiar stereographic projection of a sphere.

From line geometry we know that the points of a tangential cut of Ω at a point $C \in \Omega$ are the Klein images of all lines in 3–space which intersect the line $U = \gamma^{-1}(C)$. Since we work in $\mathbb{R}^3$ and thus neglect lines at infinity, *the following operation introduces an affine structure into $\mathcal{L}$: remove all lines which intersect some line at infinity U.* Let U be the line at infinity of the xy–plane and thus we remove the set $\mathcal{L}'$ of lines orthogonal to the z–axis to get $\mathcal{L}^o := \mathcal{L} \setminus \mathcal{L}'$.

The realization of the corresponding stereographic projection with center C is very simple: A line $A \in \mathcal{L}^o$ may be defined by its intersection point $A_0 = (a_1, a_2, 0)$ with $\pi_0 : z = 0$ and its intersection $A_1 = (a_3, a_4, 1)$ with $\pi_1 : z = 1$ (see Fig. 6). The mapping

$$\sigma : A \in \mathcal{L} \mapsto (a_1, a_2, a_3, a_4) \in \mathbb{R}^4 \tag{21}$$

from the set $\mathcal{L}^o$ onto real affine 4–space describes a *stereographic projection of the Klein quadric*. This is proved as follows. The Plücker coordinates of L and U are

$$\gamma(A) = (a_3 - a_1, a_4 - a_2, 1, a_2, -a_1, a_1 a_4 - a_2 a_3), \quad \gamma(U) = (0, \ldots, 0, 1).$$

Embedding $\mathbb{R}^4$ into P^5 by

$$\sigma(A) = (a_1, a_2, a_3, a_4) \mapsto \sigma'(A) = (a_3 - a_1, a_4 - a_2, 1, a_2, -a_1, 0),$$

it follows that the point $\sigma'(A)$ is collinear with $\gamma(A)$ and $\gamma(U) = C$, and thus it is the image of $\gamma(A)$ for projection with center C onto the hyperplane $H : x_6 = 0$ in P^5.

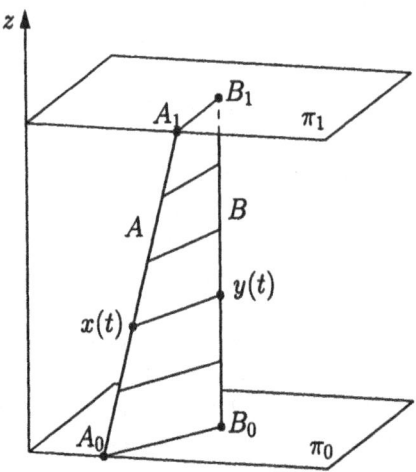

Fig. 6. Introducing a distance measure between two lines in an area of interest located between parallel planes π_0, π_1

For approximation we need a distance measure between two lines A, B. In practice, the distance within some area of interest will be important. Placing this area between the parallel planes π_0 and π_1, we may map the two lines linearly onto each other via

$$x(t) = (1 - t)A_0 + tA_1 \mapsto y(t) = (1 - t)B_0 + tB_1. \tag{22}$$

Viewing the z-axis as vertical, we may say that the line segments $x(t)y(t)$ are horizontal (cf. Fig. 6). It seems reasonable to measure the deviation of the two lines A, B in the domain between the reference planes π_0, π_1 by

$$d(A, B)^2 = \int_0^1 \|x(t) - y(t)\|^2 dt.$$

Inserting (22) we obtain

$$d(A, B)^2 = (A_0 - B_0)^2 + (A_1 - B_1)^2 + (A_0 - B_0) \cdot (A_1 - B_1)$$

$$= \sum_{i=1}^4 (a_i - b_i)^2 + (a_1 - b_1)(a_3 - b_3) + (a_2 - b_2)(a_4 - b_4). \tag{23}$$

This is the distance of their image points $\sigma(A), \sigma(B) \in \mathbb{R}^4$ in a *Euclidean metric in* $\mathbb{R}^4$, defined by positive definite quadratic form

$$\langle X, X \rangle = x_1^2 + \ldots + x_4^2 + x_1 x_3 + x_2 x_4. \tag{24}$$

For the definition of $d(A, B)$ one integrates squared distances between A, B measured in parallel planes between π_0, π_1. These distances differ from orthogonal distances to A by a factor between 1 and $1/\cos\varphi$, if φ is the angle between A and the z–axis. At least for lines whose angle with the z–axis does not exceed some tolerance dependent value $\gamma_0 < \pi/2$, (23) is a useful distance measure. Its behaviour is a counterpart to the well–known fact, that the distance distortion for stereographic projection of a sphere in $\mathbb{R}^3$ increases with the distance from the antipodal point of the projection center.

Theorem 3. *Consider two parallel planes π_0, π_1 in $\mathbb{R}^3$ and the set $\mathcal{L}^o$ of all lines which are not parallel to them. Then intersection of any line in $\mathcal{L}^o$ with π_0, π_1 gives a pair (p, q) of points, which may be considered as point in real affine 4–space $\mathbb{R}^4$. This mapping from $\mathcal{L}^o$ onto $\mathbb{R}^4$ can be interpreted as stereographic projection of the Klein quadric. The image space $\mathbb{R}^4$ can be endowed with a Euclidean metric (in an adapted coordinate system given by (24)), which corresponds to the deviation of the lines within the parallel strip bounded by planes π_0, π_1.*

This result provides a *transfer principle from approximation in line space to approximation in Euclidean 4–space*. We will illustrate it at hand of some examples.

6.1   Scattered data fitting in line space

We consider the problem of *scattered data interpolation and approximation for functions defined on line space*. Let a finite set of lines L_i (data lines) and associated real numbers f_i, obtained by some measurement or computation, be given. We would like to construct a function F, which is defined on all lines L within some domain D of interest (for example, lines with a maximum distance d to a fixed point) and exactly or approximately satisfies $F(L_i) = f_i$. One approach is the following [21]. Consider a centrally symmetric covering of the unit sphere $\Sigma \in \mathbb{R}^3$ by circular caps Γ_i, $i = 1, \ldots, m$ with rotational axes a_i and spherical radii ρ_i. For each axis a_i, let $\mathcal{L}_i$ be the set of lines that form an angle $< \rho_i$ with a_i and lie in the domain of interest. With two parallel planes that are orthogonal to a_i and enclose the domain of interest, we perform the mapping into $\mathbb{R}^4$. There, the images of data lines $L_k \in \mathcal{L}_i$ are data points with associated function values. Using the corresponding metric, they can be interpolated or approximated by any method which works in $\mathbb{R}^4$, for example radial basis functions [11]. Of course, we use the metric based on (24) instead of the canonical Euclidean metric. This gives a partial solution function F_i defined on any line in $\mathcal{L}_i$. Finally one just has to combine these partial solutions into a single one. This approach has been discussed in more detail by Peternell and Pottmann [21].

6.2   Ruled surface approximation

Theorem 3 provides a technique to interpret ruled surface approximation problems as curve approximation problems in the image space $\mathbb{R}^4$. In more detail we want to study the following. Given m lines L_i and corresponding parameter values τ_i, we will construct a ruled surface surface Φ with generators $X(t)$, such that $X(\tau_i)$ is close to L_i, in the sense of the distance function defined by (23).

Note that the distance $d(A, B)$ of two lines is not invariant under motions in $\mathbb{R}^3$, but essentially depends on the choice of the z-axis and the planes π_0, π_1. We can overcome this disadvantage by the following construction.

We have to determine a unit vector z as third coordinate axis, such that the angles formed by z and given lines L_i are as small as possible. To achieve this, the vector z is computed as solution of a regression problem to the given direction vectors (l_{i1}, l_{i2}, l_{i3}) of lines L_i. If some angle $\angle(L_i, z)$ is larger than a user defined bound $\gamma_0 < \pi/2$ one has to perform a segmentation of the data. Additionally, planes π_0, π_1 have to be chosen in such a way that they bound the domain of interest.

In order to determine $X(t)$ we minimize

$$F := \sum_{i=1}^{m} d(L_i, X(\tau_i))^2 \tag{25}$$

for an unknown ruled surface with generating lines $X(t)$. With respect to the mapping (21) we will restrict the class of ruled surfaces to those whose intersection curves with the planes π_0, π_1 are B-spline curves. Thus, the image curve of $X(t)$ in $\mathbb{R}^4$ is a B-spline curve

$$Y(t) = \sigma(X(t)) = \sum_{i=0}^{n} N_i^n(t) C_i$$

with control points C_i and B-splines of degree n as basis functions over a chosen knot vector.

Let the intersection curves of $X(t)$ with π_0, π_1 be denoted by

$$a(t) = (x_1(t), x_2(t), 0) \quad \text{and} \quad b(t) = (x_3(t), x_4(t), 1). \tag{26}$$

Since $x_i(t)$ are piecewise polynomials, F is a quadratic function in the unknown coefficients of the control points C_i. This implies that the minimization is done by solving a linear system of equations.

The ruled surface strip Φ determined by (26) possesses the following parameterization as point set in $\mathbb{R}^3$,

$$y(t, u) = (1 - u)a(t) + ub(t), \tag{27}$$

where $u \in [0, 1]$ parametrizes the line segments on the generating lines $X(t)$. The shape of $y(t, u)$ essentially depends on the chosen knot sequence and on

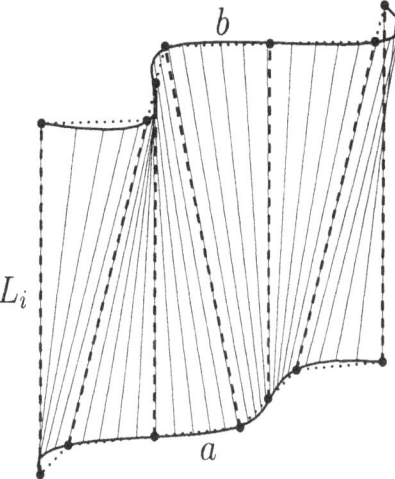

Fig. 7. Ruled surface approximating the given (dashed) lines L_i; the image $\sigma(X(t)) \subset \mathbb{R}^4$ is a cubic B-spline curve consisting of two segments

the parameter values τ_i. To improve the behaviour of the approximant Φ, we can combine (25) with the minimization of a fairness functional. A good choice, although parameter dependent, is usually the functional

$$G(y) = \int_{t_0}^{t_1} \int_0^1 (y_{tt}^2 + 2y_{tu}^2 + y_{uu}^2)\,du\,dt.$$

Inserting (27) and elaborating this we obtain a functional involving coordinate functions of $\sigma(X(t)) = Y(t) = (a, b)(t)$,

$$G(Y) = \frac{1}{3}\int_{t_0}^{t_1} (\ddot{a}^2 + \ddot{a}\ddot{b} + \ddot{b}^2)\,dt + 2\int_{t_0}^{t_1} (\dot{b}^2 - 2\dot{a}\dot{b} + \dot{a}^2)\,du.$$

Substituting $Y(t) = (a(t), b(t))$ we obtain

$$G(Y) = \frac{1}{3}\int_{t_0}^{t_1} \langle \ddot{Y}, \ddot{Y}\rangle\,dt + 2\int_{t_0}^{t_1} \dot{Y} \cdot M \cdot \dot{Y}\,dt,$$

where M is the positive semidefinite matrix

$$M = \begin{bmatrix} I & -I \\ -I & I \end{bmatrix} \text{ with } I = \begin{bmatrix} 1 & 0 \\ 0 & 1 \end{bmatrix}.$$

As an improvement of minimizing (25) we can minimize the quadratic function

$$F(X) + \lambda G(X).$$

This amounts again to the solution of a linear system in the unknown coordinates of the control points C_i.

Another possibility to improve the approximation is to apply a correction of the parameter values as mentioned in previous sections and described in [11].

6.3   Approximating line congruences

A two-parameter set of lines or more precisely, the preimage of a two-dim. surface in the Klein quadric, is called a *line congruence*. Its image under mapping σ (21) is a two-dimensional surface in $\mathbb{R}^4$. Thus, approximation problems for line congruences are transferred into surface approximation problems in Euclidean 4-space.

Approximating line congruences have applications in tool motion planning for 5-axis NC machining [31]. There, we have an additional requirement on the congruences: they shall form a fibration in some neighborhood of the surface to be machined. The inclusion of this property into the approximation algorithm is one of our current research topics.

7   Summary: A concept for approximation in spaces of geometric objects

We have treated spaces of different geometric objects: planes, lines and spheres. At first glance, the methods for approximation seem to be quite different. However, in all cases there are some basic steps involved:

- First, construct an appropriate point model for the k-dimensional space S^k of objects under consideration. If this is not yet a projective space, define – at least locally – a mapping into projective k-space P^k.
- By removing a hyperplane from the projective space P^k (removing an appropriate subset of the considered space of objects), we get an affine space A^k.
- Define an appropriate (Euclidean) metric in A^k, which is motivated by a deviation measure between two objects in S^k.
- The computation of m-parameter families of objects in S^k, which approximate some given objects from S^k, is thus transferred into an approximation problem for m-dimensional surfaces in Euclidean k–space.

Note that we presented the cases with increasing level of complication: For spheres, we immediately arrived at an affine space. For planes, we first got a projective space P^3. For lines, we first had a quadric as point model which could then be mapped onto an affine space.

The general concept still remains on a very high, not sufficiently detailed level. However, it is a nice feature that the discussed special cases fit into the following specific general framework. It concerns spaces of *affinely equivalent geometric objects*. This means, that any two objects of the considered class may be mapped into each other by an affine mapping (which needs not be unique, as we will see later).

We outline the concept for objects in $\mathbb{R}^3$, since its generalization to arbitrary dimensions is straightforward. Consider an object Γ in $\mathbb{R}^3$. We think of Γ as a curve, surface or solid. With respect to some tetrahedron $V_0, \ldots, V_3$, it may have the parametric barycentric representation

$$X(u) = \sum_{i=0}^{3} f_i(u)V_i, \quad \text{with} \quad \sum_{i=0}^{3} f_i(u) = 1. \tag{28}$$

The parameters u are from a domain in $\mathbb{R}^l$ with $l = 1$ for a curve Γ, $l = 2$ for a surface Γ and $l = 3$ for a solid Γ. The space $S = S^k$ of geometric objects is generated by the affine images of Γ. Let the tetrahedron $V_0^1, \ldots, V_3^1$ be the affine image of the tetrahedron $V_0, \ldots, V_3$. Then, the corresponding element Γ^1 of S has the representation

$$\Gamma^1: \quad X^1(u) = \sum_{i=0}^{3} f_i(u)V_i^1.$$

A simple way to get a point model for S is to interpret the 12 coordinates of the four points $V_0^1, \ldots, V_3^1$ as a point $G^1 = \sigma(\Gamma^1)$ in 12-dimensional affine space A^{12}. It is a point model for the affine maps in A^3 (see [29] for a more general point model of projective maps, which contains the present one as a subspace).

The mapping $\sigma : S \to A^{12}$ maps elements of S to points in A^{12}. We still have to introduce a metric in A^{12} based on the deviation of elements in S. This is done as follows. Two elements Γ^1, Γ^2 of S, written as

$$\Gamma^j: \quad X^j(u) = \sum_{i=0}^{3} f_i(u)V_i^j, \quad j = 1, 2,$$

may be mapped onto each other by the affine map α which maps the corresponding tetrahedra onto each other, i.e., $\alpha(V_i^1) = V_i^2$. The vectors connecting corresponding points are

$$D(u) = X^2(u) - X^1(u) = \sum_{i=0}^{3} f_i(u)(V_i^2 - V_i^1). \tag{29}$$

We view $\|D^2(u)\|$ as a function defined on the fundamental object Γ and integrate it over Γ to get a *distance* d between Γ^1 and Γ^2,

$$d^2(\Gamma^1, \Gamma^2) := \frac{1}{N} \int_{\Gamma} D^2(u) d\Gamma. \tag{30}$$

Here, we have normalized with the integral over Γ,

$$N = \int_{\Gamma} d\Gamma.$$

Inserting (29) into (30), we realize d^2 as a positive definite quadratic form in the coordinates of the difference vectors $D_i = V_i^2 - V_i^1$,

$$d(\Gamma^1, \Gamma^2) = \frac{1}{N} \left[D_0^2 \int_\Gamma f_0^2(u) d\Gamma \right. \tag{31}$$

$$\left. + 2D_0 \cdot D_1 \int_\Gamma f_0(u) f_1(u) d\Gamma + \ldots + D_3^2 \int_\Gamma f_3^2(u) d\Gamma \right].$$

Hence, we have introduced a *Euclidean metric* in A^{12} and thus have a way to solve approximation problems in S via approximation in Euclidean 12-space.

Theorem 4. *The above defined mapping σ maps elements from a space S of affinely equivalent geometric objects in $\mathbb{R}^3$ to points in A^{12}. The deviation measure (30) between objects Γ^1, Γ^2 in S introduces a Euclidean metric in the image space A^{12}.*

The occurring integrals are extended over a representative Γ of the space S. If Γ has the dimension of the space it spans (i.e., is either a line, a planar domain or a 3D solid), it does not matter which element of S has been chosen as Γ. This is so, since ratios of volume integrals are invariant under affine maps. Otherwise, this choice has an influence on the definition of the metric.

Let us briefly discuss now, in which way our previous investigations are special cases of the present more general concept.

1. For the deviation measure between planes (based on the same domain Γ of interest), we may see the affine maps between two planes realized by a projection parallel to the z-axis. The space of geometric objects is then the set of images of the domain Γ under projections parallel to the z-axis onto all planes which are not parallel to the z-axis. Clearly, we need no reference tetrahedron for the affine image of the planar figure Γ, and now the space S and the image space A^3 are only 3-dimensional. Otherwise, except for the (unnecessary) normalization, the concept for planes is precisely a special case of the general concept.

2. The deviation measure between lines defines affine maps between two lines with help of the auxiliary planes. In this way we set up the affine mapping via two points and their affine images. Clearly, we do not need more for our concept. Note that the same idea applies to the 6-dimensional space of *line segments* in $\mathbb{R}^3$ (see [3,8]).

3. Finally, for the set of spheres we have set up a special affine mapping between two spheres, namely the central similarity (compatible with the orientation). This is the only case where Γ spans 3-space and where we do not have a volume integral over the spanning space. Clearly, we could not use the whole space of affine images of a sphere, since we did not want to work with ellipsoids but with spheres only. In this way, we arrive at a 4-dimensional point model. The deviation measure is exactly the one which corresponds to the definition in the general concept.

There is still a lot of room for work in the area of approximation in spaces of geometric objects. For example, other important spaces (CSG primitives and other fundamental objects in CAD systems) could be investigated. Even the special cases of planes, lines and spheres need more research, in particular for m-parametric families of objects with $m > 1$.

There are also relations to known approaches which deserve further investigations. In particular, the present ideas are closely related to *kinematic mappings* which are used in motion design (see [30] and the references therein). It has to be investigated whether we can improve the construction of approximating motions based on our approach.

Acknowledgement

This work has been supported by grant No. P13648-MAT of the Austrian Science Fund.

References

1. R. M. C. Bodduluri, B. Ravani, Geometric design and fabrication of developable surfaces, ASME Adv. Design Autom. **2** (1992) , 243–250.
2. R. M. C. Bodduluri, B. Ravani, Design of developable surfaces using duality between plane and point geometries, Computer-Aided Design **25** (1993) , 621–632.
3. H.Y. Chen, H. Pottmann, Approximation by ruled surfaces, J. of Computational and Applied Math. **102** (1999), 143–156.
4. H. I. Choi, C. Y. Han, H. P. Moon, K. H. Roh, N.S. Wee, Medial axis transform and offset curves by Minkowski Pythagorean hodograph curves, Comp. Aided Design **31** (1999), 59–72.
5. H. Edelsbrunner, Deformable smooth surface design. Raindrop Geomagic Inc., Report 96-002, 1996.
6. G. Farin, *Curves and Surfaces for Computer Aided Geometric Design*, Academic Press, Boston 1992.
7. R. Fletcher, Practical Methods of Optimization, J. Wiley, Chichester, 1999.
8. Q. J. Ge, and B. Ravani, On representation and interpolation of line segments for computer aided geometric design, ASME Design Automation Conf., vol. 96-1 (1994) 191–198.
9. V. Hlavaty, Differential Line Geometry, P. Nordhoff Ltd., Groningen, 1953).
10. J. Hoschek, Liniengeometrie, Bibliograph. Institut, Zürich, 1971.
11. J. Hoschek, D. Lasser, *Fundamentals of Computer Aided Geometric Design*, A. K. Peters, Wellesley, Mass. 1993.
12. J. Hoschek, H. Pottmann, Interpolation and approximation with developable B-spline surfaces, in *Mathematical Methods for Curves and Surfaces*, M. Dæhlen, T. Lyche and L. L. Schumaker, eds., Vanderbilt University Press Nashville 1995, pp. 255–264.
13. J. Hoschek, M. Schneider, Interpolation and approximation with developable surfaces, in: *Curves and Surfaces with Applications in CAGD*, A. Le Méhauté, C. Rabut and L. L. Schumaker, eds., Vanderbilt University Press, Nashville 1997, pp. 185–202.

14. J. Hoschek, U. Schwanecke, Interpolation and approximation with ruled surfaces, in *The Mathematics of Surfaces VIII*, R. Cripps, ed., Information Geometers, 1998, 213–231.

15. R. Krasauskas, C. Mäurer, Studying cyclides with Laguerre geometry, Comput. Aided Geom. Design **17** (2000), 101–126.

16. C. Mäurer, Applications of sphere geometry in canal surface design, in *Curves and Surfaces*, P. J. Laurent, C. Rabut and L. L. Schumaker, eds., Vanderbilt Univ. Press, Nashville, TN, 2000.

17. M. Paluszny, W. Boehm, General cyclides, Comput. Aided Geom. Design **15** (1998), 699-710.

18. M. Peternell, H. Pottmann, Computing rational parametrizations of canal surfaces, J. Symbolic Computation **23** (1997), 255–266.

19. M. Peternell, H. Pottmann, A Laguerre geometric approach to rational offsets, Comput. Aided Geom. Design **15** (1998), 223–249.

20. M. Peternell, H. Pottmann, B. Ravani, On the computational geometry of ruled surfaces, Computer Aided Design **31** (1999), 17–32.

21. M. Peternell, H. Pottmann, Interpolating functions on lines in 3-space, in *Curves and Surfaces*, P. J. Laurent, C. Rabut and L. L. Schumaker, eds., Vanderbilt Univ. Press, Nashville, TN, 2000.

22. L. Piegl, W. Tiller, *The NURBS book*, Springer, 1995.

23. H. Pottmann, G. Farin, Developable rational Bezier and B–spline surfaces, Comput. Aided Geom. Design **12** (1995) , 513–531.

24. H. Pottmann, M. Peternell, Applications of Laguerre geometry in CAGD, Comput. Aided Geom. Design **15** (1998), 165–186.

25. H. Pottmann, M. Peternell, B. Ravani, Approximation in line space: applications in robot kinematics and surface reconstruction, in *Advances in Robot Kinematics: Analysis and Control*, J. Lenarcic and M. Husty, eds., Kluwer, 1998, 403-412.

26. H. Pottmann, M. Peternell, B. Ravani, An introduction to line geometry with applications, Computer Aided Design **31** (1999), 3–16.

27. H. Pottmann, J. Wallner, Approximation algorithms for developable surfaces. Comput. Aided Geom. Design **16** (1999), 539–556.

28. H. Pottmann, J. Wallner, B. Ravani, *Computational Line Geometry*, in preparation for publication in the Springer series "Mathematics and Visualization".

29. W. Rath, Matrix groups and kinematics in projective spaces, Abhandlungen Math. Seminar Univ. Hamburg **63** (1993), 177-196.

30. O. Röschel, Rational motion design – a survey, Computer Aided Design **30** (1998), 169-178.

31. J. Wallner, H. Pottmann, On the geometry of sculptured surface machining, in *Curves and Surfaces*, P. J. Laurent, C. Rabut and L. L. Schumaker, eds., Vanderbilt Univ. Press, Nashville, TN, 2000.

Representing the Time-Dependent Geometry of the Heart for Fluid Dynamical Analysis

C.J. Evans, M.I.G. Bloor and M.J. Wilson

Department of Applied Mathematics
University of Leeds LS2 9JT United Kingdom Email: chrise@amsta.leeds.ac.uk

Summary. A parametric model of the complex time-dependent geometry of the heart is constructed. The geometry model is created by means of a boundary value approach, solving an elliptic partial differential equation to generate a representation of the inner surface of the ventricles. The technique provides a closed-form description of the geometry and a straightforward link to analysis, facilitating the calculation of physical properties such as those relevant to fluid dynamics. As an application of this work, the geometry model is combined with commercial CFD software to analyse the blood flow in the heart. Calculations are performed at various time steps to follow the evolution of the fluid flow.

1 Introduction

The heart consists of four chambers, a right and left atrium and a right and left ventricle, and between them they drive the circulation of blood round the body. The right ventricle pumps blood through the pulmonary arteries to the lungs. Here, the blood is oxygenated and passed to the left ventricle which propels it through the aorta to the remainder of the body and finally back into the right side of the heart. The left and right atria act as auxiliary pumps, assisting the entry of blood into their corresponding ventricle. Each ventricular chamber has inlet and outlet valves which act in such a way that one closes before the other opens, so that (ideally) no backflow occurs. These four valve orifices are aligned approximately in the same plane. The two phases of the cardiac cycle are called *systole*, during which both ventricles simultaneously contract, and *diastole* when they relax again and refill with blood. More information on this can be found in any standard physiology text, such as [12].

Pedley [9] describes the left ventricle as roughly circular in cross-section, and shaped rather like a blunted arrowhead. When it contracts, there is an initial phase in which the long axis shortens slightly and the transverse cross section expands. Then the aortic valve opens and the long axis remains roughly the same size while the transverse axes shorten by around a third. The right ventricle, however, is more complex in shape. The inter-ventricular wall is functionally part of the left ventricle but the outer wall of the right ventricle is significantly thinner and has a much larger area, resulting in a crescent-like shaped cavity wrapped around the left ventricle.

Various approaches have been adopted in the past to tackle the modelling of the motion and fluid dynamics of the heart, with the most work concentrating on systolic flow in the left ventricle. Relatively recent advances in scanning technology have meant that it is possible to build up an accurate geometry model of the heart using scan data ([14], [8], [5] for example). This approach provides potentially accurate representations of the geometry of the ventricles but relies on having (processed) scan data available for the case in question. Other work such as that of Schoephoerster [4], [11] has used a simplified ventricular geometry to solve the unsteady flow problem in the left ventricle using various computational techniques.

Some of the most advanced work to date is that of Peskin and McQueen [10] who build a model encompassing both the ventricles, atria and major arteries linked to the heart, and then use their Immersed Boundary Method to examine the fluid flow. This technique treats the ventricle walls as made up of elastic fibres which occupy zero volume and move at the local fluid velocity. A set of coupled differential equations are then set up to describe the motion of the fluid, the fibers, and the interaction between the two. The calculations required to do this are extremely computationally intensive due to the complexity of the model. This technique was also used by Yoganathan et al. [15] in modelling a thin-walled left ventricle during early systole.

The aims of the work presented in this paper are as follows:

- to build a model for the geometry of the human heart that is capable of representing the complex, time-dependent changes in shape that take place during the cardiac cycle.
- to use the model to help understand the basics of fluid flow within the ventricles during the cardiac cycle.

In order to achieve these, we wish to create a model of the heart which is generic, rather than one built up from actual scan (eg. Magnetic Resonance Imaging (MRI)) data which, because of the large amount of data involved, would be much less flexible. This flexibility is desirable due not only to the large variation in ventricular shape between individuals, but also due to the wide range of disease conditions which can affect the shape and motion of the heart in many different ways.

The technique which is used to generate the geometry in this paper has been applied in the past to many other design problems, including the efficient parametrization of aircraft [1], propeller blades [3] and ship hulls [7]. Surfaces are produced by specifying boundary curves which reflect the key features of the shape to be modelled, and then forming smooth surface patches between them. A consequence of this approach is that only a small number of design parameters are required because the shape is entirely determined by the information specified around the boundary curves. Moreover, even using a small parameter space, a wide range of geometries can be generated. The current work differs from previous design work in that it uses parameters which are functions of time in order to model the changes in ventricular shape throughout a typical cardiac cycle. One of the advantages of the method over many other CAD techniques is the straightforward link to analysis, which facilitates the calculation of physical properties. This link has facilitated past work on aircraft, ship and engine schematics for which the fluid dynamics is important, as well as for other problems, where the mechanical stress is an issue. Here we will consider the fluid dynamics of the heart.

To do this, a commercial computational fluid dynamics package called Fluent is used to calculate the flow within the ventricles. Fluent comprises the main flow solver and also some preprocessing software which automatically generates a tetrahedral volume mesh from the description of the surface geometry. An idealised flow model is adopted, in which blood is taken to be a viscous incompressible Newtonian fluid. Steady calculations have been performed at various time steps to follow the evolution of the fluid flow.

In this paper, we shall describe the parametrization method, demonstrate how it is applied to the problem of the ventricles and show the resulting geometry. We will also describe the fluid dynamical results that have so far been obtained. In the following section we describe our methodology for producing the geometry model, while in sections 3 and 4 we give the fluid flow results and some discussion.

2 Method for Parametrizing the Geometry

As mentioned previously, the technique for shape parametrization that is used defines surfaces in terms of boundary curves and derivative information about the surface on these curves. A smooth surface interior is generated as the solution to an elliptic partial differential equation (PDE), which acts as an 'averaging' operator. The shape is defined by a network of curves on the surface, and the surface can be regarded as describing a smooth transition between the boundaries. Most of the work done so far has used a version of the biharmonic equation:

$$\left(\frac{\partial^2}{\partial u^2} + a^2 \frac{\partial^2}{\partial v^2} \right)^2 \boldsymbol{X}(u, v) = 0 \tag{1}$$

where $\boldsymbol{X} : (u, v) \rightarrow (x, y, z)$ with (u, v) contained in some finite subset of $\mathbb{R}^2$. a is called the smoothing parameter and controls the relative scaling between the u and v directions.

A typical surface could be formed from a rectangular domain, $(u, v) \in [0, 1] \times [0, 2\pi]$ where v is subject to periodic boundary conditions. In this case the surface would be topologically like an open cylinder, with the lines $u = 0$ and $u = 1$ forming the top and bottom edges. Suitable boundary conditions would then be of the form:

$$\boldsymbol{X}(0, v) = \boldsymbol{f}_0(v), \boldsymbol{X}(1, v) = \boldsymbol{f}_1(v) \tag{2}$$
$$\boldsymbol{X}_u(0, v) = \boldsymbol{s}_0(v), \boldsymbol{X}_u(1, v) = \boldsymbol{s}_1(v)$$

Here, f_0 and f_1 define the shapes of the edge curves parametrised in terms of the variable v. s_0 and s_1 are the corresponding derivatives, again parametrised by v. The geometric parameters of the model appear within the functions f_0, f_1, s_0, s_1 for each patch.

In general however, more than one patch of surface is needed to give a full description of the shape to be modelled, and in previous work complicated surfaces have been built up by joining multiple surface patches together. Appropriate continuity between patches is ensured by imposing consistent derivative conditions on shared boundary curves.

For example, if we take, for patch A of the left ventricle (see Fig. 2), the function f_0 which describes the boundary circle of the inflow tract, it has the following form, describing a simple circle in space.

$$f_0(v) = (x_a + r_1 \cos(v), y_a + r_1 \sin(v), z_{ab} - z_t) \qquad (3)$$

Here (x_a, y_a, z_{ab}) are the initial coordinates of the centre of the circle of the valve, r_1 is the radius of the valve and z_t describes the axial contraction of the ventricle. For this particular boundary, the only time dependence is contained in z_t .

To generate the surfaces used to model the ventricles, three patches are used in each, labelled A, B and C. Each of these patches is formed between two closed curves, with boundary conditions of the form of (2) imposed on them. Figure 1 shows the curves used to form the left ventricle surface. The derivative conditions on the curves control the direction that the surface normal points at each point on the boundary. The dotted lines in the diagram represent tangent lines to the surface on these curves, which are simply the vectors s_0 and s_1. The overall geometry produced by this method for the left and right ventricles is shown in Fig. 2. The shading shown in these pictures is generated using standard computer graphics techniques and the lines on the surfaces shown are artefacts of this.

The parameters for the left ventricle are also shown in Fig. 1 (the right ventricle situation is similar) and their time-dependence is discussed later on.

It is straightforward to express the solution to equation (1) in closed form and it has the form

$$X(u, v) = A_0(u) + \sum_{n=1}^{N} \left[A_n(u) \cos(nv) + B_n(u) \sin(nv) \right] \qquad (4)$$

where
$$A_0(u) = a_{00} + a_{01}u + a_{02}u^2 + a_{03}u^3$$
$$A_n(u) = a_{n1}e^{anu} + a_{n2}ue^{anu} + a_{n3}e^{-anu} + a_{n4}ue^{-anu}$$
$$B_n(u) = b_{n1}e^{anu} + b_{n2}ue^{anu} + b_{n3}e^{-anu} + b_{n4}ue^{-anu}$$

Here, a_{nm} and b_{nm} are all constant vectors.

To solve Eqn.(1) subject to (2), it will normally be necessary to Fourier analyse the functions f_0, f_1, s_0 and s_1. This will usually require that $N = \infty$.

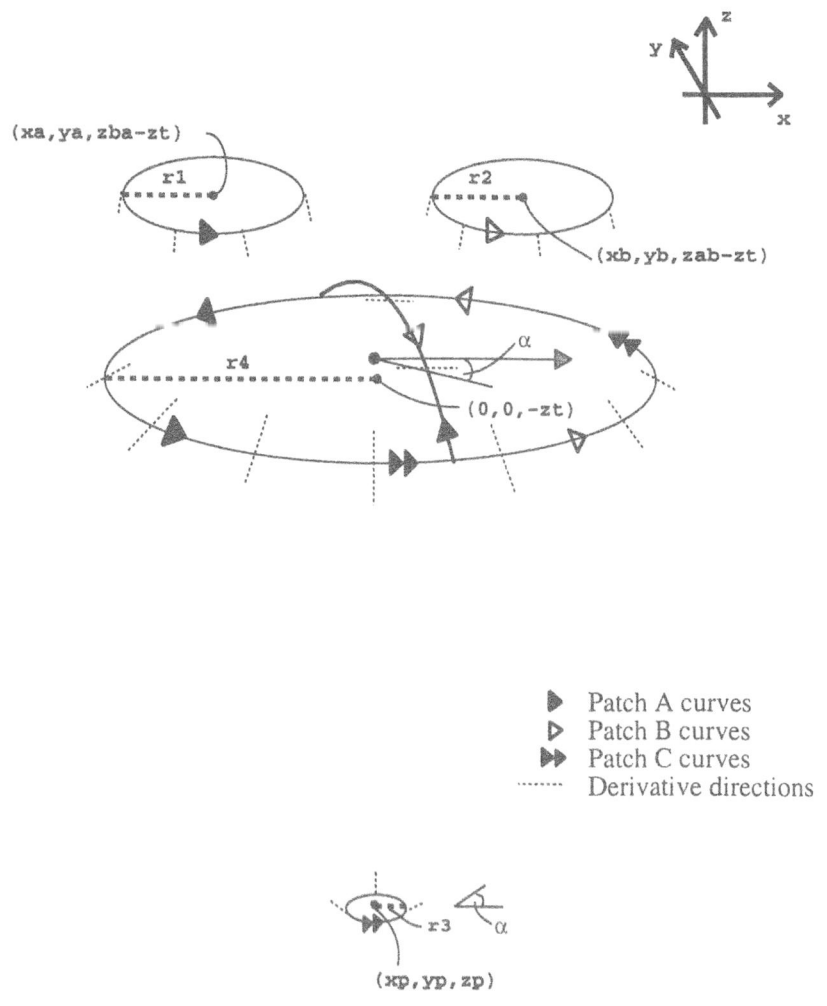

Fig. 1. Schematic diagram of the boundary curves and parameters used for the left ventricle

Practically, of course, an infinite series solution is of little value so a means of providing an approximate, but nevertheless analytic, solution is necessary.

An obvious solution to this problem would be to terminate the Fourier series after a finite number of terms. This would give an exact solution to the PDE which only matched the boundary conditions approximately. However, this would lead to holes in the surface between patches as they would not exactly meet on the adjoining curves. Also, at the outset of this process the choice of PDE was to some extent arbitrary and it is the boundary conditions which are the critical element in the modelling process. Thus a more satisfactory outcome would be to find an approximate solution to the PDE

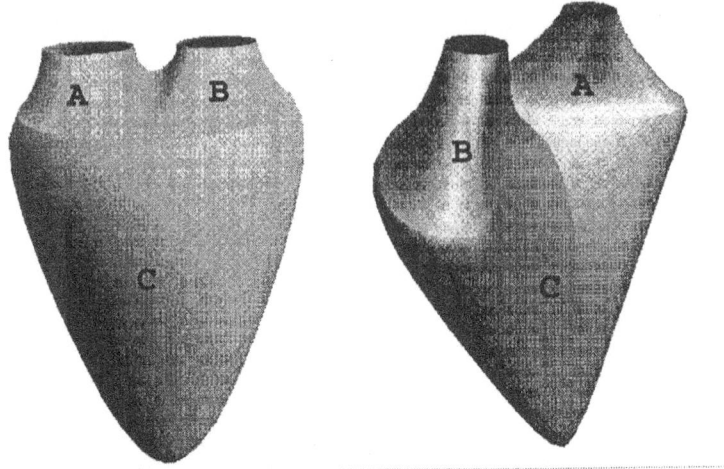

Fig. 2. Resulting geometry for the left and right ventricles.

which met the boundary conditions exactly. The method used to do this is described in [2]. This technique results in a surface that satisfies the boundary conditions precisely and is close to the exact solution of the whole problem away from the boundaries. This surface has a closed form representation.

The geometry of the left ventricle produced by the PDE method, given the boundary curves in Fig. 1 is shown (at two different time steps) in Fig. 3. Generating the geometry from values of the input parameters is a quick process, taking just a few seconds.

Changes in Time

The above process must be performed at each time step of interest. In order to change the geometry, all that need be altered in the above procedure is the boundary conditions and the smoothing parameter, a, since they completely define the shape. Information about the changes in ventricular shape during a heartbeat can be found in various physiology texts, such as [12]. Three major components of ventricular motion, as described in section 1, have been included, namely:

1. 'axial' contraction
2. 'radial' contraction and
3. twisting (see below)

In the above, the axis that is referred to stretches from the plane of the valves to the 'apex' of the ventricle. During systole, the plane of the valves moves down slightly while the apex remains roughly where it is.

Table 1. Descriptions and definitions of parameters for the left ventricle model.

Parameter	Description	Function
r_t	radius of centre circle	$r_0\left(1 - t_m \frac{\sin^2(t\pi/2)}{3}\right)$
z_t	axial contraction	$z_c t_m \sin^2(t\pi/2)$
α_t	angle of twist	$\alpha_0 t_m \sin^2(t\pi/3)$

Another important mode of contraction (expansion) that occurs during systole (diastole) is the 'tightening' ('loosening') of the muscle fibres. The fibres stretch round the ventricle in a helical-like curve and they contract as the ventricle shrinks which changes their formation. While the grid lines on the surface do not correspond to the muscle fibres in the actual ventricles, this effect is mimicked by rotating the bottom boundary curve of the main patch of surface about the axis of the each ventricle.

With these three modes of contraction included, realistic changes in volume and dimensions can be achieved. Figure 3 shows the left ventricle geometry at its extremes of size.

Some of the key time-dependent parameters that are used to implement the geometric changes described above (in the left ventricle) are as follows. The examples shown have used simple sinusoidal variation in time. Figure 1 shows the geometrical meaning of the parameters.

Here, r_0, y_0, z_c and α_0 are constants describing the amount of contraction of each type. t_m represents the time taken for one cycle to occur.

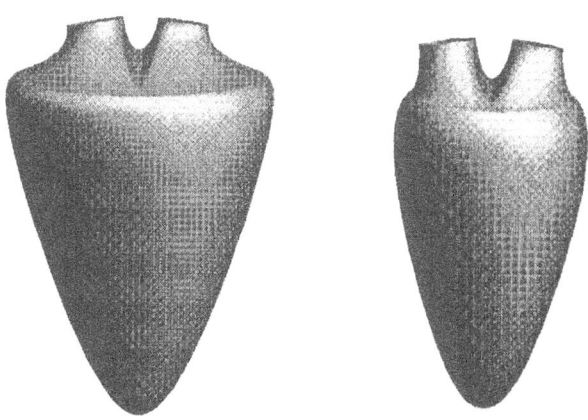

Fig. 3. The left ventricle geometry at its (a) largest and (b) smallest.

Fluid Dynamical Analysis

Steady state flow calculations were performed at various time steps throughout the cardiac cycle using a CFD package called Fluent. This software uses finite volume techniques to solve conservation equations for mass and momentum by means of a first order upwind scheme. Prior to this process, an internal tetrahedral volume mesh must be generated and this is performed automatically by programs called PreBFC and TGrid, which are preprocessing software used with Fluent.

Blood was modelled as an incompressible fluid with density $\rho = 1050 kgm^{-3}$ and viscosity $\nu = 0.004 kgm^{-1}s^{-1}$.

Boundary Conditions

In the Fluent solver, boundaries can be specified as (amongst other things) outlets, velocity inlets and walls. At velocity inlets, the velocity of the fluid in each cell on the boundary must be specified. These three types of boundary were the only ones used in the flow calculations for the ventricles. The simplest case was in systole where one valve was taken to be an outflow, the other a wall and all other regions were treated as velocity inlets. As we specify how the heart wall moves throughout the cycle, the instantaneous velocities of the grid points on the surface can easily be calculated.

The situation during diastole was slightly more complicated. Since it is the motion of the ventricle walls which our model provides, we again wished to specify the velocities of each point of the grid as boundary conditions for the flow. Therefore the majority of the surface was once again treated as a velocity inlet with negative velocity, despite the fact that the ventricle is expanding during this part of the cycle. The inflow valve was treated as an outflow (but fluid still enters through it) and the aortic valve, since it shuts for the majority of diastole, was said to be a wall.

Results of Flow Calculations

A sequence of results at various time steps throughout the cardiac cycle were obtained. The computations carried out to date have all used incompressible laminar flow models. On this basis we can model an unsteady flow as a series of steady state calculations. In future work it should be possible to perform turbulent unsteady calculations to improve the accuracy of the results.

Some pictures of the particle path lines for the left and right ventricles in systole and diastole are shown in figures 4 and 5. Fluent can also provide various other graphical means of examining the results, such as velocity profile graphs, pressure contours and others.

The resulting flows are in qualitative agreement with the observations of Taylor and Wade [13] who studied diastolic flow patterns in the left ventricle in the seventies and also borne out by more recent studies such as [6], in which doppler imaging was used to analyse left ventricular blood flow.

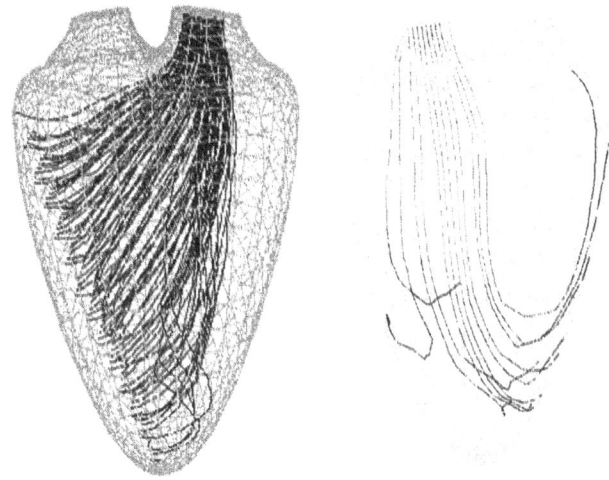

Fig. 4. Particle paths in the left ventricle during (a) systole and (b) diastole.

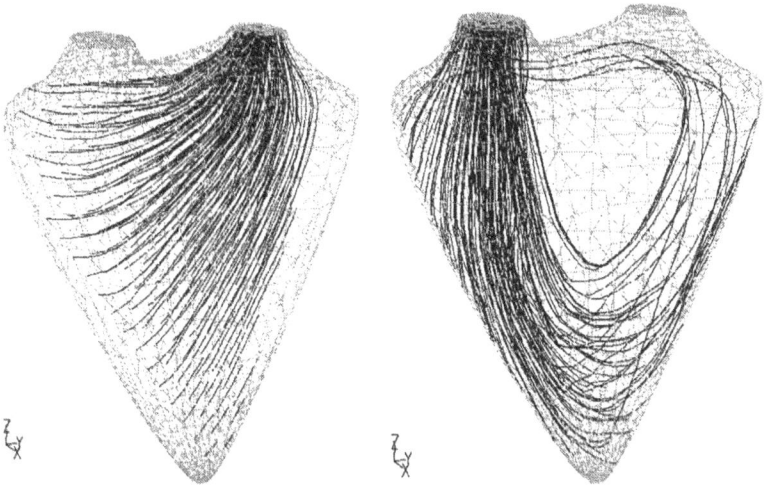

Fig. 5. Particle paths in the right ventricle during (a) systole and (b) diastole.

Discussion

A generic model of the geometry of the ventricles of the heart has been obtained using the PDE method, and time-dependence has been built in so that the geometry at any stage of the cardiac cycle can be produced. The geometry description provides a straightforward link to methods for physical analysis. For example, using this model and the Fluent computational fluid dynamics software, steady state flow calculations have been performed at stages throughout the cardiac cycle. Examples of the results obtained are shown in figures 4 and 5. The calculations can also provide velocity profiles, pressure distributions and other data.

There are many possible extensions to this work.

Firstly, as has already been mentioned, it should be possible using more advanced CFD software to perform unsteady, turbulent calculations on the same basis as described above to better approximate the conditions in the ventricles.

Secondly, the modelled geometry could be extended to include the actual geometry of the valves of the ventricles, again using the PDE method.

Finally, one major advantage of the technique described is that changes in the geometry and motion of the ventricles can be easily introduced into the model. In particular it should be possible to adapt the flow calculation in order to incorporate the effect(s) of some types of heart disease. One possibility is to examine *ischemia*, a condition in which a portion of the muscle surrounding the heart dies, affecting the way in which the ventricles contract.

3   Acknowledgments

CJE would like to acknowledge the support of a research studentship from the BBSRC (special grant number 98/B1/E/04227).

References

1. Bloor M.I.G. and Wilson M.J., "Generating Parametrizations of Wing Geometries using Partial Differential Equations", Computer Methods in Applied Mechanics and Engineering, 1997, Vol. 148, pp 125-138.
2. Bloor M.I.G. and Wilson M.J., "Spectral Approximations to PDE Surfaces", Computer-Aided Design, 1996, Vol. 28, 2, pp 145-152.
3. Dekanski C., Bloor M.I.G., Nowacki H. and Wilson M.J., "The Geometric Design of Marine Propeller Blades using the PDE Method" in Practical Design of Ships and Mobile Units, Elsevier Applied Science, London, 1992, pp 1.596-1.609.
4. Gonzalez E., Schoephoerster R.T., "A Simulation of three-dimensional Flow Dynamics in a Spherical Ventricle: Effects of abnormal wall motion", Annals of Biomedical Engineering, 1996, Vol. 24, 1, pp 48-57.
5. Haber E., Metaxas D.N. and Axel L., "Motion Analysis of the Right Ventricle from MRI Images", Lecture Notes in Computer Science, 1998, Vol. 1496, pp 177-188.
6. Kim W.Y et al., "Left-ventricular blood-flow patterns in normal subjects", Journal of the American College of Cardiology, 1995, Vol. 26, 1, pp 224-37.
7. Lowe T.W., Bloor M.I.G. and Wilson M.J., "The Automated Functional Design of Hull Surface Geometry", Journal of Ship Research, 1994, Vol. 38, 4, pp 319-328.
8. Park J., Metaxas D., Young A.A and Axel L, "Deformable Models with Parameter Functions for Cardiac Motion Analysis from Tagged MRI Data", IEEE Transactions on Medical Imaging, 1996, Vol. 15, 3.

9. Pedley T.J., "The Fluid Dynamics of Large Blood Vessels", Cambridge University Press, 1980.

10. Peskin C.S. and McQueen D.M., "Cardiac Fluid Dynamics", Critical Reviews in Biomedical Engineering, 1992, Vol. 20, 5-6, pp 451 et seq.

11. Schoephoerster R.T., Silva C.L., Ray G., "Evaluation of Left-Ventricular Function based on Simulated Systolic Flow Dynamics Computed from Regional Wall-Motion", Journal of Biomechanics, 1994, Vol. 27, 2, pp 125-136.

12. Smith J.J. and Kampine J.P., "Circulatory Physiology : the essentials", Baltimore : Williams and Wilkins, c1990, 3rd ed.

13. Taylor D.E.M. and Wade J.D., "The pattern of flow around the atrioventricular valves during diastolic ventricular filling", Journal of Physiology, 1970, Vol. 207, pp 71-2.

14. Taylor T.W. and Yamaguchi T., "Flow Patterns in 3-dimensional Left-ventricular Systolic and Diastolic flows determined from Computational Fluid-Dynamics", Biorheology, 1995, Vol. 32, 1, pp 61-71.

15. Yoganathan A.P., Lemmon J.D., Kim Y.H., Walker P.G., Levine R.A., Vesier C.C, "A Computational Study of a Thin-Walled 3-Dimensional Left Ventricle during Early Systole", Journal of Biomechanical Engineering-Transactions of the ASME, 1994, Vol. 116, 3, pp 307-314.

Application of Point-Based Smoothing to the Design of Flying Surfaces

R.M. Tookey & W. Sargeant

BAE SYSTEMS, Warton Aerodrome,
Preston, Lancashire,
PR4 1AX, England.

1   Introduction

The aerospace industry is becoming increasingly competitive and customer focused. Competitive edge is achieved by providing the customer with an improved product with quicker lead times from concept through to manufacture. This can be facilitated using a fully integrated Digital Mock-Up (DMU), i.e. a complete solid model, which allows physical activities, such as clash detection on assembly and mock air battles, to be simulated within a virtual or synthetic environment. Design changes, such as alternative wing profiles or fuselage sections, can then be incorporated into the DMU at an earlier stage allowing more 'what if ..?' and 'how to ..?' scenarios to be analysed without manufacturing any components.

Surfaces play an important role in the fully integrated DMU since they underpin the design of all the solid models. Necessarily, an aircraft is designed from the outside in. For example, the external flying surface, known as the Computed Profile Line (CPL), is designed first. This is followed by the internal skin surface, which may be a constant offset from the CPL. All contacting structures, such as bulkheads and frames, are then designed using constant offsets from this internal skin surface. In addition, tool surfaces can be constant offsets from CPL. This offset dependence requires surfaces to be of a high quality and not contain any gaps, creases, ripples or areas of tight curvature. Apart from avoiding re-work and reducing modelling time, high quality surfaces also ensure that surface blemishes are not reflected in the manufactured components. It is generally acknowledged that high quality surfaces are extremely difficult to produce using existing computer systems. In addition, more complex design requirements, such as stealth, imply that aircraft are becoming more doubly-curved (bi-directional) and fully blended. This demands even higher quality surfaces, and hence additional skills from the surface designers and additional tools from the computer systems.

The paper first reviews the traditional surface design process at BAE SYSTEMS and highlights the difficulties in producing a high quality surface. An alternative semi-automatic approach for quality surface design is then proposed. This has been test implemented within CATIA V4 and applied to the case study of the Windscreen Canopy surface for Eurofighter.

2   Review of traditional surface design process

The flowchart in Fig. 1 illustrates the traditional surface design process at BAE SYSTEMS. It contains two distinct phases, namely the Concept and Definition Phases. Initially, the customer will consult with a designer from Concept Studies. Engineers from any other relevant departments, such as Design (CATIA), Aerodynamics, Structures, Systems (mechanical), Avionics (electrical), Cost and Armaments, will then be involved in creating a series of characterising parameters representing the shape of the CPL. These parameters are optimised with respect to performance and affordability trade-offs using an in-house computer system. This part of the Concept Phase is iterative and referred to as the Concept Loop. The resulting point continuous (G^0) surface model is referred to as the Configuration Standard and can be considered as the modern day equivalent to the traditional two-dimensional component drawing in that it conveys the specifications of the component together with the surface designers intent with respect to the CPL. Component drawings are still used to specify some of the critical curve geometry. For example, it is appropriate to specify the Leading and Trailing Edges in two orthogonal views.

Since the Configuration Standard is often constructed from a sparse set of planar sections, it clearly does not contain the level of detail required to generate the Numerically-Controlled (NC) cutter-paths necessary to produce the tool surfaces. Hence, the Standard needs to be productionised. The Definition Phase is again iterative with engineers from Design, Aerodynamics, Structures and any relevant Manufacturing Departments being involved in the Definition Loop. Surface designers first ensure that the surface model has at least first order geometric contact, i.e. tangent continuity (G^1), and where possible second order continuity (G^2) which is related to the surface curvature. This is generally achieved by improving the quality of each surface by modifying the appropriate wireframe geometry. The network of input curves is manually improved until the resulting surface satisfies the quality criteria. This process is necessarily time consuming since the modification of any lateral or longitudinal curve requires all of the intersecting curves to be updated accordingly. In addition, a curve should not be modified in isolation, but should be based on the behaviour of its neighbouring curves with the result that the network of curves is harmonious and compatible with a quality surface.

Manufacturability is also built into the surface model during the Definition Loop. For example, the surface designers need to ensure the surface can be used to construct a solid model. This solid model, together with the associated surfaces, is referred to as the Production Standard and is analysed using customised Computational Fluid Dynamics (CFD) and Finite Element (FE) analysis software for aerodynamic and structural properties. Aerodynamic analyses tend to use triangular surface meshes, while structural analyses tend to use rectangular solid meshes.

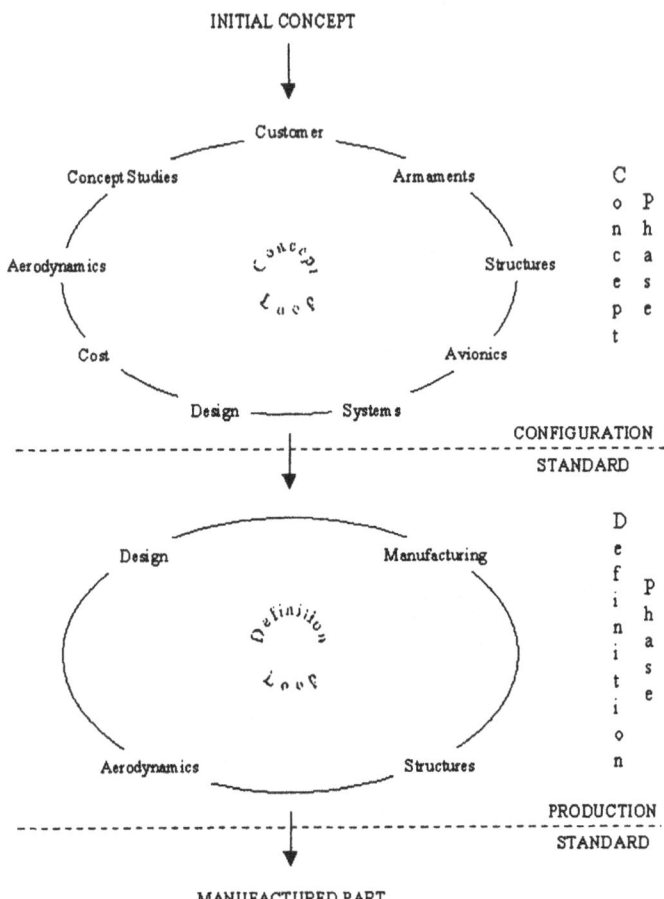

Fig. 1. Flowchart illustrating the traditional surface design process at BAE SYSTEMS

Finally, the solid model is incorporated into the DMU and can be analysed for interference with other components and visualised within Virtual Reality (VR) environments, such as CATIA's 4D Navigator. If there should be any problems with the Production Standard, then the design process can return to either the Concept or Definition Loops. It has been identified that the surface design process is both time consuming and labour intensive, relying heavily on the skills of the surface designers and their knowledge of the available curve and surface construction and analysis techniques. Alternative semi-automatic procedures are therefore required to improve the productionisation process from concept through to manufacture.

3   Background to case study

The case study is the Windscreen Canopy surface for Eurofighter (cf. Fig. 2). One Configuration and two Production Standards were developed before the surface was considered acceptable. The case study is symmetric, therefore a half model is first constructed and then symmetrised. Brief construction histories and analyses of the three surfaces are provided in the next three sub-sections.

Fig. 2. Solid visualisation of Eurofighter

3.1   Construction of Configuration Standard

The Windscreen Canopy surface was first constructed from a set of traditional two-dimensional component drawings and verbal instructions from the Aerodynamics Department. The Windscreen was required to be a cone to give certain reflection characteristics relating to the projected head-up display. The Canopy was required to have circular sections to meet structural requirements. In essence, the hoop stresses need to be uniformly distributed to manage the distortion of the Canopy and maintain its pressure seal with the Fuselage. The component drawings specified three constraint curves, namely the Windscreen Arch, Pilots Eye and Sill Line. The Windscreen Arch is a plane section curve corresponding to the join between the fixed Windscreen and the ejecting Canopy. Its location was chosen so that the associated frame would not obscure the pilots view. The Pilots Eye is a plane section curve corresponding to the minimum clearance for the pilots helmet. The Sill Line is a space curve corresponding to the intersection between the Windscreen Canopy and the Fuselage. Any surface construction must match these three constraint curves to within 0.1 mm.

An iterative procedure, based on a series of 18 circular sections spaced approximately every 200mm, was used to develop a Centre Line (also used as the upper limit curve), a lower limit curve and a spine for the surface. The three curves were designed in sympathy so that the resulting circular sweep surface, with sections normal to the spine and passing through the Centre Line, minimised the induced drag and provided sufficient clearance for the pilots helmet. The half model, which extends below the Sill Line, is illustrated in Fig. 3 and contains a strip of 34 patches of degree 8 by 5. Note that degree 8 curves are used in CATIA to approximate circular arcs. The figure also illustrates the Centre Line, Windscreen Arch and Sill Line. Note that the front of the Windscreen is towards the right of the figure, while the rear of the Canopy is towards the left.

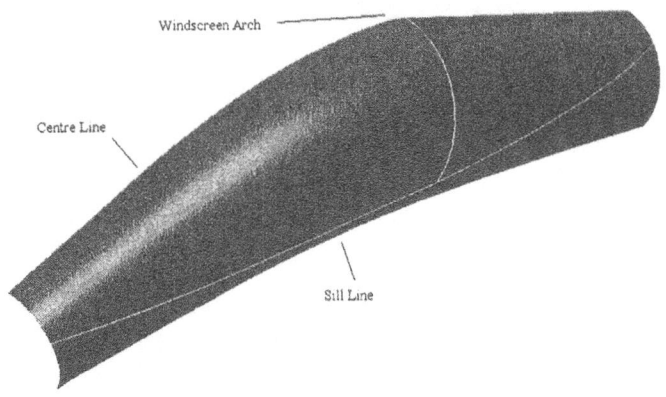

Fig. 3. Constraint curves on the Configuration Standard surface

The solid models of the Windscreen and Canopy need to be 22 mm and 8 mm thick respectively. Unfortunately, when split at the Windscreen Arch, the Canopy solid cannot be constructed. Subsequent analysis reveals that the lower surface boundary inflects in the plan view at the join between the

Windscreen and the Canopy. This results in tight curvatures, which stop the surface offsetting by the required amount.

3.2   Construction of Production Standard

In general, the modification procedure involves removing the problem area and re-designing it using additional constraint curves to force the shape. For the Windscreen Canopy, the surface designer first extracted the degree 8 patch boundaries on the Configuration Standard surface. Some of these in the problem area were then deleted and loft surfaces fitted through the rest. The two removed patches were re-designed by constructing longitudinal constraint curves and imposing tangency on the end sections with the adjacent loft surfaces. Concatenating the three surfaces into a single one results in a 7×28 array of degree 5 by 5 patches. Note that the loft surface in CATIA approximates wireframe geometry with degree 5 curves.

The Canopy surface now offsets by the required amount. In addition to the curvature discontinuities and an unnecessary rippling at the rear of the Canopy, the main problem with the initial Production Standard is the number of patches, i.e. 392 in the symmetrised model. This generally increases the model size and slows down any subsequent interrogations, such as NC machining.

The second refinement modelled each part separately and then concatenated the two surfaces. The Windscreen was constructed as a ruled surface using 2 circular sections: this results in a single patch of degree 8 by 1. The Canopy was constructed as a loft surface using 14 circular sections and specifying tangency with the Windscreen on the first section curve: this results in a 3×13 array of degree 5 patches. Concatenating all the patches into a single surface produces a 3×14 array of degree 8 by 5 patches. The surface offsets by the required amount, but the curvature is still discontinuous between the two surfaces. However, the number of patches has been significantly reduced, from 392 to 84, by adopting this second approach. This compares favourably with the 68 in the Configuration Standard.

3.3   Analysis of surfaces

Table 1 summarises the integrity of the three surface models, together with the maximum deviations from the Centre Line, Windscreen Arch and Sill Line. The table highlights the difficulty in producing an acceptable Production Standard surface since the maximum point deviations from the constraint curves should be less than 0.1 mm. However, the deviations on the Sill Line were deemed acceptable due to their location at the rear of the Canopy.

The quality of the surfaces was checked by taking planar intersects parallel to the three Cartesian axes. It is generally understood that, for a quality surface, individual intersection curves should be smooth and not ripple, while

Table 1. Comparison of surface models (* due to symmetry of half model)

Standard	Size of patch array	Max. internal point dev. (mm)	Max. internal normal dev. (°)	Max. point dev. from Centre Line (mm)	Max. point dev. from Windscreen Arch (mm)	Max. point dev. from Sill Line (mm)
Configuration	2x34	0.000	0.100*	0.010	0.009	1.111
Initial Production	14x28	0.000	0.212*	0.120	0.077	0.551
Final Production	6x14	0.009*	0.218*	0.119	0.053	0.624

as a set they should be harmonious and in sympathy. The intersects showed no obvious irregularities. Reflection curves were then constructed on the surface using light lines parallel to the three Cartesian axes. Figure 4 illustrates these reflects spaced every $1°$ with respect to the x axis on the Configuration and Final Production Standard surfaces. The waviness of the reflects towards the rear of the Configuration Canopy suggests that the surface is rippling, while the tangent discontinuities at the join between the Windscreen and the Canopy suggest that the surface is not curvature continuous. It is clear from the figure that the Final Production Canopy surface is significantly smoother.

4   Surface productionisation using point-based smoothing

The proposed process utilises an intermediate point-based representation of the model rather than a fully surface-based one. In essence, the Configuration Standard will be discretised into an array of surface points. These are then smoothed [3] so that they are consistent with a curvature continuous surface. Finally, a sub-array of these points will be interpolated to construct a quality surface. Although the points are smoothed in this paper, they could have been re-designed or optimised to meet aerodynamic or structural requirements, e.g. as a result of CFD or FE analyses.

Ideally, the proposed productionisation process will result in a high quality Production Standard surface that contains a similar number of patches to the Configuration Standard. The surface must be exactly point and tangent continuous, but needs to be only approximately curvature continuous. It should have no unnecessary ripples and must offset by the required amount whilst minimising the point deviations from the constraint curves. It is recognised that it may not be possible to achieve all of these quality requirements.

4.1   Discretisation of surface

Consider sampling a rectangular array of points from a surface and suppose the points can be automatically smoothed to within some tolerance to fulfil

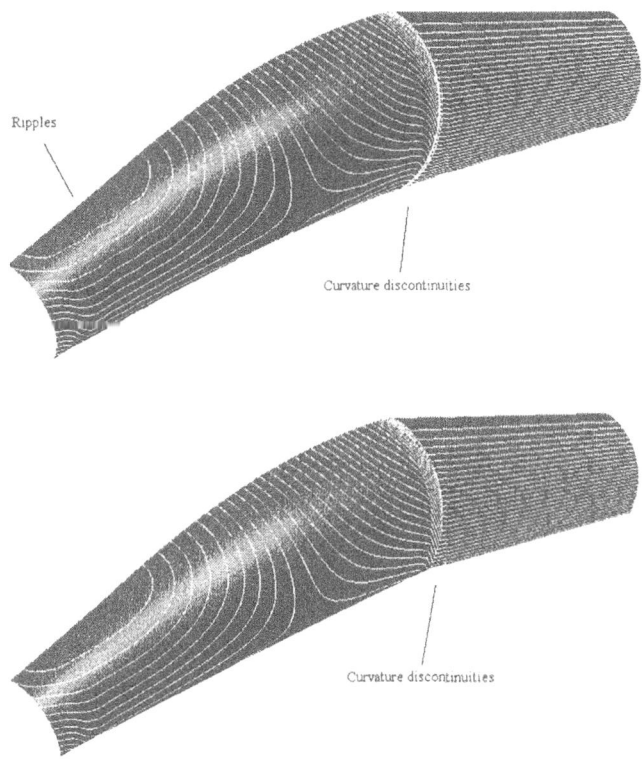

Fig. 4. Reflection curves on the (top) Configuration and (bottom) Final Production Standard surfaces

certain quality criteria. The expectation is that any re-constructed surface through the smooth points will be a quality surface [1]. This point-based approach assumes that the original array of points is sufficiently dense to characterise the shape of the surface. Any intermediate point on the surface can then be re-constructed accurately from the characterising point array.

A single surface comprising an $m \times n (m, n \geq 1)$ array of patches, having maximum degrees $d_u \geq 1$ and $d_v \geq 1$ respectively, has an array of defining control points of maximum size $(md_u + 1) \times (nd_v + 1)$. Assuming that the surface is an efficient definition, i.e. not over-defined, then it is reasonable to assume that an equivalent sized array of surface points characterises the shape of the surface. It is recognised that this is a sufficient, and not a necessary, number of points, i.e. fewer points could be used when the surface is over-defined. Therefore the 2×34 array of degree 8×5 patches in the Configuration Standard corresponds to a 17×171 array of surface points.

Having established the density of points, the spacing of them must now be determined. The performance of the point-based algorithms relies on accurate curvature estimates. Currently the smoothing algorithms use the curvature estimates of [10], which can be improved using regularly spaced points. These points are constructed by intersecting the appropriate number of boundary and isoparameter curves running in both parametric directions. It is usual for the surface-based approach to specify additional tangent and curvature information to control the shape of a surface at the boundary. Instead, the point-based approach uses additional points specified towards the boundary. Experiments suggest that an additional point at 50% of the first span (last span) on a string of points is sufficient to specify the tangent at the end of a curve. Similarly, additional points at 25% and 50% of the first span (50% and 75% of the last span) are sufficient to specify the curvature. Hence, the array of points can be augmented to provide a 21×175 array for the surface characterisation. The augmented points will only be used by the smoothing algorithm.

4.2   Point-based smoothing

The quality of the characterising points is first inspected using discrete geometric parameter curves. These are similar to isoparameter curves in that they connect opposite surface boundaries, but are based on the underlying geometry rather than the parametrisation. The concept was first introduced by Kosters [7] who developed surface reparametrisation functions based on the approximate regularisation of arc-length and normal-angle in order to optimise the display of surfaces on a computer screen. Isoparameter curves on the reparametrised surface then reflect changes in the geometric parameters across the surface. Czerkawski [5] generalised the concept to geometric sampling to visually assess the quality of a surface. Appropriate points are first determined on the surface boundaries for the underlying parametrisation. Surface curves are then constructed to connect corresponding pairs of points on opposite boundaries. Cripps and Howe [4] have since extended this concept to arrays of points.

The points are assessed using discrete geometric parameter curves with respect to arc-length. They should appear harmonious when viewed as a set. Geometric parameter curves can also be constructed with respect to the normal-angle. These are more sensitive than the arc-length parameter curves. When both sets of geometric parameter curves are harmonious, then the points are regarded as being smooth. If they are not smooth, then strings of points are removed from the array to leave a sub-array of points. This is then used to re-generate the removed points. The points are again assessed and the procedure repeated until the geometric parameter curves are harmonious [3]. Clearly, the process is iterative and dependent on the choice of sub-array. It does however provide a procedure for constructing points lying on a curvature continuous surface. The smoothing results in a 49x153 array of points characterising the full surface. This array constitutes three subdivisions from an augmented sub-array. The point-based assessment and smoothing recognises and preserves any symmetries in the point array.

The array of smoothed points was briefly analysed with respect to their deviation from the Configuration Standard surface and constraint curves. The maximum deviation of the smoothed points from the Centre Line is 6.950 mm towards the front of the Canopy, i.e. near the Windscreen Arch. Note that the productionisation process does not allow the surface designers to modify a surface by more than 0.1 mm. This is due to any tooling that has already been performed on the support panels. Although the case study has revealed that this tolerance is flexible and dependent on its location on the surface, the size of the modification is excessive and means the smoothed points are unacceptable for productionisation use.

4.3   Re-construction of surfaces

A quality surface is generally as simple as possible and constructed using the minimum number of low degree patches. The expectation therefore is that the smoothing will be consistent with reducing the number of points. Hence, the re-construction process may not require all of the points. Those that are not used can then be compared to the re-constructed surface to give a direct indication of the surface quality with respect to the smoothed points. In addition, all the re-constructed surfaces will be analysed using the traditional methods of planar intersects, reflects and surface offsets at the appropriate distances. It has already been recognised that the constraint curves will not be accurately matched.

The CATIA function for interpolating an $m \times n(m, n \geq 2)$) array of points forms an $(m - 1) \times (n - 1)$ array of degree 8 patches each connecting with $\epsilon\text{-}G^2$ continuity. Full, rather than half, models will be re-constructed using both the original and smoothed point arrays to evaluate the quality of the smoothing. Re-constructing full models ensures exact tangent and curvature continuity across the symmetry plane. The function constructs symmetric surfaces from symmetric point arrays.

Post processors generally perform isoparametric, rather than geometric, meshing for performance reasons. Therefore, the re-constructed surface must contain patches that are all rectangular and of a similar size to enable an efficient mesh to be generated. This can be achieved by interpolating a sub-array of regularly spaced points from the smoothed array (after removing the augmented points, which were only inserted for smoothing purposes). Recall that the re-constructed surfaces should contain an equivalent number of patches to the original surfaces, i.e. a 2×34 array for the Configuration Standard and a 6×14 array for the Final Production Standard. Therefore, the re-construction interpolates appropriate regularly spaced sub-arrays to provide equivalent patch arrays. Clearly, it will be difficult to specify exactly regularly spaced sub-arrays of points.

4.4   Analysis of surfaces

Table 2 summarises the integrity of the re-constructed surfaces using the original and smoothed point arrays. The maximum deviations from the appropriate point array, Centre Line, Windscreen Arch and Sill Line are also given. Note that there are no internal point or normal deviations in these degree 8 surfaces. In addition, they all offset by 22 mm, i.e. the required amount for the Windscreen. It is clear from the table that the first two re-constructed surfaces had insufficient points to match the circular sections. This resulted in a similar quality of fit for both arrays of points. However, the last two re-constructed surfaces match the sections better. An improved quality of fit is then exhibited using the smoothed rather than the original point array.

Table 2. Comparison of re-constructed surface models (* this deviation is towards the rear of the Canopy but below the Sill Line, the maximum above the Sill Line is 1.347mm)

Points fit	Size of patch array	Max. internal point dev. (mm)	Max. internal normal dev. (°)	Max. point dev. from app. point array (mm)	Max. point dev. from Centre Line (mm)	Max. point dev. from Windscreen Arch (mm)	Max. point dev. from Sill Line (mm)
1	2x34	0.000	0.000	88.593	1.986	76.961	63.818
2	2x34	0.000	0.000	88.747	6.950	78.053	63.849
3	6x14	0.000	0.000	5.434	5.719	4.376	2.908
4	6x14	0.000	0.000	2.547*	7.044	5.504	1.401

Figure 5 illustrates the reflects on the first two re-constructed surfaces. They each contain a 2×34 array of patches and use the original and smoothed points respectively. The reflects are smoother on the second re-constructed surface in the vicinity of the join between the Windscreen and the Canopy indicating that the curvature discontinuities have been decreased due to the point smoothing. However, the rippling at the rear of the Canopy is still visible. This suggests the smoothing is sensitive to discontinuous curvature profiles but not to ones that are continuous and inflecting. It is clear that a denser array of points, and hence more patches, are required for the surfaces to match the circular sections. Alternatively, tangency arrays could be specified on the lower boundary curves.

Figure 6 illustrates the reflects on the last two re-constructed surfaces. They each contain a 6×14 array of patches and use the original and smoothed points respectively. Again, the reflects are smoother on the fourth re-constructed surface in the vicinity of the join between the Windscreen and the Canopy. In addition, the rippling at the rear of the Canopy has disappeared. This is probably due to the CATIA function having fewer points to interpolate, i.e. the surface is less constrained resulting in a higher quality.

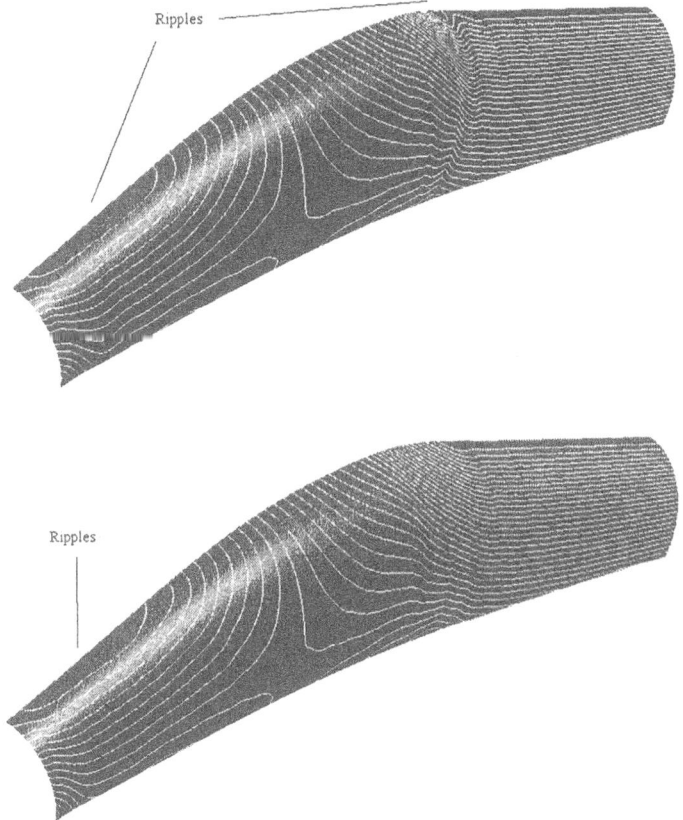

Fig. 5. Reflection curves on the (top) first and (bottom) second re-constructed surfaces

5   Aerodynamic analysis

An industrial CFD analysis was performed using FLITE3D on four of the surfaces, namely the Configuration and Final Production Standard surfaces and thethird and fourth re-constructed surfaces. The first and second re-constructed surfaces were not analysed due to their excessive deviations from the circular sections.

Each surface was mounted on a Front Fuselage assembly and the analysis simulated nose-up flight at $4\,^{\circ}$ of incidence at a speed of Mach 0.9. This typically produces a shock wave, and resulting pressure drop, over the Canopy. Table 3 lists the relative drags between the four surfaces using the Configuration Standard surface as the reference. Evidently, the Configuration and Final Production Standard surfaces are aerodynamically very similar. The

Ripples ―――――――――

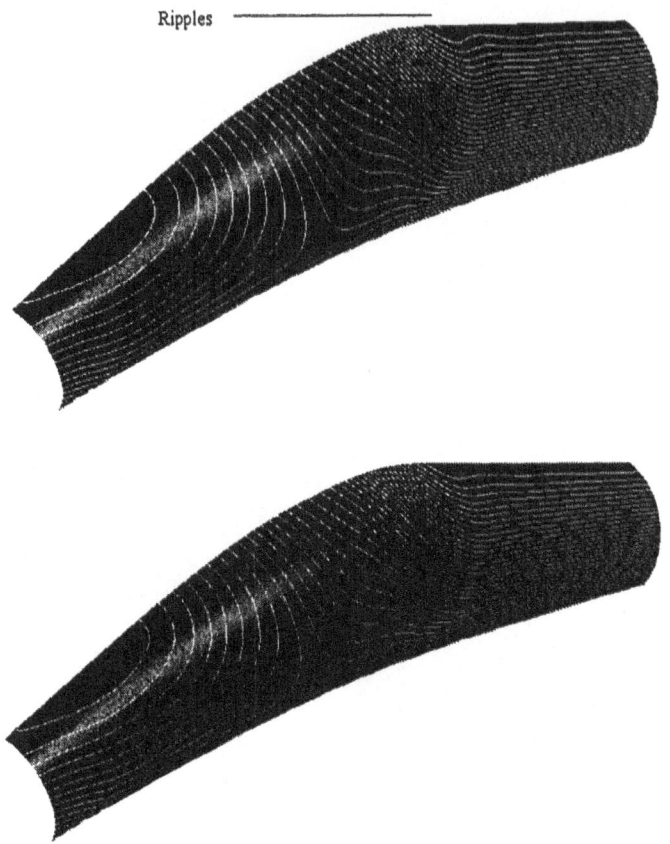

Fig. 6. Reflection curves on the (top) third and (bottom) fourth re-constructed surfaces

drag is reduced on the third re-constructed surface, and further reduced on the fourth one.

Table 3. Relative drags between the original and re-constructed surfaces

Surface	Relative drag
Configuration Standard	0.000000
Final Production Standard	-0.000003
Third re-constructed	-0.000015
Fourth re-constructed	-0.000047

Figure 7 displays Mach number contours of the airflow on the Final Production Standard and fourth re-constructed surfaces. Mach number contours are also displayed above the surfaces in the symmetry plane. The dense contours at the Windscreen Arch correspond to an expansion shock, where the airflow changes from subsonic (Mach 0.9) to supersonic (Mach 1.4), while the dense contours in the middle of the Canopy correspond to a compression shock. Between the two shocks, the air expands which results in a pressure drop. It is clear from the figures that the contours are not as dense on the fourth re-constructed surface. This improves the pilots comfort, due to less noise and buffeting, and reduces the chance of the airflow separating from the Canopy. This separation would result in a significant drop in performance. The improved drag and shock wave characteristics are a direct result of smoothing the join between the Windscreen and the Canopy.

6   Conclusions

The approach indicates a possible solution to the productionisation process. However, further work is required to overcome the conflicts between the constraint curves and the smoothing. In this paper, the points were smoothed with respect to internal curvature continuity only and did not take into account constraint curves. The analysis of the re-constructed surfaces confirms that the smoothing has improved the curvature deviations between the Windscreen and the Canopy, but has not removed the ripples towards the rear of the Canopy. Future work should include the point-based smoothing of irregular topologies, as well as regular topologies, whilst minimising the point deviations from any constraint curves, i.e. a constrained point-based optimisation.

Lin et al. [8] have demonstrated that when the point-based representation is consistent with the node points of a triangulated surface mesh, then the surface analysis and re-construction is totally integrated. This suggests that the point-based representation could be used to perform the discrete CFD and FE analyses during the Concept and Definition Loops, as well as the smoothing, resulting in high quality Configuration and Production Standards. The benefits of surface smoothing extend to manufacturing where most milling and forming techniques are dependent on the quality of the tool, and hence the component, surfaces. On-going research at The University of Birmingham suggests that surfaces could be avoided altogether and that all follow-on processes could operate on point arrays [1]. Smith [9] describes the preliminary implementation of such a system. Cook [2] and Fisher [6] respectively describe point-based approaches for surface assessment and surface machining. Future work could investigate whether biased tool surfaces, to allow for the spring-forward of Carbon Fibre Composite (CFC) parts, could be constructed directly from the smoothed point arrays representing the component.

6.1 Acknowledgements

The authors are pleased to acknowledge Karen Attwood (BAE SYSTEMS) for developing a CATIA V4 GII function for discretising a surface, Peter Cook (The University of Birmingham) for performing the point-based smoothing, and Keith Weatherill (BAE SYSTEMS) for performing the aerodynamic analysis.

References

1. Ball, A. A. (1997), CAD: master or servant of engineering?, in: Goodman, T.N.T. and Martin, R.R. (eds.), The Mathematics of Surfaces VII, Information Geometers, 17-23.
2. Cook, P.R. (2000), Point-based mathematics and graphics for CADCAM, Ph.D. thesis, The University of Birmingham, U.K.
3. Cripps, R.J. and Cook, P.R. (1999), Point-based CADCAM, in: Bramley, A.N., Mileham, A.R., Newnes, L.B. and Owen, G.W. (eds.), Advances in Manufacturing Technology XIII, Professional Engineering Publishing, 149-153.
4. Cripps, R.J. and Howe, R.E.D. (1998), Surface visualisation and assessment using geometric parameter curves, in: Baines, R.W., Taleb-Bendiab, A. and Zhao, Z. (eds.), Advances in Manufacturing Technology XII, Professional Engineering Publishing, 335-342.
5. Czerkawski, A.M. (1996), Fitting procedures for curves and surfaces, Ph.D. thesis, The University of Birmingham, U.K.
6. Fisher, M. (2000), Point-based mathematics for CAM, Ph.D. thesis, The University of Birmingham, U.K., in preparation.
7. Kosters, M. (1991), Curvature-dependent parametrization of curves and surfaces, Computer-Aided Design 23, 569-578.
8. Lin, J., Ball, A.A. and Zheng, J.J. (1998), Surface modelling and mesh generation for simulating superplastic forming, Journal of Materials Processing Technology 80-81, 613-619.
9. Smith, F. (1998), Point-based geometric modelling, in: Proceedings of SSM98 Sculptured Surface Machining, Michigan, USA.
10. Tookey, R.M. and Ball, A.A. (1997), Estimation of curvatures from planar point data, in: Goodman, T.N.T. and Martin, R.R. (eds.), The Mathematics of Surfaces VII, Information Geometers, 131-144.

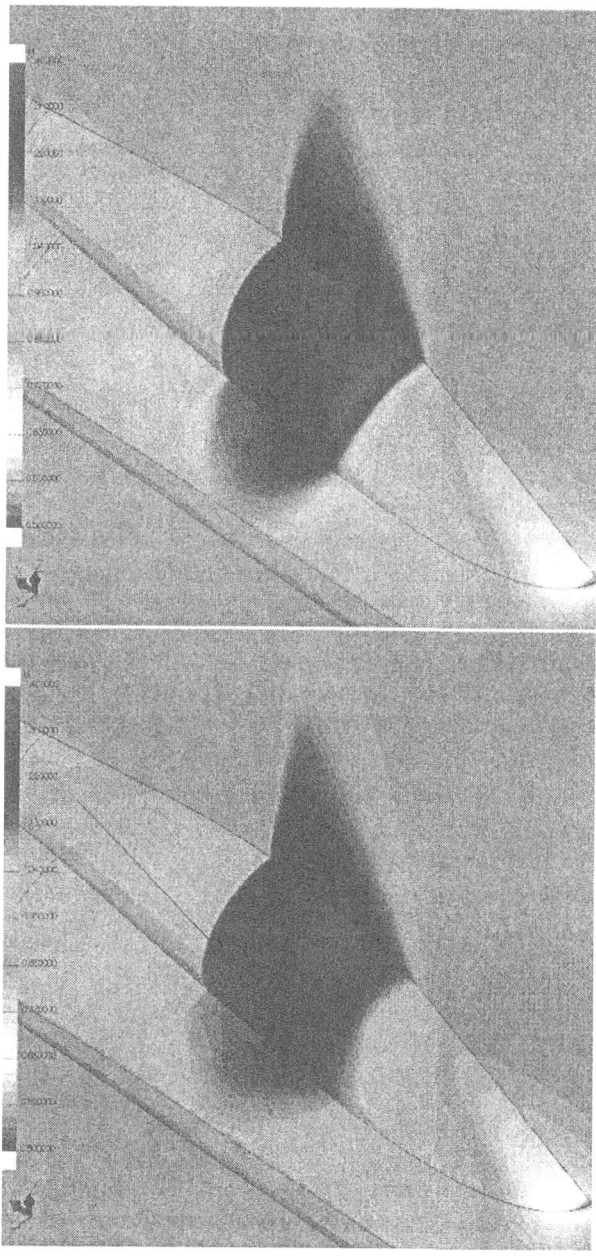

Fig. 7. Mach number contours of the airflow on the (top) Final Production Standard and (bottom) fourth re-constructed surfaces

Modelling of Material Property Variation for Layered Manufacturing

Michael J. Pratt

National Institute of Standards and Technology*
Manufacturing Systems Integration Division
100 Bureau Drive, Mail Stop 8262
Gaithersburg, MD 20899-8262, USA

Summary. Layered manufacturing (LM), alias solid freeform fabrication or rapid prototyping, is an important emerging manufacturing technique. It builds up a manufactured artefact by depositing successive layers of material under computer control. Until recently, objects manufactured by LM methods have been regarded as composed of homogeneous material. However, new methods of optimal design specify 'functionally graded' or inhomogeneous materials. Layered manufacturing provides a means for producing such variable material distributions. Furthermore, methods are under development for embedding reinforcing fibres in the deposited material, and additional means are therefore needed for the representation of material nonisotropy. The problem reduces essentially to that of parametrizing the interior of a boundary representation solid model, ideally in terms of the surfaces involved in its boundary. The paper surveys several methods with the potential for representing 3D material distributions, and examines their compatibility with ISO 10303 (STEP) an international standard for the representation of product life-cycle data. Some consideration is also given to the related problem of representing the microstructure resulting from the deposition of the material (in most LM methods) in strands or filaments.

1 Introduction

The term *layered manufacturing* (LM) is used for a family of recently developed additive manufacturing processes in which objects are constructed layer by layer, usually in a series of parallel planar laminae. Alternative names for these processes include *solid freeform fabrication* (SFF) and *rapid prototyping* (RP). The last term stems from the fact that initially such processes were used exclusively for the generation of non-functional prototype parts used for 'look, feel and fit' evaluation of component geometry during the design process. However, the use of LM is now being extended into other areas of manufacturing, in particular the production of moulds and dies, and it is likely that it will soon become used as a method for small batch production of functional components of engineering products.

---

* Also affiliated with Center for Automation Technologies, Rensselaer Polytechnic Institute, Troy, NY 12180-3590, USA.

Many LM processes currently exist, using different materials and layering methods [3]. The following classes of methods may be identified:

Photopolymer solidification (e.g., stereolithography, solid ground curing), in which a liquid resin is hardened layer by layer with a laser or ultraviolet lamp.

Material deposition (e.g., fused deposition modelling), in which drops or filaments of molten plastic or wax are deposited to construct each layer.

Powder solidification (e.g., selective laser sintering and three-dimensional printing), in which powdered material layers are solidified by adding a binder or by sintering with a laser. Parts can be built from ceramics, nylon, polycarbonate, wax or metal composites.

Lamination (laminated object manufacturing and solid ground curing). The first of these methods uses lasers to cut layers from sheets of paper, cardboard, foil or plastic, stacks them and bonds them together. The second uses a cut mask to expose regions of resin to be solidified by an ultraviolet lamp.

Weld-based approaches. Currently still at the research stage, these use welding and cladding techniques to build metal parts [12].

LM processes have several advantages over other manufacturing methods. They operate largely automatically on input data generated directly by a CAD system. Their process planning requirements are comparatively low, the primary requirement being that of determining an optimal orientation in which to build the product. They can be used to build very complex artefacts, having intricate geometry, internal voids or multiple ready-assembled components, without the use of special tooling.

However, these methods also have disadvantages. For instance, the layer by layer mode of building LM objects leads to surfaces that often have a staircase appearance, and it is at present difficult to achieve good tolerance or surface finish specifications. Poor surface quality is sometimes overcome by performing finishing operations, such as grinding or polishing, after an object is built. LM products may also have inferior material properties when compared with objects manufactured by other means. Much research is currently directed towards overcoming these and other problems.

2 Material distribution in LM objects

Although objects manufactured using LM were originally composed of single materials, the manufacture of multi-material objects is now becoming possible. LM, unlike conventional manufacturing processes, in principle permits complete 3D control over material properties, and the following possibilities are emerging:

- Entire pre-existing components may be embedded as the object is built.

- Discrete 3D regions of different materials may be deposited through the use of multiple deposition heads.
- The simultaneous use of multiple deposition heads also makes it possible to deposit continuously varying combinations of different materials.
- Reinforcing fibres may be laid down with the deposited material to obtain directional strength characteristics in the built object.

The first two of these present little difficulty from the point of view of mathematical representation, but the third and fourth raise some interesting problems. Both of them are also potentially important in practice, as explained in the following paragraphs.

New design methods such as homogenization for structural topology design [2] specify varying material properties in the interior of a designed object. As we have seen, LM provides a means for producing such artefacts. If such design practices come into productive use some means will be needed of representing the material distribution throughout the volume of the designed object, to serve as output from design and input to LM manufacturing.

Further, recent developments in LM have demonstrated the deposition of reinforcing fibres as an object is built [4]. This leads to obvious possibilities for design in terms of directional strength characteristics. Now the design representation will need to capture not just a scalar distribution of material composition but also additional non-isotropic directional information. Although this paper is primarily concerned with the representation of inhomogeneity, it is believed that all of the methods reviewed can be extended easily for the capture of material non-isotropy.

A related but slightly different material representation problem arises because in most LM processes (with the exception of those classed above as *lamination* methods) material is deposited or hardened in a strand or filament, some area-fill strategy being used within the boundaries of each layer. This being so, the structural properties of an object manufactured by LM are dependent on the particular area-fill strategy employed. For a given method they will be affected by the lengthwise strength of each filament, and by the strengths of its bonds to neighbouring filaments in the same layer and to the underlying layer on which it is deposited. It is therefore highly desirable to be able to represent the microstructure resulting from the deposition patterns within layers, so that analysis programs tailored to particular physical choices of material and process can be used to calculate strength characteristics of manufactured parts.

One important characteristic of the material representation sought is that it should be defined on a domain corresponding to the interior and boundary of a CAD model of the boundary representation type. If the boundary of the distribution coincides only approximately with the boundary of the object model, then there may be interior regions near the model boundary where no distribution is defined. This problem could be overcome by ensuring that the domain of the distribution includes the entire CAD model, but that is also non-ideal. It would lead to difficulty in the precise specification of the distribution on the object boundary itself, which is likely to be a frequent requirement from the design point of view.

The motivation behind the study reported here is that an extension is proposed for the international standard ISO 10303 (STEP) [6] to enable the standardized transfer of information between CAD and LM systems. For this purpose it is appropriate to take into account the implications of current developments on the likely future informational requirements of LM. It is almost certain that data concerning inhomogeneous and nonisotropic material characteristics will be among those requirements.

The paper is organized as follows. In Sect. 3 several possible means for the representation of material distribution on the macroscopic scale are reviewed. In the following section some further aspects of deposition strategy are explained, and the representation of material microstructure is then further considered. In Sect. 5 some brief remarks are made concerning the standard ISO 10303 mentioned above, and in particular its compatibility with the various representations discussed in the paper. Finally, a summary is given.

3   Representations for material distributions

For the representation of material macrostructure, what is ideally needed is some means of defining a function whose domain is strictly limited to the interior and boundary of a detailed boundary representation model of a 3D object. This is easy if the object is bounded by constant values of the coordinates in some 3D coordinate system, e.g., in the case of a rectangular block or a circular cylinder. More generally, it is also easy for a triparametric volume resulting from a mapping of such a coordinate system from parameter space into Euclidean space. However, for most realistic mechanical engineering objects there is no such simple parametrization.

Several well-known methods, including the finite element and boundary element methods, appear at first sight to be well suited to the problem described. However, these methods generally make use of an approximated or simplified object boundary, and therefore do not meet the requirement on accuracy of the boundary of the material representation domain. Also, such methods are often driven by data defined on the boundary of the object, whereas what is required in the material property problem is some method of specifying characteristics of the distribution in its interior.

Since no ideal method is known we must evaluate available non-ideal approaches. The choices appear to be

1. discretization of the object's volume into simple parametrizable elements,
2. use of a simple parametrization whose domain does not coincide with the object's boundaries, or

3. construction of a specific parametrization for the interior and boundary of each object, which is likely to lead to a high level of computational complexity.

These three approaches are exemplified by the methods outlined in what follows. It may be that all three approaches are needed, since the nature of the design system used for defining objects with continuously variable material properties will determine the most appropriate formulation to use.

Volume decomposition approaches

In this section several suggested decomposition methods are examined.

Voxel-based modelling. The most elementary form of cellular decomposition uses *voxels*, cuboidal cells forming a regular 3D grid [24]. Any 3D solid model may easily be converted into a voxel-based representation. Typically, the grid is imposed over the object and all cells more than 50% occupied by material are considered to be part of the object. Voxels are usually small in relation to the overall object size, and almost invariably the physical properties associated with each voxel are taken to be homogeneous in its interior. For the present application, therefore, material variation will be modelled in terms of discontinuous changes across voxel boundaries.

In one sense such a representation is well suited to layered manufacturing, since each layer in the fabrication process can be represented as a planar layer of voxels whose vertical dimension is the layer thickness. On the other hand, voxel models suffer from the same disadvantage as objects manufactured by LM, in that they exhibit surfaces with staircase effects. All faces of the voxel model are parallel to the three coordinate planes, and in general the faces of a real object can only be approximated. In general, any reorientation of the model requires recalculation of its voxel representation.

Some further important limitations of the voxel approach are that

- A voxel model is not suitable for analytical purposes, e.g. for finite element analysis. There are far too many elements, if each voxel is regarded as an element; it is only possible to modify the mesh by changing its grid characteristics, and the grid-based mesh is far from ideal for most types of analysis.
- It is not easy to associate attribute information (e.g., material conditions on the object boundary) with entire faces of an object model if those faces are discretized into assemblies of voxel facets.
- Although they are very simple, voxel-based representations are not compatible with any known standard for exchange of product data between computer-based systems.

The *octree* approach to decomposition [7,13] is more sophisticated than pure voxel decomposition. In its basic version it uses small cells in order to achieve acceptable boundary resolution in an object model, but larger ones in the interior in order to keep the overall number of cells within bounds. In the past, the use of such an approach has assumed homogeneity of the object interior, and in this case octrees are not suitable for the modelling of variable material distributions. On the other hand, since the cells are all cuboidal it is easy in principle to model material distributions inside the larger cells in terms of a Euclidean coordinate system. However, such an approach still suffers from the disadvantages listed above for the pure voxel method.

Bajaj [1] has combined the use of octrees with C^1-continuous triparametric polynomial spline functions to model spatial distributions for visualization purposes. However, this is essentially an interpolation approach. The boundary of the interpolated data set emerges from the process in a triangulated representation, whereas what is needed for LM is a fit of the spatial distribution to the pre-specified boundary of a CAD model.

Mesh-based decomposition. Irregular decompositions of the finite element type provide alternative possibilities for the modelling of material distributions. One such proposal is made by Jackson et al. at MIT [8], who mesh a solid model into tetrahedral finite elements. The material composition in the model is represented in terms of a vector-valued function which may be denoted by $M^n(x) = (m_1(x), m_2(x), \ldots, m_n(x))$, where

- x is the position vector of a point in the domain of the object,
- n is the number of different materials to be combined in the LM process, and
- $\sum_{i=1}^{n} m_i(x) = 1$, so that $m_i(x) \geq 0, i \in [1, n]$, expresses the relative concentration of material i at the point x.

In [8] the designer is provided with a convenient means of specifying the material composition in terms of distance from some geometric feature (for example, the axis of a rotational object) or from the object surface. This fixes the values of the components of $M^n(x)$ at the nodes of the mesh; the values in cell interiors are then interpolated by trivariate Bézier polynomials using the nodes as control points. Currently, only piecewise constant and piecewise linear variations have been considered. However, additional smoothness in the material distribution between cells could be attained in the future by adding further nodes to the mesh and using Bézier representations of degree two or higher.

One advantage of this approach is that the model can be directly used for finite element analysis. Against this, it may be questioned whether a mesh that is optimal for design is equally suitable for subsequent analysis.

Other disadvantages of the finite element mesh approach are as follows:

- Although this method can specify the material composition exactly at the mesh vertices, elsewhere the composition is computed by interpolation, and is therefore only approximate.
- As in the voxel approach the object boundary is also approximated, and consequently the material definition domain does not fully coincide with the object domain.

Clearly it would also be possible to decompose the object model into the hexahedral elements that are preferred by many finite element analysts. However, the same advantages and disadvantages apply in this case also.

A set-based approach. *Heterogeneous solid modeling* (HSM) is the name given to a set-based scheme developed at the University of Michigan for the representation of multi-material objects [10]. Although the mathematical formulation is slightly different, the mode of representing material properties at a point is equivalent to that described in the last section.

The notion of an r_m-set is introduced as a subset of the product space of E^3 with M^n as defined earlier. The corresponding subset of E^3 is an r-set in the usual solid modelling sense [17], and that of M^n assigns a material composition to each Euclidean point of the r-set. Within the r-set, therefore, a vector-valued function is required to specify the material distribution. This is assumed to be C^∞ continuous to ensure that material discontinuities can only occur at r-set boundaries.

The domain of the material distribution function should ideally correspond to the r-set. This is easy to achieve when the boundary of the r-set has a simple geometric configuration, but to cater for more complex cases the r_m-set is said to be undefined for points in the exterior of the r-set. The case where an r-set is occupied by a single pure material can obviously be handled by a material distribution function with $m_i = 1, m_j = 0$ for $j \neq i$.

A *material object* (r_m-object) is now defined as a finite collection of r_m-sets having the following properties:

- No pair of r_m sets in the r_m-object share any interior points.
- The material distribution function applies only to the interior of each r_m-set.
- At boundary points shared by adjacent r-sets the material composition is defined by a specialized combination operator [10].

For the representation of material variation it is proposed in [16] that a coordinate system be associated with each r_m-set for use in defining the material composition function. The best choice of coordinate system will depend on the basic shape of the underlying r-set. For example, regions that are approximately cuboidal or approximately cylindrical will most appropriately have their material distributions represented in cartesian or cylindrical coordinates respectively.

The HSM method provides material distributions that are continuous in each r-set, but not in general between the r-sets defining the shape and internal configuration of an r_m-object. It also suffers from the disadvantages that

- Except for very simple objects the domain of the material distribution function will not coincide with the domain of the object it is defined for. As an example, consider a gear-wheel. Its shape is basically cylindrical, and a cylindrical coordinate system is therefore appropriate for defining the material distribution function. However, at the periphery of the gear-wheel the teeth represent departures from the cylindricity of the configuration, and there is no flexibility in refining the material distribution function there unless further r_m-sets are defined for the teeth and used in the specification of a more complex r_m-object.
- The basic method provides no means for achieving smoothness of the material distribution function throughout the interior of an r_m-object, and so the decomposition suggested above would lead to discontinuities in the absence of further refinement of the method. A possible approach for such refinement is suggested below.

The gear-wheel example mentioned above, using a single r_m-set, prompts the thought that a natural decomposition into multiple r_m-sets may be based upon the concept of form features [19]. If a base shape and a set of imposed features all have simple geometry then r_m-sets may be defined for each of them, and it may then be possible to use a blending technique to obtain a smooth continuous overall distribution. A suitable method may be based on well-known approaches to the generation of blending surfaces in surface and solid modelling [22,23], though the dimension of the problem is now one higher. Blend boundary curves are replaced by feature boundary surfaces, and the blending is now volumetric, but the same principles can be made to apply.

An extension of the HSM approach has recently been described by Kumar et al. [11]. It is pointed out that the same basic approach can be used for modelling a variety of spatially-related product properties beyond material distribution. The cited paper notes also that the association of one or more fields of properties with a spatial model constitutes what is known to modern differential geometers as a *trivial fibre bundle* [14]. The use of fibre bundles for related purposes has also been studied by Zagajac [25].

A constructive approach

A method under development at the University of Wisconsin [18] differs from those described earlier in not being decomposition-based. It is based on two key ideas:

Shepard interpolation. This is a well-known method for the interpolation of values defined at sets of randomly distributed points in any number of dimensions. It is based on the use of functions that are radially symmetric about a central point and decrease in value with increasing distance from that point, usually according to some inverse power of the distance. The overall interpolant is a linear combination of such basis functions, one centred at each interpolated point. Shepard [21] (see also [5]) popularized the use of such methods in the 1960s, though in fact they had been used for isolated applications before that time. Essentially, they interpolate function values f_i at a corresponding set of points x_i by a function $\sum_i w_i(x)f_i$, where the $w_i(x)$ are the Shepard basis functions. These must be positive and continuous, must form a partition of unity and also satisfy the interpolation condition $w_i(x_j) = \delta_{ij}$. Subject to these conditions various formulations are available. There also exists a wider class of related *radial basis function* methods [5].

R-functions. These should not be confused with *r*-sets. An R-function is a real-valued function whose sign is completely determined by the signs of its arguments. Such functions provide analogies to the Boolean logical functions. Simple examples are provided by $\min(x_1, x_2)$ and $\max(x_1, x_2)$, which are analogous to Boolean 'and' and 'or', respectively, if we take $+$ and $-$ values of the arguments to correspond to the logical values of TRUE and FALSE. Many other R-functions are known [18], and some prove more suitable than those given above for practical applications, since they give better continuity in the generated interpolation functions.

The analogy with Boolean functions allows any shape element of a solid model, expressed in terms of Boolean combinations of half-spaces, to be defined in terms of a single implicit function, by composition of appropriate R-functions. The composition process may be performed automatically. It is shown in [20] how this may also be done for individual elements of a boundary representation model. These may then be built, through further compositions, into a function representing a full B-rep model. Then a generalization of Shepard's interpolation method can be used for interpolating continuous distributions (e.g., of material properties) specified on the boundary elements. In [18] it is shown how boundary derivative distributions may also be interpolated.

An important feature of the R-function method (RFM) is that it defines transfinite interpolants, since continuous functions are being interpolated rather than sets of discrete values. Moreover, the 'boundary' of a region may be defined very generally, allowing the specification of discrete point values or continuous distributions that do not lie on the physical boundary of the artefact. The means are therefore available for controlling the material variation in the interior of an object as well as on its exterior boundary.

The use of RFM appears to enable the representation of continuous material property distributions, without the use of any meshing or other form of decomposition, inside any volume bounded by implicitly defined surfaces.

Since conditions may also be defined in the interior of the volume, as mentioned above, there is good potential for local control anywhere it is required. RFM therefore offers strong potential for the the large-scale modelling of material properties needed for the design and manufacture of LM objects, without some of the disadvantages exhibited by the other methods surveyed. However, it is not a perfect fit for material property variation in LM, for the following reasons:

- RFM is currently proposed as a general-purpose method, not specialized for the representation of material distributions. Further work will be needed for the present application; in particular, suitable design methods for use with RFM will need to be identified.
- RFM may only be used for models whose boundary surfaces are defined by implicit surfaces. This appears to rule out its direct use for models involving parametric surfaces such as NURBS. Although in principle implicit representations may be determined for such surfaces, in practice the magnitude of the computational process would be insupportable.
- The composite R-function corresponding to a complex B-rep model will be very complex. It remains to be seen whether such functions are computationally tractable. It is possible that some means of localizing the contributions of individual boundary elements to neighbouring regions of the model may help in overcoming this problem.

4 The representation of microstructure

Space does not permit more than a few remarks on this topic, which represents a significant problem in its own right. Microstructure results in most LM objects from the filament-wise accretion of material. For the purpose of illustration, suppose that the material hardens after deposition over some finite period of time. The bonding together of adjacent strands will then be affected by the degree of hardening that has occurred in the first strand by the time its neighbour is laid down. One governing factor is the speed of deposition, and another is the area fill strategy used, which will determine the time difference between the deposition of adjacent filaments at any given point of adjacency. Furthermore, the bonding of filaments in one layer to the material in the layer below may also be affected by the hardening time and the time-periods involved, in a similar manner.

Bonding strengths between filaments will affect overall part strength. Additionally, there is likely to be material variation even within filaments as a result of the hardening process, which may occur from the inside out and therefore lead to inhomogeneity of properties at the scale of individual filament radii.

Clearly it will be a complex task to capture such small-scale variations within the entire manufactured object. Nevertheless, these variations are important in that they affect the structural properties of the artefact, and their capture is likely to be necessary for the future. For the moment it seems best to concentrate on the macroscopic variation of material and leave the microstructure for the longer term. One point that may be made is that an LM system has to generate scan paths (analogous to NC cutter paths in numerically controlled machining). It is these scan paths that give rise to the particular microstructure in a specific LM object, and they potentially provide the basis for the capture of the associated microstructure. The standard ISO 10303-42 ('Geometric and topological representation'), mentioned in Sect. 5, has the capabilities required for scan path representation, but further research is needed to find the most appropriate means of using them in modelling this small-scale aspect of material distribution.

5   International standard ISO 10303 (STEP)

The international standard ISO 10303, informally known as STEP (STandard for the Exchange of Product model data) [6,15] enables the exchange of product life-cycle data between different CAD systems or between CAD and downstream application systems. STEP covers a wide variety of different product types (electronic, electro-mechanical, mechanical, sheet metal, fiber composites, ships, architectural, process plant,...) and life-cycle stages (design, analysis, planning, manufacture,...). This range is continually expanding as new parts of the standard are issued.

The implementable parts of ISO 10303, each applicable to a particular life-cycle stage of a particular product class, are known as Application Protocols (APs). The APs themselves are constructed on the basis of a set of Integrated Resources (IRs), defining fundamental constructs that can be specialized and applied for a wide variety of purposes. Following consultation with LM vendors, users and researchers [9], it is currently proposed to develop a new STEP AP, dealing specifically with LM. The resources that will be needed include the following IRs:

ISO 10303-42 – 'Geometric and topological representation' (for modelling the product shape),

ISO 10303-44 – 'Product structure configuration' (for modelling assemblies of parts, which may be manufactured ready-assembled using LM),

ISO 10303-45 – 'Materials' (for the specification of material types),

ISO 10303-50 – 'Mathematical constructs' (for the representation of material distribution functions).

All of these are already at International Standard (IS) status with the exception of Part 50 (which is approaching the Draft IS stage at the time of writing). Part 42 provides representations of the usual implicit and parametric surfaces used in solid modelling. It will allow the precise modelling of boundary shape, in contrast with the present *de facto* faceted representation used in LM. The move to an exact geometric representation is motivated by increasingly stringent accuracy requirements for LM artefacts.

Part 50 of ISO 10303 deals with function specification and utilization in arbitrarily many dimensions. It has been developed initially with the requirements of engineering analysis in mind (specifically, finite element analysis, computational fluid dynamics and electromagnetic field analysis). The survey presented earlier in the paper leads the author to believe that the capabilities provided in this new part of ISO 10303 will be adequate for the modelling of material property variation by any of the major approaches identified.

6 Summary and conclusions

Several methods for the representation of material distributions in CAD models have been surveyed. Their primary characteristics are summarized in the table below. The third column, 'Interior continuity' relates to the basic methods as described earlier, and the fourth, 'Possible improvements' relates to ways in which the level of interior continuity of distributions may be enhanced. Although the paper has concentrated primarily on scalar distributions the methods can all be used, with only trivial extensions, for the modelling of material non-isotropy. Indeed, they have much wider potential application to the modelling of general scalar or vector fields in domains with geometrically complex boundaries.

Table 1. Summary of methods reviewed

Method	Object boundary	Interior continuity	Possible improvements
Voxels	jagged	C^0 between voxels	impossible?
Mesh	faceted	C^0 between cells	high-order basis functions
Set-based	approximated	C^0 between r-sets	blending
R-functions	exact for implicits	smooth	–

On the basis of the table, the use of voxels can be ruled out for present purposes. The mesh and set-based approaches score similarly, but neither of these methods represents the object boundary exactly (except in trivially simple cases), so that the domain of the material distribution function does not coincide with the set of points comprising the object. The R-function method does allow exact representation of the object boundary provided the elements involved are defined implicitly. It also gives smooth distributions automatically if appropriate R-functions are chosen. But the computational complexity of the method will be high in practical applications.

For data exchange purposes, the mesh-based, set-based and R-function methods can all be handled by the capabilities provided in the standard ISO 10303, both as regards geometric representation and the representation of the mathematical functions involved. In the mesh-based case it will also be necessary to use some of the standard's finite element capability, as specified in ISO 10303-104. However, the science of designing artefacts with variable material properties is in its infancy, and different methods under development are likely to lead to different approaches to the representation of material variation. It is therefore important that developments in this area are followed closely to ensure that the capabilities of ISO 10303 can be enhanced if necessary to handle the representations generated by whatever design methods become widely used in the future. Initial indications are that these capabilities are sufficient for the methods currently under development.

In the short term, it is desired to develop a new part of the ISO 10303 standard to handle LM data. The author's opinion is that this work should concentrate initially on the use of the mesh-based and set-based approaches, since these are the simplest to deal with and are closest to the type of output from current design methods. However, R-functions show good promise for the future, and in parallel with the development of the new standard their use in the design and manufacture of LM objects should be further studied, with the objective of including this approach in later versions of ISO 10303.

7   Acknowledgment

The author is grateful for financial support from the NIST project 'Systems Integration for Manufacturing Applications' (SIMA). The work has been carried out as part of the SIMA Product Engineering program.

References

1. C. L. Bajaj. Modelling physical fields for interrogative visualization. In T. N. T. Goodman and R. R. Martin, editors, *The Mathematics of Surfaces VII*. Information Geometers Ltd., Winchester, UK, 1997.
2. M. P. Bendsoe, A. Diaz, and N. Kikuchi. Topology and generalized layout optimization of elastic structures. In M.P. Bendsoe and C.A. MotaSoares, editors, *Topology Design of Structures*. Kluwer Academic Publishers, Amsterdam, 1993.
3. M. Burns. *Automated Fabrication*. Prentice-Hall, Englewood Cliffs, New Jersey, 1992.
4. R. W. Gray IV, D. G. Baird, and J. H. Bøhn. Effects of processing conditions on short TLCP fiber reinforced FDM parts. *Rapid Prototyping Journal*, **4**, 1, 14–25, 1998.
5. J. Hoschek and D. Lasser. *Computer Aided Geometric Design*. A. K. Peters, Wellesley, MA, 1993.

6. International Organization for Standardization. *Industrial Automation Systems and Integration - Product Data Representation and Exchange*, 1994. (The ISO catalogue is at http://www.iso.ch/cate/cat.html; search on 10303 for a listing of parts of the standard).

7. C. L. Jackins and S. L. Tanimoto. Oct-trees and their use in representing three-dimensional objects. *Computer Graphics & Image Processing*, **14**, 249 – 270, 1983.

8. T. R. Jackson, N. M. Patrikalakis, E. M. Sachs, and M. J. Cima. Modeling and designing components with locally controlled composition. In *Proceedings of the Solid Freeform Fabrication Symposium, Austin, TX*, 1998.

9. K. K. Jurrens. Rapid prototyping's second decade. *Rapid Prototyping (quarterly newsletter of the SME Rapid Prototyping Association)*, **4**, 1, 1 – 4, 1998.

10. V. Kumar. *Solid Modeling and Algorithms for Heterogeneous Objects*. PhD thesis, University of Michigan, Ann Arbor, MI, 1999.

11. V. Kumar, D. Burns, D. Dutta, and C. M. Hoffmann. A framework for object modeling. *Computer Aided Design*, **31**, 9, 541 – 556, 1999.

12. J. Mazumder, J. Koch, K. Nagarathnam, and J. Choi. Rapid manufacturing by laser aided direct deposition of metals. Technical report, University of Illinois, Department of Mechanical Engineering, 1996.

13. D. Meagher. Geometric modelling using octree encoding. *Computer Graphics & Image Processing*, **19**, 129 – 147, 1982.

14. C. Nash and S. Sen. *Topology and Differential Geometry for Physicists*. Academic Press, New York, NY, 1983.

15. J. Owen. *STEP: An Introduction*. Information Geometers, Winchester, UK, 2nd edition, 1997.

16. L. Patil, D. Dutta, A. D. Bhatt, K. K. Jurrens, K. W. Lyons, M. J. Pratt, and R. D.Sriram. Representation of heterogeneous objects in ISO 10303 (STEP). Submitted to ASME International Mechanical Engineering Congress and Exposition, Orlando, FL, November 2000.

17. A. A. G. Requicha. Representations of rigid solids — Theory, methods and systems. *ACM Computing Surveys*, **12**, 437 – 464, 1980.

18. V. L. Rvachev, T. I. Sheiko, V. Shapiro, and I. Tsukanov. Transfinite interpolation over implicitly defined sets. Technical Report SAL 2000 - 1, Spatial Automation Laboratory, University of Wisconsin - Madison, 2000.

19. J. J. Shah and M. Mäntylä. *Parametric and Feature-based CAD/CAM*. Wiley, 1995.

20. V. Shapiro and I. Tsukanov. Implicit functions with guaranteed differential properties. In W. F. Bronsvoort and D. C. Anderson, editors, *Proc. 5th ACM Symposium on Solid Modeling and Applications*. Association for Computing Machinery, New York, NY, 1999.

21. D. Shepard. A two-dimensional interpolation function for irregularly spaced data. In *Proceedings of 23rd ACM National Conference*, pages 517 – 524. Association for Computing Machinery, New York, NY, 1968.

22. J. Vida, R. R. Martin, and T. Várady. A survey of blending methods that use parametric surfaces. *Computer Aided Design*, **26**, 5, 341 – 365, 1994.

23. J. R. Woodwark. Blends in geometric modelling. In R. R. Martin, editor, *The Mathematics of Surfaces II*. Oxford University Press, 1987. (Proc. 2nd IMA Conf. on the Mathematics of Surfaces, Cardiff, Wales, Sept. 1986).

24. Z. Wu, S. H. Soon, and L. Feng. NURBS-based volume modeling. In *International Workshop on Volume Graphics, Swansea, Wales*, pages 321 – 330, 1999.

25. J. Zagajac. *Engineering Analysis over Subsets*. PhD thesis, Sibley School of Mechanical and Aerospace Engineering, Cornell University, Ithaca, NY, 1997.

The manufacturer's authorised representative in the EU is Springer
Nature Customer Service Centre GmbH, Europaplatz 3, 69115 Heidelberg,
Germany. If you have any concerns regarding our products, please
contact ProductSafety@springernature.com

Printed and bound by CPI Group (UK) Ltd, Croydon, CR0 4YY
23/04/2026
02095593-0001